Nonsmooth Mechanics and Convex Optimization

Nonsmooth Mechanics and Convex Optimization

Yoshihiro Kanno

CRC Press
Taylor & Francis Group
Boca Raton London New York

CRC Press is an imprint of the
Taylor & Francis Group, an **informa** business

CRC Press
Taylor & Francis Group
6000 Broken Sound Parkway NW, Suite 300
Boca Raton, FL 33487-2742

First issued in paperback 2017

ISBN 13: 978-1-138-07278-7 (pbk)
ISBN 13: 978-1-4200-9423-7 (hbk)

Library of Congress Cataloging-in-Publication Data

Kanno, Yoshihiro, 1976-
Nonsmooth mechanics and convex optimization / Yoshihiro Kanno.
p. cm.
Summary: "This book presents a methodology for comprehensive treatment of nonsmooth laws in mechanics in accordance with contemporary theory and algorithms of optimization. The author deals with theory and numeiral algorithms comprehensively, providing a new perspective n nonsmooth mechanics based on contemporary optimization. Covering linear programs; semidefinite programs; second-order cone programs; complementarity problems; optimality conditions; Fenchel and Lagrangian dualities; algorithms of operations research, and treating cable networks; membranes; masonry structures; contact problems; plasticity, this is an ideal guide of nonsmooth mechanics for graduate students and researchers in civil and mechanical engineering, and applied mathematics"--Provided by publisher.
Includes bibliographical references and index.
ISBN 978-1-4200-9423-7 (hardback)
1. Contact mechanics--Mathematics. 2. Mechanics, Applied--Mathematics. 3. Mechanics, Analytic. 4. Nonsmooth mathematical analysis. 5. Nonsmooth optimization. 6. Convex sets. 7. Duality theory (Mathematics) I. Title.

TA353.K36 2011
620.1001'51--dc22 2010045332

Visit the Taylor & Francis Web site at
http://www.taylorandfrancis.com

and the CRC Press Web site at
http://www.crcpress.com

Dedicated to M. H.

Preface

1. The principal subject of this book is to discuss how to use theory and algorithms of optimization for treating problems in applied mechanics in a comprehensive way. Particular emphasis, however, is put on the two terms involved in the title, "nonsmooth" and "convex," which distinguish the methodology of the present work from the *conventional* methods in applied and computational mechanics.

2. To see the reason the attention of "nonsmoothness" is inevitably required for dealing with problems in mechanics, consider two bodies (either deformable or rigid) that may possibly contact on their surfaces. On the boundaries of the bodies, the physical phenomena are characterized by the kinematic and static variables, say, *gap displacement* denoted by g and *reaction force* denoted by r. The so-called unilateral contact condition can be written as (see section 3.3 and Chapter 10 for details)

$$g \geq 0, \tag{1a}$$

$$\begin{cases} r = 0 & \text{if } g > 0 \quad \text{[zero reaction with gap]}, \\ r \leq 0 & \text{if } g = 0 \quad \text{[in contact with reaction]}. \end{cases} \tag{1b}$$

The relation of g and r in (1) is depicted in Figure 1(a), in which we can clearly see that the physical law changes when the threshold, either $g = 0$ or $r = 0$, is attained. Indeed, in the presence of the reaction ($r < 0$) the gap is prescribed ($g = 0$), while in the presence of the gap ($g > 0$) the reaction is prescribed ($r = 0$). As seen in Figure 1(a), the relation between g and r is nonsmooth at $(g, r) = (0, 0)$.

Within the framework of nonsmooth mechanics, we treat (1) by deriving equivalent forms involving no "if-clause"; for example, (1) is equivalent to

$$g \geq 0, \quad r \leq 0, \quad gr = 0. \tag{2}$$

The equation involved in (2), say, $gr = 0$, is called the *complementarity condition*. Thus the nonsmooth relation in Figure 1(a) is expressed by the two inequalities and the one complementarity condition. Throughout the book we frequently use the form like (2) to represent nonsmooth laws in mechanics.

In the framework of the conventional mechanics, the most popular approach to deal with the contact condition is a *gap-element*, which is a fictitious structural element with a large stiffness to prevent g from taking a negative value.

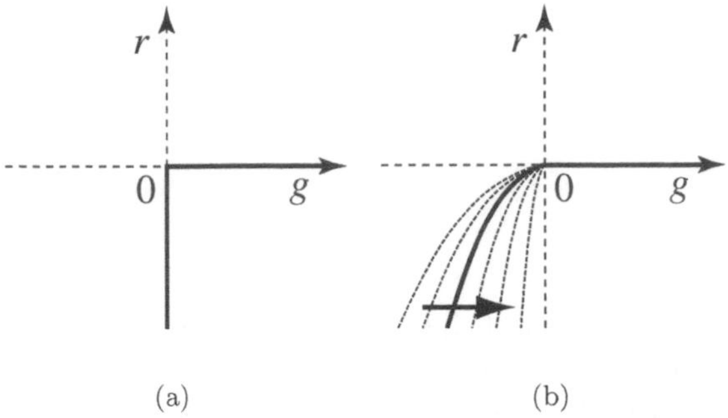

FIGURE 1: Relation between the gap and the reaction force in contact problems. (a) the nonsmooth law; (b) a regularization based on a gap-element model.

Accordingly, the constitutive relation between g and r is modeled as Figure 1(b), which intuitively converges to the law in Figure 1(a) by taking the limit of the stiffness of the gap-element.

The fundamental concept of the nonsmooth mechanics approach is to deal with the nonsmooth law without resorting to any regularization such as that in Figure 1(b). Although some mechanicians might have the opposite impression, we here assert that a nonsmooth model (in Figure 1(a)) is considered to be simpler than a regularized one (in Figure 1(b)) from the following two viewpoints:

(i) The nonsmooth model consists of the two linear half-lines, while the regularized model involves a nonlinear curve.

(ii) The nonsmooth model is symmetric with respect to exchange of the roles of g and r. This symmetry property reflects the physical law of alternatives between the gap and the reaction. In contrast, the regularized model is not symmetric in the sense that the reaction is subjected to the inequality constraint $r \leq 0$ while g can take any value.

On the other hand, the regularization certainly has been believed to provide us with efficient numerical algorithms. For example, the gap-element based contact law (in Figure 1(b)) could be treated similar to the nonlinear material law, and hence it seems that the conventional numerical algorithms designed for nonlinear materials are applicable. However, a regularized model also involves the transition of the parameter in the governing equations, such as the "zero-stiffness" and "finitely-large-stiffness" of the gap-element. For this reason, conventional numerical methods usually employ trial-and-error procedures, when they are applied to the problems arising from nonsmooth mechanics. In contrast, as stated in the succeeding section, the progress of modern optimization algorithms enables us to solve a vast class of problems with complementarity conditions effectively. Nowadays we have algorithms directly handling the complementarity conditions, without any trial-and-error procedure, with guaranteed convergence for various kinds of problems. Thus, even from a numerical point of view, we no longer need to resort to any regularization, which is just a trick for numerical computation,[1] and it is more preferable to treat the nonsmooth properties in mechanics directly.

3. As stated in section 1, we put special emphasis on "nonsmooth" and "convex" properties in mechanics. The importance of the "nonsmooth" property was discussed in section 2. Concerning "convexity," the linear optimization that is the most fundamental convex optimization has been studied extensively in the literature of mechanics. In contrast, we mostly focus on "nonlinear convex" optimization in this book. On one hand, nonlinearity brings *flexibility* of modeling considerably complicated phenomena in the real-world mechanics. On the other hand, convexity provides us with *tractability* of problems from both theoretical and computational points of view.

More specifically, a conventional way to deal with the change of physical law is to formulate the complementarity condition in the form of

$$f(x) \geq 0, \quad y \geq 0, \quad yf(x) = 0, \tag{3}$$

where f is a nonlinear function. For example, in the elastoplastic problem, f is the function defining the yield surface, x is a stress, and y is the consistency parameter corresponding to the modulus of the plastic strain rate (see Chapter 11 for details).

In many practical problems, f in (3) is a nonlinear function. In such a situation, unlike (2), the complementarity condition, say, $yf(x) = 0$, is no longer an equality constraint on a bilinear form. This point could be a source of the difficulty for dealing with (3).

Motivated by the modern optimization theory and algorithms, the methodology established in the present work is to restate (3) in the form of

$$x \in \mathcal{C}, \quad s \in \mathcal{C}, \quad \langle x, s \rangle = 0, \tag{4}$$

where $\mathcal{C}$ is a *convex cone* that has the so-called self-duality property (see Chapter 1). Indeed, the linear inequalities involved in (2) can be represented by using a *linear* convex cone $\mathbb{R}_+ = \{z \mid z \geq 0\}$ as $g \in \mathbb{R}_+$ and $-r \in \mathbb{R}_+$. Making use of *nonlinear* convex cones, however, enriches the flexibility of modeling using the unified form in (4). In fact, the present work collects

[1]Regularization is required in another instance; for example, the normal compliance regularization [300, 342] is often adopted to the unilateral contact problems of continua to ensure the existence of a weak solution. The necessity of such a regularization, certainly, is independent of the discussion made here.

various problems that were formerly stated in the form of (3) but can truly be restated in the form of either (4) or its slightly extended form

$$Ax + b \in \mathcal{C}, \quad Cs + d \in \mathcal{C}, \quad \langle Ax + b, Cs + d \rangle = 0. \tag{5}$$

It should be noted in (5), as well as in (4), that the nonlinear properties, formerly represented by f in (3), are "hidden" in the nonlinearity of $\mathcal{C}$ and that the complementarity condition is still a bilinear condition in x and s. The form of (5) enables us to use the *duality theory* (see Chapter 2), which is an elegant mathematical tool for study of optimization problems, as well as *numerical algorithms* (see Chapter 6), that are particularly designed for problems associated with convex cones.

The aim of the present work is to show, through many examples, that *nonsmooth mechanics* can enjoy the theoretical and algorithmic achievements of *modern optimization* by representing the nonsmooth laws in the form of (5).

4. This book consists of four parts: (I) the abstract framework of convex analysis for comprehensive treatment of nonsmooth mechanics (Chapters 1–3); (II) demonstration of our methodology through in-depth study of a selected class of structures (Chapters 4–5); (III) numerical algorithms for solving the problems in nonsmooth mechanics (Chapters 6–7); and (IV) the application of theoretical and numerical methodologies to the problems covering many topics in nonsmooth mechanics (Chapters 8–11).

5. A short description of the contents of the chapters is given as follows.

Chapter 1 summarizes the essential properties of a particular class of convex sets, called self-dual cones. Mathematical programming problems related to self-dual cones are introduced, which serve as fundamental tools for modeling problems in nonsmooth mechanics.

Chapter 2 introduces the modern theory of convex (or conic) optimization based on the conjugate duality. This very flexible abstract theory can be adapted to a wide variety of situations in nonsmooth mechanics.

Chapter 3 describes the application of conic optimization and convex analysis to several problems in structural engineering. Specifically, we study two topics from structural optimization (the compliance minimization and the eigenvalue optimization) and a fundamental topic from nonsmooth mechanics (the unilateral contact condition).

Chapter 4 treats the minimum principle of potential energy for cable networks, which serves as a remarkable example to demonstrate our new methodology in nonsmooth mechanics. Through the comprehensive study, this chapter shows how the framework of conic optimization (introduced in Chapter 1) is adapted to describe this particular problem arising from nonsmooth mechanics in an exact and excellent way.

Chapter 5 establishes, in continuation of the preceding chapter, the dual principle for the equilibrium of cable networks by making use of the duality theory (developed in Chapter 2). The dual principle corresponds to the minimum principle of complementary energy, which involves only the static variables. The existence and uniqueness results of solution are shown as consequences of the analysis of the duality.

Chapter 6 collects modern algorithms designed for the mathematical programming problems involving cone constraints. By virtue of these algorithms, the problems in nonsmooth mechanics can be solved numerically without any trial-and-error procedure.

Chapter 7 demonstrates the efficiency of the algorithms introduced in the preceding chapter by applying them to the problems of cable networks. The convex optimization formulations presented in Chapter 4 and Chapter 5 are solved to find the equilibrium states of cable networks.

Chapter 8 is devoted to phenomena of no-tension material from which we elaborate the equilibrium analysis of masonry structures. Following the same recipe as the study of cable networks, conic optimization formulations and their duality are investigated for the energy principles of masonry structures.

Chapter 9 considers planar membranes, in which the transition among the taut, wrinkled, and slack states are the resource of nonsmoothness property. The geometrical nonlinearity is also treated comprehensively.

Chapter 10 deals with the contact problem of deformable bodies against a rigid obstacle, in the presence of the Coulomb friction. Several formulations for this typical problem in nonsmooth mechanics are collected to see the modelling variation of a nonsmooth law.

Chapter 11 treats the elastoplastic problems, considering the flow rule with the von Mises yielding criterion. Although the problem under consideration is classic in nonsmooth mechanics, the methodology of modern convex optimization provides us with a different formulation from the conventional one, as briefly mentioned in section 3 of this preface.

6. A remarkable landmark of nonsmooth mechanics is the monograph by Duvaut–Lions [120]. The contributions by J.J. Moreau [319, 320, 322–324] on dynamic contact problems also have bridged mechanics and optimization in both theoretical and numerical aspects. The mathematical programming approach to elastoplastic analysis of structures was initiated by G. Maier [285–289]. The seminal pioneering work on variational inequalities in mechanics is due to P.D. Panagiotopoulos [353–356].

For continuum mechanics, variational inequality formulations and their theoretical consequences (e.g., the existence and uniqueness of a solution) are the principal materials dealt with in Duvaut–Lions [120]. In such a continuum setting, the reader is required to be familiar with rigorous treatment of infinite-dimensional spaces. The finite-dimensional setting is technically much simpler, which has motivated us to present the ideas for dealing with nonsmoothness issues by taking up finitely discretized structures. Although in some chapters we use the notation from continuum mechanics, to present rigorous treatment of the problems in infinite-dimensional spaces is beyond the aim of the current work; see Hlaváček–Haslinger–Nečas–Lovíšek [183], Kravchuk–

Neittaanmäki [247], and Sofonea–Matei [412] for more on contact problems of continua, and Allaire [10] and Ekeland–Temam [121] for the theory of infinite-dimensional optimization problems.

For mathematical theory of the variational and hemivariational inequalities, one may refer to Goeleven–Motreanu–Dumont–Rochdi [150] and Naniewicz–Panagiotopoulos [331], which also collect various examples, including economics, physics, as well as contact problems in mechanics. In contrast, this book is rather application oriented by describing modern algorithms with many numerical examples.

For plasticity, one of the distinguished contributions from the viewpoint of nonsmooth mechanics is from Simo–Hughes [413]. Particularly, this monograph introduces the numerical algorithm known as the return-mapping method, which has been used widely for elastoplastic analysis of structures. In contrast to the fact that the subject of Simo–Hughes [413] is restricted to plasticity, the present work collects various major topics in nonsmooth mechanics. Moreover, as already announced in section 3 of this preface, we present a formulation different from the one of the conventional plasticity theory.

The paradigm of continuous optimization has shifted drastically in the 1990s; formerly the nonlinear programming problems were distinguished from the linear programming problems from the viewpoint of tractability, and now almost all the convex nonlinear optimization problems are unified with the linear programming problems as the notion of conic optimization. The modern treatments of theory and algorithms for convex optimization are introduced by Ben-Tal–Nemirovski [46], Boyd–Vandenberghe [61], and Nesterov–Nemirovski [333]. These truly fruitful results, both in theoretical and numerical aspects, are fully enjoyed in this book.

Recently several books have been published on nonsmooth mechanics. The dynamics and bifurcation of nonsmooth mechanical systems consisting of undeformable bodies are subjects of Awrejcewicz–Lamarque [26], di Bernardo–Budd–Champneys–Kowalczyk [116], Leine–Nijmeijer [255], Leine–van de Wouw [256], and Sextro [405]. Accordingly, the major topics considered by these authors are contact, impact, and friction. For the treatment of dynamic contact problems of rigid bodies, the notion of differential inclusion is addressed by Monteiro Marques [318]; see also Deimling [105] and a survey by Stewart [420] for numerical methods. Numerical algorithms as well as theoretical results are also treated by Acary–Brogliato [1] and Brogliato [66], who also write primarily for undeformable bodies. Nevertheless, Parts II and III of Acary–Brogliato [1] serve as a good reference on numerical methods for contact mechanics. Frémond [136] gives a comprehensive treatment of various problems in nonsmooth thermo-mechanics, including the contact problems of deformable bodies with and without adhesion, mainly from the theoretical point of view.

7. After more than three decades since the work by Duvaut–Lions [120], it is hoped that the present work serves as a new bridge between nonsmooth mechanics of deformable bodies and modern convex optimization.

Although this book is primarily aimed at mechanicians, it also provides applied mathematicians with a successful case study in which achievements of modern mathematical engineering are fully applied to real-world problems. Basic and detailed exposition of the notion of complementarity and its links with convex analysis, including many examples taken from applied mechanics, may open a new door for the communities of applied and computational mechanics to a comprehensive treatment of nonsmoothness properties.

8. I am most grateful to Prof. Makoto Ohsaki, who introduced me to the fields of structural optimization and nonlinear mechanics as the thesis supervisor. Some materials in this work are originated in my dissertation and subsequent collaboration with him.

I wish to express my sincere gratitude to Prof. João A.C. Martins (*deceased*) whose profound influence has guided me to the notion of nonsmooth mechanics, and to Prof. Hirohisa Noguchi (*deceased*), who supported this book project. They sadly passed away in August 2008; in spring of that year this book project was launched.

My special appreciation goes to Prof. Kazuo Murota, who encouraged me to write this book, and gave many valuable comments on the draft. The present work has also been influenced directly and indirectly by the discussions and collaborations with Junichi Ito, Naoki Katoh, António Pinto da Costa, Izuru Takewaki, and Kazuo Yonekura. I thank them sincerely. I also appreciate that this book project was partially supported financially by the Global COE "The Research and Training Center for New Development in Mathematics."

Finally, I wish to thank Li-Ming Leong at CRC Press and Taylor & Francis for her patience, continuous encouragement, and support in the production of this book.

Tokyo, July 2010 *Yoshihiro Kanno*

Contents

I Convex Optimization over Symmetric Cone 1

1 Cones, Complementarity, and Conic Optimization 3
1.1 Proper Cones and Conic Inequalities 3
1.1.1 Convex sets and cones 3
1.1.2 Partial order induced by proper cone 5
1.2 Complementarity over Cones 6
1.2.1 Dual cones and self-duality 6
1.2.2 Complementarity problems 7
1.2.3 Variational inequalities 8
1.2.4 Complementarity over nonnegative orthant 9
1.2.5 Overview of complementarity over cones 10
1.3 Positive-Semidefinite Cone 11
1.3.1 Positive-semidefinite matrices 12
1.3.2 Inner product of matrices 16
1.3.3 Self-duality of positive-semidefinite cone 17
1.3.4 Complementarity over positive-semidefinite cone . . . 18
1.4 Second-Order Cone 19
1.4.1 Fundamentals of second-order cone 20
1.4.2 Self-duality of second-order cone 20
1.4.3 Complementarity over second-order cone 22
1.5 Conic Constraints and Their Relationship 26
1.6 Conic Optimization 29
1.6.1 Linear programming 30
1.6.2 Semidefinite programming 32
1.6.3 Second-order cone programming 33
1.7 Notes 36

2 Optimality and Duality 39
2.1 Fundamentals of Convex Analysis 39
2.1.1 Convex sets and convex functions 40
2.1.2 Monotone functions and convexity 41
2.1.3 Closed convex functions 44
2.1.4 Subdifferential 45
2.1.5 Conjugate function 47
2.2 Optimality and Duality 50
2.2.1 Dual problem 50

2.2.2 Weak duality 51
2.2.3 Strong duality 53
2.2.4 Optimality condition 54
2.2.5 Fenchel duality 55
2.2.6 Lagrangian duality 58
2.2.7 KKT conditions 61
2.3 Application to Semidefinite Programming 63
2.3.1 Fenchel dual problem of SDP 63
2.3.2 Duality and optimality of SDP 66
2.3.3 Lagrangian duality of SDP 69
2.4 Notes 72

3 Applications in Structural Engineering 73
3.1 Compliance Optimization 73
3.1.1 Definition of compliance 74
3.1.2 Compliance minimization 76
3.1.3 Worst-case compliance and robust optimization 79
3.2 Eigenvalue Optimization 81
3.2.1 Eigenvalue optimization of structures 81
3.2.2 SDP formulation 82
3.2.3 Optimality condition 84
3.3 Set-Valued Constitutive Law 86
3.3.1 Constitutive law 86
3.3.2 Linear elasticity and Legendre transformation 88
3.3.3 Inversion via Fenchel transformation 89
3.3.4 Unilateral contact law and Fenchel transformation . . 91
3.4 Notes 94

II Cable Networks: An Example in Nonsmooth Mechanics 97

4 Principles of Potential Energy for Cable Networks 99
4.1 Constitutive law 99
4.1.1 No-compression model 100
4.1.2 Inclusion form 101
4.1.3 Variational form 103
4.1.4 Complementarity form 104
4.2 Potential Energy Principles in Convex Optimization Forms . 108
4.2.1 Principle of potential energy in general form 108
4.2.2 Principle for large strain 112
4.2.3 Principle for linear strain 116
4.2.4 Principle for the Green–Lagrange strain 117
4.3 More on Cable Networks: Nonlinear Material Law 119
4.3.1 Piecewise-linear law 120
4.3.2 Piecewise-quadratic law 124

4.4 Notes 127

5 Duality in Cable Networks: Principles of Complementary Energy 129

5.1 Duality in Cable Networks (1): Large Strain 130
5.1.1 Embedding to Fenchel form 130
5.1.2 Dual problem 131
5.1.3 Duality and optimality 135
5.1.4 Principle of complementary energy 139
5.1.5 Existence and uniqueness of solution 145
5.2 Duality in Cable Networks (2): Linear Strain 147
5.2.1 Embedding to Fenchel form 148
5.2.2 Dual problem 149
5.2.3 Duality and optimality 150
5.2.4 Principle of complementary energy 152
5.3 Duality in Cable Networks (3): Green–Lagrange Strain . . . 153
5.3.1 Embedding to Fenchel form 153
5.3.2 Dual problem 155
5.3.3 Duality and optimality 157
5.3.4 Principle of complementary energy 161
5.4 Notes 163

III Numerical Methods 165

6 Algorithms for Conic Optimization 167

6.1 Primal-Dual Interior-Point Method 167
6.1.1 Outline of interior-point methods 167
6.1.2 Interior-point method for linear programming 168
6.1.3 Interior-point method for semidefinite programming . 173
6.2 Reformulation and Smoothing Method 177
6.2.1 Reformulation method 177
6.2.2 Smoothing method 180
6.2.3 Extensions to conic complementarity problems 181
6.3 Notes 183

7 Numerical Analysis of Cable Networks 185

7.1 Cable Networks with Pin-Joints 185
7.2 Cable Networks with Sliding Joints 195
7.3 Form-Finding of Cable Networks 200
7.3.1 Form-finding with specified axial forces 201
7.3.2 Special cases 202
7.4 Notes 206

IV Problems in Nonsmooth Mechanics 209

8 Masonry Structures **211**
8.1 Introduction 211
8.1.1 Notation 213
8.2 Principle of Potential Energy for Masonry Structures 214
8.2.1 Principle of potential energy 214
8.2.2 Constitutive law 216
8.2.3 Conic optimization formulation 221
8.3 Principle of Complementary Energy for Masonry Structures 225
8.3.1 Embedding to Fenchel form 225
8.3.2 Dual problem 228
8.3.3 Duality and optimality 232
8.3.4 Principle of complementary energy 235
8.4 Numerical Aspects 237
8.4.1 Spatial discretization 237
8.4.2 Examples 243
8.5 Notes 249

9 Planar Membranes **253**
9.1 Introduction 253
9.2 Analysis in Small Deformation 255
9.2.1 Principle of potential energy in small deformation . . 255
9.2.2 Conic optimization formulation 259
9.2.3 Principle of complementary energy in small deformation 261
9.3 Principle of Potential Energy for Membranes 264
9.3.1 Constitutive law 264
9.3.2 Principle of potential energy 273
9.4 Principle of Complementary Energy for Membranes 274
9.4.1 Embedding to Fenchel form 275
9.4.2 Dual problem 276
9.4.3 Duality and optimality 280
9.4.4 Principle of complementary energy 288
9.5 Numerical Aspects 291
9.5.1 Spatial discretization 291
9.5.2 Examples 295
9.6 Notes 305

10 Frictional Contact Problems **311**
10.1 Friction Law 311
10.1.1 Coulomb's law 312
10.1.2 Second-order cone complementarity formulation 314
10.2 Incremental Problem 317
10.2.1 Friction law in incremental problems 318
10.2.2 Contact kinematics 318
10.2.3 Problem formulation 321
10.3 Discussions on Various Complementarity Forms 329

10.3.1 On auxiliary variables 329
10.3.2 Maximum dissipation law and its optimality conditions 330
10.3.3 A formulation using projection operator 339
10.3.4 Friction law and normality rule 340
10.4 Notes 348

11 Plasticity 351
11.1 Fundamentals of Plasticity 351
11.2 Perfect Plasticity 356
11.2.1 Classical formulation of flow rule in perfect plasticity . 356
11.2.2 Second-order cone complementarity formulation 358
11.3 Plasticity with Isotropic Hardening 362
11.3.1 Linear isotropic hardening law 363
11.3.2 Second-order cone complementarity formulation 364
11.3.3 Incremental problem 367
11.3.4 SOCP formulation of incremental problem 370
11.4 Plasticity with Kinematic Hardening 373
11.4.1 Linear kinematic hardening 374
11.4.2 Second-order cone complementarity formulation 375
11.4.3 SOCP formulation of incremental problem 377
11.5 Notes 379

References 381

Index 417

About the Author 425

Part I

Convex Optimization over Symmetric Cone

Chapter 1

Cones, Complementarity, and Conic Optimization

Orientation In this chapter we introduce the concept of complementarity conditions defined over convex cones, which plays a key role in investigating the various problems in nonsmooth mechanics throughout the book. Following the definition of proper cone (which is a closed, convex, solid, and pointed cone) provided in section 1.1, section 1.2 introduces the complementarity condition over a proper cone, together with related problems such as the complementarity problem and the variational inequality. The nonnegative orthant, such as a nonnegative half-line, a quadrant, or an octant, is the simplest (but still important) example of proper cone. Besides the nonnegative orthant, the positive-semidefinite cone and the second-order cone possess distinguished importance in nonsmooth mechanics, which are studied in section 1.3 and section 1.4, respectively. Section 1.5 describes the constraints associated with these three cones, called the conic constraints, and their relationship. As the optimization problem subjected to a conic constraint, the notion of conic optimization is introduced in section 1.6. In the latter chapters of the book, we see that various problems in nonsmooth mechanics can be treated as conic optimization problems and that such a treatment successfully provides us with a unified perspective of nonsmooth mechanics.

1.1 Proper Cones and Conic Inequalities

Within the framework of nonsmooth mechanics, a mechanical problem is typically governed not by *equations* but by *inclusions*, i.e., relations such that some functions of state variables are included in certain sets. As a class of sets, particularly, convex sets are important for practical and theoretical reasons.

1.1.1 Convex sets and cones

Let $\mathbb{V}$ denote a Euclidean space, by which we mean a finite-dimensional vector space over the reals $\mathbb{R}$, equiped with an inner product $\langle \cdot, \cdot \rangle$. Indeed, we may consider only $\mathbb{R}^n$ without loss of generality, although a more abstract notation $\mathbb{V}$ is often more versatile. As an example, we often consider $\mathbb{S}^{n'}$, i.e., the space of $n' \times n'$ symmetric real matrices[1] (which is shortly studied in section 1.3 comprehensively), which seems to be very different from $\mathbb{R}^n$. However, the reader is suggested to put simply $\mathbb{V} = \mathbb{R}^n$, if the abstract notation seems complicated.[2]

A set $\mathcal{S} \subseteq \mathbb{V}$ is said to be *convex* if the line segment between any two points in $\mathcal{S}$ is contained in $\mathcal{S}$, i.e., if we have

$$x_1, x_2 \in \mathcal{S},\ 0 \le \tau \le 1 \quad \Rightarrow \quad \tau x_1 + (1-\tau) x_2 \in \mathcal{S},$$

where $\tau \in \mathbb{R}$. The empty set $\emptyset$ and the space $\mathbb{R}^n$ are extreme examples of convex sets. A half-space in $\mathbb{R}^n$ is another example of convex set. For $\boldsymbol{a} \in \mathbb{R}^n$ ($\boldsymbol{a} \neq 0$) and $b \in \mathbb{R}$, the set

$$\{\boldsymbol{x} \in \mathbb{R}^n \mid \boldsymbol{a}^{\mathrm{T}} \boldsymbol{x} \le b\}$$

is called a *closed half-space*, while the set

$$\{\boldsymbol{x} \in \mathbb{R}^n \mid \boldsymbol{a}^{\mathrm{T}} \boldsymbol{x} < b\}$$

is called an *open half-space*. It is easy to see that all the closed and open half-spaces are nonempty convex sets.

Define the *norm* of $x \in \mathbb{V}$ by $\|x\| = \sqrt{\langle x, x \rangle}$. For a set $\mathcal{S} \subseteq \mathbb{V}$, we say that $x \in \mathcal{S}$ is an *interior point* of $\mathcal{S}$ if there exists an $\epsilon > 0$ for which

$$\{y \in \mathbb{V} \mid \|y - x\| \le \epsilon\} \subseteq \mathcal{S}$$

holds. The set of all interior points of $\mathcal{S}$ is called the *interior* of $\mathcal{S}$ and is denoted by $\mathrm{int}\,\mathcal{S}$. A set $\mathcal{S}$ is said to be *open* if $\mathrm{int}\,\mathcal{S} = \mathcal{S}$.[3] We say that $\mathcal{S} \subseteq \mathbb{V}$ is *closed* if its complement, $\mathbb{V} \setminus \mathcal{S}$, is open.[4] The *closure* of $\mathcal{S}$, denoted $\mathrm{cl}\,\mathcal{S}$, is the smallest closed set that contains $\mathcal{S}$. We easily see that

$$\mathrm{cl}\,\mathcal{S} = \mathbb{V} \setminus \mathrm{int}(\mathbb{V} \setminus \mathcal{S}).$$

The *boundary* of $\mathcal{S}$, denoted $\mathrm{bd}\,\mathcal{S}$, is the set defined by

$$\mathrm{bd}\,\mathcal{S} = \mathrm{cl}\,\mathcal{S} \setminus \mathrm{int}\,\mathcal{S}.$$

[1] In fact, the case of $\mathbb{V} = \mathbb{S}^{n'}$ can be related to $\mathbb{R}^n$ with $n = n'(n'+1)/2$. Note that a $n' \times n'$ symmetric matrix is uniquely determined by specifying only the upper triangle components, the number of which is $n'(n'+1)/2$.

[2] If we put $\mathbb{V} = \mathbb{R}^n$, then for $\boldsymbol{x}$, $\boldsymbol{y} \in \mathbb{R}^n$ we have $\langle \boldsymbol{x}, \boldsymbol{y} \rangle = \boldsymbol{x}^{\mathrm{T}} \boldsymbol{y}$, which is the standard inner product of real n-dimensional vectors.

[3] This definition of open set is equivalent to requiring for $\mathcal{S} \subseteq \mathbb{V}$ that

$$x \in \mathcal{S} \quad \Rightarrow \quad \exists \epsilon > 0 : \{y \in \mathbb{V} \mid \|y - x\| \le \epsilon\} \subseteq \mathcal{S},$$

which means that for any $x \in \mathcal{S}$ there exists a closed ball that is centered at x and lies entirely in $\mathcal{S}$.

[4] Both the empty set $\emptyset$ and the space $\mathbb{R}^n$ themselves are examples of open and closed sets.

A set $\mathcal{C} \subseteq \mathbb{V}$ is called a *cone* if it satisfies

$$x \in \mathcal{C},\ \tau \geq 0 \quad \Rightarrow \quad \tau x \in \mathcal{C}.$$

We say that $\mathcal{C}$ is a *convex cone* if it is convex and a cone, which means that $\tau_1 x_1 + \tau_2 x_2 \in \mathcal{C}$ holds for any $x_1,\ x_2 \in \mathcal{C}$ and any $\tau_1,\ \tau_2 \geq 0$.

A cone $\mathcal{C}$ is said to be *pointed* if $\mathcal{C} \cap (-\mathcal{C}) = \{0\}$.[5] We say that a cone $\mathcal{C}$ is *solid* if it has a nonempty interior, i.e. if $\operatorname{int} \mathcal{C} \neq \emptyset$. A cone $\mathcal{C}$ is called a *proper cone* if it is closed, convex, solid, and pointed.

Example 1.1 (nonnegative orthant)
The *nonnegative orthant*, denoted $\mathbb{R}^n_+$, means the set of nonnegative vectors in $\mathbb{R}^n$, i.e.,

$$\mathbb{R}^n_+ = \{\boldsymbol{x} \in \mathbb{R}^n \mid x_i \geq 0\ (i = 1, \ldots, n)\}.$$

In particular, $\mathbb{R}_+$ denotes the set of nonnegative numbers; i.e., $x \in \mathbb{R}_+$ if and only if $x \geq 0$. Moreover, $\mathbb{R}^2_+$ and $\mathbb{R}^3_+$ correspond to the *quadrant* and the *octant*, respectively. It is easy to see that $\mathbb{R}^n_+$ is a proper cone in $\mathbb{R}^n$. □

1.1.2 Partial order induced by proper cone

Let $\mathcal{C}$ be a proper cone in $\mathbb{V}$. Then $\mathcal{C}$ can induce a partial order "$\preceq_\mathcal{C}$" on $\mathbb{V}$ by defining

$$x \preceq_\mathcal{C} y \quad \Leftrightarrow \quad y - x \in \mathcal{C}$$

for $x,\ y \in \mathbb{V}$. This partial order relation may be regarded as a "generalized inequality" between x and y, and hence $x \preceq_\mathcal{C} y$ is referred to as a *conic inequality* associated with the cone $\mathcal{C}$. Similarly, we introduce a strict partial order by defining

$$x \prec_\mathcal{C} y \quad \Leftrightarrow \quad y - x \in \operatorname{int} \mathcal{C}.$$

Example 1.2 ($\mathbb{R}^n_+$ and componentwise inequality)
In continuation of Example 1.1, consider the nonnegative orthant $\mathbb{R}^n_+$ as an example of $\mathcal{C}$, where $\mathbb{V} = \mathbb{R}^n$. For $\boldsymbol{x},\ \boldsymbol{y} \in \mathbb{R}^n$, the associated conic inequality $\boldsymbol{x} \preceq_{\mathbb{R}^n_+} \boldsymbol{y}$ corresponds to the componentwise inequality between $\boldsymbol{x}$ and $\boldsymbol{y}$, i.e.,

$$\boldsymbol{x} \preceq_{\mathbb{R}^n_+} \boldsymbol{y} \quad \Leftrightarrow \quad x_i \leq y_i\ (i = 1, \ldots, n).$$

Similarly, $\boldsymbol{x} \prec_{\mathbb{R}^n_+} \boldsymbol{y}$ means $x_i < y_i$ $(i = 1, \ldots, n)$, because $\operatorname{int} \mathbb{R}^n_+ = \{\boldsymbol{x} \in \mathbb{R}^n \mid x_i > 0\ (i = 1, \ldots, n)\}$.

As an extreme case, consider $n = 1$. Then conic inequality $x \preceq_{\mathbb{R}_+} y$ naturally coincides with the usual inequality $x \leq y$ on $\mathbb{R}$. Similarly, $x \prec_{\mathbb{R}_+} y$ is nothing but the usual strict inequality $x < y$.

[5] By the notation $\mathcal{C} \cap (-\mathcal{C}) = \{0\}$ we mean "$x \in \mathcal{C},\ -x \in \mathcal{C} \Rightarrow x = 0$". This also means that $\mathcal{C}$ contains no line.

Since the componentwise inequality of two vectors arise so frequently, in this book we usually write $\boldsymbol{x} \le \boldsymbol{y}$ instead of $\boldsymbol{x} \preceq_{\mathbb{R}^n_+} \boldsymbol{y}$. We also write $\boldsymbol{y} \ge \boldsymbol{x}$ for $\boldsymbol{x} \le \boldsymbol{y}$. In particular, $\boldsymbol{x} \ge \boldsymbol{0}$ means $x_i \ge 0$ $(i = 1, \dots, n)$. Similarly, we usually write $\boldsymbol{y} < \boldsymbol{x}$ or $\boldsymbol{x} > \boldsymbol{y}$ instead of $\boldsymbol{x} \prec_{\mathbb{R}^n_+} \boldsymbol{y}$. □

1.2 Complementarity over Cones

Complementarity is key in dealing with nonsmooth mechanics, both from theoretical and numerical points of view. Problems in nonsmooth mechanics usually involve two or more physical phases. As a simple example, consider the unilateral contact problem, which consist of the *contact* phase and the *free* phase; see section 3.3.4 for details. The existence of reaction force means that two bodies are in contact, while the existence of gap means that they are free; hence, non-zero reaction and non-zero gap cannot exist simultaneously. Such an *alternative* property is comprehensively treated in terms of a complementarity condition. In this section we focus on complementarity conditions over convex cones.

1.2.1 Dual cones and self-duality

For a given cone $\mathcal{C} \subseteq \mathbb{V}$, the *dual cone* of $\mathcal{C}$ is defined by

$$\mathcal{C}^* = \{s \in \mathbb{V} \mid \langle x, s \rangle \ge 0 \ (\forall x \in \mathcal{C})\}. \tag{1.1}$$

In other words, $\mathcal{C}^*$ is the set of all vectors that make a non-obtuse angle with any vector in $\mathcal{C}$. As the name suggests, $\mathcal{C}^*$ is a cone. Moreover, $\mathcal{C}^*$ is always convex and closed, even if $\mathcal{C}$ is not.

We write $\mathcal{C}^{**}$ to mean $(\mathcal{C}^*)^*$. If $\mathcal{C}$ is a proper cone, then so is $\mathcal{C}^*$, and hence $\mathcal{C}^{**} = \mathcal{C}$. A cone $\mathcal{C}$ is said to be *self-dual* if $\mathcal{C}^* = \mathcal{C}$.

Example 1.3 (self-duality of $\mathbb{R}^n_+$)
In continuation of Example 1.1 and Example 1.2, we consider the dual cone of the nonnegative orthant $\mathbb{R}^n_+$. Specifically, consider (1.1) with $\mathbb{V} = \mathbb{R}^n$ and $\langle \boldsymbol{x}, \boldsymbol{s} \rangle = \boldsymbol{x}^{\mathrm{T}}\boldsymbol{s}$ for $\boldsymbol{x}, \boldsymbol{s} \in \mathbb{R}^n$:

$$(\mathbb{R}^n_+)^* = \{\boldsymbol{s} \in \mathbb{R}^n \mid \boldsymbol{x}^{\mathrm{T}}\boldsymbol{s} \ge 0 \ (\forall \boldsymbol{x} \ge \boldsymbol{0})\}.$$

Suppose $\boldsymbol{s} \notin \mathbb{R}^n_+$, which means that there exists $\hat{\imath} \in \{1, \dots, n\}$ such that $s_{\hat{\imath}} < 0$. Define $\hat{\boldsymbol{x}} \in \mathbb{R}^n$ by

$$\hat{x}_i = \begin{cases} 1 & \text{if } i = \hat{\imath}, \\ 0 & \text{if } i \neq \hat{\imath}, \end{cases}$$

where $\hat{\boldsymbol{x}} \geq \mathbf{0}$. Then we see that $\hat{\boldsymbol{x}}^{\mathrm{T}}\boldsymbol{s} = s_{\hat{i}} < 0$, which implies that

$$\boldsymbol{x}^{\mathrm{T}}\boldsymbol{s} \geq 0 \ (\forall \boldsymbol{x} \geq 0) \quad \Leftrightarrow \quad \boldsymbol{s} \geq \mathbf{0}.$$

Consequently, we obtain

$$(\mathbb{R}^n_+)^* = \{\boldsymbol{s} \in \mathbb{R}^n \mid \boldsymbol{s} \geq \mathbf{0}\} = \mathbb{R}^n_+.$$

Thus the nonnegative orthant $\mathbb{R}^n_+$ is a self-dual cone. □

1.2.2 Complementarity problems

This section introduces a complementarity problem associated with a proper cone, which is regarded as a generalization of optimization problems. The complementarity problem provides a broad unifying setting for studying most problems in nonsmooth mechanics. In its general form, a complementarity problem is formally defined as follows.[6]

Definition 1.2.1 (CP). Given a proper cone $\mathcal{C} \subseteq \mathbb{V}$ and a mapping $G : \mathcal{C} \to \mathbb{V}$, the *complementarity problem* (CP) is to find $x \in \mathbb{V}$ satisfying

$$\mathrm{CP}: \quad x \in \mathcal{C}, \quad G(x) \in \mathcal{C}^*, \quad \langle x, G(x) \rangle = 0.$$

Equivalently, we also simply write

$$0 \preceq_{\mathcal{C}} x \perp G(x) \succeq_{\mathcal{C}^*} 0.$$ ■

The complementarity problem, CP, is characterized by G and $\mathcal{C}$. A slight generalization of CP is the mixed complementarity problem, which consists of two different types of variables in $\mathbb{V}_1$ and $\mathbb{V}_2$: we suppose that the variables $x_1 \in \mathbb{V}_1$ should be included in the cone $\mathcal{C}$ ($\subseteq \mathbb{V}_1$), while the variables $x_2 \in \mathbb{V}_2$ are free from the constraints. The relation between x_1 and x_2 is given by a system of equations.[7] Thus we arrive at the following definition.

Definition 1.2.2 (MiCP). Given a proper cone $\mathcal{C} \subseteq \mathbb{V}_1$ and mappings G_1, $G_2 : \mathcal{C} \times \mathbb{V}_2 \to \mathbb{V}_1 \times \mathbb{V}_2$, the *mixed complementarity problem* (MiCP) is to find $(x_1, x_2) \in \mathbb{V}_1 \times \mathbb{V}_2$ satisfying

$$\begin{aligned} \mathrm{MiCP}: \quad & G_2(x_1, x_2) = 0, \\ & 0 \preceq_{\mathcal{C}} x_1 \perp G_1(x_1, x_2) \succeq_{\mathcal{C}^*} 0. \end{aligned}$$ ■

[6] More generally, the cone $\mathcal{C}$ in Definition 1.2.1 is not necessarily proper. Throughout the book, however, we are interested only in the situation where $\mathcal{C}$ is a proper cone.

[7] The MiCP in Definition 1.2.2 is reduced to CP if $G_2(x_1, x_2) = 0$ is solvable with respect to x_2, i.e., if it is equivalently rewritten as $x_2 = H(x_1)$ with a function H.

For example, the optimality condition of a conic optimization problem can be written in the form of MiCP, as we see shall later (see section 2.3.2). In particular, the optimality condition of a linear programming problem corresponds to the case of MiCP, where both G_1 and G_2 are affine and $\mathcal{C} = \mathcal{C}^* = \mathbb{R}^n_+$. Throughout most of this book, we are interested in MiCPs, i.e., problems including some variables that are not directly subjected to conic inequalities. Hence, we often call MiCP as CP for simplicity, although in their rigorous definitions above CP is just a particular case of MiCP.

Clearly, a key feature distinguishing CP (also MiCP) from the conventional system of equations and inequalities is the condition in the form of

$$0 \preceq_{\mathcal{C}} x \perp s \succeq_{\mathcal{C}^*} 0, \tag{1.2}$$

where we consider CP with auxiliary variables $s = G(x)$. Indeed, we see in Part II and Part IV that various nonsmoothness properties in mechanics can be represented in the form of (1.2). The condition (1.2) is called the complementarity (or the perpendicularity) condition over a pair of cones $\mathcal{C}$ and $\mathcal{C}^*$. We are particularly interested in the cases in which $\mathcal{C}^* = \mathcal{C}$: we study the cases of $\mathcal{C} := \mathbb{R}^n_+$, $\mathbb{S}^n_+$, and $\mathbb{L}^n$ in section 1.2.4, section 1.3.4, and section 1.4.3, respectively.[8] See also section 1.2.5 for an overview of complementarity conditions over $\mathbb{R}^n_+$, $\mathbb{S}^n_+$, and $\mathbb{L}^n$.

1.2.3 Variational inequalities

A generalization of complementarity problems is the variational inequality, which is defined formally as follows.

Definition 1.2.3 (VI). Given $\mathcal{K} \subseteq \mathbb{V}$ and a mapping $G : \mathcal{K} \to \mathbb{V}$, the *variational inequality* (VI) is to find a vector $x \in \mathbb{V}$ satisfying

$$\begin{aligned} \mathrm{VI}: \quad & x \in \mathcal{K}, \\ & (y - x)^{\mathrm{T}} G(x) \geq 0, \quad \forall y \in \mathcal{K}. \end{aligned}$$ ■

Remark 1.2.4. Several books on nonsmooth mechanics are based on variational inequalities, rather than complementarity problems; see Goeleven–Motreanu–Dumont–Rochdi [150], Kravchuk–Neittaanmäki [247], Naniewicz–Panagiotopoulos [331], and Sofonea–Matei [412], some of whom also deal with a broader class, such as *hemivariational inequality*. In contrast, in this book we principally treat problems in nonsmooth mechanics as complementarity problems, because we focus on the problems in which $\mathcal{K}$ in VI is a proper cone. ■

[8] For the definition of each cone, see the corresponding section. In short, the positive-semidefinite cone $\mathbb{S}^n_+$ is the set of all $n \times n$ symmetric positive-semidefinite matrices, while the second-order cone $\mathbb{L}^n$ is the set of vectors $(x_0, x_1, \dots x_{n-1}) \in \mathbb{R}^n$ satisfying $x_0 \geq \sqrt{x_1^2 + \cdots + x_n^2}$.

If $\mathcal{K}$ is a proper cone, then VI becomes equivalent to CP in Definition 1.2.1 with $\mathcal{C} = \mathcal{K}$ (see [125, Proposition 1.1.3]). As another interesting situation, suppose that G is written as $G = \nabla f$ for some differentiable $f : \mathcal{K} \to \mathbb{R}$, where $\nabla f(x)$ is the gradient of f. Then VI reads

$$\forall x \in \mathcal{K}, \quad \nabla f(x)^{\mathrm{T}}(y - x) \geq 0, \quad \forall y \in \mathcal{K}, \tag{1.3}$$

while

$$f(y) - f(x) = \nabla f(x)^{\mathrm{T}}(y - x) + o(\|y - x\|).$$

Hence, (1.3) corresponds to the (first-order) optimality condition of the optimization problem

$$\min_{x}\{f(x) \mid x \in \mathcal{K}\}.$$

In particular, if f and $\mathcal{K}$ are convex, then solving this optimization problem is equivalent to VI in Definition 1.2.3.

1.2.4 Complementarity over nonnegative orthant

In section 1.2.2, we introduced CP associated with a proper cone $\mathcal{C}$ in $\mathbb{V}$; see Definition 1.2.1. This section provides a further study of a particular case in which $\mathbb{V} = \mathbb{R}^n$ and $\mathcal{C} = \mathbb{R}^n_+$. By virtue of the self-duality of $\mathbb{R}^n_+$ (see Example 1.3), CP associated with $\mathbb{R}^n_+$ is reduced to the following form:

$$\mathbf{0} \leq \boldsymbol{x} \perp \boldsymbol{g}(\boldsymbol{x}) \geq \mathbf{0}, \tag{1.4}$$

where $\boldsymbol{g} : \mathbb{R}^n_+ \to \mathbb{R}^n$ is a given mapping. Problem (1.4) is called a *nonlinear complementarity problem* (NCP). In particular, if $\boldsymbol{g}$ is affine, then we call NCP the *linear complementarity problem* (LCP).[9]

Analogous to a general CP investigated in section 1.2.2, a key feature characterizing LCP and NCP is the condition in the form of

$$\mathbf{0} \leq \boldsymbol{x} \perp \boldsymbol{s} \geq \mathbf{0}, \tag{1.5}$$

where we put $\boldsymbol{s} = \boldsymbol{g}(\boldsymbol{x})$ in (1.4). By writing the perpendicular condition $\boldsymbol{x} \perp \boldsymbol{s}$ over $\mathbb{R}^n$ explicitly, (1.5) is equivalently rewritten as

$$\boldsymbol{x} \geq \mathbf{0}, \quad \boldsymbol{s} \geq \mathbf{0}, \quad \boldsymbol{x}^{\mathrm{T}}\boldsymbol{s} = 0. \tag{1.6}$$

Since the inequalities $x_i \geq 0$ and $s_i \geq 0$ imply $s_i x_i \geq 0$, (1.6) holds if and only if the perpendicular conditions are satisfied componentwise, i.e.,

$$x_i \geq 0, \quad s_i \geq 0, \quad x_i s_i = 0 \quad (i = 1, \dots, n). \tag{1.7}$$

[9] See also section 1.6.1.

TABLE 1.1: The implication of the complementarity condition over the nonnegative orthant. The condition (1.6) holds if and only if any one of the following three states is true for each $i = 1, \dots, n$.

	$\boldsymbol{x} \in \mathbb{R}^n$		$\boldsymbol{s} \in \mathbb{R}^n$	
	inclusion	inequality	inclusion	inequality
(i)	$x_i \in \operatorname{int} \mathbb{R}_+$	$(x_i > 0)$	$s_i = 0$	
(ii)	$x_i = 0$		$s_i \in \operatorname{int} \mathbb{R}_+$	$(s_i > 0)$
(iii)	$x_i = 0$		$s_i = 0$	

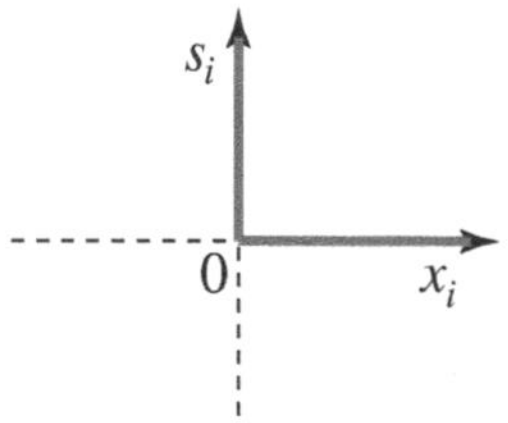

FIGURE 1.1: Solution set of (1.6).

In (1.7),

$$x_i s_i = 0 \tag{1.8}$$

is called the *complementarity condition*. The complementarity condition (1.8) is equivalent to

$$\begin{cases} x_i \neq 0 & \Rightarrow s_i = 0, \\ s_i \neq 0 & \Rightarrow x_i = 0. \end{cases} \tag{1.9}$$

The term *complementarity* stems from (1.9): If one of x_i and s_i is nonzero, then the other must be zero. Thus, (1.6) holds if and only if for each $i = 1, \dots, n$ any one of the three states listed in Table 1.1 is satisfied. The solution set of (1.6) is illustrated in Figure 1.1. We say that $\boldsymbol{x}$ and $\boldsymbol{s}$ satisfy the *strict complementarity condition* if either the case (i) or (ii) in Table 1.1 is true for each $i = 1, \dots, n$. This condition is also equivalent to $\boldsymbol{x} + \boldsymbol{s} > \boldsymbol{0}$. In contrast, if for some i the case (iii) holds, the strict complementarity is not satisfied.

1.2.5 Overview of complementarity over cones

We close this section with a brief overview of generalization of the complementarity condition.

Concerning CP introduced in Definition 1.2.1, we focus on the condition (1.2). The property of the solution set of (1.2) depends on the choice of the

TABLE 1.2: Self-dual cones and related optimization problems. $\mathbb{R}^n_+$: nonnegative orthant; $\mathbb{L}^n$: second-order cone; $\mathbb{S}^n_+$: positive-semidefinite cone of symmetric matrices.

vector space	inner product	cone	complementarity	optimization
$\mathbb{V} = \mathbb{R}^n$	$\boldsymbol{s}^{\mathrm{T}}\boldsymbol{x}$	$\mathbb{R}^n_+$	$x_i s_i = 0 \ (\forall i)$	LP
$\mathbb{V} = \mathbb{R}^n$	$\boldsymbol{s}^{\mathrm{T}}\boldsymbol{x}$	$\mathbb{L}^n$	$\boldsymbol{x} \circ \boldsymbol{s} = \boldsymbol{0}$	SOCP
$\mathbb{V} = \mathbb{S}^n$	$S \bullet X$	$\mathbb{S}^n_+$	$XS = O$	SDP

proper cone $\mathcal{C}$. A particular case in which $\mathcal{C} = \mathbb{R}^n_+$ was studied in section 1.2.4; the results are summarized in Table 1.2.

The nonnegative orthant $\mathbb{R}^n_+$ is indeed a polyhedral cone and is the simplest example of self-dual proper cones. In what follows, we generalize the notion of complementarity condition to the ones associated with non-polyhedral cones, with which we can establish more elegant models of various phenomena in nonsmooth mechanics. More specifically, we are interested in the following non-polyhedral self-dual proper cones:

- Positive-semidefinite cone, $\mathbb{S}^n_+$, in section 1.3
- Second-order cone, $\mathbb{L}^n$, in section 1.4

The semidefinite programming (SDP), which is an optimization problem associated with $\mathbb{S}^n_+$, is introduced in section 1.6.2, and section 1.6.3 is devoted to the introduction of the second-order cone programming (SOCP) associated with $\mathbb{L}^n$. The linear programming (LP), which is the more standard optimization problem associated with $\mathbb{R}^n_+$, is formally stated in section 1.6.1.

The cones $\mathbb{S}^n_+$ and $\mathbb{L}^n$, together with $\mathbb{R}^n_+$, play important roles throughout the book both from the theoretical and computational points of view. The key results of study of complementarity relations over $\mathbb{R}^n_+$, $\mathbb{S}^n_+$, and $\mathbb{L}^n$ are summarized in Table 1.2. Note that SOCP includes LP as a particular case, while SDP includes both LP and SOCP as particular cases. However, we treat LP, SOCP, and SDP individually, because their properties seem to be quite different, as shown in Table 1.2. Although LP and SOCP problems can be solved as SDP problems, doing so is not advisable both from theoretical and numerical points of view. Analogously, it is not recommended to solve LP problems as SOCP problems.

1.3 Positive-Semidefinite Cone

Example 1.1, Example 1.2, and Example 1.3 show that the nonnegative orthant, denoted $\mathbb{R}^n_+$, is a (polyhedral) proper cone. The conic inequality induced by $\mathbb{R}^n_+$ coincides with the componentwise inequality of n-dimensional vectors. Evidently, $\mathbb{R}^n_+$ is the simplest example of proper cones, although important applied problems in mechanics and structural engineering can be profitably formulated as optimization or complementarity problems over $\mathbb{R}^n_+$. Indeed, most of the classical achievements in nonsmooth mechanics rely upon the complementarity conditions over $\mathbb{R}^n_+$.

In this section, we introduce a particular *non-polyhedral* cone, which also enjoys the favorable properties. Specifically, we treat the cone of symmetric positive-semidefinite matrices. Another example of non-polyhedral cone is the second-order cone, which is studied in section 1.4. As we see in the latter parts of the book, introducing non-polyhedral proper cones opens the door to a new perspective on theory and algorithms of nonsmooth mechanics.

1.3.1 Positive-semidefinite matrices

Let $\mathbb{S}^n$ denote the set of $n \times n$ symmetric real matrices. We investigate subsets of $\mathbb{S}^n$ characterized by positive-definite and positive-semidefinite matrices, which are defined in Definition 1.3.2.

Remark 1.3.1. A few words about $\mathbb{S}^n$, concerning the notation $\mathbb{V}$ introduced in the beginning of section 1.1.1. For $n \times n$ symmetric matrix

$$X = \begin{bmatrix} X_{11} & X_{21} & \cdots & X_{n1} \\ X_{21} & X_{22} & \cdots & X_{n2} \\ \vdots & \vdots & \ddots & \vdots \\ X_{n1} & X_{n2} & \cdots & X_{nn} \end{bmatrix},$$

define $\mathbf{vec}(X) \in \mathbb{R}^{n(n+1)/2}$ by

$$\mathbf{vec}(X) = (X_{11}, X_{21}, \dots, X_{n1}, X_{22}, \dots, X_{n2}, \dots, X_{nn})^{\mathrm{T}}.$$

Then we can identify $\mathbb{S}^n$ with $\mathbb{R}^{n(n+1)/2}$ as

$$\{\mathbf{vec}(X) \mid X \in \mathbb{S}^n\} = \{\boldsymbol{x} \mid \boldsymbol{x} \in \mathbb{R}^{n(n+1)/2}\}.$$

Thus in this section (and anywhere we treat symmetric matrices) we consider $\mathbb{S}^n$ as the vector space $\mathbb{V} = \mathbb{R}^m$ with $m = n(n+1)/2$. ■

Definition 1.3.2 (Positive (semi)definiteness).

- $A \in \mathbb{S}^n$ is said to be positive semidefinite if its quadratic form, $\boldsymbol{z}^{\mathrm{T}} A \boldsymbol{z}$, satisfies $\boldsymbol{z}^{\mathrm{T}} A \boldsymbol{z} \geq 0$ for any $\boldsymbol{z} \in \mathbb{R}^n$.
  - (a) We write $A \succeq O$ if $A \in \mathbb{S}^n$ is positive semidefinite.
  - (b) We denote by $\mathbb{S}^n_+$ the set of $n \times n$ positive-semidefinite symmetric matrices. Hence, we also write $A \in \mathbb{S}^n_+$ for positive-semidefinite A.
- $A \in \mathbb{S}^n$ is said to be positive definite if $\boldsymbol{z}^{\mathrm{T}} A \boldsymbol{z} > 0$ for any $\boldsymbol{z} \in \mathbb{R}^n$, $\boldsymbol{z} \neq \boldsymbol{0}$.

(a) We write $A \succ O$ if $A \in \mathbb{S}^n$ is positive definite.

(b) We denote by $\mathbb{S}^n_{++}$ the set of $n \times n$ positive-definite symmetric matrices. Hence, we also write $A \in \mathbb{S}^n_{++}$ for positive-definite A. ■

Clearly, $\mathbb{S}^n_+$ is a closed convex cone in $\mathbb{S}^n$: If $X_1, X_2 \in \mathbb{S}^n_+$, so is $\tau_1 X_1 + \tau_2 X_2$ for all scalars $\tau_1, \tau_2 \geq 0$. Hence, we call $\mathbb{S}^n_+$ the *positive-semidefinite cone.* Moreover, $\mathbb{S}^n_+$ is pointed: $\mathbb{S}^n_+ \cap (-\mathbb{S}^n_+) = \{O\}$. Since $\operatorname{int} \mathbb{S}^n_+ = \mathbb{S}^n_{++} \neq \emptyset$, $\mathbb{S}^n_+$ is solid.[10] Thus $\mathbb{S}^n_+$ is a proper cone.

In section 1.1.2, we introduced the conic inequality by considering the partial order induced by a proper cone. The notations $\succeq$ and $\succ$ in Definition 1.3.2 are actually motivated by the conic inequality associated with $\mathbb{S}^n_+$, which is a proper cone. The partial order $A \preceq_{\mathbb{S}^n_+} B$ on $\mathbb{S}^n$ induced by $\mathbb{S}^n_+$ means $B - A \in \mathbb{S}^n_+$. However, since the conic inequality associated with $\mathbb{S}^n_+$ arises very frequently, we drop the subscript; i.e., for symmetric matrices we write simply $A \preceq B$ or $B \succeq A$ if the matrix $B - A$ is positive semidefinite. Similarly, we write $A \prec B$ and $B \succ A$ if the matrix $B - A$ is positive definite. Particularly, we write $A \succeq O$ ($\succ O$) to mean A is positive semidefinite (positive definite). Throughout most of this book, we use the notations $A \succeq O$ and $A \succ O$ only for symmetric matrices unless there is an explicit statement otherwise.

Some equivalent characterizations of positive-semidefinite matrices include the following.

Fact 1.3.3. For $A \in \mathbb{S}^n$, the following three statements are equivalent:

(a) $A \succeq O$, i.e., $\boldsymbol{z}^{\mathrm{T}} A \boldsymbol{z} \geq 0$ ($\forall \boldsymbol{z} \in \mathbb{R}^n$).

(b) All the eigenvalues of A are nonnegative.

(c) There exits $C \in \mathbb{R}^{m \times n}$ such that $A = C^{\mathrm{T}} C$. Any such C satisfies $\operatorname{rank} C = \operatorname{rank} A$. ■

Every $A \in \mathbb{S}^n$ can be diagonalized as

$$A = Q \Lambda Q^{\mathrm{T}}, \tag{1.10}$$

where $Q \in \mathbb{R}^{n \times n}$ is orthogonal, and $\Lambda \in \mathbb{R}^{n \times n}$ is diagonal. From $AQ = Q\Lambda$ we see that the columns of Q are the eigenvectors of A. The diagonal entries of Λ are the corresponding eigenvalues, i.e., if we write $\Lambda = \operatorname{diag}(\lambda_1, \ldots, \lambda_n)$, $\lambda_1, \ldots, \lambda_n$ are the eigenvalues of A. Therefore, the orthogonal transformation in the form of (1.10) is called the eigenvalue decomposition. As an immediate consequence of Fact 1.3.3, if $A \succeq O$, then the diagonal entries of Λ are nonnegative. This naturally leads to the following definition of a square root of $A \succeq O$.

[10] In contrast, $\mathbb{S}^n_{++}$ is an open convex cone.

Definition 1.3.4 (Square root of positive-semidefinite matrix)**.** For $A \succeq O$, its positive-semidefinite symmetric square root $A^{1/2}$ is defined by $A^{1/2} = Q \operatorname{diag}(\sqrt{\lambda_1}, \dots, \sqrt{\lambda_n}) Q^{\mathrm{T}}$ with expression (1.10). ■

Note that $A^{1/2}$ in Definition 1.3.4 is determined uniquely, although Q in (1.10) is not necessarily unique.

A variational characterization of eigenvalues of $A \in \mathbb{S}^n$ is provided by the Courant–Fischer theorem. In particular, the largest and smallest eigenvalues are related to the extremal values of the Rayleigh quotient as follows.

Fact 1.3.5. For $A \in \mathbb{S}^n$, we denote by $\lambda_{\max}(A)$ and $\lambda_{\min}(A)$ its largest and smallest eigenvalues. Then

$$\lambda_{\max}(A) = \max_{\boldsymbol{z} \neq \boldsymbol{0}} \frac{\boldsymbol{z}^{\mathrm{T}} A \boldsymbol{z}}{\boldsymbol{z}^{\mathrm{T}} \boldsymbol{z}},$$

$$\lambda_{\min}(A) = \min_{\boldsymbol{z} \neq \boldsymbol{0}} \frac{\boldsymbol{z}^{\mathrm{T}} A \boldsymbol{z}}{\boldsymbol{z}^{\mathrm{T}} \boldsymbol{z}}.$$ ■

Remark 1.3.6. Clearly, by using the notation in Fact 1.3.5, the positive-semidefinite constraint is alternatively written as

$$A \succeq O \quad \Leftrightarrow \quad \lambda_{\min}(A) \geq 0.$$

However, throughout the book we prefer the expression $A \succeq O$, which sheds light on the treatment of positive-semidefinite constraint from the viewpoint of the unified framework of conic inequalities. ■

Concerning Fact 1.3.5, observe

$$\frac{\boldsymbol{z}^{\mathrm{T}}(A - tI)\boldsymbol{z}}{\boldsymbol{z}^{\mathrm{T}}\boldsymbol{z}} = \frac{\boldsymbol{z}^{\mathrm{T}} A \boldsymbol{z}}{\boldsymbol{z}^{\mathrm{T}} \boldsymbol{z}} - t,$$

from which we obtain

$$A - tI \succeq O \quad \Leftrightarrow \quad \lambda_{\min}(A) \geq t. \tag{1.11}$$

Hence, we have that

$$\lambda_{\min}(A) = \max_{t \in \mathbb{R}} \{ t : A \succeq tI \},$$

which is an alternative variational characterization of the smallest eigenvalue. The assertion (1.11) is employed in section 3.2 to present an example of semidefinite programming used in structural optimization.

Almost analogous to Fact 1.3.3 for positive-semidefinite matrices, some equivalent conditions for positive definiteness of symmetric matrices are presented in Fact 1.3.7.

Fact 1.3.7. For $A \in \mathbb{S}^n$, the following three statements are equivalent:

(a) $A \succ O$, i.e., $\boldsymbol{z}^{\mathrm{T}} A \boldsymbol{z} > 0$ $(\forall \boldsymbol{z} \in \mathbb{R}^n \setminus \{\boldsymbol{0}\})$.

(b) All the eigenvalues of A are positive.

(c) $A = C^{\mathrm{T}}C$ for some $n \times n$ nonsingular matrix C. ■

It is immediately obvious that if $A \succ O$, then A is nonsingular, and $A^{-1} = Q\Lambda^{-1}Q^{\mathrm{T}}$, where $\Lambda^{-1} = \mathrm{diag}(1/\lambda_1, \ldots, 1/\lambda_n)$ with expression (1.10). Since the eigenvalues of A^{-1} are $1/\lambda_1, \ldots, 1/\lambda_n$, we have that $A^{-1} \succ O$ if and only if $A \succ O$.

Besides the square root $A^{1/2}$, any positive-definite A has a factorization in the following form.

Fact 1.3.8 (Cholesky factorization). $A \in \mathbb{S}^n$ satisfies $A \succ O$ if and only if $A = LL^{\mathrm{T}}$ for some nonsingular lower triangular matrix L. ■

The positive definiteness and the positive semidefiniteness are preserved through the following transformation.

Fact 1.3.9. Let $A \in \mathbb{S}^n$.

(i) For $P \in \mathbb{R}^{m \times n}$, $PAP^{\mathrm{T}} \succeq O$ if $A \succeq O$.

(ii) For a nonsingular $P \in \mathbb{R}^{n \times n}$, $A \succ O$ ($\succeq O$) if and only if $PAP^{\mathrm{T}} \succ O$ ($\succeq O$).[11] ■

A useful and standard tool for investigating properties of symmetric matrices is the result below, which states the positive-semidefiniteness (and the positive-definiteness) of a block-partitioned matrix.

Fact 1.3.10 (Lemma on Schur complement). Suppose

$$W = \begin{bmatrix} A & C \\ C^{\mathrm{T}} & B \end{bmatrix}$$

with $A \in \mathbb{S}^m$, $B \in \mathbb{S}^n$, $C \in \mathbb{R}^{m \times n}$, and $A \succ O$. Then the following two statements are equivalent:

(a) $W \succeq O$ ($\succ O$).

(b) $B - C^{\mathrm{T}}A^{-1}C \succeq O$ ($\succ O$).

The matrix $B - C^{\mathrm{T}}A^{-1}C$ in (b) is called the *Schur complement* of A in W. ■

To show Fact 1.3.10, consider the congruence transformation

$$\begin{bmatrix} I & O \\ -C^{\mathrm{T}}A^{-1} & I \end{bmatrix} \begin{bmatrix} A & C \\ C^{\mathrm{T}} & B \end{bmatrix} \begin{bmatrix} I & -A^{-1}C \\ O & I \end{bmatrix} = \left[\begin{array}{c|c} A & O \\ \hline O & B - C^{\mathrm{T}}A^{-1}C \end{array}\right]$$

of W, and apply Fact 1.3.9 (ii) to the obtained block-diagonal matrix. Fact 1.3.10 has several generalizations to the case in which A is singular. A useful example is presented next.[12]

[11] For $A \in \mathbb{R}^{n \times n}$, a transformation in the form $P \mapsto PAP^{\mathrm{T}}$ with a nonsingular P is called a *congruence transformation*.

[12] For a proof, see Boyd–El Ghaoui–Feron–Balakrishnan [60, p. 28].

Fact 1.3.11. Suppose $A \succeq O$ in the situation of Fact 1.3.10. Then W satisfies $W \succeq O$ if and only if

$$(I - AA^{+})C = O, \quad B - C^{\mathrm{T}} A^{+} C \succeq O,$$

where A^{+} is the (Moore–Penrose) pseudo-inverse of A. ∎

1.3.2 Inner product of matrices

A natural inner product between two matrices is defined as follows.

Definition 1.3.12. For A, $B \in \mathbb{R}^{m \times n}$, we denote by $A \bullet B$ the inner product between A and B defined by

$$A \bullet B = \mathrm{tr}(A^{\mathrm{T}} B).$$

We also write $\langle A, B \rangle$ to mean $A \bullet B$. ∎

From the definition above we immediately see that

$$A \bullet B = \sum_{i=1}^{m} \sum_{j=1}^{n} A_{ij} B_{ij}$$

with $A = (A_{ij})$ and $B = (B_{ij})$, and hence $A \bullet B = B \bullet A$. Moreover, if A, $B \in \mathbb{S}^{n}$, then $A \bullet B = \mathrm{tr}(AB)$. The norm associated with the inner product in Definition 1.3.12 is the *Frobenius norm*, written as $\|A\|_{\mathrm{F}} = \sqrt{A \bullet A}$. The following equality holds concerning the inner product and the (usual) product of matrices.

Fact 1.3.13. If $P \in \mathbb{R}^{m \times n}$, $Q \in \mathbb{R}^{m \times l}$, and $R \in \mathbb{R}^{l \times n}$, then $P \bullet (QR) = (Q^{\mathrm{T}} P) \bullet R$. ∎

As immediate consequences of Fact 1.3.13, we obtain the following.

Fact 1.3.14. For A, $B \in \mathbb{S}^{n}$, the following statements hold:

(i) $\boldsymbol{z}^{\mathrm{T}} A \boldsymbol{z} = A \bullet (\boldsymbol{z}\boldsymbol{z}^{\mathrm{T}})$ for a vector $\boldsymbol{z} \in \mathbb{R}^{n}$.

(ii) If A, $B \succeq O$, then $A \bullet B = \mathrm{tr}(A^{1/2} B A^{1/2})$. ∎

Fact 1.3.14 (i) gives the relation between quadratic form and matrix inner product. To see Fact 1.3.14 (ii), note that the matrix $A \succeq O$ has a symmetric square root $A^{1/2}$, from which, combined with Fact 1.3.13, we obtain

$$A \bullet B = \mathrm{tr}(AB) = \mathrm{tr}(A^{1/2} A^{1/2} B) = \mathrm{tr}(A^{1/2} B A^{1/2}).$$

1.3.3 Self-duality of positive-semidefinite cone

This section investigates some properties of the set $\mathbb{S}^n_+$. As mentioned before, $\mathbb{S}^n_+$ is a proper cone, i.e., it is a closed, convex, pointed, and solid cone. In contrast, $\mathbb{S}^n_{++}$ is an open convex cone.

Our next concern is the dual cone, introduced in (1.1), of $\mathbb{S}^n_+$. As established in Fact 1.3.16 below, $\mathbb{S}^n_+$ enjoys the *self-duality* property, which means that the dual cone of $\mathbb{S}^n_+$ coincides with $\mathbb{S}^n_+$ itself. We first present a simple property of the inner product over the positive-semidefinite cone.

Fact 1.3.15. Let $A, B \in \mathbb{S}^n$. If $A, B \succeq O$, then $A \bullet B \geq 0$. ∎

Proof. From Fact 1.3.14 (ii), we know that $A \bullet B = A^{1/2}BA^{1/2}$, which is the sum of the eigenvalues of $A^{1/2}BA^{1/2}$. By using Fact 1.3.9 (i), the assumption $B \succeq O$ implies $A^{1/2}BA^{1/2} \succeq O$, and hence all the eigenvalues of $A^{1/2}BA^{1/2}$ are nonnegative. □

Now we formally state the self-dual property of the positive-semidefinite cone.

Fact 1.3.16 (Self-duality of $\mathbb{S}^n_+$)**.** The dual cone $(\mathbb{S}^n_+)^*$, defined by

$$(\mathbb{S}^n_+)^* = \{S \in \mathbb{S}^n \mid X \bullet S \geq 0 \ (\forall X \in \mathbb{S}^n_+)\},$$

satisfies $(\mathbb{S}^n_+)^* = \mathbb{S}^n_+$. ∎

Proof. (A) $\mathbb{S}^n_+ \subseteq (\mathbb{S}^n_+)^*$: This immediately follows from the fact that $S \bullet X \geq 0$ for any $X, S \succeq O$ (see Fact 1.3.15).

(B) $\mathbb{S}^n_+ \supseteq (\mathbb{S}^n_+)^*$: To show this, we prove that $S \notin \mathbb{S}^n_+$ implies $S \notin (\mathbb{S}^n_+)^*$. Indeed, for $S \notin \mathbb{S}^n_+$, we have $\boldsymbol{z}^{\mathrm{T}}S\boldsymbol{z} < 0$ for some $\boldsymbol{z} \in \mathbb{R}^n$, and hence $S \bullet (\boldsymbol{z}\boldsymbol{z}^{\mathrm{T}}) < 0$, which shows $S \notin (\mathbb{S}^n_+)^*$. □

As an immediate consequence of Fact 1.3.16, the following result gives an alternative characterization of positive-semidefinite matrix.[13]

Fact 1.3.17. For $S \in \mathbb{S}^n$, the following statements hold:

(i) $S \succeq O$ if and only if $X \bullet S \geq 0 \ (\forall X \in \mathbb{S}^n_+)$.

(ii) $\inf_{X \in \mathbb{S}^n} \{S \bullet X \mid X \succeq O\} = \begin{cases} 0 & \text{if } S \succeq O, \\ -\infty & \text{otherwise.} \end{cases}$ ∎

[13] Fact 1.3.17 (ii) is obtained as follows. If $S \succeq O$, then Fact 1.3.17 (i) implies that $\inf_{X \in \mathbb{S}^n} \{S \bullet X \mid X \succeq O\} \geq 0$, and that $S \bullet X = 0$ is attained with $X = O$. Now suppose $S \not\succeq O$. Then Fact 1.3.17 (i) implies that there exists an $\hat{X} \succeq O$ such that $S \bullet \hat{X} < 0$, and hence for $\alpha \in \mathbb{R}$ we have that $S \bullet (\alpha \hat{X}) \to -\infty$ when $\alpha \to \infty$.

1.3.4 Complementarity over positive-semidefinite cone

In this section we study the property of the solution set of the complementarity condition over the positive-semidefinite cone, i.e.,

$$\mathbb{S}^n_+ \ni X \perp S \in \mathbb{S}^n_+,$$

where we consider (1.2) with $\mathcal{C} = \mathbb{S}^n_+$ and have used the self-duality of $\mathbb{S}^n_+$ (see Fact 1.3.16). More explicitly, the condition above is written as

$$X \succeq O, \quad S \succeq O, \quad X \bullet S = 0$$

with the aid of the notion of conic inequality associated with $\mathbb{S}^n_+$.

We first present the following result, which shows the relation between the perpendicular condition on the positive-semidefinite cone and the (usual) matrix multiplication.

Fact 1.3.18. Let X, $S \in \mathbb{S}^n$. For X, $S \succeq O$, $X \bullet S = 0$ if and only if $SX = O$. ■

Proof. It follows from Fact 1.3.14 (ii) that $X \bullet S = 0$ is equivalent to $\mathrm{tr}(X^{1/2}SX^{1/2}) = 0$. This condition is also equivalent to $X^{1/2}SX^{1/2} = O$, because we know that $X^{1/2}SX^{1/2} \succeq O$. Since

$$X^{1/2}SX^{1/2} = (S^{1/2}X^{1/2})^{\mathrm{T}}(S^{1/2}X^{1/2}),$$

we conclude $S^{1/2}X^{1/2} = O$. Premultiplication of $S^{1/2}$ and postmultiplication of $X^{1/2}$ result in $SX = O$. □

Notice here that $SX = O$ implies that X and S commute, which motivates us to recall the following necessary and sufficient condition for commutativity in multiplication of symmetric matrices (see, e.g., [186, Theorem 1.3.12] for a proof).

Fact 1.3.19. Let X, $S \in \mathbb{S}^n$. X and S commute (i.e., $SX = XS$) if and only if X and S are simultaneously diagonalizable (i.e., they have eigenvalue decompositions with the same orthogonal Q). ■

Thereby, X and S in Fact 1.3.18 share a common system of eigenvalues, from which we obtain Fact 1.3.20.

Fact 1.3.20. $X \in \mathbb{S}^n$ and $S \in \mathbb{S}^n$ satisfy

$$X \succeq O, \quad S \succeq O, \quad X \bullet S = 0 \tag{1.12}$$

if and only if there exists an orthogonal matrix Q satisfying

$$X = Q\,\mathrm{diag}(\lambda_1, \dots, \lambda_n)Q^{\mathrm{T}}, \quad S = Q\,\mathrm{diag}(\omega_1, \dots, \omega_n)Q^{\mathrm{T}}$$

and the condition

$$\lambda_r \geq 0, \quad \omega_r \geq 0, \quad \lambda_r\omega_r = 0 \quad (r = 1, \dots, n) \tag{1.13}$$

holds. ■

TABLE 1.3: The implication of the complementarity condition over the positive-semidefinite cone. The condition (1.12) holds if and only if there exists an orthogonal Q satisfying $X = Q\Lambda Q^{\mathrm{T}}$ and $S = Q\Omega Q^{\mathrm{T}}$ with $\Lambda = \mathrm{diag}(\lambda_1, \ldots, \lambda_n)$ and $\Omega = \mathrm{diag}(\omega_1, \ldots, \omega_n)$, and any one of the following three states is true for each $r = 1, \ldots, n$.

	$\mathbb{S}^n \ni X = Q\Lambda Q^{\mathrm{T}}$		$\mathbb{S}^n \ni S = Q\Omega Q^{\mathrm{T}}$	
	inclusion	inequality	inclusion	inequality
(i)	$\lambda_r \in \mathrm{int}\,\mathbb{R}_+$	$(\lambda_r > 0)$	$\omega_r = 0$	
(ii)	$\lambda_r = 0$		$\omega_r \in \mathrm{int}\,\mathbb{R}_+$	$(\omega_r > 0)$
(iii)	$\lambda_r = 0$		$\omega_r = 0$	

Fact 1.3.20 is a key result for the complementarity condition between two positive-semidefinite matrices, which asserts that condition (1.12) holds if and only if X and S are simultaneously diagonalizable and each pair of their eigenvalues, $\{\lambda_r, \omega_r\}$, satisfies the complementarity condition (1.13). This implies that $\mathrm{rank}\, X + \mathrm{rank}\, S \leq n$. We say that X and S in (1.12) satisfies the *strict complementarity condition* if $\mathrm{rank}\, X + \mathrm{rank}\, S = n$, or, equivalently, $X + S \succ O$, i.e., if exactly one of the two conditions $\lambda_r = 0$ or $\omega_r = 0$ holds for each $r = 1, \ldots, n$. The upshot of Fact 1.3.20 is summarized in Table 1.3 (see also Table 1.1 for comparison with the case of $\mathbb{R}^n_+$). If there exists an $r \in \{1, \ldots, n\}$ satisfying the case (iii) in Table 1.3, then the strict complementarity does not hold for such X and S. Otherwise, the strict complementarity is satisfied. Thus, the complementarity condition over $\mathbb{S}^n_+$ can be reduced to the existence of simultaneous diagonalization and the complementarity condition of eigenvalues over $\mathbb{R}^n_+$.

1.4 Second-Order Cone

Besides the positive-semidefinite cone studied in section 1.3, an important non-polyhedral convex cone is the second-order cone, which is also known under several names, namely, it is also called the quadratic cone, Lorentz cone, or ice-cream cone. Throughout this section we assume $n \geq 2$ to avoid triviality.

1.4.1 Fundamentals of second-order cone

Definition 1.4.1 (Second-order cone). The n-dimensional second-order (or Lorentz, or quadratic) cone $\mathbb{L}^n$ is defined by

$$\mathbb{L}^n = \left\{ (x_0, x_1, \dots, x_{n-1})^{\mathrm{T}} \in \mathbb{R}^n \mid x_0 \geq \sqrt{x_1^2 + \cdots + x_{n-1}^2} \right\}. \qquad \blacksquare$$

For any $\boldsymbol{z} \in \mathbb{R}^n$ we denote by $\|\boldsymbol{z}\|$ its standard Euclidean norm, i.e., $\|\boldsymbol{z}\| = (\boldsymbol{z}^{\mathrm{T}}\boldsymbol{z})^{1/2}$. Besides the notation in Definition 1.4.1, we also write

$$\mathbb{L}^n = \{\boldsymbol{x} = (x_0, \boldsymbol{x}_1) \in \mathbb{R}^n \mid x_0 \geq \|\boldsymbol{x}_1\|\},$$

where the n-dimensional vector $\boldsymbol{x}$ is partitioned as

$$\boldsymbol{x} = \begin{bmatrix} x_0 \\ \boldsymbol{x}_1 \end{bmatrix}, \quad x_0 \in \mathbb{R}, \quad \boldsymbol{x}_1 \in \mathbb{R}^{n-1},$$

and we often write $(x_0, \boldsymbol{x}_1)$ instead of $(x_0, \boldsymbol{x}_1^{\mathrm{T}})^{\mathrm{T}}$.

As the name suggests, $\mathbb{L}^n$ is a cone in $\mathbb{R}^n$. Its interior and boundary are explicitly given by

$$\begin{aligned} \operatorname{int} \mathbb{L}^n &= \{(x_0, \boldsymbol{x}_1) \in \mathbb{R} \times \mathbb{R}^{n-1} \mid x_0 > \|\boldsymbol{x}_1\|\}, \\ \operatorname{bd} \mathbb{L}^n &= \{(x_0, \boldsymbol{x}_1) \in \mathbb{R} \times \mathbb{R}^{n-1} \mid x_0 = \|\boldsymbol{x}_1\|\}. \end{aligned}$$

Thus $\mathbb{L}^n$ has a nonempty interior. Moreover, $\mathbb{L}^n \cap (-\mathbb{L}^n) = \{\boldsymbol{0}\}$ is satisfied. Consequently, $\mathbb{L}^n$ is a proper cone. This implies that, as we saw in section 1.1.2, $\mathbb{L}^n$ can induce a partial order on $\mathbb{R}^n$, in which, for $\boldsymbol{x}$, $\boldsymbol{y} \in \mathbb{R}^n$, the conic inequality $\boldsymbol{x} \preceq_{\mathbb{L}^n} \boldsymbol{y}$ means $\boldsymbol{y} - \boldsymbol{x} \in \mathbb{L}^n$. Thus we write $\boldsymbol{x} \succeq_{\mathbb{L}^n} \boldsymbol{0}$ if $\boldsymbol{x} \in \mathbb{L}^n$, although in this book we use the notation $x_0 \geq \|\boldsymbol{x}_1\|$ or $\boldsymbol{x} \in \mathbb{L}^n$ more frequently.

1.4.2 Self-duality of second-order cone

The inner product associated with $\mathbb{L}^n$ is the standard one for n-dimensional real vector, which means that for $\boldsymbol{x} = (x_0, \boldsymbol{x}_1) \in \mathbb{L}^n$ and $\boldsymbol{s} = (s_0, \boldsymbol{s}_1) \in \mathbb{L}^n$ we consider

$$\langle \boldsymbol{x}, \boldsymbol{s} \rangle = \boldsymbol{x}^{\mathrm{T}}\boldsymbol{s} = x_0 s_0 + \boldsymbol{x}_1^{\mathrm{T}}\boldsymbol{s}_1.$$

Similar to $\mathbb{S}_+^n$ (see Fact 1.3.15), the inner product of any two vectors belonging to $\mathbb{L}^n$ is nonnegative, as shown below.

Fact 1.4.2. If $\boldsymbol{x}$, $\boldsymbol{s} \in \mathbb{L}^n$, then $\boldsymbol{x}^{\mathrm{T}}\boldsymbol{s} \geq 0$.

Proof. From the Cauchy–Schwarz inequality and the assumptions that $x_0 \geq \|\boldsymbol{x}_1\|$ and $s_0 \geq \|\boldsymbol{s}_1\|$, we obtain

$$\boldsymbol{x}_1^{\mathrm{T}}\boldsymbol{s}_1 \geq -\|\boldsymbol{x}_1\|\|\boldsymbol{s}_1\| \geq -x_0 s_0,$$

hence $\boldsymbol{x}^{\mathrm{T}}\boldsymbol{s} = x_0 s_0 + \boldsymbol{x}_1^{\mathrm{T}}\boldsymbol{s}_1 \geq 0$. $\square$

According to (1.1), the dual cone of $\mathbb{L}^n$ is defined by

$$(\mathbb{L}^n)^* = \{\boldsymbol{s} \in \mathbb{R}^n \mid \boldsymbol{x}^{\mathrm{T}}\boldsymbol{s} \geq 0 \ (\forall \boldsymbol{x} \in \mathbb{L}^n)\}.$$

Analogous to the case of $\mathbb{S}^n_+$ (see Fact 1.3.16), the second-order cone has the self-duality property stated in the following.

Fact 1.4.3 (Self-duality of $\mathbb{L}^n$). $(\mathbb{L}^n)^* = \mathbb{L}^n$.

Proof. (A) $\mathbb{L}^n \subseteq (\mathbb{L}^n)^*$: This immediately follows from Fact 1.4.2.

(B) $\mathbb{L}^n \supseteq (\mathbb{L}^n)^*$: To show this, we prove that $\boldsymbol{s} \notin \mathbb{L}^n$ implies $\boldsymbol{s} \notin (\mathbb{L}^n)^*$. Let $\boldsymbol{s} = (s_0, \boldsymbol{s}_1) \notin \mathbb{L}^n$, which means $s_0 < \|\boldsymbol{s}_1\|$. Suppose $\boldsymbol{s}_1 \neq \boldsymbol{0}$. Choose

$$\hat{\boldsymbol{x}} = \begin{bmatrix} \|\boldsymbol{s}_1\| \\ -\boldsymbol{s}_1 \end{bmatrix}$$

to see $\hat{\boldsymbol{x}} \in \mathbb{L}^n$, while we obtain

$$\hat{\boldsymbol{x}}^{\mathrm{T}}\boldsymbol{s} = \|\boldsymbol{s}_1\| s_0 - \boldsymbol{s}_1^{\mathrm{T}}\boldsymbol{s}_1 < \|\boldsymbol{s}_1\|^2 - \boldsymbol{s}_1^{\mathrm{T}}\boldsymbol{s}_1 = 0.$$

This inequality implies $\boldsymbol{s} \notin (\mathbb{L}^n)^*$. Alternatively, suppose $\boldsymbol{s}_1 = \boldsymbol{0}$. Then $\boldsymbol{s} \notin \mathbb{L}^n$ implies $s_0 < 0$. Choose $\hat{\boldsymbol{x}} \in \mathbb{L}^n$ as

$$\hat{\boldsymbol{x}} = \begin{bmatrix} 1 \\ \boldsymbol{0} \end{bmatrix},$$

while we have that

$$\hat{\boldsymbol{x}}^{\mathrm{T}}\boldsymbol{s} = s_0 < 0.$$

Therefore, $\boldsymbol{s} \notin (\mathbb{L}^n)^*$. □

As an immediate consequence of Fact 1.4.3, we obtain a variational characterization of the second-order cone as follows.[14]

Fact 1.4.4. For an n-dimensional vector $\boldsymbol{s} \in \mathbb{R}^n$, the following statements hold:

(i) $\boldsymbol{s} \in \mathbb{L}^n$ if and only if $\boldsymbol{x}^{\mathrm{T}}\boldsymbol{s} \geq 0 \ (\forall \boldsymbol{x} \in \mathbb{L}^n)$.

(ii) $\displaystyle\inf_{\boldsymbol{x} \in \mathbb{R}^n} \{\boldsymbol{s}^{\mathrm{T}}\boldsymbol{x} \mid \boldsymbol{x} \in \mathbb{L}^n\} = \begin{cases} 0 & \text{if } \boldsymbol{s} \in \mathbb{L}^n, \\ -\infty & \text{otherwise.} \end{cases}$ ■

[14] Fact 1.4.4 (ii) can be proved in a manner similar to Fact 1.3.17 (ii).

1.4.3 Complementarity over second-order cone

Proceeding in parallel with the discussion in section 1.3.4 on the complementarity over $\mathbb{S}^n_+$, we here study the property of the solution set of the complementarity condition over the second-order cone, i.e.,

$$\mathbb{L}^n \ni \boldsymbol{x} \perp \boldsymbol{s} \in \mathbb{L}^n,$$

where we consider (1.2) with $C = \mathbb{L}^n$ and have used the self-duality of $\mathbb{L}^n$ (see Fact 1.4.3). We observed in Fact 1.3.18 that for $X,\ S \in \mathbb{S}^n_+$ we have the perpendicular condition $X \perp S$ if and only if their product vanishes, i.e., $XS = 0$. Likewise, in Fact 1.4.6 we see that $\boldsymbol{x} \in \mathbb{L}^n$ and $\boldsymbol{s} \in \mathbb{L}^n$ satisfy $\boldsymbol{x} \perp \boldsymbol{s}$ if and only if their "product" vanishes, namely, $\boldsymbol{x} \circ \boldsymbol{s} = \boldsymbol{0}$, with a particularly defined "multiplication operator $\circ$." To establish this, following [126, 140], we first introduce the Euclidean Jordan algebra associated with the second-order cone.

For $\boldsymbol{x} = (x_0, \boldsymbol{x}_1) \in \mathbb{R} \times \mathbb{R}^{n-1}$ and $\boldsymbol{s} = (s_0, \boldsymbol{s}_1) \in \mathbb{R} \times \mathbb{R}^{n-1}$, we define their *Jordan product* by

$$\boldsymbol{x} \circ \boldsymbol{s} = \begin{bmatrix} \boldsymbol{x}^{\mathrm{T}} \boldsymbol{s} \\ s_0 \boldsymbol{x}_1 + x_0 \boldsymbol{s}_1 \end{bmatrix}. \tag{1.14}$$

It is easy to see that this product enjoys the commutativity and satisfies

$$\boldsymbol{x} \circ \boldsymbol{s} = \boldsymbol{s} \circ \boldsymbol{x} = \begin{bmatrix} x_0 & \boldsymbol{x}_1^{\mathrm{T}} \\ \boldsymbol{x}_1 & x_0 I \end{bmatrix} \begin{bmatrix} s_0 \\ \boldsymbol{s}_1 \end{bmatrix}.$$

We consider the algebra, which consists of the product $\circ$ defined above and the vector space endowed with $\circ$. The vector $\boldsymbol{e} = (1, 0, \dots, 0)^{\mathrm{T}} \in \mathbb{R}^n$ serves as a unique identity element, i.e., $\boldsymbol{x} \circ \boldsymbol{e} = \boldsymbol{x}$ for any $\boldsymbol{x} \in \mathbb{R}^n$. The second-order cone itself can be characterized by the set of "squares" of vectors with respect to $\circ$ as follows.[15]

Fact 1.4.5. $\mathbb{L}^n = \{\boldsymbol{x} \circ \boldsymbol{x} \mid \boldsymbol{x} \in \mathbb{R}^n\}$.

Proof. (A) $\mathbb{L}^n \supseteq \{\boldsymbol{x} \circ \boldsymbol{x} \mid \boldsymbol{x} \in \mathbb{R}^n\}$: For any $\boldsymbol{x} \in \mathbb{R}^n$, let

$$\begin{bmatrix} y_0 \\ \boldsymbol{y}_1 \end{bmatrix} := \boldsymbol{x} \circ \boldsymbol{x} = \begin{bmatrix} x_0^2 + \boldsymbol{x}_1^{\mathrm{T}} \boldsymbol{x}_1 \\ 2 x_0 \boldsymbol{x}_1 \end{bmatrix}.$$

Then we have that

$$y_0^2 - \boldsymbol{y}_1^{\mathrm{T}} \boldsymbol{y}_1 = (x_0^2 + \boldsymbol{x}_1^{\mathrm{T}} \boldsymbol{x}_1)^2 - (2 x_0 \boldsymbol{x}_1)^2 = (x_0^2 - \boldsymbol{x}_1^{\mathrm{T}} \boldsymbol{x}_1)^2 \geq 0,$$

which verifies that $\boldsymbol{y} \in \mathbb{L}^n$.

[15]By virtue of a Euclidean Jordan algebra, we can treat algebraic properties on $\mathbb{L}^n$ and $\mathbb{S}^n_+$ in a unified way. Indeed, the pair $(\mathbb{S}^n, \circ)$ is a Euclidean Jordan algebra with the Jordan product defined by $X \circ S = (XS + SX)/2$. Analogous to the case of $\mathbb{L}^n$, we see that $\mathbb{S}^n_+ = \{X \circ X \mid X \in \mathbb{S}^n\}$. See [126, 428] for more detail.

(B) $\mathbb{L}^n \subseteq \{\boldsymbol{x} \circ \boldsymbol{x} \mid \boldsymbol{x} \in \mathbb{R}^n\}$: Let $\boldsymbol{y} \in \mathbb{L}^n$, which means that

$$y_0^2 - \boldsymbol{y}_1^{\mathrm{T}} \boldsymbol{y}_1 \geq 0. \tag{1.15}$$

It suffices to show that there exists $(x_0, \boldsymbol{x}_1)$ such that

$$x_0^2 + \boldsymbol{x}_1^{\mathrm{T}} \boldsymbol{x}_1 = y_0, \quad 2x_0 \boldsymbol{x}_1 = \boldsymbol{y}_1, \tag{1.16}$$

By eliminating $\boldsymbol{x}_1$ from (1.16), we obtain

$$4x_0^4 - 4y_0 x_0^2 + \boldsymbol{y}_1^{\mathrm{T}} \boldsymbol{y}_1 = 0. \tag{1.17}$$

It is guaranteed from (1.15) that equation (1.17) has a real root $x_0^2 > 0$, which concludes the proof. □

The relation between the inner product and the Jordan product of two vectors is given below.

Fact 1.4.6. For $\boldsymbol{x} = (x_0, \boldsymbol{x}_1) \in \mathbb{L}^n$ and $\boldsymbol{s} = (s_0, \boldsymbol{s}_1) \in \mathbb{L}^n$, the following three statements are equivalent:

(a) $\boldsymbol{x}^{\mathrm{T}} \boldsymbol{s} = 0$.

(b) $\boldsymbol{x} \circ \boldsymbol{s} = \boldsymbol{0}$.

(c) $s_0 \boldsymbol{x}_1 + x_0 \boldsymbol{s}_1 = \boldsymbol{0}$. ■

Proof. It suffices to show the assertion "(a) ⇔ (c)".

From the Cauchy–Schwarz inequality and the assumptions that

$$x_0 \geq \|\boldsymbol{x}_1\|, \quad s_0 \geq \|\boldsymbol{s}_1\|, \tag{1.18}$$

we obtain

$$\boldsymbol{x}_1^{\mathrm{T}} \boldsymbol{s}_1 \geq -\|\boldsymbol{x}_1\| \|\boldsymbol{s}_1\| \tag{1.19}$$
$$\geq -x_0 s_0. \tag{1.20}$$

Hence, $\boldsymbol{x}^{\mathrm{T}} \boldsymbol{s} = x_0 s_0 + \boldsymbol{x}_1^{\mathrm{T}} \boldsymbol{s}_1 = 0$ if and only if the inequalities both in (1.19) and (1.20) are satisfied as equalities. From (1.18) and the equality in (1.20), we obtain

$$x_0 = \|\boldsymbol{x}_1\|, \quad s_0 = \|\boldsymbol{s}_1\|. \tag{1.21}$$

The equality in (1.19) holds if and only if either

$$\begin{cases} \boldsymbol{x}_1 = \boldsymbol{0}; \\ \boldsymbol{s}_1 = \boldsymbol{0}; \\ \boldsymbol{x}_1 \neq \boldsymbol{0},\ \boldsymbol{s}_1 \neq \boldsymbol{0}, \text{ and } \boldsymbol{x}_1 = -\alpha \boldsymbol{s}_1 \ (\alpha > 0) \end{cases} \tag{1.22}$$

is satisfied.[16] Thus we see that (a) holds if and only if (1.21) and (1.22) are satisfied. but the latter condition is equivalent to (c). □

[16] The final case in (1.22) means that $\boldsymbol{x}_1$ is parallel to $\boldsymbol{s}_1$ in the opposite direction.

For $\boldsymbol{w}_1 \in \mathbb{R}^{n-1}$ satisfying $\|\boldsymbol{w}_1\| = 1/2$, consider $\boldsymbol{\phi}^{(1)}, \boldsymbol{\phi}^{(2)} \in \mathbb{R}^n$ defined by

$$\boldsymbol{\phi}^{(1)} = \begin{bmatrix} 1/2 \\ -\boldsymbol{w}_1 \end{bmatrix}, \quad \boldsymbol{\phi}^{(2)} = \begin{bmatrix} 1/2 \\ \boldsymbol{w}_1 \end{bmatrix}.$$

We call a pair of vectors $\{\boldsymbol{\phi}^{(1)}, \boldsymbol{\phi}^{(2)}\}$ a *Jordan frame*[17] with respect to $\mathbb{L}^n$. The following result is analogous to Fact 1.3.20 stated for $\mathbb{S}^n_+$.

Fact 1.4.7. $\boldsymbol{x} \in \mathbb{R}^n$ and $\boldsymbol{s} \in \mathbb{R}^n$ satisfy

$$\boldsymbol{x} \in \mathbb{L}^n, \quad \boldsymbol{s} \in \mathbb{L}^n, \quad \boldsymbol{x}^{\mathrm{T}}\boldsymbol{s} = 0$$

if and only if there exists a Jordan frame $\{\boldsymbol{q}^{(1)}, \boldsymbol{q}^{(2)}\}$ associated with $\mathbb{L}^n$ satisfying

$$\boldsymbol{x} = \sum_{r=1}^{2} \lambda_{(r)} \boldsymbol{q}^{(r)}, \quad \boldsymbol{s} = \sum_{r=1}^{2} \omega_{(r)} \boldsymbol{q}^{(r)},$$

and the condition

$$\lambda_{(r)} \geq 0, \quad \omega_{(r)} \geq 0, \quad \lambda_{(r)}\omega_{(r)} = 0 \quad (r = 1, 2)$$

holds. ■

In Fact 1.4.7 we see a condition that $\boldsymbol{x}$ and $\boldsymbol{s}$ share a common Jordan frame. On the other hand, for a given vector the corresponding Jordan frame is represented as follows. For $\boldsymbol{x} = (x_0, \boldsymbol{x}_1) \in \mathbb{R} \times \mathbb{R}^{n-1}$ $(n \geq 2)$, define $\lambda_{(r)}$ and $\boldsymbol{\phi}^{(r)}$ by

$$\lambda_{(1)} = x_0 - \|\boldsymbol{x}_1\|, \quad \boldsymbol{\phi}^{(1)} = \frac{1}{2}\begin{bmatrix} 1 \\ -\boldsymbol{x}_1/\|\boldsymbol{x}_1\| \end{bmatrix},$$
$$\lambda_{(2)} = x_0 + \|\boldsymbol{x}_1\|, \quad \boldsymbol{\phi}^{(2)} = \frac{1}{2}\begin{bmatrix} 1 \\ \boldsymbol{x}_1/\|\boldsymbol{x}_1\| \end{bmatrix}$$

in the case of $\boldsymbol{x}_1 \neq \boldsymbol{0}$. If $\boldsymbol{x}_1 = \boldsymbol{0}$, then with any $\boldsymbol{s}_1 \in \mathbb{R}^{n-1}$ we let

$$\boldsymbol{\phi}^{(1)} = \frac{1}{2}\begin{bmatrix} 1 \\ -\boldsymbol{s}_1/\|\boldsymbol{s}_1\| \end{bmatrix}, \quad \boldsymbol{\phi}^{(2)} = \frac{1}{2}\begin{bmatrix} 1 \\ \boldsymbol{s}_1/\|\boldsymbol{s}_1\| \end{bmatrix}.$$

We then obtain the identity

$$\boldsymbol{x} = \lambda_{(1)}\boldsymbol{\phi}^{(1)} + \lambda_{(2)}\boldsymbol{\phi}^{(2)},$$

[17] Every Jordan frame $\{\boldsymbol{q}^{(1)}, \boldsymbol{q}^{(2)}\}$ satisfies $\boldsymbol{q}^{(1)} \circ \boldsymbol{q}^{(2)} = \boldsymbol{0}$, $\boldsymbol{q}^{(1)} + \boldsymbol{q}^{(2)} = \boldsymbol{e}$, and $\boldsymbol{q}^{(r)} \circ \boldsymbol{q}^{(r)} = \boldsymbol{q}^{(r)}$ $(r = 1, 2)$. By comparing the eigenvalue decomposition of $A \in \mathbb{S}^n$, i.e., $A = \sum_{r=1}^{n} \lambda_r \boldsymbol{\phi}_r \boldsymbol{\phi}_r^{\mathrm{T}}$, we see that a Jordan frame plays a role of the set of rank-one matrices $\boldsymbol{\phi}_r \boldsymbol{\phi}_r^{\mathrm{T}}$ $(r = 1, \dots, n)$ formed from the orthonormal eigenvectors in symmetric matrix algebra. This observation may help us understand the similarity between Fact 1.3.20 and Fact 1.4.7.

TABLE 1.4: The implication of the complementarity condition over the second-order cone. The condition (1.23) is satisfied if and only if any one of the following six states holds.

	$\boldsymbol{x} = (x_0, \boldsymbol{x}_1) \in \mathbb{R} \times \mathbb{R}^{n-1}$		$\boldsymbol{s} = (s_0, \boldsymbol{s}_1) \in \mathbb{R} \times \mathbb{R}^{n-1}$	
	inclusion	inequality	inclusion	inequality
(i)	$\boldsymbol{x} \in \operatorname{int} \mathbb{L}^n$	$(x_0 > \lVert\boldsymbol{x}_1\rVert)$	$\boldsymbol{s} = \mathbf{0}$	
(ii)	$\boldsymbol{x} = \mathbf{0}$		$\boldsymbol{s} \in \operatorname{int} \mathbb{L}^n$	$(s_0 > \lVert\boldsymbol{s}_1\rVert)$
(iii)†	$\boldsymbol{x} \in \operatorname{bd} \mathbb{L}^n \setminus \{\mathbf{0}\}$	$(x_0 = \lVert\boldsymbol{x}_1\rVert > 0)$	$\boldsymbol{s} \in \operatorname{bd} \mathbb{L}^n \setminus \{\mathbf{0}\}$	$(s_0 = \lVert\boldsymbol{s}_1\rVert > 0)$
(iv)	$\boldsymbol{x} \in \operatorname{bd} \mathbb{L}^n \setminus \{\mathbf{0}\}$	$(x_0 = \lVert\boldsymbol{x}_1\rVert > 0)$	$\boldsymbol{s} = \mathbf{0}$	
(v)	$\boldsymbol{x} = \mathbf{0}$		$\boldsymbol{s} \in \operatorname{bd} \mathbb{L}^n \setminus \{\mathbf{0}\}$	$(s_0 = \lVert\boldsymbol{s}_1\rVert > 0)$
(vi)	$\boldsymbol{x} = \mathbf{0}$		$\boldsymbol{s} = \mathbf{0}$	

†In this case, there should exist a scalar $\alpha > 0$ satisfying $\boldsymbol{x}_1 = -\alpha \boldsymbol{s}_1$.

which is called the spectral factorization of $\boldsymbol{x}$ with respect to $\mathbb{L}^n$. Here, $\lambda_{(r)}$'s are called the spectral values of $\boldsymbol{x}$, while $\boldsymbol{\phi}^{(r)}$'s are called the spectral vectors. Thus in the spectral factorization on $\mathbb{L}^n$ each vector has two spectral values. It is easy to see that (i) $\lambda_{(1)} \leq \lambda_{(2)}$; (ii) $\boldsymbol{x} \in \mathbb{L}^n$ if and only if $\lambda_{(1)} \geq 0$; (iii) $\lVert\boldsymbol{\phi}^{(1)}\rVert = \lVert\boldsymbol{\phi}^{(2)}\rVert = \sqrt{2}$; (iv) $(\boldsymbol{\phi}^{(1)})^{\mathrm{T}}\boldsymbol{\phi}^{(2)} = 0$; and (v) $2\lVert\boldsymbol{x}\rVert^2 = \lambda_{(1)}^2 + \lambda_{(2)}^2$.

The complementarity condition over $\mathbb{L}^n$ established in Fact 1.4.7 may be captured more intuitively as follows. By summarizing Fact 1.4.6 and Fact 1.4.7, we see that $\boldsymbol{x}$ and $\boldsymbol{s}$ satisfy

$$\boldsymbol{x} \in \mathbb{L}^n, \quad \boldsymbol{s} \in \mathbb{L}^n, \quad \boldsymbol{x}^{\mathrm{T}}\boldsymbol{s} = 0 \tag{1.23}$$

if and only if they satisfy

$$x_0 \geq \lVert\boldsymbol{x}_1\rVert, \quad s_0 \geq \lVert\boldsymbol{s}_1\rVert, \tag{1.24}$$

and

$$x_0 > \lVert\boldsymbol{x}_1\rVert \quad \Rightarrow \quad s_0 = \lVert\boldsymbol{s}_1\rVert = 0, \tag{1.25a}$$

$$s_0 > \lVert\boldsymbol{s}_1\rVert \quad \Rightarrow \quad x_0 = \lVert\boldsymbol{x}_1\rVert = 0, \tag{1.25b}$$

$$x_0 = \lVert\boldsymbol{x}_1\rVert,\ s_0 = \lVert\boldsymbol{s}_1\rVert \quad \Rightarrow \quad \lVert\boldsymbol{x}_1\rVert\boldsymbol{s}_1 + \lVert\boldsymbol{s}_1\rVert\boldsymbol{x}_1 = \mathbf{0}. \tag{1.25c}$$

More specifically, for $\boldsymbol{x}$ and $\boldsymbol{s}$ satisfying (1.23), any one of the states listed in Table 1.4 is possible (see also Table 1.1 and Table 1.3 for comparison with the cases of $\mathbb{R}_+^n$ and $\mathbb{S}_+^n$). Here, in cases (i)–(iii) of Table 1.4 we say that $\boldsymbol{x}$ and $\boldsymbol{s}$ satisfy the strict complementarity condition, while in cases (iv)–(vi) the strict complementarity is not satisfied. Equivalently, in (1.23) the strict complementarity holds if and only if $\boldsymbol{x} + \boldsymbol{s} \in \operatorname{int} \mathbb{L}^n$, i.e., $\lambda_{(r)} + \omega_{(r)} > 0$ $(r = 1, 2)$ in the simultaneous spectral factorization in Fact 1.4.7.

The properties established above concerning the complementarity condition over the single second-order cone can be extended in a straightforward manner to that over several second-order cones. Suppose that the two vectors, $\boldsymbol{x}$ and $\boldsymbol{s}$, consist of several sub-vectors as

$$\boldsymbol{x} = \begin{bmatrix} \boldsymbol{x}_1 \\ \vdots \\ \boldsymbol{x}_k \end{bmatrix}, \quad \boldsymbol{s} = \begin{bmatrix} \boldsymbol{s}_1 \\ \vdots \\ \boldsymbol{s}_k \end{bmatrix},$$
$$\boldsymbol{x}_i, \boldsymbol{s}_i \in \mathbb{R}^{n_i} \quad (i = 1, \dots, k),$$

and further the components of $\boldsymbol{x}_i$ and $\boldsymbol{s}_i$ are written as

$$\boldsymbol{x}_i = \begin{bmatrix} x_{i0} \\ \boldsymbol{x}_{i1} \end{bmatrix}, \quad \boldsymbol{s}_i = \begin{bmatrix} s_{i0} \\ \boldsymbol{s}_{i1} \end{bmatrix}.$$

Each sub-vector $\boldsymbol{x}_i$ of $\boldsymbol{x}$ is supposed to belong to the n_i-dimensional second-order cone; i.e.,

$$\boldsymbol{x}_1 \in \mathbb{L}^{n_1}, \dots, \boldsymbol{x}_k \in \mathbb{L}^{n_k} \quad \Leftrightarrow \quad x_{i0} \geq \|\boldsymbol{x}_{i1}\| \ (i = 1, \dots, k).$$

Similarly, we suppose that $\boldsymbol{s}$ satisfies

$$\boldsymbol{s}_1 \in \mathbb{L}^{n_1}, \dots, \boldsymbol{s}_k \in \mathbb{L}^{n_k} \quad \Leftrightarrow \quad s_{i0} \geq \|\boldsymbol{s}_{i1}\| \ (i = 1, \dots, k).$$

Then $\boldsymbol{x}$ and $\boldsymbol{s}$ satisfy the complementarity condition, $\boldsymbol{x}^{\mathrm{T}}\boldsymbol{s} = 0$, if and only if each pair of $\boldsymbol{x}_i$ and $\boldsymbol{s}_i$ satisfies the complementarity condition. More precisely, the following assertion is obtained as an easy extension of Fact 1.4.6.

Fact 1.4.8. For $\boldsymbol{x}_i \in \mathbb{L}^{n_i}$, $\boldsymbol{s}_i \in \mathbb{L}^{n_i}$ $(i = 1, \dots, k)$, the following three statements are equivalent:

(a) $\sum_{i=1}^{k} \boldsymbol{x}_i^{\mathrm{T}} \boldsymbol{s}_i = 0$.

(b) $\boldsymbol{x}_i^{\mathrm{T}} \boldsymbol{s}_i = 0$ $(i = 1, \dots, k)$.

(c) $x_{i0}\boldsymbol{s}_{i1} + s_{i0}\boldsymbol{x}_{i1} = 0$ $(i = 1, \dots, k)$.

In section 1.6.3 we study the optimization problem associated with the second-order cone, termed the second-order cone programming, in which the variable vector is subjected to several second-order cone constraints as considered in Fact 1.4.8.

1.5 Conic Constraints and Their Relationship

As seen in section 1.1.2, a proper cone $C \subseteq \mathbb{V}$ can induce a partial order on the Euclidean space $\mathbb{V}$. This partial order can be regarded as a "generalized inequality," which is called a conic inequality. Throughout the book we treat optimization problems and complementarity conditions involving some conic inequalities as constraints. More specifically, let $\mathbb{Y}$ be another Euclidean space, and let $g : \mathbb{Y} \to \mathbb{V}$ be a (possibly, e.g., vector-valued or matrix-valued) function. Then we consider a constraint in the form of

$$g(x) \in \mathcal{C} \quad (\text{with } \mathcal{C} = \mathbb{R}^n_+, \mathbb{L}^n, \text{or } \mathbb{S}^n_+).$$

In this section we are particularly interested in the cases in which g is affine.

Associated with the nonnegative orthant, the conic constraint defined with an affine mapping g is reduced to conventional *linear inequality constraints*. Indeed, for given $A \in \mathbb{R}^{k\times n}$ and $\boldsymbol{b} \in \mathbb{R}^k$, we have that

$$A\boldsymbol{x} \geq \boldsymbol{b} \quad \Leftrightarrow \quad A\boldsymbol{x} - \boldsymbol{b} \in \mathbb{R}^k_+,$$

where $\mathbb{V} = \mathbb{R}^k$, $\mathbb{Y} = \mathbb{R}^n$, and $\mathcal{C} = \mathbb{R}^k_+$ in the expression above. We call a constraint on $\boldsymbol{x}$ in the form of

$$\boldsymbol{p}^{\mathrm{T}}\boldsymbol{x} + q \geq \|A\boldsymbol{x} + \boldsymbol{b}\|,$$

where $\boldsymbol{p} \in \mathbb{R}^n$ and $q \in \mathbb{R}$ are also given, a *second-order cone constraint*, since it is equivalent to requiring

$$\begin{bmatrix} \boldsymbol{p}^{\mathrm{T}}\boldsymbol{x} + q \\ A\boldsymbol{x} + \boldsymbol{b} \end{bmatrix} \in \mathbb{L}^{k+1}.$$

Concerning $\mathbb{S}^n_+$, we consider the *positive-semidefinite constraint*: Given $A_1, \ldots, A_m \in \mathbb{S}^n$ and $B \in \mathbb{S}^n$, the positive-semidefinite constraint is a constraint of the form

$$\sum_{i=1}^{m} x_i A_i + B \succeq O.$$

This condition is also referred to as the *linear matrix inequality* (LMI).

Our next concern is the relation among linear inequality constraints, second-order cone constraints, and positive-semidefinite constraints.

Every linear inequality constraint can be represented as a second-order cone constraint. Indeed, we have that

$$\boldsymbol{a}^{\mathrm{T}}\boldsymbol{x} \geq b \quad \Leftrightarrow \quad \boldsymbol{a}^{\mathrm{T}}\boldsymbol{x} - b \geq \|\mathbf{0}\|.$$

A set of some linear inequalities can be represented as a positive-semidefinite constraint, because

$$A\boldsymbol{x} \geq \boldsymbol{b} \quad \Leftrightarrow \quad \mathrm{diag}(A\boldsymbol{x} - \boldsymbol{b}) \succeq O,$$

where we denote by $\mathrm{diag}(\boldsymbol{d})$ the $n \times n$ diagonal matrix with a vector $\boldsymbol{d} \in \mathbb{R}^n$ on its diagonal.

We can express a set of positive-semidefinite constraints by using a block-diagonal matrix. More precisely, several positive-semidefinite constraints of varying dimensions can be formulated equivalently as a single positive-semidefinite constraint, because

$$X^{(1)} \in \mathbb{S}_+^{n_1}, X^{(2)} \in \mathbb{S}_+^{n_2}, \dots, X^{(k)} \in \mathbb{S}_+^{n_k} \quad \Leftrightarrow \quad \begin{bmatrix} X^{(1)} & O & \dots & O \\ O & X^{(2)} & \dots & O \\ \vdots & \vdots & \ddots & \vdots \\ O & O & \dots & X^{(k)} \end{bmatrix} \succeq O.$$

Thus linear inequality constraint is included in second-order cone constraint and positive-semidefinite constraint as particular cases.

Fact 1.5.1 implies that a second-order cone constraint can be expressed as a positive-semidefinite constraint.

Fact 1.5.1. For $x_0 \in \mathbb{R}$ and $\boldsymbol{x}_1 \in \mathbb{R}^{n-1}$, we have that

$$x_0 \geq \|\boldsymbol{x}_1\| \quad (x_0 > \|\boldsymbol{x}_1\|)$$

if and only if[18]

$$A(\boldsymbol{x}) := \begin{bmatrix} x_0 & \boldsymbol{x}_1^{\mathrm{T}} \\ \boldsymbol{x}_1 & x_0 I_{n-1} \end{bmatrix} \succeq O \quad (\succ O).$$

In other words, $\boldsymbol{x} \in \mathbb{L}^n$ (resp. $\in \operatorname{int} \mathbb{L}^n$) if and only if $A(\boldsymbol{x}) \in \mathbb{S}_+^n$ (resp. $\in \mathbb{S}_{++}^n$). To show, e.g., the assertion "$\boldsymbol{x} \in \mathbb{L}^n \Leftrightarrow A(\boldsymbol{x}) \in \mathbb{S}_+^n$," we first observe that each of $\boldsymbol{x} \in \mathbb{L}^n$ and $A(\boldsymbol{x}) \in \mathbb{S}_+^n$ implies that either $x_0 = 0$ or $x_0 > 0$ holds. Assuming $x_0 = 0$, we obtain $\boldsymbol{x}_1 = \boldsymbol{0}$ for both conditions (since, for $A(\boldsymbol{x}) \in \mathbb{S}_+^n$, otherwise by choosing $\boldsymbol{z} := (-1, \boldsymbol{x}_1)$ we would have $\boldsymbol{z}^{\mathrm{T}} A(\boldsymbol{x}) \boldsymbol{z} = -2\boldsymbol{x}_1^{\mathrm{T}} \boldsymbol{x}_1 < 0$). For $x_0 > 0$, from Fact 1.3.10 we see that $X \succeq O$ if and only if the Schur complement in X satisfies $x_0 - \boldsymbol{x}_1^{\mathrm{T}} (x_0 I)^{-1} \boldsymbol{x}_1 \geq 0$, i.e., $x_0 \geq \|\boldsymbol{x}_1\|$. Since $A(\boldsymbol{x})$ is linear in $\boldsymbol{x}$, the condition $A(\boldsymbol{x}) \succeq O$ is a linear matrix inequality; hence, Fact 1.5.1 implies that a second-order cone constraint can be expressed as a positive-semidefinite constraint.

It should be emphasized that positive-semidefinite constraints and second-order cone constraints can represent various nonlinear convex constraints. One of the most important examples is the convex quadratic constraint shown below.

Fact 1.5.2 (Convex quadratic constraint)**.** Let $Q \in \mathbb{S}_+^n$, $\boldsymbol{p} \in \mathbb{R}^n$, and $r \in \mathbb{R}$. Then the *convex quadratic constraint*

$$\boldsymbol{x}^{\mathrm{T}} Q \boldsymbol{x} + \boldsymbol{p}^{\mathrm{T}} \boldsymbol{x} + r \leq 0$$

is equivalent to each of the following conditions:

[18] The matrix $A(\boldsymbol{x})$ is sometimes called the *arrow-shaped matrix*; see, e.g., Alizadeh–Goldfarb [8, section 1].

(a) $\boldsymbol{x}$ satisfies the linear matrix inequality[19]

$$\begin{bmatrix} I_n & Q^{1/2}\boldsymbol{x} \\ \boldsymbol{x}^{\mathrm{T}}Q^{1/2} & -\boldsymbol{p}^{\mathrm{T}}\boldsymbol{x} - r \end{bmatrix} \succeq O.$$

(b) $\boldsymbol{x}$ satisfies the second-order cone constraint

$$-\boldsymbol{p}^{\mathrm{T}}\boldsymbol{x} - r + 1 \geq \left\| \begin{bmatrix} -\boldsymbol{p}^{\mathrm{T}}\boldsymbol{x} - r - 1 \\ 2Q^{1/2}\boldsymbol{x} \end{bmatrix} \right\|. \qquad \blacksquare$$

Fact 1.5.2 (a) immediately follows from Fact 1.3.10. Fact 1.5.2 (b) can be obtained as a special case of the following hyperbolic constraint.

Fact 1.5.3 (Hyperbolic constraint)**.** Let $x, y \in \mathbb{R}$ and $\boldsymbol{w} \in \mathbb{R}^n$. The *hyperbolic constraint*

$$xy \geq \boldsymbol{w}^{\mathrm{T}}\boldsymbol{w}, \quad x \geq 0, \quad y \geq 0$$

is satisfied if and only if the second-order cone constraint

$$x + y \geq \left\| \begin{bmatrix} x - y \\ 2\boldsymbol{w} \end{bmatrix} \right\|$$

is satisfied. ■

1.6 Conic Optimization

This section introduces the *conic optimization* (or, more precisely, *conic linear program*), which is the optimization problem of a linear function over the intersection of a proper cone and affine space. Conic optimization is also regarded as the optimization problem consisting of the conic constraints investigated in section 1.5 and the linear objective function. In Parts II and IV we shall see that the conic optimization and its extension play a key role in the study of various problems in nonsmooth mechanics.

Let $\mathcal{C} \subseteq \mathbb{V}$ be a proper cone. Given $a_i \in \mathbb{V}$, $b_i \in \mathbb{R}$ $(i = 1, \dots, m)$, and $c \in \mathbb{V}$, consider the optimization problem

$$\left.\begin{aligned} \mathrm{CO_P} : \min_{x} \ & \langle c, x \rangle \\ \mathrm{s.t.} \ & \langle a_i, x \rangle = b_i, \quad i = 1, \dots, m, \\ & x \succeq_{\mathcal{C}} 0, \end{aligned}\right\} \qquad (1.26)$$

[19]Recall that $Q^{1/2}$ denotes the symmetric square root of the positive-semidefinite Q. See Definition 1.3.4.

where $x \in \mathbb{V}$ is the variable.[20] We call $\mathrm{CO_P}$ the *linear programming problem over convex cones*, or the *conic optimization* problem. The *dual problem* of $\mathrm{CO_P}$ is the optimization problem formulated in the variables $y \in \mathbb{R}^m$ and $s \in \mathbb{V}$ as

$$\left.\begin{aligned} \mathrm{CO_D} : \max_{y,s} \ & \sum_{i=1}^{m} b_i y_i \\ \text{s.t.} \ & \sum_{i=1}^{m} y_i a_i + s = c, \\ & s \succeq_{\mathcal{C}^*} 0, \end{aligned}\right\} \tag{1.27}$$

where $\mathcal{C}^*$ is the dual cone of $\mathcal{C}$.

Note that both $\mathrm{CO_P}$ and $\mathrm{CO_D}$ are convex optimization problems. Indeed, for $\mathrm{CO_P}$ we see that the objective function $\langle c, x\rangle$ is a linear function, the equality constraints are also linear, and the conic inequality constraint, $x \succeq_{\mathcal{C}} 0$, is convex because by definition a proper cone $\mathcal{C}$ is convex.[21] Similarly, $\mathrm{CO_D}$ has an linear objective function, linear equality constraints, and the conic inequality constraint, $s \succeq_{\mathcal{C}^*} 0$, is convex since by definition of a dual cone $\mathcal{C}^*$ is convex (see section 1.2.1).

If $\mathcal{C}$ is a self-dual cone, then it is possible to reformulate $\mathrm{CO_D}$ in the form of $\mathrm{CO_P}$, and in that sense $\mathrm{CO_P}$ and $\mathrm{CO_D}$ belong to the same class of optimization problems. In sections 1.6.1–1.6.3 we study such conic optimization problems, particularly those associated with $\mathcal{C} = \mathbb{R}^n_+$, $\mathbb{L}^n$, or $\mathbb{S}^n_+$. The relation between the primal and dual problems, together with the optimality conditions, is the topic of Chapter 2.

1.6.1 Linear programming

We begin with the simplest special class of the conic optimization. Specifically, consider problems (1.26) and (1.27) with $\mathbb{V} = \mathbb{R}^n$ and $\mathcal{C} = \mathbb{R}^n_+$. Then $\langle \cdot, \cdot \rangle$ in (1.26) is identified with the usual inner product of two n-dimensional vectors. The resulting optimization problem is called the linear programming problem, which is described formally below.

Definition 1.6.1 (LP). Given $\boldsymbol{a}_1, \dots, \boldsymbol{a}_m, \boldsymbol{c} \in \mathbb{R}^n$ and $\boldsymbol{b} \in \mathbb{R}^m$, the *linear programming* (LP) problem is the optimization problem defined as

$$\left.\begin{aligned} \mathrm{LP_P} : \min_{\boldsymbol{x}} \ & \boldsymbol{c}^{\mathrm{T}} \boldsymbol{x} \\ \text{s.t.} \ & \boldsymbol{a}_i^{\mathrm{T}} \boldsymbol{x} = b_i, \quad i = 1, \dots, m, \\ & \boldsymbol{x} \geq \boldsymbol{0}, \end{aligned}\right\}$$

[20]Recall that the conic inequality $x \succeq_{\mathcal{C}} 0$ means $x \in \mathcal{C}$ (see section 1.1.2).

[21]See section 1.1.1 for the definition of a proper cone.

where $\boldsymbol{x} \in \mathbb{R}^n$ is the variable vector. The dual problem of $\mathrm{LP_P}$ is defined as

$$\left.\begin{aligned} \mathrm{LP_D} : \max_{\boldsymbol{y},\boldsymbol{s}} \ & \sum_{i=1}^{m} b_i y_i \\ \text{s.t.} \ & \sum_{i=1}^{m} y_i \boldsymbol{a}_i + \boldsymbol{s} = \boldsymbol{c}, \\ & \boldsymbol{s} \geq \boldsymbol{0}, \end{aligned}\right\}$$

where $\boldsymbol{y} \in \mathbb{R}^m$ and $\boldsymbol{s} \in \mathbb{R}^n$ are the variables. ■

In other words, the optimization problem is called an LP if its objective function is linear and constraints are some linear equalities and linear inequalities. Note that $\mathrm{LP_D}$ is also an LP and can be converted to the form of $\mathrm{LP_P}$.

Remark 1.6.2. A few words about the notation for the consistency with most textbooks on optimization. Define $A \in \mathbb{R}^{m\times n}$ by

$$A = \begin{bmatrix} \boldsymbol{a}_1^{\mathrm{T}} \\ \hline \vdots \\ \hline \boldsymbol{a}_m^{\mathrm{T}} \end{bmatrix},$$

and then the equality constraints of $\mathrm{LP_P}$ are simply written as $A\boldsymbol{x} = \boldsymbol{b}$. Therewith, the equality constraints of $\mathrm{LP_D}$ can be rewritten as $A^{\mathrm{T}}\boldsymbol{y} + \boldsymbol{s} = \boldsymbol{c}$. ■

Concerning the complementarity problem (CP), defined in Definition 1.2.1, we consider a special case in which $\mathcal{C} = \mathcal{C}^* = \mathbb{R}^n_+$, $\mathbb{V} = \mathbb{R}^n$, and G is an affine function; in this case, CP is called a linear complementarity problem, as defined rigorously below.

Definition 1.6.3 (LCP). Given a matrix $M \in \mathbb{R}^{n\times n}$ and a vector $\boldsymbol{q} \in \mathbb{R}^n$, the *linear complementarity problem* (LCP) is to find $\boldsymbol{x} \in \mathbb{R}^n$ satisfying

$$\mathrm{LCP}: \quad \boldsymbol{0} \leq \boldsymbol{x} \perp M\boldsymbol{x} + \boldsymbol{q} \geq \boldsymbol{0}.$$ ■

Similarly, as a special case of MiCP (mixed complementarity problem), defined in Definition 1.2.2, suppose that both G_1 and G_2 are affine and $\mathcal{C} = \mathcal{C}^* = \mathbb{R}^n_+$. Then the corresponding problem is called a *mixed linear complementarity problem* (MLCP). More specifically, MLCP is written in the variables $\boldsymbol{x}_1 \in \mathbb{R}^{n_1}$ and $\boldsymbol{x}_2 \in \mathbb{R}^{n_2}$ as

$$A_1\boldsymbol{x}_1 + A_2\boldsymbol{x}_2 + \boldsymbol{b} = \boldsymbol{0}, \tag{1.28a}$$

$$\boldsymbol{0} \leq \boldsymbol{x}_1 \perp M_1\boldsymbol{x}_1 + M_2\boldsymbol{x}_2 + \boldsymbol{q} \geq \boldsymbol{0}, \tag{1.28b}$$

for given matrices A_1, A_2, M_1, M_2 and vectors $\boldsymbol{b}$, $\boldsymbol{q}$. Note that the optimality condition of an LP is written in the form of the MLCP; see section 6.1.2.1.

1.6.2 Semidefinite programming

Our next concern is the conic optimization induced by the cone of positive-semidefinite matrices. More precisely, we consider problems (1.26) and (1.27) with $\mathbb{V} = \mathbb{S}^n$ and $\mathcal{C} = \mathbb{S}^n_+$. In this case, $\langle \cdot, \cdot \rangle$ in (1.26) is understood as the inner product of matrices defined in Definition 1.3.12. The optimization problem obtained is called a semidefinite programming problem, which is formally defined below.

Definition 1.6.4 (SDP). Given $A_1, \dots, A_m, C \in \mathbb{S}^n$ and $\boldsymbol{b} \in \mathbb{R}^m$, the *semidefinite programming* (SDP) problem is the optimization problem defined as

$$\left.\begin{aligned} \mathrm{SDP_P} : \min_{X} \ & C \bullet X \\ \text{s.t.} \ & A_i \bullet X = b_i, \quad i = 1, \dots, m, \\ & X \succeq O. \end{aligned}\right\} \tag{1.29}$$

where $X \in \mathbb{S}^n$ is the variable matrix. The problem dual to $\mathrm{SDP_P}$ is formulated as

$$\left.\begin{aligned} \mathrm{SDP_D} : \max_{\boldsymbol{y}, S} \ & \sum_{i=1}^{m} b_i y_i \\ \text{s.t.} \ & \sum_{i=1}^{m} y_i A_i + S = C, \\ & S \succeq O, \end{aligned}\right\} \tag{1.30}$$

where $\boldsymbol{y} \in \mathbb{R}^m$ and $S \in \mathbb{S}^n$ are the variables. ■

In other words, SDP is the optimization problem of a linear function under linear matrix inequality constraints and linear equality constraints. Since $X \succeq O$ is a nonlinear convex constraint, an SDP is a nonlinear convex optimization problem. We often call an optimization problem the SDP if it can be reduced to the form of $\mathrm{SDP_P}$ without showing the explicit reduction. Note that $\mathrm{SDP_P}$ and $\mathrm{SDP_D}$ are called the standard forms of primal and dual SDP problems, respectively.

Sometimes $\mathrm{SDP_P}$ (and also $\mathrm{SDP_D}$) is called the *linear SDP*, although it is a nonlinear optimization problem. By this terminology we distinguish the usual SDP problem from the problem including some additional nonlinear constraints other than the positive-semidefinite constraint.

As shown in section 1.5, positive-semidefinite constraint includes linear inequality constraint as a particular case. Consequently, LP is a special case of SDP.

The *quadratic programming* (QP) problem is the minimization problem of a convex quadratic function over an affine space. A QP problem can be expressed in the form of

$$\left.\begin{aligned} \mathrm{QP_P} : \min_{\boldsymbol{x} \in \mathbb{R}^n} \ & (1/2)\boldsymbol{x}^{\mathrm{T}} Q \boldsymbol{x} + \boldsymbol{c}^{\mathrm{T}} \boldsymbol{x} \\ \text{s.t.} \ & A\boldsymbol{x} = \boldsymbol{b}, \\ & \boldsymbol{x} \geq \boldsymbol{0}, \end{aligned}\right\} \tag{1.31}$$

where $Q \succeq O$.[22] If the constraints are also convex quadratic inequalities, then the problem is called the *quadratically constrained quadratic programming* (QCQP) problem, which is formulated as

$$\left.\begin{array}{ll} \mathrm{QCQP}: \min\limits_{\boldsymbol{x}\in\mathbb{R}^n} & (1/2)\boldsymbol{x}^{\mathrm{T}}Q\boldsymbol{x} + \boldsymbol{c}^{\mathrm{T}}\boldsymbol{x} \\ \qquad\quad \text{s.t.} & (1/2)\boldsymbol{x}^{\mathrm{T}}Q_l\boldsymbol{x} + \boldsymbol{p}_l^{\mathrm{T}}\boldsymbol{x} + r_l \le 0, \quad l = 1,\dots,k, \end{array}\right\} \tag{1.32}$$

where $Q, Q_1, \dots, Q_k \succeq O$. As we see in Fact 1.5.2 (a), any convex quadratic inequality can be represented as a linear matrix inequality; thus QP and QCQP are special cases of SDP. For example, the QP in (1.31) can be converted to an SDP as

$$\left.\begin{array}{ll} \min\limits_{\boldsymbol{x},t} & t \\ \text{s.t.} & \begin{bmatrix} I_n & Q^{1/2}\boldsymbol{x} \\ \boldsymbol{x}^{\mathrm{T}}Q^{1/2} & 2(t - \boldsymbol{p}^{\mathrm{T}}\boldsymbol{x}) \end{bmatrix} \succeq O, \\ & A\boldsymbol{x} = \boldsymbol{b}, \\ & \boldsymbol{x} \ge \boldsymbol{0}, \end{array}\right\}$$

where $t \in \mathbb{R}$ is an auxiliary variable.

1.6.3 Second-order cone programming

A *second-order cone programming* (SOCP) problem is a minimization of a linear objective function under some second-order cone constraint and affine constraints. As a simplest example of SOCP, consider the optimization problem

$$\left.\begin{array}{ll} \min\limits_{\boldsymbol{x}=(x_0,\boldsymbol{x}_1)} & \boldsymbol{c}^{\mathrm{T}}\boldsymbol{x} \\ \text{s.t.} & A\boldsymbol{x} = \boldsymbol{b}, \\ & x_0 \ge \|\boldsymbol{x}_1\|, \end{array}\right\} \tag{1.33}$$

where $(x_0, \boldsymbol{x}_1) \in \mathbb{R} \times \mathbb{R}^{n-1}$ is the variable vector, $\boldsymbol{b} \in \mathbb{R}^m$ and $\boldsymbol{c} \in \mathbb{R}^n$ are constant vectors, and $A \in \mathbb{R}^{m\times n}$ is a constant matrix. Note that we here consider the n-dimensional second-order cone; i.e., we consider the $\mathrm{CO_P}$ in (1.26) with $\mathcal{C} = \mathbb{L}^n$.

Problem (1.33) includes a single second-order cone constraint. More generically, we usually consider a problem with several second-order cone constraints, which is called an SOCP as well. Let $\boldsymbol{x}_1, \dots, \boldsymbol{x}_k$ be the variable vectors, where $\boldsymbol{x}_l \in \mathbb{R}^{n_l}$ $(l = 1, \dots, k)$. Suppose that each $\boldsymbol{x}_l$ should be included

[22] Problem (1.31) with $Q \not\succeq O$ is also called a quadratic programming problem by some authors. The terms "convex quadratic programming problem" and "nonconvex quadratic programming problem" are also used to distinguish problems with $Q \succeq O$ and $Q \not\succeq O$. Furthermore, problem (1.32), possibly with $Q, Q_1, \dots, Q_k \not\succeq O$, is also called a quadratic programming problem by some authors.

in $\mathbb{L}^{n_l}$, which is also written as $x_{l0} \geq \|\boldsymbol{x}_{l1}\|$ with $\boldsymbol{x}_l = (x_{l0}, \boldsymbol{x}_{l1}) \in \mathbb{R} \times \mathbb{R}^{n_l - 1}$. This situation corresponds to considering $\mathrm{CO_P}$ with $\mathbb{V} = \mathbb{R}^{n_1} \times \cdots \times \mathbb{R}^{n_k}$ and $\mathcal{C} = \mathbb{L}^{n_1} \times \cdots \times \mathbb{L}^{n_k}$. Thus SOCP in its general form is formally described below.

Definition 1.6.5 (SOCP)**.** Given $A_l \in \mathbb{R}^{m \times n_l}$, $\boldsymbol{c}_l \in \mathbb{R}^{n_l}$ $(l = 1, \dots, k)$, and $\boldsymbol{b} \in \mathbb{R}^m$, the *second-order cone programming* (SOCP) problem is the optimization problem defined as

$$\left.\begin{aligned} \mathrm{SOCP_P}: \quad & \min_{\boldsymbol{x}_1, \dots, \boldsymbol{x}_k} && \sum_{l=1}^{k} \boldsymbol{c}_l^{\mathrm{T}} \boldsymbol{x}_l \\ & \quad \mathrm{s.t.} && \sum_{l=1}^{k} A_l \boldsymbol{x}_l = \boldsymbol{b}, \\ & && x_{l0} \geq \|\boldsymbol{x}_{l1}\|, \quad l = 1, \dots, k, \end{aligned}\right\}$$

where $\boldsymbol{x}_l = (x_{l0}, \boldsymbol{x}_{l1})$ $(l = 1, \dots, k)$ are the variables. The problem dual to $\mathrm{SOCP_P}$ is formulated as

$$\left.\begin{aligned} \mathrm{SOCP_D}: \quad & \max_{\boldsymbol{y}, \boldsymbol{s}_1, \dots, \boldsymbol{s}_k} && \boldsymbol{b}^{\mathrm{T}} \boldsymbol{y} \\ & \quad \mathrm{s.t.} && A_l^{\mathrm{T}} \boldsymbol{y} + \boldsymbol{s}_l = \boldsymbol{c}_l, \quad && l = 1, \dots, k, \\ & && s_{l0} \geq \|\boldsymbol{s}_{l1}\|, && l = 1, \dots, k, \end{aligned}\right\}$$

where $\boldsymbol{y} \in \mathbb{R}^m$ and $\boldsymbol{s}_l \in \mathbb{R}^{n_l}$ $(l = 1, \dots, k)$ are the variables. ■

Thus, SOCP is an optimization problem of a linear function under second-order cone constraints and linear equality constraints. In view of this, the following problem, for example, is also an SOCP:

$$\left.\begin{aligned} & \min_{\boldsymbol{x}} \ \boldsymbol{c}^{\mathrm{T}} \boldsymbol{x} \\ & \mathrm{s.t.} \ \boldsymbol{p}_l^{\mathrm{T}} \boldsymbol{x} + q_l \geq \|A_l \boldsymbol{x} + \boldsymbol{b}_l\|, \quad l = 1, \dots, k. \end{aligned}\right\}$$

Indeed, this problem can be converted to the form of $\mathrm{SOCP_P}$.

In section 1.5 we see that the linear inequality constraint is regarded as a special case of second-order cone constraint. Hence, LP is a special case of SOCP. In contrast, SOCP is included in SDP as a special case, because Fact 1.5.1 implies that each second-order cone constraint can be represented as a positive-semidefinite constraint. As we saw in Fact 1.5.2 (b), a convex quadratic inequality can be expressed as a second-order cone constraint. Hence, SOCP includes QP and QCQP as particular cases. Indeed, the QP in (1.31) can be converted to an SOCP by introducing an extra variable t as

$$\left.\begin{aligned} & \min_{\boldsymbol{x}, t} \ t \\ & \mathrm{s.t.} \ -\boldsymbol{p}^{\mathrm{T}} \boldsymbol{x} + t + 1 \geq \left\| \begin{bmatrix} -\boldsymbol{p}^{\mathrm{T}} \boldsymbol{x} + t - 1 \\ \sqrt{2} Q^{1/2} \boldsymbol{x} \end{bmatrix} \right\|, \\ & \qquad A \boldsymbol{x} = \boldsymbol{b}, \\ & \qquad \boldsymbol{x} \geq \boldsymbol{0}. \end{aligned}\right\}$$

As seen in Fact 1.5.1, a second-order cone constraint can be expressed as a positive-semidefinite constraint. Therefore, SOCP is included in SDP as a special case. For example, $\mathrm{SOCP_P}$ in Definition 1.6.5 can be rewritten as the following SDP:

$$\left.\begin{array}{ll} \displaystyle\min_{\boldsymbol{x}_1,\dots,\boldsymbol{x}_k} & \displaystyle\sum_{l=1}^{k} \boldsymbol{c}_l^{\mathrm{T}} \boldsymbol{x}_l \\ \text{s.t.} & \displaystyle\sum_{l=1}^{k} A_l \boldsymbol{x}_l = \boldsymbol{b}, \\ & \begin{bmatrix} x_{l0} & \boldsymbol{x}_{l1}^{\mathrm{T}} \\ \boldsymbol{x}_{l1} & x_{l0} I_{n_l-1} \end{bmatrix} \succeq O, \quad l = 1, \dots, k. \end{array}\right\}$$

We often deal with the convex optimization with the several kinds of conic constraints. For example, consider a problem in the variable $\boldsymbol{x} \in \mathbb{R}^n$ in the form of

$$\left.\begin{array}{ll} \displaystyle\min_{\boldsymbol{x}} & \boldsymbol{c}^{\mathrm{T}} \boldsymbol{x} \\ \text{s.t.} & A\boldsymbol{x} + \boldsymbol{b} \geq \boldsymbol{0}, \\ & \hat{\boldsymbol{a}}_1^{\mathrm{T}} \boldsymbol{x} + \hat{b}_1 \geq \|\hat{A}_2 \boldsymbol{x} + \hat{\boldsymbol{b}}_2\|, \\ & \displaystyle\sum_{i=1}^{n} x_i F_i + F_0 \succeq O, \end{array}\right\} \tag{1.34}$$

where $A \in \mathbb{R}^{m_1 \times n}$, $\boldsymbol{b} \in \mathbb{R}^{m_1}$, $\hat{\boldsymbol{a}}_1 \in \mathbb{R}^n$, $\hat{b}_1 \in \mathbb{R}$, $\hat{A}_2 \in \mathbb{R}^{m_2 \times n}$, $\hat{\boldsymbol{b}}_2 \in \mathbb{R}^{m_2}$, and $F_0, F_1, \dots, F_n \in \mathbb{S}^{m_3}$. Problem (1.34) is then a conic optimization problem, with the linear inequality constraints, second-order cone constraint, and positive-semidefinite constraint. Since both a linear inequality constraint and a second-order cone constraint are rewritten as a positive-semidefinite constraint (see section 1.5), all the constraints of problem (1.34) can be expressed as a positive-semidefinite constraint. This implies that problem (1.34) is equivalently rewritten as an SDP problem (but, certainly, it cannot be rewritten as an LP or SOCP problem). For this reason we sometimes call problem (1.34) an SDP problem. Alternatively, consider a problem obtained by removing the constraint $\sum_{i=1}^{n} x_i F_i + F_0 \succeq O$ from (1.34). This problem is still rewritten as an SDP problem but can also be rewritten as an SOCP problem, because a linear inequality constraint is regarded as a special case of a second-cone constraint (see section 1.5). In this case the problem is called not an SDP problem but either an SOCP problem or a conic optimization problem with linear inequality constraints and a second-order cone constraint. In this way we clarify the problem class, to which the optimization problem under consideration belongs.

1.7 Notes

The form of conic optimization problems was introduced as a standard problem formulation of nonlinear convex optimization by Nesterov–Nemirovski [333]. Similar accounts are found in the textbooks by Ben-Tal–Nemirovski [46] and Boyd–Vandenberghe [61], which include details of linear programming, second-order cone programming, and semidefinite programming, as well as various applications. The duality theory of conic convex optimization and related topics are found in Luo–Sturm–Zhang [283, 284], and Sturm [425].

Linear programming certainly plays the most fundamental role in optimization theory and algorithms and is the subject of many books, including Chvátal [89], Dantzig–Thapa [103], and Schrijver [404]. Among them, with particular emphasis on the interior-point method, are Vanderbei [445] and Ye [460].

For seeing the whole perspective of semidefinite programming, Wolkowicz–Saigal–Vandenberghe [454] serves as an excellent handbook. For surveys, see Helmberg [172], Todd [434], and Vandenberghe–Boyd [444]. A survey particularly on the eigenvalue optimization is in Lewis–Overton [261]. The lemma on the Schur complement (Fact 1.3.10) is one of basic tools for investigating matrix inequalities. Some generalizations of this lemma are found in Boyd–El Ghaoui–Feron–Balakrishnan [60]. For the complementarity condition over the positive-semidefinite cone and its implication, see Alizadeh–Haeberly–Overton [9]. In section 3.1 and section 3.2, we investigate some applications of SDP in the field of structural engineering. More applications can be found in, e.g., combinatorial optimization [17, 151, 152, 453], control [60, 401], relaxations of the nonlinear optimization problems [174, 240, 250], moment problems in probability theory [51, 251, 252], and uncertainty analyses [67, 210, 212, 213, 339]. SDP and SOCP provide us with a new perspective on nonsmooth mechanics, which is certainly the principal aim of this book.

A fundamental tool with which we reformulate problems in applications as SDP problems is the Schur complement (Fact 1.3.10). Besides this, the so-called *S-lemma* is also important, although it is not discussed in this book; see Pólik–Terlaky [373] for a survey. Applications of S-lemma to derive SDP problems can be found in Ben-Tal–Nemirovski [46, section 4.10.5], Boyd–El Ghaoui–Feron–Balakrishnan [60], Calafiore–El Ghaoui [67], and Kanno–Takewaki [210, 211].

Theoretical results and applications of second-order cone programming are surveyed by Alizadeh–Goldfarb [8] and Lobo–Vandenberghe–Boyd–Lebret [272]. The Euclidean Jordan algebra associated with a proper cone is the subject dealt with by Alizadeh–Goldfarb [8], Faraut–Korányi [126], and Sun–Sun [428]. In section 3.1 we see an application of SOCP in structural optimization. Furthermore, in Parts II and IV, SOCP serves as a fundamental tool for analysis of nonsmooth mechanics. More applications in the field other than applied mechanics can be found in Lobo–Vandenberghe–Boyd–Lebret [272] and Sasakawa–Tsuchiya [399].

Fundamental references on complementarity problems and variational inequalities are Cottle–Pang–Stone [96], Facchinei–Pang [125], and Kinderlehrer–Stampacchia [222]. Various applications of these problems can be found in the surveys by Ferris–Pang [130] and Harker–Pang [161].

Main concepts treated in this chapter (e.g., convexity, monotonicity) can be extended to infinite-dimensional spaces; see, e.g., Jahn [190] and Ekeland–Temam [121]. However, special attention should be given to the differences between infinite-dimensional and finite-dimensional spaces (Borwein–Lewis [59, Chap. 10]).

Chapter 2

Optimality and Duality

Orientation This chapter introduces the notion of convex analysis and duality in convex optimization, which are two fundamental mathematical tools for investigating the nonsmooth mechanics. Following a brief introduction of the convex analysis in section 2.1, we establish in section 2.2 the Fenchel and Lagrangian dualities of convex optimizations. As an illustrative example of the use of those duality theories, the strong duality and the optimality conditions of semidefinite programming (SDP) are studied in section 2.3.

Readers who are familiar with the mechanical engineering but not with the convex analysis should refer to section 3.3 for an engineering application of the notion of convex analysis to the model of (nonsmooth) constitutive law.

2.1 Fundamentals of Convex Analysis

We say that f is a (real-valued) *function* on $\mathbb{R}^n$, denoted $f : \mathbb{R}^n \to \mathbb{R}$, if a real value $f(\boldsymbol{x})$ is associated to any $\boldsymbol{x} \in \mathbb{R}^n$. A function is also called a *mapping* (or *map*). When dealing with optimization problems, it is often convenient to consider a function f while allowing to take $+\infty$ as the value of $f(\boldsymbol{x})$. In such a case, f is called an *extended real valued function*, and we write $f : \mathbb{R}^n \to \mathbb{R} \cup \{+\infty\}$ (or, $f : \mathbb{R}^n \to (-\infty, +\infty]$). In what follows, we use $\mathbb{V}$ to denote a Euclidean space, as done in Chapter 1, although we may simply consider $\mathbb{V} = \mathbb{R}^n$ without loss of generality.

For any function $f : \mathbb{V} \to \mathbb{R} \cup \{+\infty\}$, we call the set

$$\operatorname{dom} f = \{x \in \mathbb{V} \mid f(x) < +\infty\}$$

the *effective domain* of f.

The set obtained as a translation of a vector subspace is called an affine space. An affine space $\mathcal{A} \subseteq \mathbb{V}$ can be represented by using some points $x_0, x_1, \dots, x_k \in \mathbb{V}$ as

$$\mathcal{A} = \left\{ x \in \mathbb{V} \mid x = x_0 + \sum_{i=1}^{k} \alpha_i x_i, \ \alpha_1, \dots, \alpha_k \in \mathbb{R} \right\}.$$

For a set $\mathcal{S} \subseteq \mathbb{V}$, the smallest affine set containing $\mathcal{S}$, denoted $\operatorname{aff} \mathcal{S}$, is called the *affine hull* of $\mathcal{S}$. We say that the point $x \in \mathcal{S}$ is a *relative interior point* if there exists an $\epsilon > 0$ for which

$$(\{y \in \mathbb{V} \mid \|y - x\| \leq \epsilon\} \cap \operatorname{aff} \mathcal{S}) \subseteq \mathcal{S}$$

holds. The set of all relative interior points of $\mathcal{S}$ is called the *relative interior* of $\mathcal{S}$ and is denoted by $\operatorname{ri} \mathcal{S}$. In particular, if $\mathcal{S}$ is a singleton, say, $\mathcal{S} = \{x\}$, then $\operatorname{aff} \mathcal{S} = \operatorname{ri} \mathcal{S} = \mathcal{S}$.

2.1.1 Convex sets and convex functions

Remark 2.1.1. A word about the notation $\mathbb{V}$. We may simply take $\mathbb{R}^n$ as $\mathbb{V}$, considering only a finite-dimensional space. ■

Recall that a set $\mathcal{S} \subseteq \mathbb{V}$ is said to be convex if, for all $x_1, x_2 \in \mathcal{S}$, we have

$$\tau x_1 + (1 - \tau) x_2 \in \mathcal{S}, \quad \forall \tau \in [0, 1].$$

The *closure* of $\mathcal{S}$, denoted $\operatorname{cl} \mathcal{S}$, has been defined as the smallest closed set containing $\mathcal{S}$. Analogously, the *convex hull* of $\mathcal{S}$, denoted $\operatorname{co} \mathcal{S}$, means the smallest convex set containing $\mathcal{S}$.

A function $f : \mathbb{V} \to \mathbb{R} \cup \{+\infty\}$ is said to be *convex* if, for any $x_1, x_2 \in \mathbb{V}$, it satisfies[1]

$$\tau f(x_1) + (1 - \tau) f(x_2) \geq f(\tau x_1 + (1 - \tau) x_2), \quad \forall \tau \in [0, 1].$$

The effective domain of a convex function is convex. We define f to be a *concave* function if $-f$ is convex. We call f a *strictly convex function* if, for any $x, y \in \mathbb{V}$ $(x \neq y)$, we have

$$\tau f(x) + (1 - \tau) f(y) > f(\tau x + (1 - \tau) x), \quad \forall \tau \in]0, 1[.$$

Noticc here that, for $a, b \in \mathbb{R}$ with $a < b$, we denote by $[a, b]$ and $]a, b[$ the closed and open intervals between a and b, respectively, i.e.,

$$[a, b] = \{x \in \mathbb{R} \mid a \leq x \leq b\},$$
$$]a, b[= \{x \in \mathbb{R} \mid a < x < b\}.$$

Convex set and convex function relate to each other as follows. The *graph* of a function $f : \mathbb{V} \to \mathbb{R} \cup \{+\infty\}$ is a subset of $\mathbb{V} \times \mathbb{R}$ defined by

$$\operatorname{graph} f = \{(x, Y) \in \mathbb{V} \times \mathbb{R} \mid Y = f(x)\}.$$

[1] In this inequality, we adapt the rule $+\infty \geq +\infty$ for convention. In addition, we do not consider the case in which f takes $-\infty$ at some x to avoid pathological situations; see, e.g., Ekeland–Temam [121, pp. 8–9] and Ciarlet [90, p. 186].

The *epigraph* of f, denoted $\operatorname{epi} f$, is defined as

$$\operatorname{epi} f = \{(x, Y) \in \mathbb{V} \times \mathbb{R} \mid Y \geq f(x)\},$$

which is the set of points "above the graph." Given any set $\mathcal{S} \subseteq \mathbb{V}$, we denote by $\delta_{\mathcal{S}} : \mathbb{V} \to \mathbb{R} \cup \{+\infty\}$ the *indicator function* of $\mathcal{S}$, which is defined by

$$\delta_{\mathcal{S}}(x) = \begin{cases} 0 & \text{if } x \in \mathcal{S}, \\ +\infty & \text{if } x \notin \mathcal{S}. \end{cases} \tag{2.1}$$

A function f is convex if and only if $\operatorname{epi} f$ is a convex set. Conversely, a set $\mathcal{S}$ is convex if and only if $\delta_{\mathcal{S}}$ is a convex function.

A convex function $f : \mathbb{V} \to \mathbb{R} \cup \{+\infty\}$ is said to be a *proper convex function* if $\operatorname{dom} f \neq \emptyset$. As an example, for any nonempty convex set $\mathcal{S}$, its indicator function $\delta_{\mathcal{S}}$ is proper convex.

2.1.2 Monotone functions and convexity

Let $\mathcal{S} \subseteq \mathbb{V}$ be nonempty. A mapping $F : \mathcal{S} \to \mathbb{V}$ is said to be *monotone* on $\mathcal{S}$ if it satisfies

$$\langle F(x) - F(y), x - y \rangle \geq 0, \quad \forall x, y \in \mathcal{S},$$

i.e., if we have

$$x, y \in \mathcal{S} \quad \Rightarrow \quad \langle F(x) - F(y), x - y \rangle \geq 0.$$

We say that F is *strictly monotone* on $\mathcal{S}$ if it satisfies

$$x, y \in \mathcal{S},\ x \neq y \quad \Rightarrow \quad \langle F(x) - F(y), x - y \rangle > 0.$$

When $\mathcal{S} = \mathbb{V}$, we similarly say that F is monotone and strictly monotone, respectively.

The relation between convexity and monotonicity is stated in Fact 2.1.2, in the setting of $\mathbb{V} = \mathbb{R}^n$ for simple presentation. Note that, for differentiable $f : \mathbb{R}^n \to \mathbb{R}$, we denote by $\nabla f(\boldsymbol{x})$ its *gradient*, which is a column vector defined by

$$\nabla f(\boldsymbol{x}) = \left(\frac{\partial f}{\partial x_i}(\boldsymbol{x}) \mid i = 1, \dots, n \right).$$

The *Hessian* of f, denoted by $\nabla^2 f(\boldsymbol{x})$, is the $n \times n$ symmetric matrix defined by

$$\nabla^2 f(\boldsymbol{x}) = \left(\frac{\partial^2 f}{\partial x_i \partial x_j}(\boldsymbol{x}) \mid i, j = 1, \dots, n \right),$$

provided that $f \in \mathrm{C}^2$. Now we state the relation among the convexity of f, the monotonicity of ∇f, and the positive definiteness of $\nabla^2 f$.

Fact 2.1.2. Suppose that $f : \mathbb{R}^n \to \mathbb{R}$ is continuously twice differentiable (i.e., $f \in \mathrm{C}^2$). Then

(i) the following three statements are equivalent:

(a) f is convex.

(b) ∇f is monotone.

(c) $\nabla^2 f \succeq O$.

(ii) f is strictly convex if and only if ∇f is strictly monotone.

(iii) f is strictly monotone if $\nabla^2 f \succ O$. ■

To show this, we begin by establishing an inequality characterizing differentiable convex functions.

Proposition 2.1.3. *Suppose that $f : \mathbb{R}^n \to \mathbb{R}$ is differentiable. Then f is convex if and only if it satisfies*

$$f(\boldsymbol{x}_2) \geq f(\boldsymbol{x}_1) + \nabla f(\boldsymbol{x}_1)^{\mathrm{T}}(\boldsymbol{x}_2 - \boldsymbol{x}_1) \tag{2.2}$$

for any $\boldsymbol{x}_1$, $\boldsymbol{x}_2 \in \mathbb{R}^n$.

Proof. Suppose that f is a convex function; i.e.,

$$(1-\alpha)f(\boldsymbol{x}_1) + \alpha f(\boldsymbol{x}_2) \geq f((1-\alpha)\boldsymbol{x}_1 + \alpha\boldsymbol{x}_2)$$

for any $\alpha \in]0,1[$. This inequality is equivalently rewritten as

$$\alpha f(\boldsymbol{x}_2) - \alpha f(\boldsymbol{x}_1) \geq f(\boldsymbol{x}_1 + \alpha(\boldsymbol{x}_2 - \boldsymbol{x}_1)) - f(\boldsymbol{x}_1).$$

By dividing both sides by α, we obtain

$$\begin{aligned} f(\boldsymbol{x}_2) - f(\boldsymbol{x}_1) &\geq \frac{f(\boldsymbol{x}_1 + \alpha(\boldsymbol{x}_2 - \boldsymbol{x}_1)) - f(\boldsymbol{x}_1)}{\alpha} \\ &\to \nabla f(\boldsymbol{x}_1)^{\mathrm{T}}(\boldsymbol{x}_2 - \boldsymbol{x}_1) \quad (\alpha \to +0) \end{aligned}$$

as expected.

Conversely, suppose that f satisfies (2.2). For any $\boldsymbol{x}_1$, $\boldsymbol{x}_2 \in \mathbb{R}^n$ and $\alpha \in]0,1[$, define $\boldsymbol{z}$ by

$$\boldsymbol{z} = \alpha\boldsymbol{x}_1 + (1-\alpha)\boldsymbol{x}_2.$$

Then (2.2) implies that

$$f(\boldsymbol{x}_i) \geq f(\boldsymbol{z}) + \nabla f(\boldsymbol{z})^{\mathrm{T}}(\boldsymbol{x}_i - \boldsymbol{z}), \quad i = 1, 2 \tag{2.3}$$

hold. In (2.3), consider that we multiply the inequality with $i = 1$ by α and that with $i = 2$ by $1 - \alpha$. Adding the obtained inequalities yields

$$\begin{aligned} &\alpha f(\boldsymbol{x}_1) + (1-\alpha)f(\boldsymbol{x}_2) \\ &\qquad \geq f(\boldsymbol{z}) + \nabla f(\boldsymbol{z})^{\mathrm{T}}\left[\alpha\boldsymbol{x}_1 + (1-\alpha)\boldsymbol{x}_2 - \boldsymbol{z}\right] \\ &\qquad = f(\boldsymbol{z}) \\ &\qquad = f(\alpha\boldsymbol{x}_1 + (1-\alpha)\boldsymbol{x}_2), \end{aligned}$$

which shows the convexity of f. □

With the aid of Proposition 2.1.3 we now establish the equivalence "(a) $\Leftrightarrow$ (b)" in Fact 2.1.2 (i).

Proposition 2.1.4. *Suppose that* $f : \mathbb{R}^n \to \mathbb{R}$ *is continuously differentiable (i.e.,* $f \in \mathrm{C}^1$*). Then* f *is convex if and only if* ∇f *is monotone.*

Proof. Suppose that f is convex. Then Proposition 2.1.3 asserts that

$$f(\boldsymbol{x}_2) \geq f(\boldsymbol{x}_1) + \nabla f(\boldsymbol{x}_1)^{\mathrm{T}}(\boldsymbol{x}_2 - \boldsymbol{x}_1) \tag{2.4}$$

holds for any $\boldsymbol{x}_1, \boldsymbol{x}_2 \in \mathbb{R}^n$. Exchanging $\boldsymbol{x}_1$ and $\boldsymbol{x}_2$ yields

$$f(\boldsymbol{x}_1) \geq f(\boldsymbol{x}_2) + \nabla f(\boldsymbol{x}_2)^{\mathrm{T}}(\boldsymbol{x}_1 - \boldsymbol{x}_2). \tag{2.5}$$

Adding (2.4) and (2.5) results in

$$(\nabla f(\boldsymbol{x}_1) - \nabla f(\boldsymbol{x}_2))^{\mathrm{T}}(\boldsymbol{x}_1 - \boldsymbol{x}_2) \geq 0,$$

which shows the monotonicity of ∇f.

Conversely, suppose that ∇f is monotone. For any $\boldsymbol{x}_1, \boldsymbol{x}_2 \in \mathbb{R}^n$, the mean-value theorem guarantees that we have

$$f(\boldsymbol{x}_2) - f(\boldsymbol{x}_1) = \nabla f(\boldsymbol{z})^{\mathrm{T}}(\boldsymbol{x}_2 - \boldsymbol{x}_1), \tag{2.6}$$

where $\boldsymbol{z} = \boldsymbol{x}_1 + \alpha(\boldsymbol{x}_2 - \boldsymbol{x}_1)$ for some $\alpha \in]0, 1[$. Since $\boldsymbol{x}_2 - \boldsymbol{x}_1 = (\boldsymbol{z} - \boldsymbol{x}_1)/\alpha$, we obtain

$$\begin{aligned}&(\nabla f(\boldsymbol{z}) - \nabla f(\boldsymbol{x}_1))^{\mathrm{T}}(\boldsymbol{x}_2 - \boldsymbol{x}_1)\\ &\qquad = \frac{1}{\alpha}(\nabla f(\boldsymbol{z}) - \nabla f(\boldsymbol{x}_1))^{\mathrm{T}}(\boldsymbol{z} - \boldsymbol{x}_1) \geq 0,\end{aligned} \tag{2.7}$$

where the inequality follows from the monotonicity of ∇f. By using (2.6) and (2.7), we have

$$\begin{aligned}&f(\boldsymbol{x}_2) - f(\boldsymbol{x}_1)\\ &\quad = (\nabla f(\boldsymbol{z}) - \nabla f(\boldsymbol{x}_1))^{\mathrm{T}}(\boldsymbol{x}_2 - \boldsymbol{x}_1) + \nabla f(\boldsymbol{x}_1)^{\mathrm{T}}(\boldsymbol{x}_2 - \boldsymbol{x}_1) \quad &&[\text{from (2.6)}]\\ &\quad \geq \nabla f(\boldsymbol{x}_1)^{\mathrm{T}}(\boldsymbol{x}_2 - \boldsymbol{x}_1). &&[\text{from (2.7)}]\end{aligned}$$

By virtue of Proposition 2.1.3, this inequality shows the convexity of f. □

The equivalence "(a) $\Leftrightarrow$ (c)" in Fact 2.1.2 (i) is also established as a consequence of Proposition 2.1.3 as follows.

Proposition 2.1.5. *Suppose that* $f : \mathbb{R}^n \to \mathbb{R}$ *is continuously twice differentiable (i.e.,* $f \in \mathrm{C}^2$*). Then* f *is convex if and only if it satisfies* $\nabla^2 f(\boldsymbol{x}) \succeq O$ *for any* $\boldsymbol{x} \in \mathbb{R}^n$.

Proof. Suppose that f is a convex function. For any fixed $\boldsymbol{d} \in \mathbb{R}^n$ and $\epsilon > 0$, the second-order Taylor approximation of $f(\boldsymbol{x} + \epsilon \boldsymbol{d})$ around $\boldsymbol{x}$ yields

$$f(\boldsymbol{x} + \epsilon \boldsymbol{d}) = f(\boldsymbol{x}) + \epsilon \nabla f(\boldsymbol{x})^{\mathrm{T}} \boldsymbol{d} + \epsilon^2 \frac{1}{2} \boldsymbol{d}^{\mathrm{T}} \nabla^2 f(\boldsymbol{x}) \boldsymbol{d} + o(\epsilon^2).$$

On the other hand, from Proposition 2.1.3 we obtain

$$f(\boldsymbol{x} + \epsilon \boldsymbol{d}) \geq f(\boldsymbol{x}) + \epsilon \nabla f(\boldsymbol{x})^{\mathrm{T}} \boldsymbol{d}.$$

Hence, we have that

$$\epsilon^2 \frac{1}{2} \boldsymbol{d}^{\mathrm{T}} \nabla^2 f(\boldsymbol{x}) \boldsymbol{d} + o(\epsilon^2) \geq 0. \tag{2.8}$$

By dividing (2.8) by ϵ^2 and letting $\epsilon \to +0$, we obtain

$$\boldsymbol{d}^{\mathrm{T}} \nabla^2 f(\boldsymbol{x}) \boldsymbol{d} \geq 0. \tag{2.9}$$

Since the inequality (2.9) holds for any direction $\boldsymbol{d}$, we conclude that $\nabla^2 f(\boldsymbol{x})$ is positive semidefinite.

Conversely, suppose $\nabla^2 f(\boldsymbol{x}) \succeq O$ for any $\boldsymbol{x}$. It follows from Taylor's theorem that for any $\boldsymbol{x}$, $\boldsymbol{d} \in \mathbb{R}^n$ there exists $\theta \in]0,1[$ satisfying

$$f(\boldsymbol{x} + \boldsymbol{d}) = f(\boldsymbol{x}) + \nabla f(\boldsymbol{x})^{\mathrm{T}} \boldsymbol{d} + \frac{1}{2} \boldsymbol{d}^{\mathrm{T}} \nabla^2 f(\boldsymbol{x} + \theta \boldsymbol{d}) \boldsymbol{d}.$$

Since $\nabla^2 f(\boldsymbol{x} + \theta \boldsymbol{d}) \succeq O$, we obtain

$$f(\boldsymbol{x} + \boldsymbol{d}) \geq f(\boldsymbol{x}) + \nabla f(\boldsymbol{x})^{\mathrm{T}} \boldsymbol{d}$$

for any $\boldsymbol{x}$ and $\boldsymbol{d}$. Therefore, f satisfies (2.2) with $\boldsymbol{x}_1 = \boldsymbol{x}$ and $\boldsymbol{x}_2 = \boldsymbol{x} + \boldsymbol{d}$; thus, the convexity of f follows from Proposition 2.1.3. □

As an important consequence of Fact 2.1.2 in mechanics, we see that the elastic strain energy is a convex function of strain if and only if the constitutive law is a monotone function of strain. This is also equivalent to the condition that the elastic tensor (and also, e.g., the elongation stiffness) is positive semidefinite.

2.1.3 Closed convex functions

A function $f : \mathbb{V} \to \mathbb{R} \cup \{+\infty\}$ is said to be *lower semicontinuous* if its epigraph, epi f, is a closed set.[2] We call f a *closed convex function* if it is lower semicontinuous and convex. Clearly, f is a closed convex function if and only if epi f is a closed convex set. Conversely, S is a closed convex set if and only if δ_S is a closed convex function. By a *closed proper convex function* we mean a proper convex function that is semicontinuous.

We say that f is *upper semicontinuous* if $-f$ is lower semicontinuous. The ordinary continuity of f is achieved by a combination of lower and upper

[2] This is equivalent to defining lower semicontinuity as follows. For a given $\bar{x} \in \mathbb{V}$, we say that f is lower semicontinuous at $\bar{x}$ if

$$f(\bar{x}) \leq \liminf_{k \to \infty} f(x_k) \ (= \lim_{k \to \infty} \inf_{k' \geq k} f(x_{k'}))$$

is satisfied for any sequence $\{x_k\}$ such that $x_k \to \bar{x}$. Then we can verify that f is lower semicontinuous at every $x \in \mathbb{V}$ if and only if epi f is a closed set.

TABLE 2.1: Convex functions and convex sets.

$f : \mathbb{R}^n \to \mathbb{R} \cup \{+\infty\}$		$\mathcal{S} \subseteq \mathbb{R}^n$	
function	property	set	property
f	convex	epi f	convex
$\delta_{\mathcal{S}}$	convex	$\mathcal{S}$	convex
f	proper convex	epi f	nonempty convex †
$\delta_{\mathcal{S}}$	proper convex	$\mathcal{S}$	nonempty convex
f	lower semicontinuous	epi f	closed
f	closed convex	epi f	closed convex
$\delta_{\mathcal{S}}$	closed convex	$\mathcal{S}$	closed convex
f	closed proper convex	epi f	nonempty closed convex †

†We exclude any function f which takes $f(x) = -\infty$ at some x.

semicontinuity; More precisely, f is said to be *continuous* if it is lower semicontinuous and upper semicontinuous. The classes of convex functions and sets introduced in this section, as well as their relationship, are summarized in Table 2.1.

2.1.4 Subdifferential

In general, for a vector space $\mathbb{V}$ (over $\mathbb{R}$) we denote by $\mathbb{V}^*$ its dual space (or its topological dual), which is the vector space of continuous linear mappings on $\mathbb{V}$. In particular, when $\mathbb{V}$ under consideration is supposed to be a Euclidean space,[3] such as $\mathbb{R}^n$, we then simply have $\mathbb{V}^* = \mathbb{V}$. However, in what follows we sometimes write $\mathbb{V}^*$ to distinguish it from $\mathbb{V}$.

The difference between $\mathbb{V}$ and $\mathbb{V}^*$ often reflects the difference of physical meanings of variables belonging to these spaces. For example, if $u \in \mathbb{V}$ is a displacement vector, then $f \in \mathbb{V}^*$ corresponds to a force vector. In contrast, if we write $u \in \mathbb{V}$ and $v \in \mathbb{V}$, then v is regarded as another displacement vector. Thus, roughly speaking, the duality between $\mathbb{V}$ and $\mathbb{V}^*$ corresponds to the work-conjugate relation between variables belonging to these spaces. Within the theory of continuous mechanics, the displacement-field and the force-field belong to different spaces (see, e.g., Ciarlet [90], Marsden–Hughes [295]). For $x \in \mathbb{V}$ and $x^* \in \mathbb{V}^*$, the value at x of x^* is denote by $\langle x, x^* \rangle$, which is called the *pairing* of x and x^*. Note that the pairing $\langle \cdot, \cdot \rangle$ is a bilinear mapping on $\mathbb{V}^* \times \mathbb{V}$. When $\mathbb{V} = \mathbb{R}^n$ is a Euclidean space, the pairing $\langle x, x^* \rangle$ corresponds

[3]In other words, we do not consider infinite-dimensional spaces, such as Hilbert space or Sobolev space. For fundamentals of convex analysis in an infinite-dimensional space, see [121, 390].

to the conventional inner product of two n-dimensional vectors x and x^*. The notion of pairing is related to the work in mechanics; in the example above, the work done by the displacement $u \in \mathbb{V}$ against the force $f \in \mathbb{V}^*$ is written as $\langle f, u \rangle$, while for the pair of $u \in \mathbb{V}$ and $v \in \mathbb{V}$ we do not consider their pairing. However, readers not interested in such a strict treatment can simply put $\mathbb{V} = \mathbb{V}^* = \mathbb{R}^n$ anywhere in what follows.

For a convex function $f : \mathbb{V} \to \mathbb{R} \cup \{+\infty\}$ and a point $x \in \mathbb{V}$, we call $s \in \mathbb{V}^*$ satisfying

$$f(y) \geq f(x) + \langle s, y - x \rangle \quad (\forall y \in \mathbb{V}) \tag{2.10}$$

a *subgradient* of f at x. The *subdifferential* of f at x, denoted $\partial f(x)$, means the set of all subgradients of f at x, i.e.,

$$\partial f(x) = \{s \in \mathbb{V}^* \mid f(y) \geq f(x) + \langle s, y - x \rangle \ (\forall y \in \mathbb{V})\}. \tag{2.11}$$

This definition immediately shows that a proper convex function $f : \mathbb{V} \to \mathbb{R} \cup \{+\infty\}$ satisfies

$$\partial(\lambda f)(x) = \lambda f(x), \quad \forall x \in \mathbb{V}$$

for any $\lambda > 0$. The following properties are also fundamental for calculating subgradients.

Fact 2.1.6. Let f, $g : \mathbb{V} \to \mathbb{R} \cup \{+\infty\}$ be proper convex functions. Then the following statements hold:

(i) $\partial(f + g)(x) \supseteq \partial f(x) + \partial g(x)$ $(\forall x \in \mathbb{V})$.

(ii) If $\mathrm{ri}(\mathrm{dom}\, f) \cap \cdots \cap \mathrm{ri}(\mathrm{dom}\, g) \neq \emptyset$, then $\partial(f + g)(x) = \partial f(x) + \partial g(x)$ $(\forall x \in \mathbb{V})$.[4]

Fact 2.1.7. Let $f : \mathbb{V} \to \mathbb{R} \cup \{+\infty\}$ be convex, and let $s \in \mathbb{V}^*$. Then $s \in \partial f(x)$ if and only if

$$x \in \arg\max_x \{\langle s, x \rangle - f(x) \mid x \in \mathbb{V}\} \tag{2.12}$$

holds. ■

Proof. By definition (2.10) of subgradient, $s \in \partial f(x)$ if and only if

$$\langle s, x \rangle - f(x) \geq \langle s, x' \rangle - f(x') \quad (\forall x' \in \mathbb{V}),$$

which is equivalent to (2.12). □

The subdifferential $\partial f(x)$ at a given x is a (possibly empty) convex set. If f is differentiable at x, then $\partial f(x) = \{\nabla f(x)\}$. A sufficient condition for the nonemptiness of subdifferential is given as follows.[5]

[4]See Theorem 23.8 in Rockafellar [389] for a proof.

[5]See Theorem 23.4 in Rockafellar [389] for a proof.

Fact 2.1.8. Let $f : \mathbb{V} \to \mathbb{R} \cup \{+\infty\}$ be a proper convex function. If $x \in \mathrm{ri}(\mathrm{dom}\, f)$, then $\partial f(x) \neq \emptyset$. ■

Besides $\mathbb{V}$, consider another Euclidean space $\mathbb{Y}$. For example, $\mathbb{V} := \mathbb{R}^n$ and $\mathbb{Y} := \mathbb{R}^m$, possibly with $m = n$. A *set-valued mapping* $F : \mathbb{V} \to \mathcal{P}(\mathbb{Y})$, also denoted by $\mathbb{V} \ni x \mapsto F(x) \subseteq \mathbb{Y}$, is a mapping that associates to any $x \in \mathbb{V}$ a (possibly empty) subset of $\mathbb{Y}$.[6] Here, we denote by $\mathcal{P}(\mathbb{Y})$ the *power set* of $\mathbb{Y}$, which is the set of all subsets of $\mathbb{Y}$. A set-valued mapping is also called a *multifunction* or a *point-to-set mapping*. As a particular case of set-valued mapping, $f : \mathbb{V} \to \mathcal{P}(\mathbb{R})$ is sometimes called a *set-valued function*. The notion of set-valued function (and also set-valued mapping) can be generalized to be an extended real-valued function, such as $F : \mathbb{V} \to \mathcal{P}(\mathbb{R} \cup \{+\infty\})$. The subdifferential of a function $f : \mathbb{R} \to \mathbb{R} \cup \{+\infty\}$ is a set-valued function in general. In engineering literature (see [101]), the term *multi-valued function* is sometimes used to represent a set-valued function that is "singleton-valued" almost everywhere, i.e., except at a finite number of isolated points where it is set-valued. In this book, however, we do not use the term multi-valued function to distinguish this particular case from general set-valued functions.

For a convex function $f : \mathbb{V} \to \mathbb{R} \cup \{+\infty\}$, the *(one-sided) directional derivative* of f at $x \in \mathrm{dom}\, f$ with respect to $d \in \mathbb{V}$ is defined by

$$f'(x; d) = \lim_{t \searrow 0} \frac{f(x + td) - f(x)}{t}.$$

Then we have that

$$f'(x; d) \geq \langle s, d \rangle, \quad \forall d \in \mathbb{V}$$

if and only if $s \in \partial f(x)$.[7]

2.1.5 Conjugate function

Given a function $f : \mathbb{V} \to \mathbb{R} \cup \{+\infty\}$,[8] the *conjugate function* of f is a function $f^* : \mathbb{V}^* \to \mathbb{R} \cup \{+\infty\}$ defined by

$$f^*(s) = \sup\{\langle s, x \rangle - f(x) \mid x \in \mathbb{V}\}. \tag{2.13}$$

The mapping $f \mapsto f^*$ is called the *Fenchel transformation*.[9] Since $\langle s, x \rangle - f(x)$ ($\forall x \in \mathbb{V}$) can be considered as a family of linear functions of s, (2.13) means that f^* is the pointwise supremum of the family of linear functions and therefore f^* is convex. The following inequality is immediately obtained from definition (2.13) of f^*.

---

[6] Although the notation $F : \mathbb{V} \to \mathbb{Y}$ originally means that F is a (usual) mapping, this notation is also used to denote a set-valued mapping by some authors for convenience; see Borwein–Lewis [59, p. 114] and Facchinei–Pang [125, p. 138].

[7] See Theorem 23.2 in Rockafellar [389].

[8] f is not necessarily convex, but we assume that $\mathrm{dom}\, f \neq \emptyset$.

[9] The Fenchel transformation is also called the Fenchel–Legendre transformation.

Fact 2.1.9 (Fenchel–Young inequality)**.** For $f : \mathbb{V} \to \mathbb{R} \cup \{+\infty\}$ such that $\mathrm{dom}\, f \neq \emptyset$, we have

$$f(x) + f^*(s) \geq \langle s, x \rangle, \quad \forall (x, s) \in \mathbb{V} \times \mathbb{V}^* \qquad \blacksquare$$

Necessary and sufficient conditions for the situation in which $f(x)+f^*(s) = \langle s, x\rangle$ holds are established in Proposition 2.1.12.

By repeating the procedure of defining f^* from f, we can obtain the conjugate function of f^*. The result, denoted by $(f^*)^*$ or f^{**}, is

$$f^{**}(x) = \sup\{\langle s, x\rangle - f^*(s) \mid s \in \mathbb{V}^*\}.$$

We call f^{**} the *biconjugate function* of f. Note that f^{**} is a function on $\mathbb{V}$ into $\mathbb{R} \cup \{+\infty\}$.

In Fact 2.1.10 and Fact 2.1.11, we see that f^{**} is related to f through the notion of closure of function defined below. For a function $f : \mathbb{V} \to \mathbb{R} \cup \{+\infty\}$,[10] the *closure* of f is the function $\mathrm{cl}\, f : \mathbb{V} \to \mathbb{R} \cup \{+\infty\}$ defined by

$$\mathrm{epi}(\mathrm{cl}\, f) = \mathrm{cl}(\mathrm{epi}\, f).$$

We define the *closed convex hull* of f, denoted $\mathrm{cl}\,\mathrm{co}\, f : \mathbb{V} \to \mathbb{R} \cup \{+\infty\}$, by[11]

$$\mathrm{epi}(\mathrm{cl}\,\mathrm{co}\, f) = \mathrm{cl}(\mathrm{co}\,\mathrm{epi}\, f).$$

It is easy to check that $\mathrm{cl}\, f$ is a minorant of f, i.e., $\mathrm{cl}\, f \leq f$ pointwise. More precisely, $\mathrm{cl}\, f$ is the largest minorant of f among lower semicontinuous functions. Analogously, $\mathrm{cl}\,\mathrm{co}\, f$ is the largest minorant of f among closed proper convex functions.

Fact 2.1.10. For any $f : \mathbb{V} \to \mathbb{R} \cup \{+\infty\}$ such that $\mathrm{dom}\, f \neq \emptyset$, the following statements hold:

(i) $f^{**}(x) \leq f(x)$ $(\forall x \in \mathbb{V})$.

(ii) $f^{**} = \mathrm{cl}\,\mathrm{co}\, f$.

(iii) $f^{***} = f^*$. ■

Fact 2.1.11. For $f : \mathbb{V} \to \mathbb{R} \cup \{+\infty\}$, the following statements hold:

(i) If f is a proper convex function, then $f^{**} = \mathrm{cl}\, f$.

(ii) If f is a closed proper convex function, then $f^{**} = f$. ■

---

[10] f is not necessarily convex, but we assume that $\mathrm{dom}\, f \neq \emptyset$.

[11] The closed convex hull "$\mathrm{cl}\,\mathrm{co}\, f$" of f is also written as "$\overline{\mathrm{co}} f$" by some authors.

The conjugate function of a convex function is closely related to the subdifferential as follows.

Proposition 2.1.12. *For a closed proper convex function* $f : \mathbb{V} \to \mathbb{R} \cup \{+\infty\}$*, the following three statements are equivalent:*

(a) $s \in \partial f(x)$.

(b) $f(x) + f^*(s) = \langle s, x \rangle$.

(c) $x \in \partial f^*(s)$.

Proof. (A) "(a) $\Leftrightarrow$ (b)": It follows from definition (2.11) of the subdifferential that (a) is rewritten as

$$\langle s, x \rangle - f(x) \geq \langle s, y \rangle - f(y), \quad \forall y \in \mathbb{V}.$$

By considering the maximum value of the right-hand side, the condition above is reduced to

$$\langle s, x \rangle - f(x) \geq \sup_y \{ \langle s, y \rangle - f(y) \mid y \in \mathbb{V} \}.$$

By applying definition (2.13) of the conjugate function to the right-hand side of the inequality above, we conclude that (a) is equivalent to the inequality

$$\langle s, x \rangle - f(x) \geq f^*(s).$$

At the same time we also have the Fenchel–Young inequality (Fact 2.1.9), and hence (a) is equivalent to (b).[12]

(B) "(b) $\Leftrightarrow$ (c)": Since f is assumed to be a closed proper convex function, it satisfies Fact 2.1.11 (ii). Hence, (b) can be rewritten as

$$f^*(s) + f^{**}(x) = \langle s, x \rangle.$$

Application of the equivalence "(b) $\Leftrightarrow$ (a)" yields that this equation is equivalent to (c). □

Table 2.2 is a summary of the relation between the subdifferential and the conjugate function of a closed proper convex function. As a consequence of (2.11), (2.13), Fact 2.1.7, and Fact 2.1.12, all the expressions in this table are equivalent.

The concepts of conjugate function and subdifferential, together with their relations in Table 2.2, are indispensable to analysis of nonsmooth mechanics. See section 3.3 for the use of these concept for describing the nonsmooth constitutive law, which is the most intuitive and the simplest example of applications of the convex analysis. This section also presents the close relation between the Fenchel transformation (or the conjugate transformation) and the classical Legendre transformation.

[12]Thus, to show the equivalence "(a) $\Leftrightarrow$ (b)", it suffices to assume that f is a proper convex function.

TABLE 2.2: Equivalent expressions for the subdifferential of a closed proper convex $f : \mathbb{R}^n \to \mathbb{R} \cup \{+\infty\}$.

	primal	dual
(i)	$\boldsymbol{s} \in \partial f(\boldsymbol{x})$	$\boldsymbol{x} \in \partial f^*(\boldsymbol{s})$
(ii)	$f(\boldsymbol{x}') - f(\boldsymbol{x}) \geq \langle \boldsymbol{s}, \boldsymbol{x}' - \boldsymbol{x} \rangle \ (\forall \boldsymbol{x}')$	$f^*(\boldsymbol{s}') - f^*(\boldsymbol{s}) \geq \langle \boldsymbol{s}' - \boldsymbol{s}, \boldsymbol{x} \rangle \ (\forall \boldsymbol{s}')$
(iii)	$\boldsymbol{x} \in \arg\max_{\boldsymbol{x} \in \mathbb{R}^n} \{\langle \boldsymbol{s}, \boldsymbol{x} \rangle - f(\boldsymbol{x})\}$	$\boldsymbol{s} \in \arg\max_{\boldsymbol{s} \in \mathbb{R}^n} \{\langle \boldsymbol{s}, \boldsymbol{x} \rangle - f^*(\boldsymbol{s})\}$
(iv)	$f(\boldsymbol{x}) + f^*(\boldsymbol{s}) = \langle \boldsymbol{s}, \boldsymbol{x} \rangle$	

2.2 Optimality and Duality

For the same reason given in the beginning of section 2.1.4, we denote by $\mathbb{V}^*$ the dual space of $\mathbb{V}$ and distinguish these two spaces. The elements of $\mathbb{V}$ and $\mathbb{V}^*$ are, in general, denoted by x and x^*, respectively, and $\langle x^*, x \rangle$ denotes their pairing (i.e., the inner product when $\mathbb{V}$ is a Euclidean space).

Although we shall not present a rigorous treatment of infinite-dimensional setting (see, e.g., [121, 390]), we make some comments concerning this issue. Let $\mathbb{V}$ be a vector space, possibly an infinite-dimensional space. Then the dual space $\mathbb{V}^*$ is defined as the vector space of continuous linear functionals over $\mathbb{V}$. The *pairing* (or *duality pairing*) $\langle x^*, x \rangle$, which is a bilinear functional on $\mathbb{V}^* \times \mathbb{V}$, represents the value at x of the (continuous linear) functional $x^* \in \mathbb{V}^*$. In the following we consider two pairs of spaces: the pair $\mathbb{V}$ and $\mathbb{V}^*$; and the pair $\mathbb{Y}$ and $\mathbb{Y}^*$. However, the reader who is not interested in such an abstract treatment may simply let

$$\mathbb{V} = \mathbb{V}^* = \mathbb{R}^n, \quad \mathbb{Y} = \mathbb{Y}^* = \mathbb{R}^m$$

without loss of generality in finite-dimensional setting.

2.2.1 Dual problem

Given a function $F : \mathbb{V} \to \mathbb{R} \cup \{+\infty\}$, consider the optimization problem

$$(\mathrm{P}) : \quad \inf\{F(x) \mid x \in \mathbb{V}\}, \tag{2.14}$$

which is called the *primal problem*.

Associated with F in (P), we introduce a function $\Phi : \mathbb{V} \times \mathbb{Y} \to \mathbb{R} \cup \{+\infty\}$

that satisfies[13]

$$\Phi(x, 0) = F(x), \quad \forall x \in \mathbb{V}. \tag{2.15}$$

For each $z \in \mathbb{Y}$, define the optimization problem (P_z) and the function $\theta : \mathbb{Y} \to \mathbb{R} \cup \{+\infty\}$ by

$$(\mathrm{P}_z): \quad \theta(z) := \inf\{\Phi(x, z) \mid x \in \mathbb{V}\}, \tag{2.16}$$

which is regarded as a family of optimization problems with respect to the parameter z. We call (P_z) the *perturbed problem* of (P). The function θ is called the *optimal value function* of (P). In particular, (P) itself corresponds to the case in which $z = 0$, say, $(\mathrm{P}_0) \equiv (\mathrm{P})$.

By definition, the conjugate function of Φ, denoted $\Phi^* : \mathbb{V}^* \times \mathbb{Y}^* \to \mathbb{R} \cup \{+\infty\}$, is given by

$$\Phi^*(x^*, z^*) = \sup_{x,z}\{\langle x^*, x\rangle + \langle z^*, z\rangle - \Phi(x, z) \mid (x, z) \in \mathbb{V} \times \mathbb{Y}\}. \tag{2.17}$$

Making use of Φ^*, we define the *dual problem* of (P) by

$$(\mathrm{P}^*): \quad \sup\{-\Phi^*(0, z^*) \mid z^* \in \mathbb{Y}^*\}. \tag{2.18}$$

Henceforth, we use the notations $\inf(\mathrm{P})$ and $\sup(\mathrm{P}^*)$ to mean

$$\inf(\mathrm{P}) = \theta(0) = \inf\{F(x) \mid x \in \mathbb{V}\}, \tag{2.19}$$

$$\sup(\mathrm{P}^*) = \sup\{-\Phi^*(0, z^*) \mid z^* \in \mathbb{Y}^*\} \tag{2.20}$$

for simplicity. We call $x \in \mathbb{V}$ an optimal solution of (P) if F attains $\inf(\mathrm{P})$ at x. Similarly, $z^* \in \mathbb{Y}^*$ satisfying $-\Phi^*(0, z^*) = \sup(\mathrm{P}^*)$ is said to be an optimal solution of (P^*). We write $\min(\mathrm{P})$ and $\max(\mathrm{P}^*)$ instead of $\inf(\mathrm{P})$ and $\sup(\mathrm{P}^*)$, respectively, if the corresponding problem has an optimal solution. We say that (P) is feasible if $\inf(\mathrm{P}) < +\infty$. Analogously, (P^*) is said to be feasible if $\sup(\mathrm{P}^*) > -\infty$.

2.2.2 Weak duality

As a study of the relation between (P) and (P^*), we start with establishing the inequality

$$\inf(\mathrm{P}) \geq \sup(\mathrm{P}^*), \tag{2.21}$$

which is called the *weak duality*.

Our first concern is to study the fundamental properties of θ defined in (2.16).

[13] For an example of Φ, see (2.29) in section 2.2.5, which is used for developing the Fenchel duality. The other examples are found in (2.44) in section 2.2.7 and (2.66) in section 2.3.1.

Proposition 2.2.1. *If Φ is a closed proper convex function, then θ is a convex function.*

Proof. What we have to show is that the inequality

$$\theta(\tau z_1 + (1-\tau)z_2) \leq \tau\theta(z_1) + (1-\tau)\theta(z_2), \quad \tau \in [0,1] \tag{2.22}$$

holds for any z_1, $z_2 \in \mathbb{Y}$ such that $\theta(z_1) < +\infty$ and $\theta(z_2) < +\infty$. By using definition (2.16) of θ, the left-hand side of (2.22) is reduced to

$$\theta(\tau z_1 + (1-\tau)z_2) = \inf\{\Phi(x, \tau z_1 + (1-\tau)z_2) \mid x \in \mathbb{V}\}. \tag{2.23}$$

Moreover, (2.16) implies that, for any $a_1 > \theta(z_1)$ (resp., for any $a_2 > \theta(z_2)$), there exists an $x_1 \in \mathbb{V}$ (resp., $x_2 \in \mathbb{V}$) satisfying

$$\theta(z_1) \leq \Phi(x_1, z_1) \leq a_1 \quad (\text{resp., } \theta(z_2) \leq \Phi(x_2, z_2) \leq a_2).$$

Hence, we can see that the inequalities

$$\begin{aligned} &\inf\{\Phi(x, \tau z_1 + (1-\tau)z_2) \mid x \in \mathbb{V}\} \\ &\quad \leq \Phi(\tau x_1 + (1-\tau)x_2, \tau z_1 + (1-\tau)z_2) \\ &\quad \leq \tau\Phi(x_1, z_1) + (1-\tau)\Phi(x_2, z_2) \\ &\quad \leq \tau a_1 + (1-\tau)a_2 \end{aligned} \tag{2.24}$$

hold.[14] By letting $a_1 \searrow \theta(z_1)$ and $a_2 \searrow \theta(z_2)$, (2.24) yields (2.22) in the limit. □

We identify θ^* as the objective function of (P*) as follows.

Proposition 2.2.2. $\theta^*(z^*) = \Phi^*(0, z^*)$ *for any* $z^* \in \mathbb{Y}^*$.

Proof. By using the definition of conjugate function (twice), we obtain

$$\begin{aligned} \theta^*(z^*) &= \sup_z\{\langle z^*, z\rangle - \theta(z) \mid z \in \mathbb{Y}\} \\ &= \sup_z\left\{\langle z^*, z\rangle - \inf_{x\in\mathbb{V}} \Phi(x, z) \mid z \in \mathbb{Y}\right\} \\ &= \sup_{x,z}\{\langle z^*, z\rangle + \langle 0, x\rangle - \Phi(x, z) \mid (x, z) \in \mathbb{V} \times \mathbb{Y}\} \\ &= \Phi^*(0, z^*). \end{aligned}$$

□

By virtue of the following result, we can associate the optimal value of (P*) with θ^{**}.

Proposition 2.2.3. $\sup(\mathrm{P}^*) = \theta^{**}(0)$.

Proof. From definition (2.20) of $\sup(\mathrm{P}^*)$ together with Proposition 2.2.2, it is straightforward to obtain

$$\begin{aligned} \sup(\mathrm{P}^*) &= \sup\{-\theta^*(z^*) \mid z^* \in \mathbb{Y}^*\} \\ &= \sup\{\langle 0, z^*\rangle - \theta^*(z^*) \mid z^* \in \mathbb{Y}^*\} \\ &= \theta^{**}(0). \end{aligned}$$

□

---

[14] The second inequality in (2.24) follows from the assumption of the convexity of Φ.

From (2.19) and Proposition 2.2.3, the difference of the primal and optimal values, which is called the *duality gap*, can be written as

$$\inf(\mathrm{P}) - \sup(\mathrm{P}^*) = \theta(0) - \theta^{**}(0).$$

The weak duality below asserts that the duality gap is nonnegative.

Theorem 2.2.4 (Weak duality). $+\infty \geq \inf(\mathrm{P}) \geq \sup(\mathrm{P}^*) \geq -\infty$.

Proof. The conclusion follows from the inequality $\theta(0) \geq \theta^{**}(0)$ (see Fact 2.1.10 (i)), Proposition 2.2.3, and (2.19). □

2.2.3 Strong duality

Our next concern is the case in which the inequality $\inf(\mathrm{P}) \geq \sup(\mathrm{P}^*)$ in Theorem 2.2.4 is satisfied as equality. We begin with an immediate corollary of the weak duality.

Proposition 2.2.5. $\inf(\mathrm{P}) = \sup(\mathrm{P}^*)$ *if and only if* $\theta^{**}(0) = \theta(0)$.

Proof. This is an immediate consequence of Theorem 2.2.4, (2.19), and Proposition 2.2.3. □

The situation in Proposition 2.2.5 is satisfied if, for example, θ is a closed proper convex function (see Fact 2.1.11 (ii)).

Note that the assertion of Proposition 2.2.5 is not related to the boundedness of $\inf(\mathrm{P})$ and $\sup(\mathrm{P}^*)$, as well as the existence of optimal solutions. A necessary and sufficient condition for guaranteeing the boundedness of $\inf(\mathrm{P})$ and $\sup(\mathrm{P}^*)$ and the existence of optimal solutions is established in the so-called strong duality theorem (Theorem 2.2.6).

If (P^*) has an optimal solution, then the set of optimal solutions of (P^*) is $\partial\theta^{**}(0)$.[15] It is known that if $\partial\theta(z) \neq \emptyset$ then $\theta(z) = \theta^{**}(z)$.[16] Moreover, if $\theta(z) = \theta^{**}(z)$, then $\partial\theta(z) = \partial\theta^{**}(z)$.[17] Thus, if $\partial\theta(0) \neq \emptyset$, the set of optimal solutions of (P^*) is $\partial\theta(0)$ (and obviously (P^*) has an optimal solution). This observation is formally stated as follows:

Theorem 2.2.6 (Strong duality). *The following two statements are equivalent:*

---

[15]In Proposition 2.2.2 we see that (P^*) is a maximization problem of $-\theta^*$ in $\mathbb{Y}^*$. Hence, $\bar{z}^* \in \mathbb{Y}^*$ is an optimal solution of (P^*) if and only if it satisfies $-\theta^*(\bar{z}^*) \geq -\theta^*(z^*)$ $(\forall z^* \in \mathbb{Y}^*)$. The latter condition yields

$$-\theta^*(\bar{z}^*) = \sup\{\langle 0, z^*\rangle - \theta^*(z^*) \mid z^* \in \mathbb{Y}^*\} = \theta^{**}(0).$$

Thus, we obtain $\theta^*(\bar{z}^*) + \theta^{**}(0) = \langle 0, \bar{z}^*\rangle$, which implies $\bar{z}^* \in \partial\theta^{**}(0)$ (see Table 2.2).

[16]See, e.g., Ekeland–Temam [121, Eq. I.(5.3)].

[17]See, e.g., Ekeland–Temam [121, Eq. I.(5.4)].

(a) $\inf(\mathrm{P})$ *is finite, and* $\partial\theta(0) \neq \emptyset$.[18]

(b) $\inf(\mathrm{P}) = \max(\mathrm{P}^*)$.[19]

Proof. From (2.19), we see that assertion (a) means that $\theta(0)$ is finite and there exists a $\bar{z}^* \in \partial\theta(0)$. By definition, $\bar{z}^* \in \partial\theta(0)$ if and only if

$$\theta(z) \geq \theta(0) + \langle \bar{z}^*, z \rangle, \quad \forall z \in \mathbb{Y},$$

which is equivalent to

$$\begin{aligned}\theta(0) &= \min\{\theta(z) - \langle \bar{z}^*, z \rangle \mid z \in \mathbb{Y}\} \\ &= -\theta^*(\bar{z}^*).\end{aligned} \tag{2.25}$$

By substituting (2.19) and the conclusion of Proposition 2.2.2 into (2.25), we obtain

$$\inf(\mathrm{P}) = -\Phi^*(0, \bar{z}^*). \tag{2.26}$$

Together with the weak duality (Theorem 2.2.4), (2.26) implies that $\bar{z}^*$ is optimal for (P^*), and hence $\max(\mathrm{P}^*) = -\Phi^*(0, \bar{z}^*)$. □

2.2.4 Optimality condition

We here establish the optimality conditions for (P) and (P^*).

Theorem 2.2.7. *Suppose that both* (P) *and* (P^*) *have optimal solutions and that*

$$-\infty < \min(\mathrm{P}) = \max(\mathrm{P}^*) < +\infty \tag{2.27}$$

is satisfied. Then, for $\bar{x} \in \mathbb{V}$ *and* $\bar{z}^* \in \mathbb{Y}^*$, *the following three statements are equivalent:*

(a) $\bar{x}$ *and* $\bar{z}^*$ *are optimal for* (P) *and* (P^*), *respectively.*

(b) $\bar{x}$ *and* $\bar{z}^*$ *satisfy* $\Phi(\bar{x}, 0) + \Phi^*(0, \bar{z}^*) = 0$.

(c) $\bar{x}$ *and* $\bar{z}^*$ *satisfy* $(0, \bar{z}^*) \in \partial\Phi(\bar{x}, 0)$.

Proof. Firstly, observe that the equation in (b) can be rewritten as

$$\Phi(\bar{x}, 0) + \Phi^*(0, \bar{z}^*) = \langle (\bar{x}, 0), (0, \bar{z}^*) \rangle,$$

which is therein equivalent to (c).[20] Thus it suffices to show the equivalence of (a) and (b).

[18] We say that (P) is *stable* if (a) is satisfied.

[19] The assertion (b) means that (P^*) has an optimal solution and its optimal value is finite.

[20] See the equivalence of (i) and (iv) in Table 2.2.

Assume (a), i.e., assume that $\bar{x}$ and $\bar{z}^*$ satisfy

$$\Phi(\bar{x},0) = \inf(\mathrm{P}),$$
$$-\Phi^*(0,\bar{z}^*) = \sup(\mathrm{P}^*).$$

Then assertion (b) follows from hypothesis (2.27).

Conversely, assume that $\bar{x}$ and $\bar{z}^*$ satisfy (b). Then, by using the definition of conjugate function, we obtain

$$\begin{aligned}
-\Phi(\bar{x},0) &= \Phi^*(0,\bar{z}^*) \\
&= \sup_{x,z}\{\langle \bar{z}^*, z\rangle - \Phi(x,z) \mid (x,z) \in \mathbb{V}\times\mathbb{Y}\} \\
&\geq \sup_{x}\{\langle \bar{z}^*, 0\rangle - \Phi(x,0) \mid x \in \mathbb{V}\} \\
&= -\inf\{\Phi(x,0) \mid x \in \mathbb{V}\} \\
&= -\inf(\mathrm{P}),
\end{aligned}$$

i.e., $\inf(\mathrm{P}) \geq \Phi(\bar{x},0)$, but this implies that $\inf(\mathrm{P}) = \Phi(\bar{x},0)$ and that $\bar{x}$ is optimal for (P). Similarly, we also see that

$$-\Phi^*(0,\bar{z}^*) = -\sup\{\Phi^*(0,z^*) \mid z^* \in \mathbb{Y}^*\},$$

i.e., $\bar{z}^*$ is optimal for (P*). Thus we have assertion (a), which concludes the proof. □

2.2.5 Fenchel duality

Given closed proper convex functions $f : \mathbb{V} \to \mathbb{R}\cup\{+\infty\}$, $g : \mathbb{Y} \to \mathbb{R}\cup\{+\infty\}$, and a continuous linear mapping $\Lambda : \mathbb{V} \to \mathbb{Y}$, consider the optimization problem[21]

$$(\mathrm{P_F}): \quad \inf_x\{f(x) + g(\Lambda x) \mid x \in \mathbb{V}\}, \tag{2.28}$$

which is regarded as the primal problem in the framework of Fenchel duality. This situation corresponds to considering (P) in (2.14) with F given by

$$F(x) = f(x) + g(\Lambda x).$$

We define Φ in (2.15) by

$$\Phi(x,z) = f(x) + g(\Lambda x - z). \tag{2.29}$$

We first derive the problem dual to ($\mathrm{P_F}$), by applying Φ in (2.29) to the pair of (P) in (2.14) and (P*) in (2.18). Note that the *adjoint operator* of Λ, denoted $\Lambda^* : \mathbb{Y}^* \to \mathbb{V}^*$, is defined by[22]

$$\langle z^*, \Lambda x\rangle = \langle \Lambda^* z^*, x\rangle, \quad \forall x \in \mathbb{V},\ \forall z^* \in \mathbb{Y}^*. \tag{2.30}$$

---

[21]As a simple example, let $\mathbb{V} = \mathbb{R}^n$ and $\mathbb{Y} = \mathbb{R}^m$. Then Λ is interpreted as an $m \times n$ real matrix.

[22]For example, let $\mathbb{V} = \mathbb{R}^n$ and $\mathbb{Y} = \mathbb{R}^m$, then $\mathbb{V}^* = \mathbb{R}^n$ and $\mathbb{Y}^* = \mathbb{R}^m$. In this case, $\Lambda \in \mathbb{R}^{m\times n}$ is interpreted as a conventional matrix; hence, Λ^* is identified with the transpose of Λ, i.e., $\Lambda^{\mathrm{T}} \in \mathbb{R}^{n\times m}$.

Proposition 2.2.8. *For* $(\mathrm{P_F})$ *defined by* (2.28)*, its dual problem is*

$$(\mathrm{P_F^*}): \quad \sup\{-f^*(\Lambda^* z^*) - g^*(-z^*) \mid z^* \in \mathbb{Y}^*\}. \tag{2.31}$$

Proof. Application of the definition of the conjugate function to Φ yields

$$\begin{aligned}\Phi^*(0, z^*) &= \sup_{x,z}\{\langle 0, x\rangle + \langle z^*, z\rangle - \Phi(x, z) \mid (x, z) \in \mathbb{V} \times \mathbb{Y}\}\\ &= \sup_{x,z}\{\langle z^*, z\rangle - f(x) - g(\Lambda x - z) \mid (x, z) \in \mathbb{V} \times \mathbb{Y}\}.\end{aligned}$$

For fixed x we introduce a new variable $t := \Lambda x - z$, which results in

$$\begin{aligned}\Phi^*(0, z^*) &= \sup_{x\in\mathbb{V}}\left\{\sup_{t\in\mathbb{Y}}\{\langle z^*, \Lambda x - t\rangle - f(x) - g(t)\}\right\}\\ &= \sup_{x\in\mathbb{V}}\left\{\langle z^*, \Lambda x\rangle - f(x) + \sup_{t\in\mathbb{Y}}\{\langle -z^*, t\rangle - g(t)\}\right\}\\ &= \sup_{x\in\mathbb{V}}\{\langle \Lambda^* z^*, x\rangle - f(x)\} + g^*(-z^*)\\ &= f^*(\Lambda^* z^*) + g^*(-z^*).\end{aligned}$$

□

Application of Theorem 2.2.7 yields the optimality conditions for $(\mathrm{P_F})$ and $(\mathrm{P_F^*})$ as follows.

Proposition 2.2.9. *Suppose that both* $(\mathrm{P_F})$ *and* $(\mathrm{P_F^*})$ *have optimal solutions and that*

$$-\infty < \min(\mathrm{P_F}) = \max(\mathrm{P_F^*}) < +\infty$$

is satisfied. Then, for $\bar{x} \in \mathbb{V}$ *and* $\bar{z}^* \in \mathbb{Y}^*$*, the following three statements are equivalent:*

(a) $\bar{x}$ *and* $\bar{z}^*$ *are optimal for* $(\mathrm{P_F})$ *and* $(\mathrm{P_F^*})$*, respectively.*

(b) $\bar{x}$ *and* $\bar{z}^*$ *satisfy*

$$\begin{aligned}f(\bar{x}) + f^*(\Lambda^* \bar{z}^*) &= \langle \Lambda^* \bar{z}^*, \bar{x}\rangle,\\ g(\Lambda\bar{x}) + g^*(-\bar{z}^*) &= \langle -\bar{z}^*, \Lambda\bar{x}\rangle.\end{aligned}$$

(c) $\bar{x}$ *and* $\bar{z}^*$ *satisfy*

$$\begin{aligned}\Lambda^* \bar{z}^* &\in \partial f(\bar{x}),\\ -\bar{z}^* &\in \partial g(\Lambda\bar{x}).\end{aligned}$$

Proof. We prove this assertion by applying Theorem 2.2.7 to the pair of $(\mathrm{P_F})$ and $(\mathrm{P_F^*})$ with Φ given in (2.29).

The condition of Theorem 2.2.7 (b) is equivalently written for Φ in (2.29) as

$$\begin{aligned}0 &= \Phi(\bar{x}, 0) + \Phi^*(0, \bar{z}^*)\\ &= \big(f(\bar{x}) + g(\Lambda\bar{x})\big) + \big(f^*(\Lambda^*\bar{z}^*) + g^*(-\bar{z}^*)\big)\\ &= \big[f(\bar{x}) + f^*(\Lambda^*\bar{z}^*) - \langle \Lambda^*\bar{z}^*, \bar{x}\rangle\big] + \big[g(\Lambda\bar{x}) + g^*(-\bar{z}^*) - \langle -\bar{z}^*, \Lambda\bar{x}\rangle\big].\end{aligned}$$

In the last expression above, both terms in the square brackets are nonnegative because of Fact 2.1.9. Thus we see that the part (b) of the present proposition is equivalent to Theorem 2.2.7 (b).

To see that assertion (c) is equivalent to assertion (b), we use the equivalence of Proposition 2.1.12 (a) and (b); by putting

$$x = \bar{x}, \quad s = \Lambda^* \bar{z}^*$$

in the notation of Proposition 2.1.12 we obtain

$$f(\bar{x}) + f^*(\Lambda^* \bar{z}^*) = \langle \Lambda^* \bar{z}^*, \bar{x} \rangle \quad \Leftrightarrow \quad \Lambda^* \bar{z}^* \in \partial f(\bar{x}),$$

while by putting

$$x = \Lambda \bar{x}, \quad s = -\bar{z}^*, \quad f = g$$

we obtain

$$g(\Lambda \bar{x}) + g^*(-\bar{z}^*) = \langle -\bar{z}^*, \Lambda \bar{x} \rangle \quad \Leftrightarrow \quad -\bar{z}^* \in \partial g(\Lambda \bar{x}),$$

which concludes the proof. □

The following result, which is aligned with Theorem 2.2.6 (the strong duality), provides us with a sufficient condition that guarantees the existence of optimal solution for $(\mathrm{P}_\mathrm{F}^*)$. Recall that f and g are assumed to be closed proper convex functions.

Proposition 2.2.10. *Suppose that the following conditions are satisfied:*

(i) $\inf(\mathrm{P}_\mathrm{F})$ *is finite.*

(ii) *There exists an* $x \in \mathrm{ri}(\mathrm{dom}\, f)$ *satisfying* $\Lambda x \in \mathrm{ri}(\mathrm{dom}\, g)$.

Then $(\mathrm{P}_\mathrm{F}^*)$ *has an optimal solution, and* $\inf(\mathrm{P}_\mathrm{F}) = \max(\mathrm{P}_\mathrm{F}^*)$ *holds.*

Proof. It suffices to show that (P_F) satisfies Theorem 2.2.6 (a). Since we here assume (i), we shall show $\partial\theta(0) \neq \emptyset$, where, from (2.16) and (2.29), θ is defined by

$$\theta(z) = \inf\{f(x) + g(\Lambda x - z) \mid x \in \mathbb{V}\}.$$

We first give an observation

$$\begin{aligned} &\theta(z) < +\infty \\ &\quad \Leftrightarrow \quad \{x \in \mathbb{V} \mid f(x) < +\infty,\ g(\Lambda x - z) < +\infty\} \neq \emptyset. \end{aligned} \tag{2.32}$$

As the next observation, for

$$x \in \mathrm{dom}\, f, \quad y \in \mathrm{dom}\, g,$$

let $z = \Lambda x - y$. Then we obtain $(\Lambda x - z) \in \mathrm{dom}\, g$. Thus the right-hand side of (2.32) is satisfied, and hence $z \in \mathrm{dom}\, \theta$. Proceeding in parallel with this observation, choose

$$x \in \mathrm{ri}(\mathrm{dom}\, f), \quad y \in \mathrm{ri}(\mathrm{dom}\, g), \tag{2.33}$$

and let $z = \Lambda x - y$. Then we obtain $(\Lambda x - z) \in \mathrm{ri}(\mathrm{dom} g)$, and hence $z \in \mathrm{ri}(\mathrm{dom}\, \theta)$. The assumption (ii) implies that we can choose $y = \Lambda x$ in (2.33), which yields $z = 0$, and hence $0 \in \mathrm{ri}(\mathrm{dom}\, \theta)$. Since f and g are closed proper convex functions, θ is a proper convex function. Hence, by Fact 2.1.8, $0 \in \mathrm{ri}(\mathrm{dom}\, \theta)$ implies $\partial\theta(0) \neq \emptyset$. □

As a counterpart of Proposition 2.2.10, the existence of optimal solution for $(\mathrm{P_F})$ is guaranteed in the following manner.

Proposition 2.2.11. *Suppose that the following conditions are satisfied:*

(i) $\sup(\mathrm{P_F^*})$ *is finite.*

(ii) *There exists a* $z^* \in \mathrm{ri}(\mathrm{dom}\, g^*)$ *satisfying* $-\Lambda^* z^* \in \mathrm{ri}(\mathrm{dom}\, f^*)$.

Then $(\mathrm{P_F})$ *has an optimal solution, and* $\min(\mathrm{P_F}) = \sup(\mathrm{P_F^*})$ *holds.*

Proof. The proof is conducted in a manner similar to that of Proposition 2.2.10, but we here regard $(\mathrm{P_F^*})$ as a primal problem. To this end, we begin by rewriting $(\mathrm{P_F^*})$ as

$$\begin{aligned}&\sup\{-f^*(\Lambda^* z^*) - g^*(-z^*) \mid z^* \in \mathbb{Y}^*\}\\&\qquad = -\inf\{g^*(-z^*) + f^*(\Lambda^* z^*) \mid -z^* \in \mathbb{Y}^*\},\end{aligned}$$

which is in the form of $(\mathrm{P_F})$ in (2.28). Thus $-z^*$ plays the role of x in Proposition 2.2.10, and hence assumption (ii) required in Proposition 2.2.10 reads

$$\exists -z^* \in \mathrm{ri}(\mathrm{dom}\, g^*) : \quad \Lambda^* z^* \in \mathrm{ri}(\mathrm{dom}\, f^*).$$

This condition is equivalent to (ii) in the present assertion. Moreover, it is obvious that assumption (i) in Proposition 2.2.10 corresponds to the boundedness of $-\sup(\mathrm{P_F^*})$, which concludes the proof. □

As our final result concerning the Fenchel duality, we here assert a sufficient condition for the existence of optimal solutions both of $(\mathrm{P_F})$ and $(\mathrm{P_F^*})$.

Proposition 2.2.12. *Suppose that the following conditions are satisfied:*

(i) *There exists an* $x \in \mathrm{ri}(\mathrm{dom}\, f)$ *satisfying* $\Lambda x \in \mathrm{ri}(\mathrm{dom}\, g)$.

(ii) *There exists a* $z^* \in \mathrm{ri}(\mathrm{dom}\, g^*)$ *satisfying* $-\Lambda^* z^* \in \mathrm{ri}(\mathrm{dom}\, f^*)$.

Then each of $(\mathrm{P_F})$ *and* $(\mathrm{P_F^*})$ *has an optimal solution, and* $\min(\mathrm{P_F}) = \max(\mathrm{P_F^*})$ *holds.*

Proof. Since we have Proposition 2.2.10 and Proposition 2.2.11, it suffices to show that $\inf(\mathrm{P_F})$ and $\sup(\mathrm{P_F^*})$ are finite.

The feasibility of each problem yields $\inf(\mathrm{P_F}) < +\infty$ and $\sup(\mathrm{P_F^*}) > -\infty$. Moreover, we also have the weak duality, say, $\sup(\mathrm{P_F^*}) \le \inf(\mathrm{P_F})$ (see Theorem 2.2.4). Thus $\inf(\mathrm{P_F})$ and $\sup(\mathrm{P_F^*})$ are finite. □

2.2.6 Lagrangian duality

In section 2.2.1 through section 2.2.5 we work with four types of variables: x, x^*, z, and z^*. Among them, x is the variable in (P), while z^* is that in (P*). We use z to introduce the perturbations on (P) as seen in (2.16), but z does not appear in the optimization problems explicitly. Analogously,

x^* introduced in (2.17) is regarded as the parameters only for introducing the perturbations on (P*). The Lagrangian duality, which we study here, discusses the duality and optimality only with x and z^*. The fundamental tool is the Lagrangian, which is defined by "eliminating" z from Φ as seen below.

Definition 2.2.13 (Lagrangian of (P)). For problem (P) defined in (2.14), consider a function Φ satisfying (2.15). Then the function $L : \mathbb{V} \times \mathbb{Y}^* \to \mathbb{R} \cup \{+\infty\}$ defined by

$$L(x, z^*) = -\sup_{z}\{\langle z^*, z\rangle - \Phi(x, z) \mid z \in \mathbb{Y}\} \tag{2.34}$$

is called the *Lagrangian* of (P). The variables $z^* \in \mathbb{Y}^*$ are called the *Lagrange multipliers*. ■

We first establish the following result indicating that (P*), which is the dual problem of (P) as introduced in (2.18), can be expressed in terms of the Lagrangian L.

Proposition 2.2.14. *The optimization problem*

$$(\mathrm{P}_{\mathrm{L}}^*) : \quad \sup_{z^* \in \mathbb{Y}^*} \left\{ \inf_{x}\{L(x, z^*) \mid x \in \mathbb{V}\} \right\} \tag{2.35}$$

is equivalent to (P*) *defined by* (2.18).

Proof. It follows from definition (2.17) of Φ^* and definition (2.34) of L that we obtain

$$\begin{aligned} \Phi^*(x^*, z^*) &= \sup_{x \in \mathbb{V}} \left\{ \langle x^*, x\rangle + \sup_{z}\{\langle z^*, z\rangle - \Phi(x, z) \mid z \in \mathbb{Y}\} \right\} \\ &= \sup_{x \in \mathbb{V}}\{\langle x^*, x\rangle - L(x, z^*)\}. \end{aligned}$$

Substitution of $x^* = 0$ to the equation above yields

$$\Phi^*(0, z^*) = -\inf_{x}\{L(x, z^*) \mid x \in \mathbb{V}\}, \tag{2.36}$$

which concludes the proof. □

Similarly, under the assumption stated below, we can express (P) in terms of L.

Proposition 2.2.15. *Suppose that* $\Phi(x, \cdot) : \mathbb{Y} \to \mathbb{R} \cup \{+\infty\}$ *is a closed proper convex function for each* $x \in \mathbb{V}$. *Then the optimization problem*

$$(\mathrm{P}_{\mathrm{L}}) : \quad \inf_{x \in \mathbb{V}} \left\{ \sup_{z^*}\{L(x, z^*) \mid z^* \in \mathbb{Y}^*\} \right\} \tag{2.37}$$

is equivalent to (P) *given in* (2.14).

Proof. We write $\Phi_x(\cdot) = \Phi(x, \cdot)$ for simplicity. The assumption on Φ allows us to apply Fact 2.1.11 (ii), from which we obtain

$$\begin{aligned}\Phi(x,z) &= \Phi_x^{**}(z) \\ &= \sup_{z^* \in \mathbb{Y}^*} \left\{ \langle z^*, z\rangle - \sup_z \{\langle z^*, z\rangle - \Phi(x,z) \mid z \in \mathbb{Y}\} \right\} \\ &= \sup_{z^* \in \mathbb{Y}^*} \{\langle z, z^*\rangle + L(x, z^*)\}.\end{aligned}$$

Thus the substitution of $z = 0$ yields

$$\Phi(x,0) = \sup_{z^*}\{L(x,z^*) \mid z^* \in \mathbb{Y}^*\}, \tag{2.38}$$

where $\Phi(x,0) = F(x)$. This concludes the proof. □

As a consequence of Proposition 2.2.14 and Proposition 2.2.15, we can claim that the weak duality (Theorem 2.2.4) established between (P) and (P*) can be also regarded as an implication of the inequality

$$\inf_{x \in \mathbb{V}} \sup_{z^* \in \mathbb{Y}^*} L(x, z^*) \geq \sup_{z^* \in \mathbb{Y}^*} \inf_{x \in \mathbb{V}} L(x, z^*).$$

The point at which the inequality above hold as equalities is characterized as follows.

Definition 2.2.16 (Saddle point)**.** Consider a Lagrangian L introduced in Definition 2.2.13. A point $(\bar{x}, \bar{z}^*) \in \mathbb{V} \times \mathbb{Y}^*$ is called a *saddle point* of L if it satisfies

$$L(\bar{x}, z^*) \leq L(\bar{x}, \bar{z}^*) \leq L(x, \bar{z}^*), \quad \forall x \in \mathbb{V},\ \forall z^* \in \mathbb{Y}^*. \qquad \blacksquare$$

The optimal solutions of $(\mathrm{P_L})$ and $(\mathrm{P_L^*})$ (and hence those of (P) and (P*)) are now linked with the saddle point of the Lagrangian L.

Proposition 2.2.17. *Suppose that* $\Phi : \mathbb{V} \times \mathbb{Y} \to \mathbb{R} \cup \{+\infty\}$ *is a closed proper convex function. Then the following two statements are equivalent:*

(a) $(\bar{x}, \bar{z}^*)$ *is a saddle point of* L.

(b) $\bar{x}$ *and* $\bar{z}^*$ *are optimal solutions of* $(\mathrm{P_L})$ *and* $(\mathrm{P_L^*})$*, respectively, and* $\min(\mathrm{P_L}) = \max(\mathrm{P_L^*})$ *holds.*

Proof. Recall that we have shown in the proofs of Proposition 2.2.14 and Proposition 2.2.15 that (see (2.36) and (2.38))

$$-\Phi^*(0, z^*) = \inf_{x \in \mathbb{V}} L(x, z^*), \tag{2.39}$$

$$\Phi(x, 0) = \sup_{z^* \in \mathbb{Y}^*} L(x, z^*). \tag{2.40}$$

First, assume (a). By virtue of (2.39), Definition 2.2.16 of saddle point yields

$$\begin{aligned}L(\bar{x}, \bar{z}^*) &= \inf_x \{L(x, \bar{z}^*) \mid x \in \mathbb{V}\} \\ &= -\Phi^*(0, \bar{z}^*).\end{aligned}$$

Similarly, by virtue of (2.40), Definition 2.2.16 yields

$$\begin{aligned} L(\bar{x}, \bar{z}^*) &= \sup_{z^*}\{L(\bar{x}, z^*) \mid z^* \in \mathbb{Y}^*\} \\ &= \Phi(\bar{x}, 0). \end{aligned}$$

Thus we conclude

$$\Phi(\bar{x}, 0) + \Phi^*(0, \bar{z}^*) = 0,$$

which is nothing but the optimality condition (Theorem 2.2.7 (b)), and hence assertion (b) is obtained.

Conversely, assume (b). By using (2.39) and (2.40), the optimality of $\bar{x}$ and $\bar{z}^*$ implies

$$\begin{aligned} -\Phi^*(0, \bar{z}^*) &= \inf_{x \in \mathbb{V}} L(x, \bar{z}^*) \leq L(\bar{x}, \bar{z}^*), \\ \Phi(\bar{x}, 0) &= \sup_{z^* \in \mathbb{Y}^*} L(\bar{x}, z^*) \geq L(\bar{x}, \bar{z}^*), \end{aligned}$$

from which we obtain

$$\Phi(\bar{x}, 0) = \sup_{z^* \in \mathbb{Y}^*} L(\bar{x}, z^*) \geq L(\bar{x}, \bar{z}^*) \geq \inf_{x \in \mathbb{V}} L(x, \bar{z}^*) = -\Phi^*(0, \bar{z}^*). \tag{2.41}$$

On the other hand, the assumption of min $(\mathrm{P_L})$ = max $(\mathrm{P_L^*})$ means that

$$-\Phi^*(0, \bar{z}^*) = \Phi(\bar{x}, 0),$$

i.e., the rightmost term of (2.41) is equal to the leftmost term of (2.41). We thus obtain

$$\sup_{z^* \in \mathbb{Y}^*} L(\bar{x}, z^*) = L(\bar{x}, \bar{z}^*) = \inf_{x \in \mathbb{V}} L(x, \bar{z}^*).$$

Consequently, by Definition 2.2.16, $(\bar{x}, \bar{z}^*)$ is a saddle point of L. □

2.2.7 KKT conditions

Based on the Lagrangian duality introduced above, we next consider the optimality condition for the conventional *nonlinear programming* (NLP) problem.

Let $f_0, f_1, \dots, f_m : \mathbb{R}^n \to \mathbb{R}$ and $h_1, \dots, h_k : \mathbb{R}^n \to \mathbb{R}$ be differentiable functions. Then a nonlinear programming problem in the standard form is formulated as

$$\left.\begin{aligned} \text{(NLP)}: \quad & \min_{\boldsymbol{x}} f_0(\boldsymbol{x}) \\ & \text{s.t. } f_j(\boldsymbol{x}) \leq 0, \quad j = 1, \dots, m, \\ & \qquad h_l(\boldsymbol{x}) = 0, \quad l = 1, \dots, k. \end{aligned}\right\} \tag{2.42}$$

In what follows, we assume that $f_0, f_1, \dots, f_m$ are convex and that $h_1, \dots, h_k$ are affine, i.e., they can be written in the forms of

$$h_l(\boldsymbol{x}) = \boldsymbol{a}_l^{\mathrm{T}} \boldsymbol{x} - b_l, \quad l = 1, \dots, k.$$

In this setting, problem (2.42) is a convex optimization problem.

To revert to the situation considered in section 2.2.1, we set the primal variables and the associated spaces as

$$x = \boldsymbol{x}, \qquad z = (\boldsymbol{\lambda}, \boldsymbol{\nu}), \tag{2.43a}$$

$$\mathbb{V} = \mathbb{R}^n, \quad \mathbb{Y} = \mathbb{R}^m \times \mathbb{R}^k, \tag{2.43b}$$

where $\boldsymbol{\lambda} \in \mathbb{R}^m$ and $\boldsymbol{\nu} \in \mathbb{R}^k$. Define $\Phi : \mathbb{V} \times \mathbb{Y} \to \mathbb{R} \cup \{+\infty\}$ by

$$\Phi(\boldsymbol{x}; \boldsymbol{\lambda}, \boldsymbol{\nu}) = \begin{cases} f_0(x) & \text{if } f_j(x) + z_j \leq 0 \ \ (j = 1, \dots, m), \\ & \qquad h_l(x) + \nu_l = 0 \ \ (l = 1, \dots, k), \\ +\infty & \text{otherwise}, \end{cases} \tag{2.44}$$

which is a closed proper convex function.[23] Then (NLP) in (2.42) is equivalently rewritten as

$$\inf\{\Phi(\boldsymbol{x}; \mathbf{0}, \mathbf{0}) \mid \boldsymbol{x} \in \mathbb{R}^n\}.$$

Thus, by using Φ in (2.44), (NLP) is represented in the form of (P) in (2.14) with (2.15).

The Lagrangian of (NLP) is now obtained by applying Definition 2.2.13 to Φ in (2.44). Direct calculation, using (2.43) and (2.44), yields

$$\begin{aligned} L(&\boldsymbol{x}; \boldsymbol{\lambda}^*, \boldsymbol{\nu}^*) \\ &= -\sup_{\boldsymbol{\lambda}, \boldsymbol{\nu}}\{\langle \boldsymbol{\lambda}^*, \boldsymbol{\lambda}\rangle + \langle \boldsymbol{\nu}^*, \boldsymbol{\nu}\rangle - \Phi(\boldsymbol{x}; \boldsymbol{\lambda}, \boldsymbol{\nu})\} \\ &= f_0(\boldsymbol{x}) - \sup_{\boldsymbol{\lambda}}\{\langle \boldsymbol{\lambda}^*, \boldsymbol{\lambda}\rangle \mid f_j(\boldsymbol{x}) \leq -\lambda_j \ (j = 1, \dots, m)\} \\ &\qquad - \sup_{\boldsymbol{\nu}}\{\langle \boldsymbol{\nu}^*, \boldsymbol{\nu}\rangle \mid h_l(\boldsymbol{x}) = -\nu_l \ (l = 1, \dots, k)\} \\ &= \begin{cases} f_0(\boldsymbol{x}) + \displaystyle\sum_{j=1}^{m} \lambda_j^* g_j(\boldsymbol{x}) + \sum_{l=1}^{k} \nu_l^* h_l(\boldsymbol{x}) & \text{if } \boldsymbol{\lambda}^* \geq \mathbf{0}, \\ -\infty & \text{otherwise}. \end{cases} \end{aligned} \tag{2.45}$$

According to Definition 2.2.16, a saddle point $(\bar{\boldsymbol{x}}; \bar{\boldsymbol{\lambda}}, \bar{\boldsymbol{\mu}})$ of L in (2.45) should satisfy

$$f_j(\bar{\boldsymbol{x}}) \leq 0, \ \bar{\lambda}_j^* \geq 0, \ \bar{\lambda}_j^* f_j(\bar{\boldsymbol{x}}) = 0, \qquad j = 1, \dots, m, \tag{2.46a}$$

$$h_l(\bar{\boldsymbol{x}}) = 0, \qquad l = 1, \dots, k, \tag{2.46b}$$

$$\nabla f_0(\bar{\boldsymbol{x}}) + \sum_{j=1}^{m} \bar{\lambda}_j^* \nabla f_j(\bar{\boldsymbol{x}}) + \sum_{l=1}^{k} \bar{\nu}_l^* \nabla h_l(\bar{\boldsymbol{x}}) = 0, \tag{2.46c}$$

[23] Slightly differently from the notation $\Phi(x, z)$ used above, we here write $\Phi(x; z) = \Phi(\boldsymbol{x}; \boldsymbol{\lambda}, \boldsymbol{\nu})$ to distinguish the primal variables $\boldsymbol{x}$ and dual variables $(\boldsymbol{\lambda}, \boldsymbol{\mu})$ clearly.

which are called the *Karush–Kuhn–Tucker* (KKT) *conditions*. Thus, if there exist $\bar{\boldsymbol{x}}$, $\bar{\boldsymbol{\lambda}}^*$, and $\bar{\boldsymbol{\mu}}^*$ satisfying the KKT conditions (2.46), then $\bar{\boldsymbol{x}}$ is the optimal solution of (NLP).

To ensure that the KKT conditions also provide the necessary condition for the optimality of (NLP),[24] we need to assume that (NLP) satisfies an appropriate constraint qualification. A simple one is the *Slater constraint qualification*: We say that (NLP) satisfies the Slater constraint qualification if there exists a vector $\boldsymbol{x}$ satisfying

$$\begin{aligned} f_j(\boldsymbol{x}) &< 0, \quad j = 1, \dots, m, \\ h_l(\boldsymbol{x}) &= 0, \quad l = 1, \dots, k. \end{aligned}$$

Under the Slater constraint qualification, one can show that $\bar{\boldsymbol{x}}$ is optimal if and only if there exist $\bar{\boldsymbol{\lambda}}^*$ and $\bar{\boldsymbol{\mu}}^*$ satisfying the KKT conditions (2.46); see, e.g., Avriel [25, section 4.5], Boyd–Vandenberghe [61, section 5.5], and Nocedal–Wright [340, Chap. 12] for more detail.

2.3 Application to Semidefinite Programming

As an example of the usage of the duality theory, we here apply the Fenchel and Lagrangian duality theories to the semidefinite programming (SDP) problem, which was introduced briefly in section 1.6.2. The Fenchel and Lagrangian dual problems of SDP are derived in section 2.3.1 and section 2.3.3, respectively, and the optimality conditions for SDP are investigated in section 2.3.2.

2.3.1 Fenchel dual problem of SDP

Recall that the (primal) standard form of SDP is given by (1.29), i.e., given $A_1, \dots, A_m, C \in \mathbb{S}^n$ and $\boldsymbol{b} \in \mathbb{R}^m$, the SDP problem is formulated as

$$(\mathrm{P_{SDP}}): \quad \left. \begin{aligned} &\min_X \ \langle C, X \rangle \\ &\text{s.t.} \ \langle A_i, X \rangle = b_i, \quad i = 1, \dots, m, \\ &\qquad \mathbb{S}^n \ni X \succeq O. \end{aligned} \right\} \tag{2.47}$$

---

[24] It should be clear that we here consider a convex smooth nonlinear programming problem.

The dual problem has been introduced, without detailed exposition of derivation, as (1.30), i.e.,

$$
(\mathrm{D_{SDP}}): \quad \left. \begin{aligned} & \max_{\boldsymbol{y}, S} \sum_{i=1}^{m} b_i y_i \\ & \text{s. t.} \ \sum_{i=1}^{m} y_i A_i + S = C, \\ & \qquad S \succeq O. \end{aligned} \right\} \tag{2.48}
$$

In this section, we show that $(\mathrm{D_{SDP}})$ can be derived by applying the framework of Fenchel duality to $(\mathrm{P_{SDP}})$. The derivation of $(\mathrm{D_{SDP}})$ by using the Lagrangian duality is investigated in section 2.3.3.

We begin by embedding problem (2.47) into the form of (2.28), which plays a role of primal problem in the Fenchel duality theory. To this end, we set

$$
\mathbb{V} = \mathbb{V}^* = \mathbb{S}^n, \quad \mathbb{Y} = \mathbb{Y}^* = \mathbb{R}^m.
$$

Define $f : \mathbb{S}^n \to \mathbb{R} \cup \{+\infty\}$ and $g : \mathbb{R}^m \to \mathbb{R} \cup \{+\infty\}$ by

$$
f(X) = \begin{cases} \langle C, X \rangle & \text{if } X \succeq O, \\ +\infty & \text{otherwise,} \end{cases} \tag{2.49}
$$

$$
g(\boldsymbol{z}) = \begin{cases} 0 & \text{if } \boldsymbol{z} = \boldsymbol{b}, \\ +\infty & \text{otherwise,} \end{cases} \tag{2.50}
$$

which are closed proper convex functions. We also define the operation of the linear mapping $\Lambda : \mathbb{S}^n \to \mathbb{R}^m$ on X by

$$
\Lambda X = \begin{bmatrix} \langle A_1, X \rangle \\ \vdots \\ \langle A_m, X \rangle \end{bmatrix}. \tag{2.51}
$$

Consequently, the primal problem $(\mathrm{P_{SDP}})$ in (2.47) is adapted to the situation of the Fenchel primal problem (2.28) as

$$
(\mathrm{P_{SDP\text{-}F}}): \quad \inf_{X} \{ f(X) + g(\Lambda X) \mid X \in \mathbb{S}^n \}. \tag{2.52}
$$

From (2.31), the Fenchel dual problem is given by

$$
(\mathrm{P^*_{SDP\text{-}F}}): \quad \sup_{\boldsymbol{z}^*} \{ -f^*(\Lambda^* \boldsymbol{z}^*) - g^*(-\boldsymbol{z}^*) \mid \boldsymbol{z}^* \in \mathbb{R}^m \}. \tag{2.53}
$$

To write (2.53) explicitly, we prepare the following fundamental facts.

Proposition 2.3.1. *For f, g, and Λ defined by* (2.49), (2.50), *and* (2.51), *respectively, we have*

$$f^*(X^*) = \begin{cases} 0 & \text{if } C - X^* \succeq O, \\ +\infty & \text{otherwise,} \end{cases} \tag{2.54}$$

$$g^*(\boldsymbol{z}^*) = \boldsymbol{b}^{\mathrm{T}}\boldsymbol{z}^*, \tag{2.55}$$

$$\Lambda^*\boldsymbol{z}^* = \sum_{i=1}^{m} z_i^* A_i. \tag{2.56}$$

Proof. By directly applying the definition of conjugate function to f in (2.49), we obtain

$$\begin{aligned} f^*(X^*) &= \sup\{\langle X^*, X\rangle - f(X) \mid X \in \mathbb{S}^n\} \\ &= \sup\{\langle X^*, X\rangle - f(X) \mid X \in \operatorname{dom} f\} \\ &= \sup\{\langle X^* - C, X\rangle \mid X \succeq O\} \\ &= -\inf\{\langle C - X^*, X\rangle \mid X \succeq O\}. \end{aligned}$$

Then (2.54) follows from Fact 1.3.17 (ii). Definition (2.50) of g and some direct calculations yield (2.55) as

$$\begin{aligned} g^*(\boldsymbol{z}^*) &= \sup\{\langle \boldsymbol{z}^*, \boldsymbol{z}\rangle - g(\boldsymbol{z}) \mid \boldsymbol{z} \in \mathbb{R}^m\} \\ &= \sup\{\langle \boldsymbol{z}^*, \boldsymbol{z}\rangle - g(\boldsymbol{z}) \mid \boldsymbol{z} \in \operatorname{dom} g\} \\ &= \sup\{\langle \boldsymbol{z}^*, \boldsymbol{z}\rangle - 0 \mid \boldsymbol{z} = \boldsymbol{b}\} \\ &= \boldsymbol{b}^{\mathrm{T}}\boldsymbol{z}^*. \end{aligned}$$

Application of the definition of an adjoint operator, say, (2.30), to (2.51) yields

$$\langle \boldsymbol{z}^*, \Lambda X\rangle = \sum_{i=1}^{m} z_i^* \langle A_i, X\rangle = \left\langle \sum_{i=1}^{m} z_i^* A_i, X \right\rangle = \langle \Lambda^*\boldsymbol{z}^*, X\rangle,$$

which means (2.56). □

Turning toward the Fenchel dual problem (2.53), we immediately see from Proposition 2.3.1 that

$$-f^*(\Lambda^*\boldsymbol{z}^*) = \begin{cases} 0 & \text{if } C - \sum_{i=1}^{m} z_i^* A_i \succeq O, \\ -\infty & \text{otherwise,} \end{cases}$$

$$-g^*(-\boldsymbol{z}^*) = \boldsymbol{b}^{\mathrm{T}}\boldsymbol{z}^*.$$

Consequently, problem (2.53) is written explicitly as

$$(\mathrm{P}^*_{\text{SDP-F}}): \quad \left.\begin{aligned} &\max_{\boldsymbol{z}^*} \boldsymbol{b}^{\mathrm{T}}\boldsymbol{z}^* \\ &\text{s.t. } C - \sum_{i=1}^{m} z_i^* A_i \succeq O. \end{aligned}\right\} \tag{2.57}$$

By setting

$$y = z^*, \quad S = C - \sum_{i=1}^{m} z_i^* A_i,$$

we can confirm that problem (2.57) is indeed reduced to ($\mathrm{D_{SDP}}$) in (2.48).

2.3.2 Duality and optimality of SDP

The preceding section showed that the Fenchel dual problem of ($\mathrm{P_{SDP}}$) in (2.47) can be formulated as ($\mathrm{P^*_{SDP\text{-}F}}$) in (2.57), which coincides with ($\mathrm{D_{SDP}}$) in (2.48). In this section we derive the optimality condition of those problems based on the Fenchel duality introduced in section 2.2.4.

For the primal problem ($\mathrm{P_{SDP}}$) in (2.47), a matrix $X \in \mathbb{S}^n$ is called a *strictly feasible solution* if it satisfies

$$\langle A_i, X\rangle = b_i \ (i = 1, \dots, m), \quad X \succ O. \tag{2.58}$$

Analogously, a vector $z^* \in \mathbb{R}^m$ satisfying

$$C - \sum_{i=1}^{m} z_i^* A_i \succ O \tag{2.59}$$

is called a strictly feasible solution of the dual problem ($\mathrm{P^*_{SDP\text{-}F}}$) in (2.57). The following result, which establishes *strong duality* of SDP, is obtained as a direct application of Proposition 2.2.12.

Proposition 2.3.2. *Suppose that both* ($\mathrm{P_{SDP}}$) *and* ($\mathrm{P^*_{SDP\text{-}F}}$) *have strictly feasible solutions. Then each problem has an optimal solution, and* $\min(\mathrm{P_{SDP}}) = \max(\mathrm{P^*_{SDP\text{-}F}})$ *holds.*

Proof. It suffices to show that ($\mathrm{P_{SDP}}$) and ($\mathrm{P^*_{SDP\text{-}F}}$) satisfy the assumptions required in Proposition 2.2.12, where ($\mathrm{P_{SDP}}$) and ($\mathrm{P^*_{SDP\text{-}F}}$) correspond to $\mathrm{P_F}$ and $\mathrm{P_F^*}$, respectively.

From the definitions of f and g (see (2.49) and (2.50)), we obtain

$$\mathrm{ri}(\mathrm{dom}\, f) = \{X \in \mathbb{S}^n \mid X \succ O\},$$
$$\mathrm{ri}(\mathrm{dom}\, g) = \{z \in \mathbb{R}^m \mid z = b\}.$$

On the other hand, it follows from (2.58) that ($\mathrm{P_{SDP}}$) has a strictly feasible solution if and only if

$$\{X \in \mathbb{S}^n \mid \Lambda X = b,\ X \succ O\} \neq \emptyset.$$

Hence, the assumption of the strict feasibility of ($\mathrm{P_{SDP}}$) means

$$\exists X \in \mathrm{ri}(\mathrm{dom}\, f) : \quad \Lambda X \in \mathrm{ri}(\mathrm{dom}\, g).$$

Consequently, ($\mathrm{P_{SDP}}$) satisfies assumption (i) required in Proposition 2.2.12.

The strict feasibility of $(\mathrm{P}^*_{\text{SDP-F}})$ is equivalent to

$$\left\{ \boldsymbol{z}^* \in \mathbb{R}^m \mid C - \sum_{i=1}^{m} z_i^* A_i \succ O \right\} \neq \emptyset.$$

On the other hand, from the definitions of f^* and g^* (see Proposition 2.3.1), we obtain

$$\begin{aligned} \mathrm{ri}(\mathrm{dom}\, f^*) &= \{X^* \in \mathbb{S}^n \mid C - X^* \succ O\}, \\ \mathrm{ri}(\mathrm{dom}\, g^*) &= \mathbb{R}^m. \end{aligned}$$

Thus the strict feasibility of $(\mathrm{P}^*_{\text{SDP-F}})$ means that assumption (ii) required in Proposition 2.2.12 is satisfied. □

By applying Proposition 2.2.9 to the pair of $(\mathrm{P}_{\text{SDP}})$ and $(\mathrm{P}^*_{\text{SDP-F}})$, we obtain the optimality conditions for SDP as follows.

Proposition 2.3.3. *Suppose that both* $(\mathrm{P}_{\text{SDP}})$ *and* $(\mathrm{P}^*_{\text{SDP-F}})$ *have strictly feasible solutions. Then* $\bar{X} \in \mathbb{S}^n$ *and* $\bar{\boldsymbol{z}}^* \in \mathbb{R}^m$ *are optimal solutions of* $(\mathrm{P}_{\text{SDP}})$ *and* $(\mathrm{P}^*_{\text{SDP-F}})$*, respectively, if and only if there exists an* $\bar{S} \in \mathbb{S}^n$ *satisfying*

$$A_i \bullet \bar{X} = b_i, \quad i = 1, \dots, m, \tag{2.60a}$$

$$\bar{S} + \sum_{i=1}^{m} \bar{z}_i^* A_i = C, \tag{2.60b}$$

$$\bar{X} \succeq O, \quad \bar{S} \succeq O, \quad \bar{S} \bullet \bar{X} = 0. \tag{2.60c}$$

Proof. We begin by noting that the assumptions required in Proposition 2.2.9 are guaranteed to be satisfied by Proposition 2.3.2. Then Proposition 2.2.9 (c) asserts that $\bar{X}$ and $\bar{\boldsymbol{z}}^*$ are optimal for $(\mathrm{P}_{\text{SDP}})$ and $(\mathrm{P}^*_{\text{SDP-F}})$, respectively, if and only if they satisfy

$$\Lambda^* \bar{\boldsymbol{z}}^* \in \partial f(\bar{X}), \tag{2.61a}$$

$$-\bar{\boldsymbol{z}}^* \in \partial g(\Lambda \bar{X}). \tag{2.61b}$$

To see that (2.60) is equivalent to (2.61), we derive the following expressions:

$$\partial f(\bar{X}) = \begin{cases} \{X^* \in \mathbb{S}^n \mid \langle C - X^*, \bar{X} \rangle = 0,\ C - X^* \succeq O\} & \text{if } \bar{X} \succeq O, \\ \emptyset & \text{otherwise,} \end{cases} \tag{2.62}$$

$$\partial g(\Lambda \bar{X}) = \begin{cases} \mathbb{R}^m & \text{if } \langle A_i, \bar{X} \rangle = b_i \ (i = 1, \dots, m), \\ \emptyset & \text{otherwise.} \end{cases} \tag{2.63}$$

To obtain (2.62), we make use of the relation that $X^* \in \partial f(\bar{X})$ if and only if[25]

$$f(\bar{X}) + f^*(X^*) = \langle X^*, \bar{X} \rangle. \tag{2.64}$$

[25] See Table 2.2 and Proposition 2.1.12.

From (2.49) and (2.54), the left-hand side of (2.64) is finite if and only if

$$\bar{X} \succeq O, \quad C - X^* \succeq O$$

is satisfied. In this case, equation (2.64) is reduced to

$$\langle C, \bar{X} \rangle = \langle X^*, \bar{X} \rangle,$$

i.e., $\langle C - X^*, \bar{X} \rangle = 0$, and hence we obtain (2.62).

Concerning (2.63), we also make use of the fact that $\boldsymbol{z}^* \in \partial g(\bar{\boldsymbol{z}})$ if and only if[26]

$$g(\bar{\boldsymbol{z}}) + g^*(\boldsymbol{z}^*) = \langle \boldsymbol{z}^*, \bar{\boldsymbol{z}} \rangle \tag{2.65}$$

holds. By using (2.50) and (2.55), we see that the left-hand side of (2.65) is finite if and only if

$$\bar{\boldsymbol{z}} = \boldsymbol{b}$$

is satisfied. In this case, (2.65) holds identically. Thus we obtain

$$\partial g(\bar{\boldsymbol{z}}) = \begin{cases} \mathbb{R}^m & \text{if } \bar{\boldsymbol{z}} = \boldsymbol{b}, \\ \emptyset & \text{otherwise,} \end{cases}$$

from which and (2.51) we arrive at (2.63).

Consequently, by using (2.62) and (2.63), together with (2.56), we see that (2.61) can be written as

$$\sum_{i=1}^{m} \bar{z}_i^* A_i \in \begin{cases} \{X^* \in \mathbb{S}^n \mid \langle C - X^*, \bar{X} \rangle = 0, \ C - X^* \succeq O\}, & \text{if } \bar{X} \succeq O, \\ \emptyset & \text{otherwise,} \end{cases}$$

$$-\bar{\boldsymbol{z}}^* \in \begin{cases} \mathbb{R}^m & \text{if } \langle A_i, \bar{X} \rangle = b_i \ (i = 1, \dots, m), \\ \emptyset & \text{otherwise.} \end{cases}$$

Now the conditions in (2.60) are obtained by denoting $\bar{S} = C - X^*$. □

Note that the optimality conditions (2.60) for SDP can be considered to be in the form of the mixed complementarity problem; see Definition 1.2.1 in section 1.2.2, with $\mathcal{C} = \mathbb{S}^n$. In (2.60a)–(2.60c), we take particular notice of (2.60c), which is nothing but the complementarity on the positive-semidefinite cone studied in section 1.3.4. By virtue of Fact 1.3.20,[27] we can see that, at the optimal solution of ($\mathrm{P}_{\mathrm{SDP}}$) and ($\mathrm{P}^*_{\mathrm{SDP\text{-}F}}$), the primal variable $\bar{X}$ and the dual slack variable $\bar{S}$ share a common system of eigenvectors, and each pair of their eigenvalues satisfies the complementarity condition.

[26] See Table 2.2 and Proposition 2.1.12.

[27] See also Table 1.3.

2.3.3 Lagrangian duality of SDP

In this section we apply the notion of the Lagrangian duality, studied in section 2.2.6, to the SDP problem ($\mathrm{P_{SDP}}$) in (2.47), and show that the Lagrangian dual problem of ($\mathrm{P_{SDP}}$) coincides with ($\mathrm{D_{SDP}}$) in (2.48). To this end, we set

$$\mathbb{V} = \mathbb{V}^* = \mathbb{S}^n, \quad \mathbb{Y} = \mathbb{Y}^* = \mathbb{R}^m \times \mathbb{S}^n.$$

For $z \in \mathbb{Y}$ and $z^* \in \mathbb{Y}^*$, we write

$$\begin{aligned} z &= (\boldsymbol{z}_1, Z_2), \quad \boldsymbol{z}_1 \in \mathbb{R}^m, \quad Z_2 \in \mathbb{S}^n, \\ z^* &= (\boldsymbol{z}_1, Z_2), \quad \boldsymbol{z}_1^* \in \mathbb{R}^m, \quad Z_2^* \in \mathbb{S}^n \end{aligned}$$

for the sake of convenience.

Define $\Phi : \mathbb{V} \times \mathbb{Y} \to \mathbb{R} \cup \{+\infty\}$ by

$$\Phi(X, \boldsymbol{z}_1, Z_2) = \begin{cases} C \bullet X & \text{if } \mathcal{A}X - \boldsymbol{b} = \boldsymbol{z}_1, \; X \succeq Z_2, \\ +\infty & \text{otherwise,} \end{cases} \tag{2.66}$$

where $\mathcal{A}X$ is defined by (2.51). Since

$$\Phi(X, \mathbf{0}, O) = \begin{cases} C \bullet X & \text{if } \mathcal{A}X = \boldsymbol{b}, \; X \succeq O, \\ +\infty & \text{otherwise,} \end{cases} \tag{2.67}$$

we see that the primal problem ($\mathrm{P_{SDP}}$) can be rewritten as

$$(\mathrm{P}_{(\mathbf{0},O)}) : \quad \inf_X \{\Phi(X, \mathbf{0}, O) \mid X \in \mathbb{V}\}, \tag{2.68}$$

which corresponds to the form of (P_z) in (2.16) with $z = 0$. Thus Φ defined in (2.66) satisfies the condition (2.15). Moreover, we easily see that $\Phi(X, \cdot, \cdot) : \mathbb{R}^m \times \mathbb{S}^n \to \mathbb{R} \cup \{+\infty\}$ is a closed proper convex function for each $X \in \mathbb{S}^n$, i.e., Φ in (2.66) also satisfies the assumption required in Proposition 2.2.15. Thus the setting for studying the Lagrangian duality is adjusted immaculately. We first derive the Lagrangian as follows.

Proposition 2.3.4. *The Lagrangian of* ($\mathrm{P}_{(\mathbf{0},O)}$)*, related to the perturbations* (2.66)*, is*

$$L(X, \boldsymbol{z}_1^*, Z_2^*) = \begin{cases} C \bullet X - (\boldsymbol{z}_1^*)^{\mathrm{T}}(\mathcal{A}X - \boldsymbol{b}) - Z_2^* \bullet X & \text{if } Z_2^* \succeq O, \\ -\infty & \text{otherwise,} \end{cases} \tag{2.69}$$

where $\boldsymbol{z}_1^* \in \mathbb{R}^m$ and $Z_2^* \in \mathbb{S}^n$ are the Lagrange multipliers.

Proof. From Definition 2.2.13 the Lagrangian of ($\mathrm{P}_{(\mathbf{0},O)}$) associated with the perturbations considered in (2.66) is defined by

$$\begin{aligned} &L(X, \boldsymbol{z}_1^*, Z_2^*) \\ &\quad = - \sup_{\boldsymbol{z}_1, Z_2} \{\langle \boldsymbol{z}_1^*, \boldsymbol{z}_1 \rangle + \langle Z_2^*, Z_2 \rangle - \Phi(X, \boldsymbol{z}_1, Z_2) \mid (\boldsymbol{z}_1, Z_2) \in \mathbb{Y}\}. \end{aligned}$$

By straightforward calculations we proceed as

$$
\begin{aligned}
&L(X, \boldsymbol{z}_1^*, Z_2^*) \\
&\quad = -\sup_{\boldsymbol{z}_1, Z_2} \{(\boldsymbol{z}_1^*)^{\mathrm{T}} \boldsymbol{z}_1 + Z_2^* \bullet Z_2 - \Phi(X, \boldsymbol{z}_1, Z_2) \mid (\boldsymbol{z}_1, Z_2) \in \operatorname{dom} \Phi(X, \cdot, \cdot)\} \\
&\quad = -\sup_{\boldsymbol{z}_1, Z_2} \{(\boldsymbol{z}_1^*)^{\mathrm{T}} (\Lambda X - \boldsymbol{b}) + Z_2^* \bullet Z_2 - C \bullet X \mid X \succeq Z_2\} \\
&\quad = C \bullet X - (\boldsymbol{z}_1^*)^{\mathrm{T}} (\Lambda X - \boldsymbol{b}) - \sup_{Z_2} \{Z_2^* \bullet Z_2 \mid X \succeq Z_2\}.
\end{aligned}
$$

To conclude the proof, it suffices to note that, by Fact 1.3.17 (ii), we have

$$
\begin{aligned}
&-\sup_{Z_2} \{Z_2^* \bullet Z_2 \mid X \succeq Z_2\} \\
&\quad = \inf_{Z_2} \{-Z_2^* \bullet X + Z_2^* \bullet (X - Z_2) \mid X \succeq Z_2\} \\
&\quad = -Z_2^* \bullet X + \inf_{Z_2} \{Z_2^* \bullet (X - Z_2) \mid X - Z_2 \succeq O\} \\
&\quad = \begin{cases} -X \bullet Z_2^* & \text{if } Z_2^* \succeq O, \\ -\infty & \text{otherwise.} \end{cases}
\end{aligned}
$$

□

Within the framework of the Lagrangian duality, the primal problem, introduced as $(\mathrm{P_L})$ in Proposition 2.2.15 for a general setting, is defined by using the Lagrangian L by

$$
(\mathrm{P_{SDP\text{-}L}}): \quad \inf_{X \in \mathbb{S}^n} \sup\{L(X, \boldsymbol{z}_1^*, Z_2^*) \mid (\boldsymbol{z}_1^*, Z_2^*) \in \mathbb{R}^m \times \mathbb{S}^n\}.
$$

We next confirm that $(\mathrm{P_{SDP\text{-}L}})$ truly coincides with $(\mathrm{P_{SDP}})$, by showing the following fact.

Proposition 2.3.5. *For $\Phi(X, \mathbf{0}, O)$ and L given in* (2.67) *and* (2.69), *respectively, we have that*

$$
\sup_{\boldsymbol{z}_1^*, Z_2^*} \{L(X, \boldsymbol{z}_1^*, Z_2^*) \mid (\boldsymbol{z}_1^*, Z_2^*) \in \mathbb{R}^m \times \mathbb{S}^n\} = \Phi(X, \mathbf{0}, O).
$$

Proof. Direct calculations yield

$$
\begin{aligned}
&\sup\{L(X, \boldsymbol{z}_1^*, Z_2^*) \mid (\boldsymbol{z}_1^*, Z_2^*) \in \mathbb{R}^m \times \mathbb{S}^n\} \\
&\quad = \sup\{C \bullet X - (\boldsymbol{z}_1^*)^{\mathrm{T}} (\Lambda X - \boldsymbol{b}) - Z_2^* \bullet X \mid (\boldsymbol{z}_1^*, Z_2^*) \in \operatorname{dom} L(X, \cdot, \cdot)\} \\
&\quad = C \bullet X - \inf_{\boldsymbol{z}_1^*} \{(\Lambda X - \boldsymbol{b})^{\mathrm{T}} \boldsymbol{z}_1^* \mid \boldsymbol{z}_1^* \in \mathbb{R}^m\} - \inf_{Z_2^*} \{X \bullet Z_2^* \mid Z_2^* \in \mathbb{S}^n\}.
\end{aligned}
$$

The assertion follows from the observation

$$
-\inf_{\boldsymbol{z}_1^*} \{(\Lambda X - \boldsymbol{b})^{\mathrm{T}} \boldsymbol{z}_1^* \mid \boldsymbol{z}_1^* \in \mathbb{R}^m\} = \begin{cases} 0 & \text{if } \Lambda X - \boldsymbol{b} = \mathbf{0}, \\ +\infty & \text{otherwise,} \end{cases}
$$

and Fact 1.3.17 (ii). □

It is straightforward to see from Proposition 2.3.5 that ($\mathrm{P_{SDP\text{-}L}}$) coincides with ($\mathrm{P}_{(\mathbf{0},O)}$) in (2.68), and hence ($\mathrm{P_{SDP}}$) in (2.47) also. Thus it is confirmed that the primal SDP problem ($\mathrm{P_{SDP}}$) can be written in the form of the primal problem of the Lagrangian duality theory by using L in (2.69).

According to Proposition 2.2.14, the Lagrangian dual problem of ($\mathrm{P_{SDP\text{-}L}}$) is defined by

$$(\mathrm{P^*_{SDP\text{-}L}}): \qquad \sup_{(\boldsymbol{z}_1^*, Z_2^*) \in \mathbb{R}^m \times \mathbb{S}^n} \inf\{L(X, \boldsymbol{z}_1^*, Z_2^*) \mid X \in \mathbb{S}^n\}.$$

For writing ($\mathrm{P^*_{SDP\text{-}L}}$) in an explicit form, we establish the following result.

Proposition 2.3.6. *For L given in* (2.69), *we have*

$$\inf_X\{L(X, \boldsymbol{z}_1^*, Z_2^*) \mid X \in \mathbb{S}^n\} = \begin{cases} \boldsymbol{b}^{\mathrm{T}}\boldsymbol{z}_1^* & \text{if } \Lambda^*\boldsymbol{z}_1^* + Z_2^* = C, \ \ Z_2^* \succeq O, \\ -\infty & \text{otherwise.} \end{cases}$$

Proof. If $Z_2^* \not\succeq O$, then nothing to be proved because $L(X, \boldsymbol{z}_1^*, Z_2^*) = -\infty$ implies $\inf\{L(X, \boldsymbol{z}_1^*, Z_2^*) \mid X \in \mathbb{S}^n\} = -\infty$. Hence, we consider only the case of $Z_2^* \succeq O$. Then by some direct calculations we have

$$\begin{aligned} &\inf_X\{L(X, \boldsymbol{z}_1^*, Z_2^*) \mid X \in \mathbb{S}^n\} \\ &\quad = \inf_X\{C \bullet X - (\Lambda X - \boldsymbol{b})^{\mathrm{T}}\boldsymbol{z}_1^* - X \bullet Z_2^* \mid X \in \mathbb{S}^n\} \\ &\quad = \boldsymbol{b}^{\mathrm{T}}\boldsymbol{z}_1^* + \inf_X\{(C - \Lambda^*\boldsymbol{z}_1^* - Z_2^*) \bullet X \mid X \in \mathbb{S}^n\} \\ &\quad = \begin{cases} \boldsymbol{b}^{\mathrm{T}}\boldsymbol{z}_1^* & \text{if } C - \Lambda^*\boldsymbol{z}_1^* - Z_2^* = O, \\ -\infty & \text{otherwise,} \end{cases} \end{aligned}$$

which concludes the proof. □

From Proposition 2.3.6, we can write the Lagrangian dual problem ($\mathrm{P^*_{SDP\text{-}L}}$) explicitly as

$$\left.\begin{aligned} \max_{\boldsymbol{z}_1^*, Z_2^*} \ & \boldsymbol{b}^{\mathrm{T}}\boldsymbol{z}_1^* \\ \text{s.t.} \ & \Lambda^*\boldsymbol{z}_1^* + Z_2^* = C, \\ & Z_2^* \succeq O. \end{aligned}\right\} \tag{2.70}$$

By setting

$$\boldsymbol{y} = \boldsymbol{z}_1^*, \quad S = Z_2^*,$$

we can see that problem (2.70) coincides with ($\mathrm{D_{SDP}}$) in (2.48). Thus, as a consequence of the investigation in section 2.3.1 and section 2.3.3, we conclude that both the Fenchel and Lagrangian dual problems of the primal standard form ($\mathrm{P_{SDP}}$) of SDP in (2.47) coincide with the dual standard form ($\mathrm{D_{SDP}}$) of SDP in (2.48).

2.4 Notes

The fundamental reference on convex analysis is Rockafellar [389]. Other monographs on convex analysis include Barvinok [35], Borwein–Lewis [59], Ekeland–Temam [121], Giorgi–Guerraggio–Thierfelder [147], Hiriart-Urruty–Lemaréchal [179], Jahn [190], and Tuy [440], some of which are more or less oriented to specific optimization problems. The description of the duality theory in section 2.2 follows mainly Ekeland–Temam [121], with some simplifications. Although this chapter is restricted to finite-dimensional cases, the standard way the conjugate duality is developed in an infinite-dimensional setting is found in Ekeland–Temam [121] and Rockafellar [390]; see also Allaire [10], Aubin–Ekeland [24], Borwein–Lewis [59, Chap. 10], Jahn [190], and Luenberger [282]. For fundamentals of functional analysis, see Oden–Demkowicz [341] and Reddy [383].

The strong duality of SDP was shown in section 2.3 based on the Fenchel duality. For more details and other proofs on the duality theorem of conic optimization; see Ben-Tal–Nemirovski [46] and Boyd–Vandenberghe [61], which also include the duality results of LP and SOCP. Related topics on the duality of SDP, as well as conic optimization, are found in Luo–Sturm–Zhang [283, 284], Pataki [359], Ramana [381], Ramana–Tunçel–Wolkowicz [382], and Sturm [425].

Regarding the duality of optimization problems in an infinite-dimensional setting, see Anderson–Nash [11], Grinold [155], Levinson [259], and Tyndall [441] for LP and Shapiro [408, 409] for conic optimization; see also Faybusovich–Moore [127] and Faybusovich–Tsuchiya [128] for interior-point algorithms for infinite-dimensional optimization problems.

Chapter 3

Applications in Structural Engineering

Orientation In this chapter we discuss three topics, that serve as illustrative examples of conic optimization and convex analysis introduced in Chapter 1 and Chapter 2, respectively. The first two applications are taken from the structural optimization: in section 3.1 the compliance minimization of a truss is shown to be reduced to semidefinite programming (SDP) and second-order cone programming (SOCP), while the truss optimization with the specified fundamental frequency is dealt with in section 3.2. In section 3.3 we link the Fenchel transformation (or the conjugate transformation) to the notion of complementary strain energy in mechanics via a study on the unilateral contact condition, which also serves as an introduction to the frictional contact problem investigated exhaustively in Chapter 10.

3.1 Compliance Optimization

When the linear elastic response is assumed, the compliance of a structure means the external work $\boldsymbol{p}^{\mathrm{T}}\boldsymbol{u}$ done by the static load, where $\boldsymbol{p}$ is the vector of given external forces, and $\boldsymbol{u}$ is the vector of displacements at the equilibrium state. The compliance is regarded as a global measure of the stiffness of a structure; if the compliance is small, the structure will be stiff against the given load.

The compliance is one of the most representative measures dealt with in the structural optimization. In particular, we take up the optimization of a truss by minimizing the compliance. Consider a truss with the given coordinates of nodes and the candidate members. For optimizing such a truss, it is natural to choose the design variables, denoted $\boldsymbol{x} \in \mathbb{R}^m$, as the member cross-sectional areas, where m is the number of members. Then the problem is to find the set of existing members, as well as their cross-sectional areas, to minimize the compliance. As shown below, this optimization problem is formulated as a conic optimization problem.

3.1.1 Definition of compliance

Let $K \in \mathbb{S}^n_+$ denote the stiffness matrix of a structure. The order n of K is the number of degrees of freedom of displacements of the structure. Let $\boldsymbol{x} \in \mathbb{R}^m$ denote the vector of design variables, and we consider K as a function of $\boldsymbol{x}$. For example, for a truss, we usually denote by x_i the member cross-sectional area, and m is the number of members. Then the stiffness matrix K of the truss is positive semidefinite for any $\boldsymbol{x} \geq \boldsymbol{0}$, as far as small deformation is assumed.

Let $\boldsymbol{u} \in \mathbb{R}^n$ denote the displacement vector. The internal energy of the structure, stored by $\boldsymbol{u}$, is written as

$$w_{\boldsymbol{x}}(\boldsymbol{u}) = \frac{1}{2}\boldsymbol{u}^{\mathrm{T}} K(\boldsymbol{x})\boldsymbol{u}.$$

The *compliance* of the structure is the function $\pi^{\mathrm{c}}(\boldsymbol{x};\cdot) : \mathbb{R}^n \to \mathbb{R} \cup \{+\infty\}$ defined by

$$\begin{aligned} \pi^{\mathrm{c}}(\boldsymbol{x};\boldsymbol{p}) &= 2w^*_{\boldsymbol{x}}(\boldsymbol{p}) \\ &= \sup_{\boldsymbol{u}}\{2\boldsymbol{p}^{\mathrm{T}}\boldsymbol{u} - \boldsymbol{u}^{\mathrm{T}} K(\boldsymbol{x})\boldsymbol{u} \mid \boldsymbol{u} \in \mathbb{R}^n\}. \end{aligned} \tag{3.1}$$

Remark 3.1.1. The compliance is interpreted here intuitively from the mechanical point of view. To begin, observe that the total potential energy function, denoted $\Phi(\boldsymbol{x};\cdot) : \mathbb{R}^n \to \mathbb{R}$, corresponding to the given external load $\boldsymbol{p} \in \mathbb{R}^n$ is defined by

$$\Phi(\boldsymbol{x};\boldsymbol{u}) = w_{\boldsymbol{x}}(\boldsymbol{u}) - \boldsymbol{p}^{\mathrm{T}}\boldsymbol{u}, \tag{3.2}$$

with which (3.1) can be rewritten as

$$\pi^{\mathrm{c}}(\boldsymbol{x};\boldsymbol{p}) = -2\inf_{\boldsymbol{u}} \Phi(\boldsymbol{x};\boldsymbol{u}). \tag{3.3}$$

Suppose that $\boldsymbol{p} \in \operatorname{Im} K(\boldsymbol{x})$. Then the stationarity condition of $\Phi(\boldsymbol{x};\cdot)$ is written as

$$K(\boldsymbol{x})\boldsymbol{u} = \boldsymbol{p}, \tag{3.4}$$

which is the equilibrium equation. Since $K(\boldsymbol{x})$ is positive semidefinite, (3.3) is realized at the stationary point of $\Phi(\boldsymbol{x};\cdot)$, which yields

$$\pi^{\mathrm{c}}(\boldsymbol{x};\boldsymbol{p}) = \boldsymbol{p}^{\mathrm{T}}\bar{\boldsymbol{u}} = \bar{\boldsymbol{u}}^{\mathrm{T}} K(\boldsymbol{x})\bar{\boldsymbol{u}}, \quad \text{with } \bar{\boldsymbol{u}} : K(\boldsymbol{x})\bar{\boldsymbol{u}} = \boldsymbol{p}. \tag{3.5}$$

In (3.5) we see that the compliance coincides with the external work done by $\boldsymbol{p}$, which is also equivalent to the twice of the stored energy, at the equilibrium state. Alternatively, if $\boldsymbol{p} \notin \operatorname{Im} K(\boldsymbol{x})$, then from (3.1) it follows that $\pi^{\mathrm{c}}(\boldsymbol{x};\boldsymbol{p}) = +\infty$.

Thus $\pi^{\mathrm{c}}(\boldsymbol{x};\boldsymbol{p})$ is finite if and only if the structure has an equilibrium state for a given $\boldsymbol{p}$. In particular, if $K(\boldsymbol{x})$ is nonsingular, then (3.5) is reduced to

$$\pi^{\mathrm{c}}(\boldsymbol{x};\boldsymbol{p}) = \boldsymbol{p}^{\mathrm{T}} K(\boldsymbol{x})^{-1}\boldsymbol{p}. \tag{3.6}$$

■

As seen in Remark 3.1.1, the compliance is equivalent to the external work done by the specified external load, and hence it can be regarded as a global performance measure of flexibility. The stiffness of a structure is to be increased by reducing the compliance. The compliance is often used for formulating structural optimization problems because of its simplicity. In particular, the compliance is a convex function with respect to the design variables $\boldsymbol{x}$ for a broad class of structures as shown below.

Proposition 3.1.2. *Let $\mathcal{X} \subseteq \mathbb{R}^m$ be a feasible set of design variables $\boldsymbol{x}$. Suppose that $\boldsymbol{u}^{\mathrm{T}}K(\cdot)\boldsymbol{u} : \mathcal{X} \to \mathbb{R}$ is concave on $\mathcal{X}$ for any $\boldsymbol{u} \in \mathbb{R}^n$. Then $\pi^{\mathrm{c}}(\cdot;\boldsymbol{p}) : \mathcal{X} \to \mathbb{R} \cup \{+\infty\}$ is convex on $\mathcal{X}$.*

Proof. It suffices to show that the inequality[1]

$$\pi^{\mathrm{c}}((\boldsymbol{x}_1 + \boldsymbol{x}_2)/2; \boldsymbol{p}) \leq \frac{1}{2}(\pi^{\mathrm{c}}(\boldsymbol{x}_1; \boldsymbol{p}) + \pi^{\mathrm{c}}(\boldsymbol{x}_2; \boldsymbol{p})) \tag{3.8}$$

holds for any $\boldsymbol{x}_1$, $\boldsymbol{x}_2 \in \mathcal{X}$.

The assumption on the concavity of $\boldsymbol{u}^{\mathrm{T}}K(\cdot)\boldsymbol{u}$ means that, for any $\boldsymbol{x}_1$, $\boldsymbol{x}_2 \in \mathcal{X}$, the inequality

$$\boldsymbol{u}^{\mathrm{T}}K((\boldsymbol{x}_1 + \boldsymbol{x}_2)/2)\boldsymbol{u} \geq \frac{1}{2}(\boldsymbol{u}^{\mathrm{T}}K(\boldsymbol{x}_1)\boldsymbol{u} + \boldsymbol{u}^{\mathrm{T}}K(\boldsymbol{x}_2)\boldsymbol{u})$$

holds; from this and definition (3.2) of Φ we obtain

$$2\Phi((\boldsymbol{x}_1 + \boldsymbol{x}_2)/2; \boldsymbol{u}) \geq \Phi(\boldsymbol{x}_1; \boldsymbol{u}) + \Phi(\boldsymbol{x}_2; \boldsymbol{u}).$$

With the use of this inequality and (3.3), we see that the left-hand side of (3.8) satisfies

$$\begin{aligned} &\pi^{\mathrm{c}}((\boldsymbol{x}_1 + \boldsymbol{x}_2)/2; \boldsymbol{p}) \\ &\quad = -2 \inf_{\boldsymbol{u}} \Phi((\boldsymbol{x}_1 + \boldsymbol{x}_2)/2; \boldsymbol{u}) \\ &\quad \leq -\inf_{\boldsymbol{u}} \{\Phi(\boldsymbol{x}_1; \boldsymbol{u}) + \Phi(\boldsymbol{x}_2; \boldsymbol{u})\} \\ &\quad = \sup_{\boldsymbol{u}} \{-\Phi(\boldsymbol{x}_1; \boldsymbol{u}) - \Phi(\boldsymbol{x}_2; \boldsymbol{u})\}. \end{aligned} \tag{3.9}$$

On the other hand, substituting (3.3) into the right-hand side of (3.8) yields

$$\begin{aligned} &\frac{1}{2}(\pi^{\mathrm{c}}(\boldsymbol{x}_1; \boldsymbol{p}) + \pi^{\mathrm{c}}(\boldsymbol{x}_2; \boldsymbol{p})) \\ &\quad = -\inf_{\boldsymbol{u}_1} \Phi(\boldsymbol{x}_1; \boldsymbol{u}_1) - \inf_{\boldsymbol{u}_2} \Phi(\boldsymbol{x}_2; \boldsymbol{u}_2) \\ &\quad = \sup_{\boldsymbol{u}_1, \boldsymbol{u}_2} \{-\Phi(\boldsymbol{x}_1; \boldsymbol{u}_1) - \Phi(\boldsymbol{x}_2; \boldsymbol{u}_2)\}. \end{aligned} \tag{3.10}$$

Compare (3.9) and (3.10) to see (3.8), which concludes the proof. □

[1]By definition, π^{c} is said to be convex on $\mathcal{X}$ if it satisfies

$$\pi^{\mathrm{c}}(\tau \boldsymbol{x}_1 + (1-\tau)\boldsymbol{x}_2) \leq \tau \pi^{\mathrm{c}}(\boldsymbol{x}_1) + (1-\tau)\pi^{\mathrm{c}}(\boldsymbol{x}_2), \quad \forall \tau \in [0, 1] \tag{3.7}$$

for any $\boldsymbol{x}_1$, $\boldsymbol{x}_2 \in \mathcal{X}$. Inequality (3.8) corresponds to the case of $\tau = 1/2$ in (3.7); indeed, for a continuous function it suffices to show that (3.8) is satisfied for any $\boldsymbol{x}_1$, $\boldsymbol{x}_2 \in \mathcal{X}$.

3.1.2 Compliance minimization

Let v_i denote the structural volume for unit value of x_i. We denote by V_0 the upper bound of the structural volume. For example, for a truss, x_i and v_i correspond to the cross-sectional area and the length of the ith member, respectively. Typically, the feasible set of design variables is given as

$$\mathcal{X} = \{\boldsymbol{x} \in \mathbb{R}^m \mid \boldsymbol{x} \geq \underline{\boldsymbol{x}},\ \boldsymbol{v}^{\mathrm{T}}\boldsymbol{x} \leq V_0\},$$

where $\underline{x}_i$ denotes the nonnegative lower bound specified for x_i. Note that we set $\boldsymbol{x} = \boldsymbol{0}$ for optimizing the topology, which means the member connectivity, of a truss. Given an external load $\boldsymbol{p}$, the minimization problem of compliance is now formulated as

$$\left.\begin{aligned} (\mathrm{MinC}) : \min_{\boldsymbol{x}} \ & \pi^{\mathrm{c}}(\boldsymbol{x};\boldsymbol{p}) \\ \mathrm{s.t.}\ & \boldsymbol{x} \geq \underline{\boldsymbol{x}}, \\ & \boldsymbol{v}^{\mathrm{T}}\boldsymbol{x} \leq V_0. \end{aligned}\right\} \tag{3.11}$$

We easily see that problem (3.11) is convex if the assumption of Proposition 3.1.2 holds. An extreme example is the case of a truss, in which the stiffness matrix can be written as

$$K(\boldsymbol{x}) = \sum_{i=1}^{m} x_i K_i \tag{3.12}$$

$$= \sum_{i=1}^{m} \frac{E x_i}{v_i} \boldsymbol{\beta}_i \boldsymbol{\beta}_i^{\mathrm{T}}, \tag{3.13}$$

where $K_1, \dots, K_m \in \mathbb{S}^n_+$ are constant positive-semidefinite matrices, $\boldsymbol{\beta}_1, \dots, \boldsymbol{\beta}_m \in \mathbb{R}^n$ are constant vectors, and E is Young's modulus (or the elastic modulus).

3.1.2.1 SDP formulation

Suppose for a moment that $K(\boldsymbol{x})$ is nonsingular for simplicity. Then $\pi^{\mathrm{c}}(\boldsymbol{x};\boldsymbol{p})$ is finite, and (3.6) means that $\tau \geq \pi^{\mathrm{c}}(\boldsymbol{x};\boldsymbol{p})$ if and only if $\tau - \boldsymbol{p}^{\mathrm{T}}K(\boldsymbol{x})^{-1}\boldsymbol{p} \geq 0$. Fact 1.3.10 (a lemma on the Schur complement) asserts that this condition is equivalent to

$$\begin{bmatrix} \tau & \boldsymbol{p}^{\mathrm{T}} \\ \boldsymbol{p} & K(\boldsymbol{x}) \end{bmatrix} \succeq O. \tag{3.14}$$

In contrast, for a singular $K(\boldsymbol{x})$, suppose $\boldsymbol{p} \notin \mathrm{Im}\, K(\boldsymbol{x})$ as an extreme case.[2] Then the definition in (3.1) leads to $\pi^{\mathrm{c}}(\boldsymbol{x};\boldsymbol{p}) = +\infty$. Such a situation is also involved in (3.14) appropriately, as stated formally as follows.

[2] From the mechanical point of view, this assumption implies that the structure is kinematically indeterminate (unstable). Therefore, the structure is regarded to have *no* stiffness against such a $\boldsymbol{p}$, which agrees with the consequence of calculation of (3.1), say, $\pi^{\mathrm{c}}(\boldsymbol{x};\boldsymbol{p}) = +\infty$.

Proposition 3.1.3. *There exists $\tau \in \mathbb{R}$ satisfying* (3.14) *if and only if* $\pi^{\mathrm{c}}(\boldsymbol{x};\boldsymbol{p}) < +\infty$. *Moreover, if* $\pi^{\mathrm{c}}(\boldsymbol{x};\boldsymbol{p})$ *is finite, then the inequality* $\tau \geq \pi^{\mathrm{c}}(\boldsymbol{x};\boldsymbol{p})$ *is equivalent to* (3.14).

As an immediate consequence of Proposition 3.1.3, problem (3.11) is equivalently rewritten as

$$\left.\begin{array}{ll} \displaystyle\min_{\tau,\boldsymbol{x}} & \tau \\ \mathrm{s.\,t.} & \begin{bmatrix} \tau & \boldsymbol{p}^{\mathrm{T}} \\ \boldsymbol{p} & K(\boldsymbol{x}) \end{bmatrix} \succeq O, \\ & \boldsymbol{x} - \underline{\boldsymbol{x}} \geq \boldsymbol{0}, \\ & V_0 - \boldsymbol{v}^{\mathrm{T}}\boldsymbol{x} \geq 0, \end{array}\right\} \qquad (3.15)$$

where τ is an auxiliary variable to be optimized.

In particular, for a truss $K(\boldsymbol{x})$ is given by (3.12), and thence problem (3.15) is reduced to an SDP problem.[3] To see this more explicitly, we claim that $\boldsymbol{x} - \underline{\boldsymbol{x}} \geq \boldsymbol{0}$ in (3.15) is equivalent to the positive-semidefinite constraint of the matrix $\mathrm{diag}(\boldsymbol{x} - \underline{\boldsymbol{x}})$, and hence problem (3.15) is reduced to

$$\left.\begin{array}{ll} \displaystyle\min_{\tau,\boldsymbol{x}} & \tau \\ \mathrm{s.\,t.} & \left[\begin{array}{cc|c|c} \tau & \boldsymbol{p}^{\mathrm{T}} & & \\ \boldsymbol{p} & \displaystyle\sum_{i=1}^{m} x_i K_i & & \\ \hline & & \mathrm{diag}(\boldsymbol{x} - \underline{\boldsymbol{x}}) & \\ \hline & & & V_0 - \boldsymbol{v}^{\mathrm{T}}\boldsymbol{x} \end{array}\right] \succeq O. \end{array}\right\}$$

Thus the set of the constraints of (3.15) is rewritten as a positive-semidefinite constraint on a block-diagonal matrix, as is the constraint of the standard form of SDP.

Remark 3.1.4. As shown in Remark 3.1.1, the compliance $\pi^{\mathrm{c}}(\boldsymbol{x};\boldsymbol{p})$ is equal to $\boldsymbol{p}^{\mathrm{T}}\boldsymbol{u}$, if $\boldsymbol{u}$ satisfies the equilibrium equation (3.4). Therefore, (MinC) in (3.11) is also formulated as

$$\left.\begin{array}{ll} \displaystyle\min_{\boldsymbol{x},\boldsymbol{u}} & \boldsymbol{p}^{\mathrm{T}}\boldsymbol{u} \\ \mathrm{s.\,t.} & K(\boldsymbol{x})\boldsymbol{u} = \boldsymbol{p}, \\ & \boldsymbol{x} - \underline{\boldsymbol{x}} \geq \boldsymbol{0}, \\ & V_0 - \boldsymbol{v}^{\mathrm{T}}\boldsymbol{x} \geq 0. \end{array}\right\} \qquad (3.16)$$

Compared with this formulation, the SDP formulation has an advantage in that problem (3.15) can be solved without difficulty even if $K(\boldsymbol{x})$ becomes singular. Indeed, for optimizing the topology of a truss we set $\underline{\boldsymbol{x}} = \boldsymbol{0}$, and it is often the case that the optimal solution has very few members compared with the initial truss. In such a case $K(\boldsymbol{x})$ often becomes singular, which makes it difficult to deal with

[3] See section 1.6.2 for the definition of SDP. We here show that problem (3.15) is reduced to the dual standard form (1.30) of SDP.

$K(\boldsymbol{x})\boldsymbol{u} = \boldsymbol{p}$ in (3.16) as equality constraints. In conventional methods, then, a small value is given to $\underline{x}_i$ as an ad hoc means to prevent the numerical instability. However, by using the primal-dual interior-point method, the SDP formulation (3.15) can be solved efficiently without resorting to any regularization. ■

3.1.2.2 SOCP formulation

Besides an SDP reformulation, it is also possible to reduce the compliance minimization problem (3.11) for a truss as an SOCP problem.[4]

First, observe that the member elongation c_i of a truss is written as $c_i = \boldsymbol{\beta}_i^{\mathrm{T}}\boldsymbol{u}$ by using the vector $\boldsymbol{\beta}_i$ given in (3.13). For simplicity, suppose that $x_i > 0$ meanwhile. The stored energy of the ith member can be written as

$$w_i = \frac{1}{2}\frac{Ex_i}{v_i}c_i^2 \tag{3.17}$$

$$= \frac{v_i}{2E}\frac{q_i^2}{x_i}, \tag{3.18}$$

where $q_i = (E/v_i)x_i c_i$ corresponds to the axial force. In (3.5) we see that $\pi^{\mathrm{c}}(\boldsymbol{x};\boldsymbol{p})$ coincides with twice $\sum_{i=1}^m w_i$ at the equilibrium state. The force-balance equation, required to be satisfied by the internal forces $q_1, \dots, q_m$ for the the given external forces $\boldsymbol{p}$, is written as

$$\sum_{i=1}^{m} q_i \boldsymbol{\beta}_i = \boldsymbol{p}. \tag{3.19}$$

Consequently, minimizing the compliance is equivalent to minimizing $\sum_{i=1}^m w_i$ in expression (3.18) under constraint (3.19),[5] where w_i, q_i, and x_i are the variables to be optimized.

In a minimization problem of $\sum_{i=1}^m w_i$, we can replace the equality constraint (3.18) with the inequality constraint $w_i \geq (v_i/2E)(q_i^2/x_i)$ without changing the optimal solution. Since $x_i \geq 0$, this inequality is further rewritten as

$$w_i x_i \geq (v_i/2E)q_i^2,$$

which is also valid in the case of $x_i = 0$. This observation yields for prob-

[4]See section 1.6.3 for the definition of SOCP. We here show that problem (3.15) is reduced to the dual standard form of SOCP in Definition 1.6.5.

[5]Because the compliance is $2\sum_{i=1}^m w_i$ and minimizing this is equivalent to minimizing $\sum_{i=1}^m w_i$.

lem (3.11) the equivalent reformulation

$$\left.\begin{array}{ll}\displaystyle\min_{\boldsymbol{x},\boldsymbol{q},\boldsymbol{w}} & \displaystyle\sum_{i=1}^{m} 2w_i \\ \text{s.t.} & w_i x_i \geq (v_i/2E)q_i^2, \quad i=1,\ldots,m, \\ & \displaystyle\sum_{i=1}^{m} q_i \boldsymbol{\beta}_i = \boldsymbol{p}, \\ & \boldsymbol{x} - \underline{\boldsymbol{x}} \geq \mathbf{0}, \\ & V_0 - \boldsymbol{v}^{\mathrm{T}}\boldsymbol{x} \geq 0. \end{array}\right\}$$

Finally, it is easy to see that this problem can be converted to an SOCP problem as follows: a linear inequality constraint is regarded as a special case of the second-order cone constraint,[6] and, by using Fact 1.5.3, each hyperbolic constraint in this problem is reduced to a second-order cone constraint as

$$w_i x_i \geq (v_i/2E)q_i^2 \quad \Leftrightarrow \quad w_i + x_i \geq \left\| \begin{bmatrix} w_i - x_i \\ \sqrt{2v_i/E}q_i \end{bmatrix} \right\|.$$

3.1.3 Worst-case compliance and robust optimization

A multiple-load case is considered in this section. We denote by $\boldsymbol{p}_1, \ldots, \boldsymbol{p}_k \in \mathbb{R}^n$ the linearly independent load vectors, where k is the number of the load scenarios. The set of loads under consideration is written as

$$\mathcal{P} = \{\boldsymbol{p}_1, \ldots, \boldsymbol{p}_k\}, \tag{3.20}$$

which is called the *uncertainty set* of the external load. Since the larger value of the compliance means less stiffness of a structure, the worst case is associated with the maximal compliance corresponding to the loads among $\boldsymbol{p} \in \mathcal{P}$. Thus the *worst-case* (or the worst-scenario or the critical) load is defined by

$$\boldsymbol{p}^{\mathrm{wc}} \in \arg\max_{\boldsymbol{p}\in\mathcal{P}} \pi^{\mathrm{c}}(\boldsymbol{x};\boldsymbol{p}).$$

This optimization problem is called the *worst-case detection* problem. The *robust optimization* is the problem to minimize the compliance in the worst case, i.e.,

$$\min_{\boldsymbol{x}\in\mathcal{X}} \sup_{\boldsymbol{p}\in\mathcal{P}} \pi^{\mathrm{c}}(\boldsymbol{x};\boldsymbol{p}). \tag{3.21}$$

[6] See section 1.5.

For $\mathcal{P}$ given by (3.20), we easily see that problem (3.21) is reduced to

$$\left.\begin{array}{ll} \min\limits_{\tau,\boldsymbol{x}} \tau & \\ \text{s.t. } \tau \geq \pi^{\mathrm{c}}(\boldsymbol{x};\boldsymbol{p}_j), & j=1,\dots,k, \\ \quad\ \boldsymbol{x} \geq \underline{\boldsymbol{x}}, & \\ \quad\ \boldsymbol{v}^{\mathrm{T}}\boldsymbol{x} \leq V_0, & \end{array}\right\}$$

where τ is an extra variable. By using Proposition 3.1.3, this problem is rewritten as

$$\left.\begin{array}{ll} \min\limits_{\tau,\boldsymbol{x}} \tau & \\ \text{s.t. } \begin{bmatrix} \tau & \boldsymbol{p}_j^{\mathrm{T}} \\ \boldsymbol{p}_j & K(\boldsymbol{x}) \end{bmatrix} \succeq O, & j=1,\dots,k, \\ \quad\ \boldsymbol{x} - \underline{\boldsymbol{x}} \geq \boldsymbol{0}, & \\ \quad\ V_0 - \boldsymbol{v}^{\mathrm{T}}\boldsymbol{x} \geq 0, & \end{array}\right\}$$

which is reduced to an SDP problem if $K(\boldsymbol{x})$ is given by (3.13).

A more sophisticated model for uncertainty in the external load may be considered as follows. For $\boldsymbol{p}_j$ $(j = 1,\dots,k)$ in (3.20), define $Q \in \mathbb{R}^{n\times k}$ by

$$Q = (\boldsymbol{p}_1,\dots,\boldsymbol{p}_k).$$

Consider the ellipsoidal uncertainty set of $\boldsymbol{p}$ given by

$$\mathcal{P} = \{Q\boldsymbol{\zeta} \mid \|\boldsymbol{\zeta}\| \leq 1\}, \tag{3.22}$$

where $\boldsymbol{\zeta} \in \mathbb{R}^k$ is called the vector of *uncertain parameters* (or unknown-but-bounded parameters). Then the robust optimization problem is again defined by (3.21), although infinitely many loads are considered in (3.22) while only a finite number of loads are treated in (3.20). In a manner similar to Proposition 3.1.3, an upper bound for the worst-case compliance, corresponding to uncertainty set (3.22), can be given by using a linear matrix inequality as follows.

Proposition 3.1.5. *There exists* $\tau \in \mathbb{R}$ *satisfying*

$$\begin{bmatrix} \tau I_k & Q^{\mathrm{T}} \\ Q & K(\boldsymbol{x}) \end{bmatrix} \succeq O \tag{3.23}$$

if and only if $\sup\{\pi^{\mathrm{c}}(\boldsymbol{x};\boldsymbol{p}) \mid \boldsymbol{p} \in \mathcal{P}\} < +\infty$. *Moreover, if* $\sup\{\pi^{\mathrm{c}}(\boldsymbol{x};\boldsymbol{p}) \mid \boldsymbol{p} \in \mathcal{P}\}$ *is finite, then the inequality* $\tau \geq \max\{\pi^{\mathrm{c}}(\boldsymbol{x};\boldsymbol{p}) \mid \boldsymbol{p} \in \mathcal{P}\}$ *is equivalent to* (3.23).

Proof. By using (3.1) and (3.22), the compliance in the worst-case is written as

$$\begin{aligned} &\sup_{\boldsymbol{p}\in\mathcal{P}} \pi^{\mathrm{c}}(\boldsymbol{x};\boldsymbol{p}) \\ &\quad = \sup_{\boldsymbol{p}\in\mathcal{P}} \left\{ \sup_{\boldsymbol{u}\in\mathbb{R}^n} \{2\boldsymbol{p}^{\mathrm{T}}\boldsymbol{u} - \boldsymbol{u}^{\mathrm{T}}K(\boldsymbol{x})\boldsymbol{u}\} \right\} \\ &\quad = \sup_{\boldsymbol{u}\in\mathbb{R}^n,\ \boldsymbol{\zeta}\in\mathbb{R}^k} \{2\boldsymbol{p}^{\mathrm{T}}\boldsymbol{u} - \boldsymbol{u}^{\mathrm{T}}K(\boldsymbol{x})\boldsymbol{u} \mid \boldsymbol{p} = Q\boldsymbol{\zeta},\ \|\boldsymbol{\zeta}\| \leq 1\}. \end{aligned}$$

Hence, $\tau \geq \sup_{\boldsymbol{p} \in \mathcal{P}} \pi^{\mathrm{c}}(\boldsymbol{x}; \boldsymbol{p})$ holds if and only if

$$\tau - 2(Q\boldsymbol{\zeta})^{\mathrm{T}} \boldsymbol{u} + \boldsymbol{u}^{\mathrm{T}} K \boldsymbol{u} \geq 0 \quad (\forall \boldsymbol{u} \in \mathbb{R}^{n},\ \forall \boldsymbol{\zeta} \in \mathbb{R}^{k} : \|\boldsymbol{\zeta}\| \leq 1) \tag{3.24}$$

is satisfied. Since the left-hand side of (3.24) attains the minimum when $\|\boldsymbol{\zeta}\| = 1$, it suffices to consider only the case in which $\|\boldsymbol{\zeta}\| = 1$ holds. Then (3.24) becomes equivalent to

$$\tau \boldsymbol{\zeta}^{\mathrm{T}} \boldsymbol{\zeta} - 2(Q\boldsymbol{\zeta})^{\mathrm{T}} \boldsymbol{u} + \boldsymbol{u}^{\mathrm{T}} K(\boldsymbol{x}) \boldsymbol{u} \geq 0 \quad (\forall \boldsymbol{u},\ \forall \boldsymbol{\zeta} : \|\boldsymbol{\zeta}\| = 1).$$

This condition is rewritten in terms of the quadratic form of a symmetric matrix as

$$[\boldsymbol{\zeta}^{\mathrm{T}}\ \boldsymbol{u}^{\mathrm{T}}] \begin{bmatrix} \tau I & -Q^{\mathrm{T}} \\ -Q & K(\boldsymbol{x}) \end{bmatrix} \begin{bmatrix} \boldsymbol{\zeta} \\ \boldsymbol{u} \end{bmatrix} \geq 0 \quad (\forall \boldsymbol{u},\ \forall \boldsymbol{\zeta} : \|\boldsymbol{\zeta}\| = 1),$$

which is equivalent to (3.23). □

As a consequence of Proposition 3.1.5, the robust optimization problem (3.21) is reformulated as

$$\left. \begin{aligned} & \min_{\tau, \boldsymbol{x}} \ \tau \\ & \text{s.t.} \ \begin{bmatrix} \tau I_k & Q^{\mathrm{T}} \\ Q & K(\boldsymbol{x}) \end{bmatrix} \succeq O, \\ & \qquad \boldsymbol{x} - \underline{\boldsymbol{x}} \geq \boldsymbol{0}, \\ & \qquad V_0 - \boldsymbol{c}^{\mathrm{T}} \boldsymbol{x} \geq 0. \end{aligned} \right\}$$

Thus the robust optimization problem of trusses is again reduced to an SDP problem, because the stiffness matrix of a truss is a linear function of $\boldsymbol{x}$ as seen in (3.12).

3.2 Eigenvalue Optimization

For a broad class of structures, the fundamental (i.e., minimal) natural frequency is regarded as a primary performance measure that reflects the global stiffness against dynamic loads. Therefore, it is natural to attempt to maximize the fundamental natural frequency for the specified structural volume, or equivalently, to minimize the structural volume for the specified fundamental natural frequency. The latter problem is formulated as an SDP problem in this section.

3.2.1 Eigenvalue optimization of structures

Consider an m-member truss, where the locations of nodes and candidate members are specified, like the one in section 3.1. Let $K \in \mathbb{S}^{n}_{+}$ denote the stiffness matrix, where n is the number of degrees of freedom of displacements. We denote by $M \in \mathbb{S}^n_+$ the mass matrix, due to both the structural and the non-structural masses. Note that K and M are the functions of the design variables, denoted by $\boldsymbol{x} \in \mathbb{R}^m$, which are the member cross-sectional areas in our context.

The generalized eigenvalue problem for the free vibration is formulated as

$$K\boldsymbol{\phi}_j = \Omega_j M \boldsymbol{\phi}_j, \quad j = 1, \dots, n, \tag{3.25}$$

where $\Omega_j \in \mathbb{R}$ and $\boldsymbol{\phi}_j \in \mathbb{R}^n$ are the eigenvalue and the corresponding eigenvector, respectively, which are functions of $\boldsymbol{x}$. Note that $\sqrt{\Omega_j}$ corresponds to the jth undamped natural circular frequency of the structure.[7]

For a while we assume that $x_i > 0$ $(i = 1, \dots, m)$, and thence $K(\boldsymbol{x})$ and $M(\boldsymbol{x})$ are positive definite. Then all the eigenvalues defined by (3.25) are positive. The smallest eigenvalue in (3.25), denoted $\Omega_{\min}(\boldsymbol{x})$, is the *fundamental eigenvalue* of the structure. Given a lower bound $\underline{\Omega} > 0$ for the fundamental eigenvalue, we attempt to find $\boldsymbol{x}$ minimizing the structural volume. This optimization problem is formulated as

$$\left.\begin{aligned} (\text{EigOpt}) : \min_{\boldsymbol{x}} \ & \boldsymbol{v}^{\mathrm{T}}\boldsymbol{x} \\ \text{s.t.} \ & \Omega_{\min}(\boldsymbol{x}) \geq \underline{\Omega}, \\ & \boldsymbol{x} \geq \underline{\boldsymbol{x}}. \end{aligned}\right\} \tag{3.26}$$

Since $\Omega_{\min}(\boldsymbol{x})$ is defined as the minimum value of a finite number of functions of $\boldsymbol{x}$, i.e., the minimum value of $\Omega_1(\boldsymbol{x}), \dots, \Omega_n(\boldsymbol{x})$, it is a nonsmooth function of $\boldsymbol{x}$. Indeed, it is known that $\Omega_{\min}(\boldsymbol{x})$ is differentiable if the multiplicity of the minimal eigenvalue of (3.25) is one (i.e., if the minimal eigenvalue is *simple*); otherwise it is not differentiable in general. We say that $\Omega_{\min}(\boldsymbol{x})$ is *multiple* if the multiplicity of the minimal eigenvalue of (3.25) is greater than one. It is also well known that the minimal eigenvalue of the optimal solution of problem (3.26) is often multiple [80]. Therefore, it is often that $\Omega_{\min}(\boldsymbol{x})$ is not differentiable at the optimal solution $\boldsymbol{x}$; hence, particularly sophisticated treatments are required both from theoretical and numerical points of view. It is difficult, in general, to solve problem (3.26) by using a gradient-based nonlinear programming approach based on standard sensitivity analysis. Applying the methodology of conic optimization resolves this issue entirely, as seen in the next section.

3.2.2 SDP formulation

To see that the eigenvalue optimization problem (3.26) is reformulated as an SDP problem, we first show that the lower bound constraint on the minimum

---

[7]The natural circular frequency is also called the natural angular frequency.

eigenvalue,

$$\Omega_{\min}(\boldsymbol{x}) \geq \underline{\Omega}, \tag{3.27}$$

in (3.26) is reduced to the linear matrix inequality.

If $\boldsymbol{x}$ satisfies $x_i > 0$ $(i = 1, \dots, m)$, then $M(\boldsymbol{x})$ is positive definite. In this case, application of (a slightly generalized) Rayleigh's principle (Fact 1.3.5) to the eigenvalue problem (3.25) yields the variational characterization

$$\Omega_{\min}(\boldsymbol{x}) = \inf_{\boldsymbol{\phi}: M(\boldsymbol{x})\boldsymbol{\phi} \neq \mathbf{0}} \frac{\boldsymbol{\phi}^{\mathrm{T}} K(\boldsymbol{x}) \boldsymbol{\phi}}{\boldsymbol{\phi}^{\mathrm{T}} M(\boldsymbol{x}) \boldsymbol{\phi}}. \tag{3.28}$$

By substituting this relation, the eigenvalue constraint (3.27) is reduced to

$$\boldsymbol{\phi}^{\mathrm{T}}(K(\boldsymbol{x}) - \underline{\Omega} M(\boldsymbol{x}))\boldsymbol{\phi} \geq 0 \quad (\forall \boldsymbol{\phi} \in \mathbb{R}^n),$$

which is equivalent to

$$K(\boldsymbol{x}) - \underline{\Omega} M(\boldsymbol{x}) \succeq O. \tag{3.29}$$

If we allow x_i to become equal to zero (i.e., if we put $\underline{\boldsymbol{x}} = \mathbf{0}$ in problem (3.26)), then $M(\boldsymbol{x})$ (and $K(\boldsymbol{x})$ also) is not necessarily positive definite, although it is always positive semidefinite. If $M(\boldsymbol{x}) \not\succ O$, then the meaning of the smallest eigenvalue in (3.25) is not clear. In this case, we define the fundamental eigenvalue of the structure by (3.28); see [3]. Note that $M(\boldsymbol{x}) \not\succ O$ means that some nodes are removed from the initial truss, due to the vanishing of some members in the course of optimization. Then the set of displacements corresponding to the existing nodes coincides with $\operatorname{Im} M(\boldsymbol{x})$, and hence we consider the infimum in (3.28) only for $\boldsymbol{\phi} \in \operatorname{Im} M(\boldsymbol{x}) \setminus \{\mathbf{0}\}$, where the number of eigenvalues considered is $\operatorname{rank} M(\boldsymbol{x})$. With this definition of $\Omega_{\min}(\boldsymbol{x})$ for $M(\boldsymbol{x}) \succeq O$, one can show that (3.27) is equivalent to (3.29) [3].

In (3.29), the mass matrix is additively decomposed as

$$M(\boldsymbol{x}) = M_{\mathrm{S}}(\boldsymbol{x}) + M_0,$$

where $M_{\mathrm{S}} \in \mathbb{S}^n_+$ reflects the contributions of structural masses, and $M_0 \in \mathbb{S}^n_+$ represents a constant non-structural mass. Particularly, for a truss $M_{\mathrm{S}}(\boldsymbol{x})$ can be written in the form of

$$M_{\mathrm{S}}(\boldsymbol{x}) = \sum_{i=1}^{m} x_i M_i,$$

where $M_1, \dots, M_m \in \mathbb{S}^n$ are constant matrices that are positive semidefinite. Recall that the stiffness matrix $K(\boldsymbol{x})$ also depends on $\boldsymbol{x}$ linearly as seen in (3.12). Therefore, (3.29) is rewritten as

$$\sum_{i=1}^{m} x_i (K_i - \underline{\Omega} M_i) - \underline{\Omega} M_0 \succeq O,$$

which is a linear matrix inequality. As a consequence, problem (3.26) is equivalently reformulated as

$$\left.\begin{array}{ll}\min\limits_{\boldsymbol{x}} & \boldsymbol{v}^{\mathrm{T}}\boldsymbol{x} \\ \mathrm{s.\,t.} & \sum\limits_{i=1}^{m} x_i(K_i - \underline{\Omega}M_i) - \underline{\Omega}M_0 \succeq O, \\ & \boldsymbol{x} - \underline{\boldsymbol{x}} \geq \mathbf{0}, \end{array}\right\} \tag{3.30}$$

which is an SDP problem.[8]

3.2.3 Optimality condition

The optimality condition for the eigenvalue optimization problem in (3.26) is derived within the framework of the strong duality of SDP.

As seen above, the eigenvalue optimization problem (3.26) for trusses is reformulated as an SDP problem in (3.30). On the other hand, recall that the duality and optimality of SDP are established in section 2.3, where problems (2.47) and (2.48) are shown to be the primal-dual pair of SDP problems. Therefore, the problem dual to (3.30), which is in the form of (2.48), can be obtained by adopting (2.47) as

$$\left.\begin{array}{ll}\max\limits_{Y,\boldsymbol{\eta}} & \underline{\Omega}M_0 \bullet Y + \underline{\boldsymbol{x}}^{\mathrm{T}}\boldsymbol{\eta} \\ \mathrm{s.\,t.} & (K_i - \underline{\Omega}M_i) \bullet Y + \eta_i = v_i, \quad i = 1,\dots,m, \\ & \boldsymbol{\eta} \geq \mathbf{0}, \\ & \mathbb{S}^n \ni Y \succeq O, \end{array}\right\} \tag{3.31}$$

where Y and $\boldsymbol{\eta}$ are the variables.[9]

The optimality condition for the SDP problems in the standard form is given in Proposition 2.3.3. For the pair (3.30) and (3.31), this condition is

[8] Problem (3.30) corresponds to the dual standard form (1.30) of SDP introduced in section 1.6.2. To see this more explicitly, we claim that the constraints of (3.30) are reduced to the positive-semidefinite constraint on a block-diagonal matrix as

$$\sum_{i=1}^{m} x_i \left[\begin{array}{c|c} -(K_i - \underline{\Omega}M_i) & O \\ \hline O & -\operatorname{diag}(\boldsymbol{e}_i) \end{array}\right] + S = \left[\begin{array}{c|c} -\underline{\Omega}M_0 & O \\ \hline O & -\operatorname{diag}(\underline{\boldsymbol{x}}) \end{array}\right], \quad S \succeq O,$$

where $\boldsymbol{e}_i \in \mathbb{R}^m$ is the ith column vector of the identity matrix I_m.

[9] Problem (3.31) corresponds to the primal standard form (1.29) of SDP introduced in section 1.6.2.

written as

$$\bar{S} = \sum_{i=1}^{m} \bar{x}_i (K_i - \underline{\Omega} M_i) - \underline{\Omega} M_0, \tag{3.32a}$$

$$(K_i - \underline{\Omega} M_i) \bullet \bar{Y} + \bar{\eta}_i = v_i, \qquad i = 1, \dots, m, \tag{3.32b}$$

$$\bar{S} \succeq O, \ \bar{Y} \succeq O, \ \bar{S} \bullet \bar{Y} = O, \tag{3.32c}$$

$$\bar{x}_i \geq \underline{x}_i, \ \bar{\eta}_i \geq 0, \ (\bar{x}_i - \underline{x}_i)\bar{\eta}_i = 0, \quad i = 1, \dots, m. \tag{3.32d}$$

More precisely, $\bar{\boldsymbol{x}}$ and $(\bar{Y}, \bar{\boldsymbol{\eta}})$ are optimal for (3.30) and (3.31), respectively, if and only if they satisfy (3.32) with $\bar{S} \in \mathbb{S}^n$ is defined by (3.32a).

In the optimality condition (3.32), it is noted that (3.32c) is the complementarity condition over positive-semidefinite cones. From Fact 1.3.20 (see also Table 1.3 for a summary), $\bar{S}$ and $\bar{Y}$ in (3.32) share a common system of eigenvectors, and the complementarity condition holds for each pair of eigenvalues. That is, $\bar{S}$ and $\bar{Y}$ are written as

$$\bar{S} = \sum_{j=1}^{n} \omega_j \boldsymbol{\psi}_j \boldsymbol{\psi}_j^{\mathrm{T}}, \tag{3.33a}$$

$$\bar{Y} = \sum_{j=1}^{n} \lambda_j \boldsymbol{\psi}_j \boldsymbol{\psi}_j^{\mathrm{T}}, \tag{3.33b}$$

$$\omega_j \geq 0, \quad \lambda_j \geq 0, \quad \omega_j \lambda_j = 0 \quad (j = 1, \dots, n), \tag{3.33c}$$

where $\boldsymbol{\psi}_1, \dots, \boldsymbol{\psi}_n$ are orthonormal eigenvectors of S and Y. We assume without loss of generality that ω_j's and λ_j's, respectively, are arranged in the non-descending and non-ascending orders, i.e., $\omega_1 \leq \dots \leq \omega_n$ and $\lambda_1 \geq \dots \geq \lambda_n$.

At the optimal solution $\bar{\boldsymbol{x}}$ of (3.26), let t denote the multiplicity of the fundamental (i.e., smallest) eigenvalues of the structure. As the result of optimization, the fundamental eigenvalue becomes equal to $\underline{\Omega}$. In the eigenvalue problem (3.25), we denote by $\boldsymbol{\phi}_j$ $(j = 1, \dots, t)$ the eigenvectors corresponding to $\underline{\Omega}$, where $\boldsymbol{\phi}_1, \dots, \boldsymbol{\phi}_t$ are chosen to be orthogonal. It then follows from (3.25) and (3.32a) that $\bar{S}\boldsymbol{\phi}_j = \mathbf{0}$ $(j = 1, \dots, t)$. This implies in (3.33a) that S has zero eigenvalues, i.e., $0 = \omega_1 = \dots = \omega_t < \omega_{t+1}$. Use (3.33c) to see $\lambda_{t+1} = \dots = \lambda_n = 0$, from which (3.33b) is reduced to

$$\bar{Y} = \sum_{j=1}^{t} \lambda_j \boldsymbol{\psi}_j \boldsymbol{\psi}_j^{\mathrm{T}}. \tag{3.34}$$

Since the subspace spanned by $\boldsymbol{\psi}_1, \dots, \boldsymbol{\psi}_t$ is the same as that by $\boldsymbol{\phi}_1, \dots, \boldsymbol{\phi}_t$, (3.34) is further rewritten as

$$\bar{Y} = \sum_{j=1}^{t} \sum_{k=1}^{t} H_{jk} \boldsymbol{\phi}_j \boldsymbol{\phi}_k^{\mathrm{T}}. \tag{3.35}$$

Here, $H = (H_{jk})$ should be a symmetric matrix, because so is $\bar{Y}$. Moreover, it is easy to see that $\bar{Y} \succeq O$ in (3.32c) is satisfied if and only if $H \succeq O$. Substitution of (3.35) into (3.32b) results in

$$\sum_{j=1}^{t} \sum_{k=1}^{t} H_{jk} \boldsymbol{\phi}_j^{\mathrm{T}} (K_i - \underline{\Omega} M_i) \boldsymbol{\phi}_k + \bar{\eta}_i = v_i$$

for each $i = 1, \dots, m$.

As a consequence of the discussion above, the optimality condition (3.32) is satisfied if and only if there exists $H \in \mathbb{S}^t$ satisfying

$$\bar{x}_i \geq \underline{x}_i \quad (i = 1, \dots, m), \tag{3.36a}$$

$$\begin{cases} \hat{v}_i = v_i & \text{if } \bar{x}_i > \underline{x}_i \\ \hat{v}_i \leq v_i & \text{if } \bar{x}_i = \underline{x}_i \end{cases} \quad (i = 1, \dots, m), \tag{3.36b}$$

$$\bar{S} \succeq O, \quad H \succeq O, \tag{3.36c}$$

where $\bar{S}$ is defined by (3.32a), $\hat{v}_i$ is defined by

$$\hat{v}_i = \sum_{j=1}^{t} \sum_{k=1}^{t} H_{jk} \boldsymbol{\phi}_j^{\mathrm{T}} (K_i - \underline{\Omega} M_i) \boldsymbol{\phi}_k,$$

and $\boldsymbol{\phi}_1, \dots, \boldsymbol{\phi}_t$ are the eigenvectors corresponding to the fundamental eigenvalue of the structure. The optimality condition (3.36) for problem (3.30) coincides with the traditional form found in the literature of structural optimization (see, e.g., [407]), analyses of which are usually based on the perturbation technique.

3.3 Set-Valued Constitutive Law

As an application of the notion of convex analysis introduced in section 2.1, this section investigates a constitutive law and its inverse represented as set-valued mappings. Specifically, it is shown that a constitutive law in nonsmooth mechanics is dealt with via the subdifferentials and conjugate functions of a stored energy function.

3.3.1 Constitutive law

As a very simple example of structures, consider a bar (or a spring) shown in Figure 3.1. Let c denote the elongation of the bar, and q is the axial force (internal force of the bar) corresponding to c.

FIGURE 3.1: A bar subjected to the elongation c and the corresponding axial force q.

A material of the bar is called *elastic* if there exists a function $\hat{q} : \mathbb{R} \to \mathbb{R}$, which relates the elongation to the axial force as

$$q = \hat{q}(c). \tag{3.37}$$

In other words, q of the elastic bar is uniquely determined solely by c. Equation (3.37) is called the *constitutive law*. We call the function $\hat{q}$ the *response function* for the axial force.

Suppose for a while that $\hat{q}$ is continuously differentiable. The *elongation stiffness* of the bar is a function defined by

$$\hat{k}(c) = \frac{\mathrm{d}}{\mathrm{d}c}\hat{q}(c). \tag{3.38}$$

For a given c, we often write $k = \hat{k}(c)$, where k is called the *tangent stiffness* at c. The *stored energy* (or *strain energy*) function $\hat{w} : \mathbb{R} \to \mathbb{R}$ is defined by[10]

$$\hat{w}(c) = \int_0^c \hat{q}(c)\mathrm{d}c. \tag{3.39}$$

We often write $w = \hat{w}(c)$ to represent the value of stored energy. The response function is certainly retrieved from the stored energy function as

$$\hat{q}(c) = \frac{\mathrm{d}}{\mathrm{d}c}\hat{w}(c). \tag{3.40}$$

Suppose, for example, that $\hat{w}$ is convex. Apply Fact 2.1.2 to the relations in (3.38) and (3.39) to see that the convexity of $\hat{w}$ is equivalent to the monotonicity of $\hat{q}$, and to $\hat{k}(c) \geq 0$ $(\forall c \in \mathbb{R})$ also. Table 3.1 summarizes such results on the constitutive relation. The *linear elasticity*, or the linear elastic material, means that there exists a constant $k_0 > 0$ such that $\hat{k}(c) = k_0$ $(\forall c \in \mathbb{R})$. If $\hat{q}$

[10]By the *strain energy function* we mean the stored energy function in (3.39) divided by the volume of the bar (in the reference configuration). Also, this strain energy function is called the *stored energy function* by some authors. In this section we do not distinguish these two terms; c can be regarded as either the elongation or the strain, and at the same time q is either the axial force or the strain, k is either the elongation stiffness or the Young modulus. Although some authors call the strain energy function (i.e., the stored energy per volume) the *strain energy density* function, we do not use this term in the book.

TABLE 3.1: The implication of the convexity of the (twice continuously differentiable) stored energy function $\hat{w}$.

$\hat{w}$	$\hat{q}$	$\hat{k}$
convex	monotone	nonnegative
strictly convex	strictly monotone	positive

is strictly monotone (not limited to the linearly elastic case), then it admits an inversion, and the inverse of the constitutive law (3.37) is immediately obtained as

$$c = \hat{q}^{-1}(q),$$

where $\hat{q}^{-1}$ denotes the the inverse of function $\hat{q}$. Thus the strict convexity of $\hat{w}$ is important in the elementary sense, although we can deal with a broader class of constitutive law by virtue of the methodology of convex analysis as seen in the subsequent sections.

3.3.2 Linear elasticity and Legendre transformation

As the simplest case of constitutive law, we here take up the linear elastic material to recall the relation among the strain energy, the complementary energy, and the constitutive law.

Let $k > 0$ be a constant, according to the assumption of linear elasticity. Write the stored energy function as

$$\hat{w}(c) = \frac{1}{2}kc^2 \tag{3.41}$$

to see that (3.40) yields the corresponding constitutive law

$$\hat{q}(c) = kc. \tag{3.42}$$

Since $k > 0$, $\hat{q}$ has the inverse function, and thence the inverse constitutive law is written as

$$c = \hat{q}^{-1}(q) = \frac{1}{k}q. \tag{3.43}$$

The *complementary energy* function, denoted $\hat{w}^{\mathrm{c}}$, is conventionally obtained by

$$\hat{w}^{\mathrm{c}}(q) = qc - \hat{w}(c), \quad \text{where } q = \hat{q}(c). \tag{3.44}$$

The mapping $\hat{w} \mapsto \hat{w}^{\mathrm{c}}$ is called the Legendre transformation; see, e.g., [18, 149, 384].[11] Provided that $\hat{w}$ is differentiable and strictly convex, there exists an inverse $\hat{q}^{-1}$ of $\hat{q}$, and $\hat{w}^{\mathrm{c}}(q)$ in (3.44) can be expressed in terms of q as

$$\hat{w}^{\mathrm{c}}(q) = q\hat{q}^{-1}(q) - \hat{w}(\hat{q}^{-1}(q)). \tag{3.45}$$

In the case of linear elasticity, substitution of (3.43) into (3.45) yields

$$\hat{w}^{\mathrm{c}}(q) = \frac{1}{2}\frac{q^2}{k}. \tag{3.46}$$

The inverse constitutive law, say, (3.43), is then retrieved as

$$c = \frac{\mathrm{d}\hat{w}^{\mathrm{c}}}{\mathrm{d}q}(q). \tag{3.47}$$

The analysis carried out above is represented in Figure 3.2. We begin with the strain energy function $\hat{w}$ in (3.41), then the constitutive law in (3.42) is obtained by differentiating $\hat{w}$; the inverse constitutive law (3.43) is computed, which is to be integrated to obtain the complementary energy function $\hat{w}^{\mathrm{c}}$ in (3.46). Thus, in the procedure above, we make use of $\hat{q}^{-1}$, the inverse of the constitutive law, to derive an explicit form of the result of the classical Legendre transformation[12] in (3.44), although the existence of a (single-valued) inverse of $\hat{q}$ is not necessarily guaranteed when we encounter a more complicated situation than the linear elasticity. In contrast, in section 3.3.3 we see that the inversion of the constitutive law is pursued as a conclusion of the Fenchel transformation. In section 3.3.4, this new procedure of inversion is shown to be applicable to obtain an inverse of a set-valued constitutive law, where a problem in nonsmooth mechanics often involves a set-valued constitutive law.

3.3.3 Inversion via Fenchel transformation

This section shows that the relation in Figure 3.2 can be recovered through the Fenchel transformation (or the conjugate transformation).

Recall that the Fenchel transformation is defined by (2.13). The conjugate function of $\hat{w}$, denoted $\hat{w}^*$, is defined by

$$\begin{aligned} \hat{w}^*(q) &= \sup_c \{qc - \hat{w}(c) \mid c \in \mathbb{R}\} \\ &= \sup_c \left\{qc - \frac{1}{2}kc^2 \mid c \in \mathbb{R}\right\}, \end{aligned} \tag{3.48}$$

[11]The Legendre transformation generally results in a set-valued function.

[12]It was mentioned in section 2.1.5 that the term *Legendre transformation* is used to represent the Fenchel transformation (i.e., the conjugate transformation) by some authors. However, we distinguish these two words in a conventional way found in the literature of applied mechanics.

$$\hat{w}(c) = \frac{1}{2}kc^2 \quad \xleftrightarrow[\text{transformation}]{\text{Legendre}} \quad \hat{w}^{\mathrm{c}}(q) = \frac{1}{2}\frac{q^2}{k}$$

differentiation $\downarrow$ $\qquad\qquad$ $\uparrow$ integration

$$q = kc \quad \xleftrightarrow[\text{inversion}]{} \quad c = \frac{1}{k}q$$

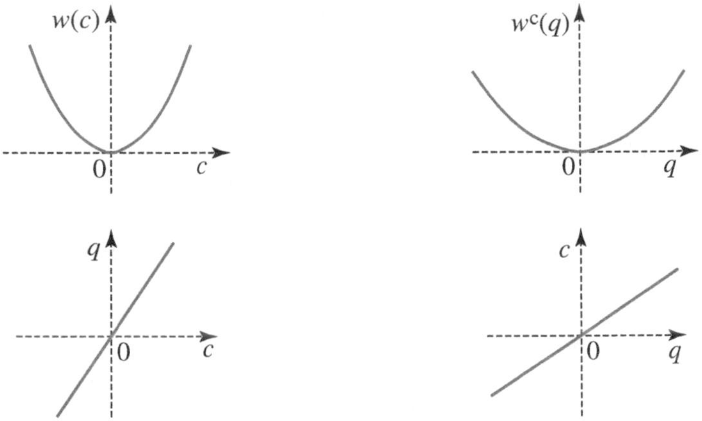

FIGURE 3.2: Schematic exposition of the classical derivation of the complementary energy from the strain energy in the linear elasticity.

where (3.41) was used. From the stationarity condition, the supremum in (3.48) is attained at $q = kc$, with which c is eliminated from (3.48) to see that

$$\hat{w}^*(q) = \frac{1}{2}\frac{q^2}{k}. \tag{3.49}$$

Thus the conjugate function $\hat{w}^*$ coincides with the complementary energy function $\hat{w}^{\mathrm{c}}$ in (3.46). The inverse constitutive law is then obtained by differentiating $\hat{w}^*$, which certainly results in (3.43).

The analysis carried out in this section is summarized in Figure 3.3. The complementary energy function is obtained as the conjugate function of the strain energy function $\hat{w}$; then the constitutive law and its inverse are obtained by (sub)differentiating $\hat{w}$ and $\hat{w}^*$, respectively. Thus by using the Fenchel transformation we retrieve the *inversion* relation, but it is now noted that the subdifferentials in Figure 3.3 can be defined for nonsmooth $\hat{w}$ and $\hat{w}^*$. The subdifferential of a nonsmooth function is a set-valued function in general, and if this is the case $\partial\hat{w}$ does not have an inverse in the conventional sense. This means that the procedure in Figure 3.3 can lead to a *generalization* of the inversion, which enables us to deal with the "inversion" of the unilateral contact law as discussed in the following section.

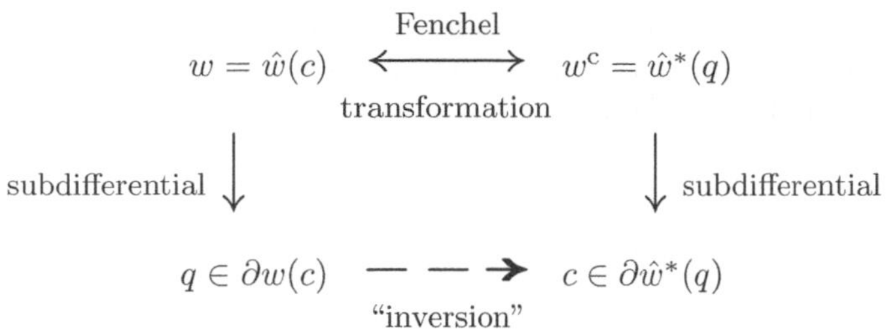

FIGURE 3.3: Relation among the constitutive law, strain energy, complementary energy, and inverse of the constitutive law within the framework of convex analysis.

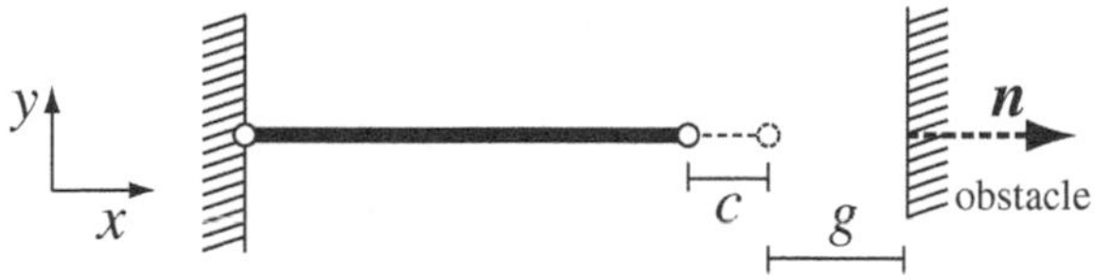

FIGURE 3.4: A bar subjected to the elongation c and the corresponding axial force q.

3.3.4 Unilateral contact law and Fenchel transformation

Consider the fixed rigid obstacle shown in Figure 3.4, which possibly make contact with the free node of the bar. Such a node is called the *contact candidate node*. To see an engineering use of the methodology of convex analysis, we establish the constitutive relation between the kinematic and static quantities associated with the contact candidate node. See Chapter 10 for a comprehensive discussion of the contact problem with the Coulomb friction; in this section only the frictionless contacts are considered.

Let g denote the *gap* (or the distance) between the contact candidate node and the obstacle. The kinematic contact states are classified by the sign of the contact gap g as follows:

- $g > 0$: the node is free.
- $g = 0$: the node is in contact.
- $g < 0$: the node is in penetration.

Since we assume that the obstacle is a rigid body, the node is not allowed to penetrate into the obstacle, and hence $g \geq 0$ should be satisfied.

We next consider the constraint on the static quantity. The force that acts from the surface of the obstacle to the node is called the *contact force* or the *reaction*. To maintain the constraint on the reaction, introduce the normal direction $\boldsymbol{n}$ inward to the obstacle surface as shown in Figure 3.4. As a physical characteristic of frictionless contact, the reaction, denoted r, should be in the direction of $\boldsymbol{n}$. Furthermore, the static contact states are classified by the sign of r as follows:

- $r < 0$: in compression
- $r = 0$: the zero reaction
- $r > 0$: in tension

Here, the tensile reaction can take place if we assume the existence of adhesion. As a simple physical modelling, we assume the absence of adhesion. Then the contact candidate node cannot be pulled by the obstacle, and thence only the compressive reaction is admissible. Summarizing the observation above, we see that g and r should satisfy the inequalities

$$g \geq 0 \quad : \text{[non-penetration]}, \tag{3.50a}$$
$$r \leq 0 \quad : \text{[non-adhesion]}. \tag{3.50b}$$

Besides the inequality constraints (3.50), observe that the existence of gap implies the zero reaction, while the existence of reaction means that the node is in contact. This relation is summarized as the following alternative conditions:

$$\begin{cases} g > 0 & \Rightarrow \quad r = 0, \\ r < 0 & \Rightarrow \quad g = 0, \end{cases}$$

which is equivalently rewritten as

$$gr = 0 \quad : \text{[complementarity]}. \tag{3.51}$$

Consequently, the contact law is completely given by the system of (3.50) and (3.51), which is known as the (*Signorini*) *unilateral contact* condition.

We next establish the "constitutive law," which is compatible with the unilateral contact law in (3.50) and (3.51), by following the procedure presented in Figure 3.3 of section 3.3.3. Since $r = 0$ for $g > 0$, the corresponding stored energy, denoted $\hat{w}(g)$, vanishes for $g > 0$. Alternatively, since $g < 0$ is not allowed, $\hat{w}(g)$ takes infinitely large value for $g < 0$.[13] Thus the stored energy associated with the contact condition is written as (see Figure 3.5)

$$\begin{aligned} \hat{w}(g) &= \delta_{\mathbb{R}_+}(g) \\ &= \begin{cases} 0 & \text{if } g \geq 0, \\ +\infty & \text{if } g < 0, \end{cases} \end{aligned} \tag{3.52}$$

[13] More intuitively, suppose that there exists a *fictitious* elastic spring behind the surface of the obstacle, and consider the potential energy stored in the spring. By increasing the stiffness of the fictitious spring to an infinitely large value, we can realize the non-penetration condition, and at the same time the stored energy also takes the infinite value.

$$w = \delta_{\mathbb{R}_+}(g) \quad \xleftrightarrow{\text{Fenchel}}_{\text{transformation}} \quad w^c = \delta_{\mathbb{R}_-}(r)$$

subdifferential $\downarrow$ $\qquad\qquad$ $\downarrow$ subdifferential

$$r \in \partial\delta_{\mathbb{R}_+}(g) \quad \dashrightarrow \quad g \in \partial\delta_{\mathbb{R}_-}(r)$$

"inversion"

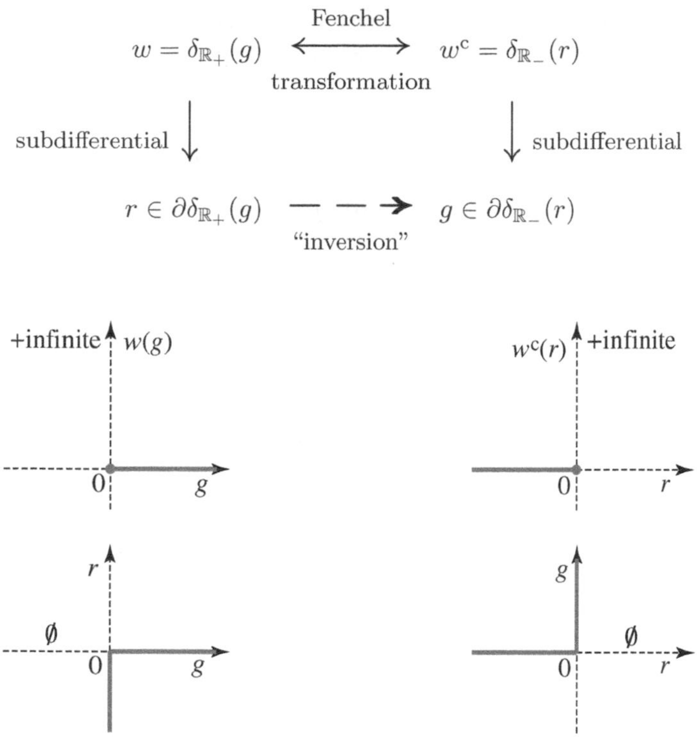

FIGURE 3.5: Schematic exposition of the derivation of the inverse of the unilateral contact law.

which is the indicator function of the nonnegative orthant (half-line), say, $\mathbb{R}_+ = \{g \mid g \ge 0\}$. The constitutive law, which expresses r in terms of g, is then obtained via the subdifferential of $\hat{w}$ as the inclusion

$$r \in \partial\delta_{\mathbb{R}_+}(g). \tag{3.53}$$

Direct calculations, according to definition (2.11) of the subdifferential, yield

$$\partial\delta_{\mathbb{R}_+}(g) = \begin{cases} \{0\} & \text{if } g > 0, \\ ]-\infty, 0] & \text{if } g = 0, \\ \emptyset & \text{if } g < 0. \end{cases}$$

As seen in the graph depicted in Figure 3.5, (3.53) is equivalent to the unilateral contact law in (3.50) and (3.51).

Although the constitutive law is once obtained as (3.53), $\partial\delta_{\mathbb{R}_+}$ is a set-valued function that is not invertible in a usual sense. Therefore, the methodology based on the Legendre transformation (in section 3.3.2) is not applicable, but we use the one based on the Fenchel transformation (in section 3.3.3).

The complementary energy function is defined as the conjugate function of $\hat{w}$, which is calculated as

$$\begin{aligned}\hat{w}^{\mathrm{c}}(r) &= \hat{w}^{*}(r) \\ &= \sup_{g}\{rg - \hat{w}(g) \mid g \in \mathbb{R}\} \\ &= \sup_{g}\{rg \mid g \geq 0\} \\ &= \begin{cases} +\infty & \text{if } r > 0, \\ 0 & \text{if } r \leq 0, \end{cases}\end{aligned}$$

or, in short, we can write

$$\hat{w}^{\mathrm{c}}(r) = \delta_{\mathbb{R}_{-}}(r), \tag{3.54}$$

where $\mathbb{R}_{-}$ denotes the nonpositive half-line, i.e.,

$$\mathbb{R}_{-} = \{x \in \mathbb{R} \mid x \leq 0\}.$$

The inverse constitutive law is then obtained by (sub)differentiating $\hat{w}^{\mathrm{c}}$ as

$$g \in \partial\delta_{\mathbb{R}_{-}}(r), \tag{3.55}$$

where

$$\partial\delta_{\mathbb{R}_{-}}(r) = \begin{cases} \emptyset & \text{if } r > 0, \\ [0, +\infty[& \text{if } r = 0, \\ \{0\} & \text{if } r < 0. \end{cases}$$

As clearly seen in Figure 3.5, relation (3.55) is regarded as an inverse of (3.52), both of which are set-valued functions. Thus the notion of convex analysis is indispensable to the study of nonsmooth mechanics.

3.4 Notes

1. The compliance minimization is the most fundamental problem in the structural optimization; see, e.g., Bendsøe–Sigmund [37], Christensen–Klarbring [86], and Rozvany–Bendsøe–Kirsch [396] for its fundamentals. Although we focus only on the SDP and SOCP formulations (in section 3.1.2.1 and section 3.1.2.2, respectively), various equivalent formulations are known for the compliance optimization of trusses; see Achtziger–Bendsøe–Ben-Tal–Zowe [2], Beckers–Fleury [36], Ben-Tal–Bendsøe [40], and Jarre–Kočvara–Zowe [193].

The robust optimization concerning the worst-case compliance and its SDP reformulation, analyzed in section 3.1.1, are from Ben-Tal–Nemirovski [44] (see also Ben-Tal–Nemirovski [46, section 4.8]). This problem is regarded as one of applications of *robust convex optimization* due to Ben-Tal–Nemirovski [45]. Robust convex optimization is a methodology dealing with the robust counterparts of uncertain conic optimization problems, where the data of a conic optimization are assumed to run through a given bounded set; see Ben-Tal–Boyd–Nemirovski [42], Ben-Tal–El Ghaoui–Nemirovski [43], and Ben-Tal–Nemirovski [47, 48] for more comprehensive surveys of the robust convex optimization. For finding the *global* optimal topology of a truss against the worst-case compliance, a branch-and-bound method in conjunction with the SDP formulation was presented by Yonekura–Kanno [461]. As further applications of SDP to robust optimization problems in structural engineering, see Guo–Bai–Zhang–Gao [158] and Kanno–Takewaki [211] for the robust optimization of structures, and Guo–Bai–Zhang [157] and Kanno–Takewaki [210, 212, 213] for the worst-case analysis of structures. For general treatments of the worst-case analysis, also called the *worst scenario method*, the *unknown-but-bounded uncertainty approach*, or the *guaranteed performance approach*, see monographs by Ben-Haim [38], Ben-Haim–Elishakoff [39], and Hlaváček–Chleboun–Babuška [182].

2. Eigenvalue optimization of the free vibration, treated in section 3.2, has been studied very extensively in the field of structural optimization, see, e.g., [111, 220, 304, 330, 361]. The first example of optimal structure with multiple eigenvalue is from Olhoff–Rasmussen [345], in which optimal columns with double buckling load factors are found. Subsequently, the optimality conditions of structures with multiple eigenvalues have been discussed by many authors; see Bratus–Seyranian [64, 65], Cox–Overton [97], Masur [303], Rodorigues–Guedes–Bendsøe [394], and Seyranian–Lund–Olhoff [407]. Computational approaches to sensitivity analysis of multiple eigenvalues are found in Choi–Kim [80], Haug–Choi [169], and Seyranian [406].

The SDP formulation of the eigenvalue optimization of trusses, presented in section 3.2.2, is from Ohsaki–Fujisawa–Katoh–Kanno [343]. Further discussions on this formulation are found in Achtziger–Kočvara [3, 4]. The derivation of the optimality condition using the SDP duality was originally considered by Kanno–Ohsaki [203]. Traditionally, the analysis of optimality condition has mostly been based on the perturbation technique [407]. In section 3.2.3 we made use of the complementarity condition over positive-semidefinite cone, which provides us with a simpler derivation than the original proof in [203].

As a related topic to the optimization concerning the fundamental frequency, it is noted that the optimization of structures considering the constraint on the linear buckling load factor is also formulated as an eigenvalue optimization. Specifically, the lower-bound constraint on the linear buckling load factor is written as a positive-semidefinite constraint of a certain symmetric matrix. The optimization problem then turns out to be a *non-*

linear SDP problem, which involves nonlinear equality constraints, as well as the linear matrix inequality. Nonlinear SDP formulations for the optimization against the linear buckling load factor was independently presented by Kanno–Ohsaki–Katoh [209] and Ben-Tal–Jarre–Kočvara–Nemirovski–Zowe [41]. For further discussions on the numerical aspect, see Kočvara [233] and Kočvara–Stingl [234].

3. The topic in section 3.3.4 concerning the unilateral contact law from the perspective of the convex analysis is classical; see Duvaut–Lions [120], Moreau [320], and Panagiotopoulos [355]. Much of the exposition given in section 3.3 is inspired by Curnier [101]; see also Glocker [149], Klarbring [227], and Stavroulakis [414]. As shown in section 3.3.4, the relation between the contact gap and reaction, subjected to the unilateral contact condition, is expressed as a complementarity condition over the nonnegative orthants. Accordingly, for the frictionless contact problem in the linear elasticity, the minimization problem of the potential energy can be formulated as a QP problem.[14] Early contributions concerning the QP formulation for the frictionless contacts include Chand–Haug–Rim [81], de Saxcé–Nguyen-Dang [110], Demkowicz [115], Haug–Chand–Pan [168], Klarbring [225], Moreau [319], Nguyen-Dang–de Saxcé [338], and Panagiotopoulos [354]. For treatment of the frictional cases, see Chapter 10.

[14]Which means a convex quadratic programming problem; see section 1.6.2.

Part II

Cable Networks: An Example in Nonsmooth Mechanics

Chapter 4

Principles of Potential Energy for Cable Networks

Orientation In this and the succeeding chapters, we take up the static analysis of cable networks, which serves as a simple but interesting and rich example of nonsmooth mechanics. A cable network is an assembly of a finite number of cable members, which can transmit only tensile forces. If we try to apply a compression force to a cable member, then the elongation stiffness suddenly disappears. Thus a nonsmoothness property appears in the response, or the constitutive relation, of each cable member. We shall see how this nonsmoothness property is dealt with by using the conic constraints and the complementarity conditions introduced in Part I. It is rather surprising that the minimization problem of the potential energy for cable networks can be formulated as a conic (and hence a convex) optimization problem even in the large deformation setting.

4.1 Constitutive law

A cable network is a structure realized as an assembly of cable members. One remarkable example that can be regarded as a "pure" cable network, i.e., a structure consisting of only cable members, is the roof of the Arena for the Munich Olympics in 1972 [402]. Besides these pure cable networks, it is often that a cable member is employed together with other structural elements, particularly in large-span civil engineering structures, because its light-weight property and its flexible realizability of curved surface given by designers. Such a class of structures, called *cable-supported structure*, includes cable-suspended roofs, cable domes, cable-supported membranes, and tensegrity structures; see, e.g., Bradshaw–Campbell–Gargari–Mirmiran–Tripeny [62], Harris–Li [162], Lewis [263], Motro [327], Narayanan [332], Pellegrino [362], and Skelton–de Olivira [411].

Since the cable member is a one-dimensional structural element, it can serve as a simple example of materials dealt with in nonsmooth mechanics, while the analyses below, particularly in the large deformation setting, provide nontrivial results. We begin the investigation of cable networks with the constitutive law of a cable member, which possesses the nonsmoothness property arising from the slackening behavior. Readers unfamiliar with mechanics should begin with the constitutive law of a truss member in the beginning of section 3.3, which serves as the simplest example of constitutive law.

4.1.1 No-compression model

A cable in structural engineering is a one-dimensional structural element that possesses very small compressive and flexural stiffnesses but can transmit the tensile force well. A drastic, but quite practically acceptable, idealization of such an element is a *no-compression* model, i.e., a truss element that cannot sustain compressive forces at all but for tensile forces behaves elastically. More specifically, when we aim to apply a compressive axial force, a cable member suddenly loses the elongation stiffness. In such a situation we say that the cable is in the *slack state*. Otherwise, the cable is said to be in the *taut state*, in the presence of a tensile axial force.

In this way, the elongation stiffness of a cable is supposed to be positive in a taut state and zero in a slack state. The elongation stiffness is thence a discontinuous function of the elongation, which is a property distinguishing the cable element from the conventional truss element. As a consequence, the nonsmoothness property appears in the constitutive law of the cable element; i.e., the axial force is a nonsmooth function of the elongation.

Let c and q denote the elongation and the axial force of the cable member, respectively. Suppose that the behavior of the cable in a taut state is linearly elastic. Then the response function, denoted $\hat{q}_{\rm cab} : \mathbb{R} \to \mathbb{R}$, is written as

$$\hat{q}_{\rm cab}(c) = \begin{cases} k_{\rm e} c & \text{if } c \geq 0, \\ 0 & \text{otherwise,} \end{cases} \tag{4.1}$$

which gives the constitutive law in the form of

$$q = \hat{q}_{\rm cab}(c). \tag{4.2}$$

In (4.1), the elongation stiffness $k_{\rm e}$ in the taut region is a constant given by

$$k_{\rm e} = \frac{Ea}{l}, \tag{4.3}$$

where a is the member cross-sectional area, l is the member length, and E is the Young modulus.

Integration of $\hat{q}_{\rm cab}$ in (4.1) yields the stored energy function, denoted $\hat{w}_{\rm cab} : \mathbb{R} \to \mathbb{R}$, as

$$\hat{w}_{\rm cab}(c) = \begin{cases} \dfrac{1}{2} k_{\rm e} c^2 & \text{if } c \geq 0, \\ 0 & \text{otherwise.} \end{cases} \tag{4.4}$$

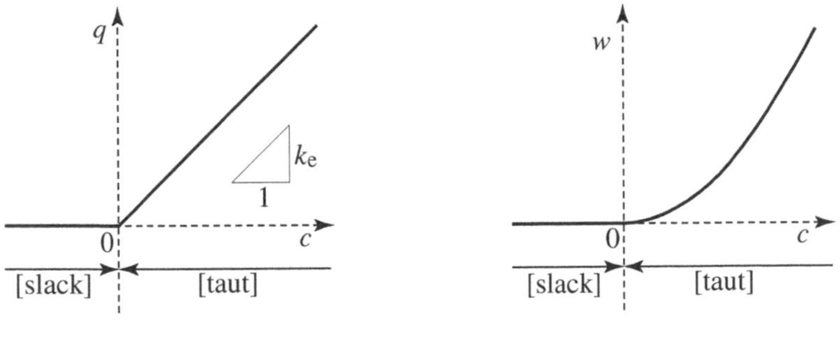

FIGURE 4.1: A model of linearly elastic cable. (a) constitutive law; (b) strain energy function.

It is also observed in (4.1) that the tangent stiffness k vanishes when the cable is slack, while k is equal to k_e when the cable is taut, i.e.,[1]

$$k = \begin{cases} k_e & \text{if } c \geq 0, \\ 0 & \text{if } c < 0. \end{cases} \tag{4.5}$$

Figure 4.1 depicts the graphs of $\hat{q}_{cab}$ and $\hat{w}_{cab}$, where we clearly see that the axial force $q = \hat{q}_{cab}(c)$ is a nonsmooth function. This nonsmoothness property distinguishes a cable element from a conventional truss element investigated in sections 3.3.1–3.3.2.

4.1.2 Inclusion form

In a manner similar to section 3.3, an "inversion" of the constitutive law (4.1) can be obtained in a consistent way within the framework of Fenchel transformation as follows.

We begin by recalling the definition of the stored energy $\hat{w}_{cab}(c)$ in (4.4). The constitutive law, say, (4.1), is alternatively written as

$$q \in \partial \hat{w}_{cab}(c). \tag{4.6}$$

Here, $\hat{w}_{cab}$ is continuously differentiable, and thence (4.6) is reduced to

$$q = \frac{d\hat{w}_{cab}}{dc}(c) = \begin{cases} k_e q & \text{if } q \geq 0, \\ 0 & \text{if } q < 0, \end{cases}$$

[1] A mathematically rigorous expression of tangent stiffness, as a set-valued function, is given in section 4.1.2; see (4.8).

which coincides with expression (4.2) together with (4.1). Thus $\partial\hat{w}_{\rm cab}$ is a (single-valued) function, and (4.6) is read as an equation. The tangent stiffness k at c is obtained by differentiating $\hat{q}_{\rm cab}$ as

$$k \in \hat{k}_{\rm cab}(c) := \partial\hat{q}_{\rm cab}(c), \tag{4.7}$$

where

$$\partial\hat{q}_{\rm cab}(c) = \begin{cases} \{k_{\rm e}\} & \text{if } c > 0, \\ [0, k_{\rm e}] & \text{if } c = 0, \\ \{0\} & \text{if } c < 0. \end{cases} \tag{4.8}$$

This is the rigorous definition of the tangent stiffness of the cable within the framework of convex analysis, which is to be compared with an intuitive expression given in (4.5). Note that (4.8) is a set-valued function, and in (4.7) the tangent stiffness is defined in the form of an inclusion.

We next apply the Fenchel transformation to $\hat{w}_{\rm cab}$ to deduce an inverse of the constitutive law for the cable; see Figure 4.2 for a summary of the procedure.

The conjugate function of $\hat{w}_{\rm cab}$ in (4.4), denoted $\hat{w}^*_{\rm cab} : \mathbb{R} \to \mathbb{R} \cup \{+\infty\}$, is obtained as

$$\begin{aligned} \hat{w}^*_{\rm cab}(q) &= \sup_{c\in\mathbb{R}}\{qc - \hat{w}_{\rm cab}(c)\} \\ &= \max\left\{ \sup_{c\geq 0}\left\{qc - \frac{1}{2}k_{\rm e}c^2\right\}, \sup_{c<0}\{qc\} \right\} \\ &= \begin{cases} \sup\left\{qc - \dfrac{1}{2}k_{\rm e}c^2 \mid c \geq 0\right\} & \text{if } q \geq 0, \\ \sup\{qc \mid c < 0\} & \text{if } q < 0, \end{cases} \\ &= \begin{cases} \dfrac{1}{2}\dfrac{q^2}{k_{\rm e}} & \text{if } q \geq 0, \\ +\infty & \text{if } q < 0, \end{cases} \end{aligned} \tag{4.9}$$

which corresponds to the complementary stored energy function. Then the elongation is expressed in terms of the axial force as

$$c \in \partial\hat{w}^*_{\rm cab}(q), \tag{4.10}$$

where

$$\partial\hat{w}^*_{\rm cab}(q) = \begin{cases} \left\{\dfrac{q}{k_{\rm e}}\right\} & \text{if } q > 0, \\ ]-\infty, 0] & \text{if } q = 0, \\ \emptyset & \text{if } q < 0. \end{cases} \tag{4.11}$$

$$\underset{\text{transformation}}{w = \hat{w}_{\text{cab}}(c) \quad \overset{\text{Fenchel}}{\longleftrightarrow} \quad w^{c} = \hat{w}^{*}_{\text{cab}}(q)}$$

subdifferential $\downarrow$ $\qquad\qquad\qquad\qquad$ $\downarrow$ subdifferential

$$q \in \hat{w}_{\text{cab}}(c) \quad \underset{\text{``inversion''}}{- - \rightarrow} \quad c \in \partial\hat{w}^{*}_{\text{cab}}(q)$$

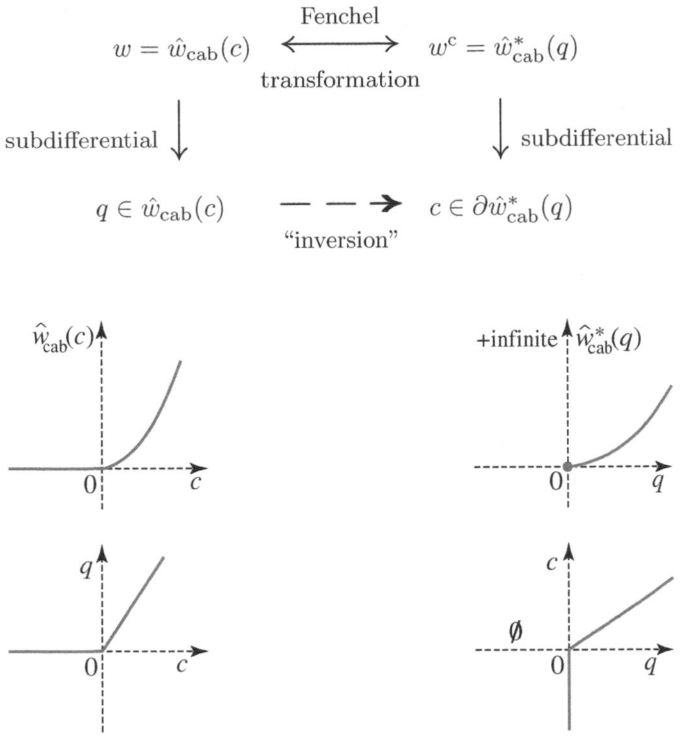

FIGURE 4.2: Relation among the constitutive law, strain energy, complementary energy, and inverse of the constitutive law for a cable member.

As depicted in Figure 4.2, $\partial\hat{w}^{*}_{\text{cab}}$ in (4.11) is a set-valued function. From the mechanical point of view, the elongation cannot be determined uniquely when the cable is slack, i.e., when $q = 0$. The inclusion (4.10) is regarded as the inversion of the constitutive law in (4.6).

4.1.3 Variational form

For $\hat{w}_{\text{cab}}$ defined in (4.4), we here investigate an alternative equivalent expression, which plays a key role in formulating the minimization problem of the potential energy for cable networks in the form of conic optimization.

Proposition 4.1.1. *Consider the optimization problem*

$$\left.\begin{array}{c} \underset{y}{\min} \ \dfrac{1}{2}k_{e}y^{2} \\ \text{s.\,t. } \ y \ge c, \end{array}\right\} \tag{4.12}$$

in the variable $y \in \mathbb{R}$. *Then* $\bar{y} \in \mathbb{R}$ *is optimal if and only if*

$$\bar{y} = \begin{cases} c & \text{if } c \geq 0, \\ 0 & \text{if } c < 0. \end{cases} \tag{4.13}$$

Moreover, the optimal value of (4.12) *is equal to* $\hat{w}_{\rm cab}(c)$, *where* $\hat{w}_{\rm cab}$ *is defined by* (4.4).

Proof. Problem (4.12) is a QP (quadratic programming) problem (with nonempty interior feasible set), the optimality condition of which is derived as follows.

The Lagrangian of (4.12) is given by

$$L(y, z^*) = \begin{cases} \frac{1}{2} k_{\rm e} y^2 - z^*(y - c) & \text{if } z^* \geq 0, \\ -\infty & \text{otherwise,} \end{cases} \tag{4.14}$$

where $z^* \in \mathbb{R}$ is the Lagrangian multiplier.[2] It then follows from Proposition 2.2.17 that $\bar{y}$ is optimal for problem (4.12) if and only if $(\bar{y}, \bar{z}^*)$ is a saddle point of L, i.e.,

$$L(\bar{y}, z^*) \leq L(\bar{y}, \bar{z}^*) \leq L(y, \bar{z}^*), \quad \forall y, \ \forall z^*.$$

This saddle-point condition is explicitly written as the following KKT conditions:

$$k_{\rm e}\bar{y} - \bar{z}^* = 0, \tag{4.15a}$$

$$\bar{y} - c \geq 0, \quad \bar{z}^* \geq 0, \quad \bar{z}^*(\bar{y} - c) = 0. \tag{4.15b}$$

By eliminating $\bar{z}^*$ from (4.15), we obtain

$$\bar{y} - c \geq 0, \quad \bar{y} \geq 0, \quad \bar{y}(\bar{y} - c) = 0, \tag{4.16}$$

where $k_{\rm e} > 0$ has been used. The complementarity condition in (4.16) means

$$\bar{y} = c \quad \text{or} \quad \bar{y} = 0,$$

from which the two inequalities in (4.16) we conclude (4.13).

It is straightforward to see from (4.4) that $\frac{1}{2} k_{\rm e} \bar{y} = \hat{w}_{\rm cab}(c)$ holds if $\bar{y}$ satisfies (4.13). □

4.1.4 Complementarity form

It has been shown in the proof of Proposition 4.1.1 that the optimality condition for problem (4.12) is written as (4.15). In this section, the condition (4.15) can be derived naturally through the mechanical consideration of the property of cables.

[2]Set $x = y$, $z^* = z^*$, and $\mathbb{V} = \mathbb{Y} = \mathbb{R}$ to revert to the notation used in section 2.2.6.

We first make clear that the slackening behavior is assumed to be *non-dissipative*; i.e., no energy dissipates through the slackening behavior. In other words, the slackening behavior is assumed to be *reversible*. In addition, the response in the taut state is assumed to be elastic. Therefore, the strain energy of a cable is totally reversible. This means that the constitutive law, illustrated in Figure 4.1(a), does not vary depending on the loading history that a cable has experienced.

Now we state the physical hypotheses formally, which are postulated on the response $q = \hat{q}_{\rm cab}(c)$ of a cable for a given elongation c.

(i) Any compressive axial force cannot be sustained.

(ii) The elongation c can be additively decomposed to the elastic part and slackening part.

(iii) The slack can take place only when no tensile axial force exists.

(iv) The tensile axial force is caused elastically solely by the elastic elongation.

Hypothesis (i) means that the axial force q should always satisfy

$$q \geq 0. \tag{4.17}$$

In accordance with hypothesis (ii), we decompose the elongation c as[3]

$$c = c_{\rm e} + c_{\rm s}, \tag{4.18}$$

where $c_{\rm e}$ is the elastic elongation, and $c_{\rm s}$ is the slackening elongation. Moreover, since the slackening part corresponds to the quantity of shortening of the member length, we have that

$$c_{\rm s} \leq 0. \tag{4.19}$$

It follows from hypothesis (iii) that the existence of tensile force implies the absence of slack, i.e.,

$$q > 0 \quad \Rightarrow \quad c_{\rm s} = 0. \tag{4.20}$$

Conversely, the existence of slack implies that the axial force is not tensile, say, $q \leq 0$. From this inequality and (4.17), we obtain

$$c_{\rm s} < 0 \quad \Rightarrow \quad q = 0. \tag{4.21}$$

Hypothesis (iv) means that q is a function depending only on $c_{\rm e}$. Since the elongation stiffness in the taut state is given by (4.3), we obtain

$$q = k_{\rm e} c_{\rm e}. \tag{4.22}$$

[3]We shall see right below that this decomposition is determined uniquely.

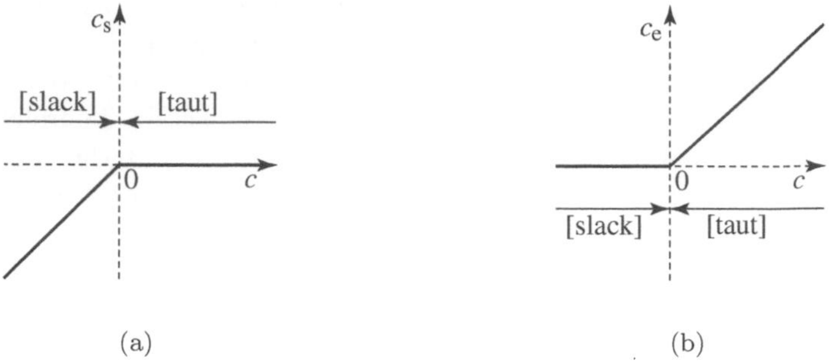

FIGURE 4.3: Decomposition (4.23a) of the elongation c to $c_{\rm e}$ and $c_{\rm s}$. Condition (4.23) implies (4.24).

Thus these hypotheses are formulated as (4.17)–(4.22).

We next observe that (4.20) and (4.21), in conjunction with (4.17) and (4.19), imply the complementarity condition

$$qc_{\rm s} = 0.$$

Therefore, conditions (4.17)–(4.22) above are equivalently rewritten as

$$c = c_{\rm e} + c_{\rm s}, \tag{4.23a}$$

$$q = k_{\rm e} c_{\rm e}, \tag{4.23b}$$

$$c_{\rm s} \le 0, \quad q \ge 0, \quad qc_{\rm s} = 0. \tag{4.23c}$$

By putting

$$\bar{y} = c_{\rm e}, \quad \bar{z}^* = q,$$

and by eliminating $c_{\rm s}$, (4.23) is reduced to (4.15), where the latter condition is the optimality condition for problem (4.12) as shown in the proof of Proposition 4.1.1. Therefore, (4.23) implies (4.13), in which $\bar{y}$ is to be replaced with $c_{\rm e}$, i.e.,

$$c_{\rm e} = \begin{cases} c & \text{if } c \ge 0, \\ 0 & \text{if } c < 0. \end{cases} \tag{4.24}$$

Thus $c_{\rm e}$ in (4.23) is determined uniquely, and so is $c_{\rm s}$. Thus the decomposition of c in (4.23a) is determined uniquely, as illustrated in Figure 4.3. Furthermore, from (4.23b) and (4.24), we can see that q in (4.23) satisfies the constitutive law

$$q = \hat{q}_{\rm cab}(c), \tag{4.25}$$

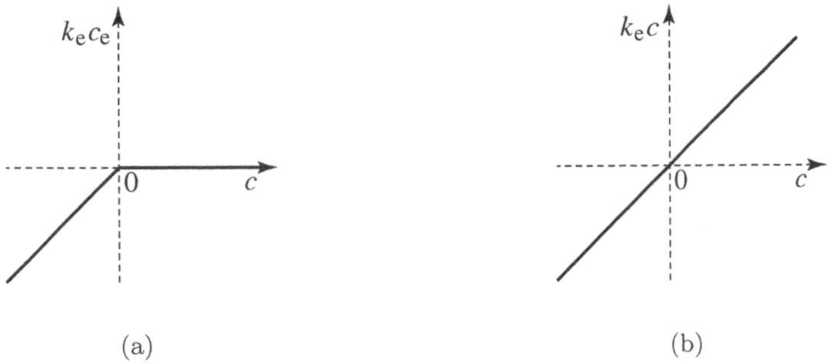

FIGURE 4.4: The cable member and its reference elastic body. (a) the constitutive law of the cable in (4.26); (b) the constitutive law of the reference elastic body in (4.27).

where $\hat{q}_{\rm cab}$ is defined by (4.1). We thus conclude that, for a given c, q satisfies (4.25) if and only if there exist $c_{\rm e}$ and $c_{\rm s}$ satisfying (4.23).

In what follows, for clarifying the physical meaning of variables, we use the notation $c_{\rm e}$ instead of y.

Condition (4.23b) suggests to introduce a concept of *reference elastic body* for the cable. As mentioned before, when $c_{\rm e}$, $c_{\rm s}$, and q satisfy (4.23), then the constitutive law of the cable is written as

$$q = k_{\rm e} c_{\rm e}. \tag{4.26}$$

Note that $c_{\rm e}$ is considered to be a function of c. The reference elastic body is a fictitious structural element, the constitutive law of which is given by

$$q = k_{\rm e} c. \tag{4.27}$$

Thus the reference elastic body associated with the cable is a linear elastic truss element. The relation of these two structural elements is illustrated in Figure 4.4. In other words, the reference elastic body is a structural element that can be obtained by "removing" the slackening behavior from the cable element. Note that the slackening property of a cable is not due to its material but to its shape (or geometry); a cable is often realized as a stranded cable, and has a relatively small cross-sectional area compared with length. The model of a reference elastic body is obtained by assuming that the material property is "inherited" even when the cable is subjected to a compression force. Thus, as seen in (4.27) in conjunction with (4.3), the reference elastic body can be interpreted as a truss member, which has the same material property E and the same geometry (l, a) as the cable member. It is noted that reference elastic body is also used in the variational formulation (4.12). The objective function of problem (4.12) is the strain energy function of the reference elastic body defined by (4.27).

The reference elastic body plays an important role in investigating the constitutive laws of masonry structures (Chapter 8) and membranes (Chapter 9).

4.2 Potential Energy Principles in Convex Optimization Forms

Consider a cable network in the physical three-dimensional space. The cable network consists of a finite number of one-dimensional truss elements, whose compressive and flexural stiffnesses are supposed to be negligibly small. We call each truss element a *cable* hereafter, the constitutive law of which was investigated in section 4.1.

4.2.1 Principle of potential energy in general form

Suppose that a cable network consists of m cable members and that no distributed loads are applied on each cable member. As a consequence, the deformed configuration of the cable network is uniquely determined by specifying the locations of all nodes. We denote by $\boldsymbol{u} \in \mathbb{R}^d$ the vector of the nodal displacements, where d is the number of degrees of freedom, including the degrees of the prescribed displacements.

Let c_i denote the elongation of the ith member. Since c_i is determined uniquely by specifying $\boldsymbol{u}$, we write

$$c_i = \hat{c}_i(\boldsymbol{u}),$$

which is the compatibility relation. For concrete expressions of $\hat{c}_i : \mathbb{R}^d \to \mathbb{R}$, see section 4.2.2 through section 4.2.4.

For the set of indices of the degrees of freedom, denoted $\Gamma = \{1, \dots, d\}$, consider a partition $\Gamma = \Gamma_{\mathrm{N}} \cup \Gamma_{\mathrm{D}}$ into the two disjoint subsets. Suppose that for the degrees of freedom included in Γ_{N} the external loads are given, while for those included in Γ_{D} the displacements are prescribed. We denote by $\underline{p}_j \in \mathbb{R}$ $(j \in \Gamma_{\mathrm{N}})$ the specified external forces and by $\underline{u}_j \in \mathbb{R}$ $(j \in \Gamma_{\mathrm{D}})$ the prescribed displacements. We assume that the applied external forces are conservative, which means path independent.[4]

By *kinematic variables* we mean state variables representing the deformation of a structure. For example, the displacement $\boldsymbol{u}$ and the elongation c_i

[4]For example, a pressure load caused by a gas applied to the surface of a deformable body depends on the deformed configuration of the body, and hence it is path dependent. Throughout the book we consider only conservative external forces. Note that a friction force dealt with in Chapter 10 is non-conservative, but it is a *reaction* force (which means that a force related to the presence of kinematic constraints), which should be distinguished from the external force.

are kinematic variables, which represent the global and local deformations.[5] The constraints restricting the kinematic variables are called kinematic constraints. For the present cable network, the compatibility conditions and the Dirichlet conditions, i.e.,

$$c_i = \hat{c}_i(\boldsymbol{u}), \quad i = 1, \dots, m,$$
$$u_j = \underline{u}_j, \quad j \in \Gamma_{\mathrm{D}},$$

are the kinematic constraints. We say that $(\boldsymbol{u}, \boldsymbol{c})$ is kinematically admissible if it satisfies the kinematic constraints. In contrast, the variables representing the forces are called static variables. For example, the axial force and the nodal force are static variables.[6] The constraints restricting static variables are called the static constraints. Typical static constraints are the force-balance equation and the Neumann conditions. The static variables are said to be statically admissible if the static constraints are satisfied. It is noted that the constitutive law states the relation between internal kinematic variables and internal static variables.

The minimization problem of the potential energy is formulated only in terms of kinematic variables, which is the minimization of the total potential energy under the kinematic constraints. In contrast, the minimization problem of the complementary energy is formulated only in terms of static variables, which is the minimization of the total complementary energy under the static constraints. See Reddy [384] and Washizu [452] for more background.

For the cable network, the minimization problem of the potential energy is formulated as

$$\left.\begin{aligned} \text{(PE)} : \min_{\boldsymbol{u}, \boldsymbol{c}} \ & \sum_{i=1}^{m} \hat{w}_{\mathrm{cab}}^{i}(c_i) - \sum_{j \in \Gamma_{\mathrm{N}}} \underline{p}_j u_j \\ \text{s.t.} \ & c_i = \hat{c}_i(\boldsymbol{u}), \quad i = 1, \dots, m, \\ & u_j = \underline{u}_j, \quad j \in \Gamma_{\mathrm{D}}, \end{aligned}\right\} \tag{4.28}$$

which is the problem to minimize the total potential energy among kinematically admissible variables. It should be clear that $\hat{w}_{\mathrm{cab}}^{i}$ involved in the objective function is the stored energy function for the ith cable member, defined as (4.4). The nonsmoothness property of cables is essentially inherited by $\hat{w}_{\mathrm{cab}}^{i}$ as the disjunction in (4.4); see also Figure 4.1.

The key idea to handle the nonsmoothness property in a tractable way is to transfer it to the *inequality constraint*. This is attained by applying

[5]The elongation is regarded as a generalized strain, where the strain is also a kinematic variable.

[6]The axial force is regarded as a generalized stress, and certainly the stress is also a static variable.

Proposition 4.1.1 to the objective function of (PE). Then (PE) is equivalently reduced to

$$\left.\begin{aligned} (\overline{\mathrm{PE}}) : \min_{\boldsymbol{u}, \boldsymbol{c}_{\mathrm{e}}} \ & \sum_{i=1}^{m} \frac{1}{2} k_{\mathrm{e}i} c_{\mathrm{e}i}^2 - \sum_{j \in \Gamma_{\mathrm{N}}} \underline{p}_j u_j \\ \text{s.t.} \ & c_{\mathrm{e}i} \geq \hat{c}_i(\boldsymbol{u}), \quad i = 1, \dots, m, \\ & u_j = \underline{u}_j, \quad j \in \Gamma_{\mathrm{D}}, \end{aligned}\right\} \tag{4.29}$$

which no longer includes disjunction conditions. More precisely, the optimal solutions of (PE) in (4.28) and $(\overline{\mathrm{PE}})$ in (4.29) are related as stated below.

Proposition 4.2.1. $(\overline{\mathrm{PE}})$ *is equivalent to* (PE) *in the following sense:*

(i) *Let* $(\bar{\boldsymbol{u}}, \bar{\boldsymbol{c}})$ *be optimal for* (PE)*, and define a vector* $\bar{\boldsymbol{c}}_{\mathrm{e}}$ *by*

$$\bar{c}_{\mathrm{e}i} = \begin{cases} \bar{c}_i & \text{if } \bar{c}_i \geq 0, \\ 0 & \text{otherwise,} \end{cases} \qquad i = 1, \dots, m. \tag{4.30}$$

Then $(\bar{\boldsymbol{u}}, \bar{\boldsymbol{c}}_{\mathrm{e}})$ *is optimal for* $(\overline{\mathrm{PE}})$.

(ii) *If* $(\bar{\boldsymbol{u}}, \bar{\boldsymbol{c}}_{\mathrm{e}})$ *is optimal for* $(\overline{\mathrm{PE}})$*, then*

$$\bar{c}_{\mathrm{e}i} = \begin{cases} \hat{c}_i(\bar{\boldsymbol{u}}) & \text{if } \hat{c}_i(\bar{\boldsymbol{u}}) \geq 0, \\ 0 & \text{otherwise,} \end{cases} \qquad i = 1, \dots, m. \tag{4.31}$$

Moreover, if we define $\bar{\boldsymbol{c}}$ *by*

$$\bar{c}_i = \hat{c}_i(\bar{\boldsymbol{u}}), \quad i = 1, \dots, m, \tag{4.32}$$

then $(\bar{\boldsymbol{u}}, \bar{\boldsymbol{c}})$ *is optimal for* (PE).

(iii) $\min(\mathrm{PE}) = \min(\overline{\mathrm{PE}})$.

Proof. We begin by showing assertion (i). By using Proposition 4.1.1, particularly that the optimal value of (4.12) coincides with $\hat{w}_{\mathrm{cab}}(c)$, we see that (PE) can be restated as

$$\left.\begin{aligned} \min_{\boldsymbol{u}, \boldsymbol{c}_{\mathrm{e}}, \boldsymbol{c}} \ & \sum_{i=1}^{m} \min_{c_{\mathrm{e}i}} \left\{ \frac{1}{2} k_{\mathrm{e}i} c_{\mathrm{e}i}^2 \mid c_{\mathrm{e}i} \geq c_i \right\} - \sum_{j \in \Gamma_{\mathrm{N}}} \underline{p}_j u_j \\ \text{s.t.} \ & c_i = \hat{c}_i(\boldsymbol{u}), \quad i = 1, \dots, m, \end{aligned}\right\} \tag{4.33}$$

without changing the optimal solution and the optimal value. In this problem, $c_{\mathrm{e}i}$ is subjected solely to the constraint $c_{\mathrm{e}i} \geq c_i$, which is involved in the inner minimization of the objective function. Hence, problem (4.33) can be further rewritten equivalently as

$$\left.\begin{aligned} \min_{\boldsymbol{u}, \boldsymbol{c}_{\mathrm{e}}, \boldsymbol{c}} \ & \sum_{i=1}^{m} \frac{1}{2} k_{\mathrm{e}i} c_{\mathrm{e}i}^2 - \sum_{j \in \Gamma_{\mathrm{N}}} \underline{p}_j u_j \\ \text{s.t.} \ & c_{\mathrm{e}i} \geq c_i, \quad i = 1, \dots, m, \\ & c_i = \hat{c}_i(\boldsymbol{u}), \quad i = 1, \dots, m. \end{aligned}\right\} \tag{4.34}$$

Elimination of $\boldsymbol{c}$ from problem (4.34) results in ($\overline{\mathrm{PE}}$). Thus (PE) and ($\overline{\mathrm{PE}}$) share the same optimal value. Moreover, if $(\bar{\boldsymbol{u}}, \bar{\boldsymbol{c}})$ is an optimal solution of (PE), then $\bar{\boldsymbol{u}}$ is also optimal for ($\overline{\mathrm{PE}}$).

The remaining variables of ($\overline{\mathrm{PE}}$), say, $\boldsymbol{c}_{\mathrm{e}}$, are determined as follows. Since $(\bar{\boldsymbol{c}}, \bar{\boldsymbol{u}})$ is assumed to be optimal for (PE), it satisfies $\bar{c}_i = \hat{c}_i(\bar{\boldsymbol{u}})$ $(i = 1, \dots, m)$. For the inner problem involved in the objective function of (4.33), the optimality condition is given by (4.13) of Proposition 4.1.1, which should be also satisfied by $\boldsymbol{c}$, $\boldsymbol{c}_{\mathrm{e}}$, and $\boldsymbol{u}$ at the optimal solution of (4.33). Consequently, (4.30) is satisfied at the optimal solution of problem (4.34).

The discussion above (particularly the optimality condition in Proposition 4.1.1 for the inner problem of (4.33)) also means that the optimal solution $(\bar{\boldsymbol{c}}_{\mathrm{e}}, \bar{\boldsymbol{u}})$ of ($\overline{\mathrm{PE}}$) satisfies (4.31) in assertion (ii). Therefore, definition (4.32) of $\bar{\boldsymbol{c}}$ means

$$\bar{c}_i = \begin{cases} \bar{c}_{\mathrm{e}i} & \text{if } \bar{c}_{\mathrm{e}i} \geq 0, \\ \hat{c}_i(\bar{\boldsymbol{u}}) \ (< 0) & \text{otherwise,} \end{cases} \qquad i = 1, \dots, m.$$

Hence, by using (4.4) we obtain $\hat{w}^i_{\mathrm{cab}}(\bar{c}_i) = (1/2) k_{\mathrm{e}i} \bar{c}_{\mathrm{e}i}$, and the optimality of $(\bar{\boldsymbol{c}}, \bar{\boldsymbol{u}})$ for (PE) is guaranteed if assertion (iii) holds, which is to be shown below.

Concerning assertion (iii), it follows from (4.4) that condition (4.30) in assertion (i) implies $\hat{w}^i_{\mathrm{cab}}(\bar{c}_{\mathrm{e}i}) = \hat{w}^i_{\mathrm{cab}}(\bar{c}_i)$. Therefore, assertion (iii) follows from the claim of optimality in (i). □

Remark 4.2.2. In general, in the geometrically nonlinear theory an equilibrium state does not correspond to the (global) minimum solution of the minimization problem of the potential energy; any stationary solution is in equilibrium. By contrast, in Proposition 4.2.1 we focus only on the (global) minimum solution of (PE). However, no equilibrium state is excluded by solving ($\overline{\mathrm{PE}}$) instead of (PE). We postpone this issue for a while; we shall show that the *minimum* principle truly holds for cable networks in section 5.1.4 by virtue of the duality theory. ■

Remark 4.2.3. Compared with (PE) and ($\overline{\mathrm{PE}}$) for a cable network, the minimization problem of the total potential energy for a truss is formulated as

$$\left.\begin{aligned} (\mathrm{PE})_{\mathrm{truss}} : \min_{\boldsymbol{u}, \boldsymbol{c}} \ & \sum_{i=1}^{m} \frac{1}{2} k_{\mathrm{e}i} c_i^2 - \sum_{j \in \Gamma_{\mathrm{N}}} \underline{p}_j u_j \\ \mathrm{s.\,t.} \ & c_i = \hat{c}_i(\boldsymbol{u}), \quad i = 1, \dots, m, \\ & u_j = \underline{u}_j, \quad j \in \Gamma_{\mathrm{D}}. \end{aligned}\right\} \tag{4.35}$$

Within the geometrically linear theory, $\hat{c}_i$ is an affine function (see section 4.2.3), and thence the problem above is convex (specifically, it is a QP problem). In contrast, if the geometrical nonlinearity is considered, then $\hat{c}_i$ is a nonlinear function (as seen right below), and hence $(\mathrm{PE})_{\mathrm{truss}}$ is nonconvex. In the following section, we see, rather surprisingly, that ($\overline{\mathrm{PE}}$) is convex even in the geometrically nonlinear theory. ■

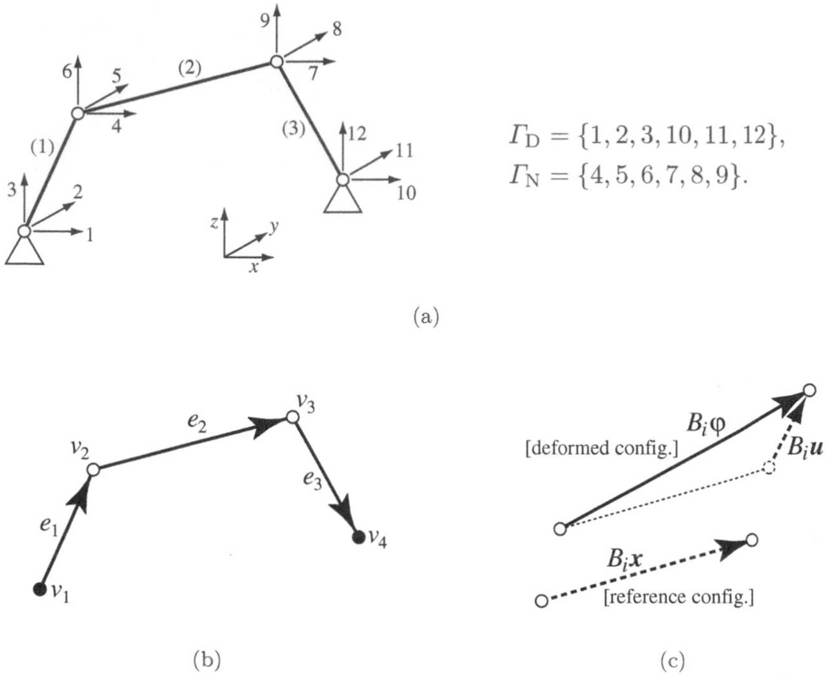

$$\Gamma_{\mathrm{D}} = \{1, 2, 3, 10, 11, 12\},$$
$$\Gamma_{\mathrm{N}} = \{4, 5, 6, 7, 8, 9\}.$$

FIGURE 4.5: An example of cable network and compatibility conditions. (a) the indices of members and degrees of freedom of a cable network; (b) the directed graph $\mathcal{G} = (\mathcal{V}, \mathcal{E})$ induced from the cable network in (a); (c) the member i at the reference and the deformed configurations.

4.2.2 Principle for large strain

Suppose that the cable network under consideration is subjected to large deformation. Although it might be simpler to begin our analysis with the geometrically linear theory, we postpone the linear case meanwhile. With a view to applying the complementarity condition over convex cones, the geometrically nonlinear theory serves as a proper example to see advantages of using *nonlinear* convex cones. Indeed, the large deformation case investigated in this section is described in an excellent way by virtue of the second-order cone inequality, while the geometrically linear problem is associated with the linear inequality. For this reason we first analyze the case of large deformation with large strain.

At the reference configuration, we denote by $\boldsymbol{x} \in \mathbb{R}^d$ the position vector of the nodes with respect to a fixed orthonormal reference coordinate system, where d is the number of the degrees of freedom. Note again that $d = |\Gamma_{\mathrm{N}}| + |\Gamma_{\mathrm{D}}|$. For example, we have $d = 3N$ for a cable network placed in the three-dimensional space, where N is the total number of nodes. In the case of the cable network depicted in Figure 4.5(a), we see that $N = 4$ and $d = 12$.

Let $\boldsymbol{\varphi} \in \mathbb{R}^d$ denote the position vector of the nodes at the deformed configuration. We denote by

$$\boldsymbol{u} = \boldsymbol{\varphi} - \boldsymbol{x} \tag{4.36}$$

the vector of the nodal displacements.

We next consider the compatibility relation between the nodal positions and the member elongation. For the given deformation $\boldsymbol{\varphi}$, we denote by $l_i(\boldsymbol{\varphi}) \in \mathbb{R}_+$ the length of the ith member. Note that the member length at the initial configuration, denoted $\underline{l}_i = l_i(\boldsymbol{x})$, is prescribed for each member. The member elongation, denoted c_i, is then written as

$$c_i = l_i(\boldsymbol{\varphi}) - l_i(\boldsymbol{x}) \tag{4.37}$$

$$= \|B_i\boldsymbol{\varphi}\| - \underline{l}_i, \tag{4.38}$$

where $B_i \in \mathbb{R}^{3\times d}$ is a constant matrix, each element of which is either 1, −1, or 0. For more account on B_i's, see Example 4.1 and Remark 4.2.5.

Example 4.1

As an example, consider B_i's, Γ_{D}, and Γ_{N} for the cable network illustrated in Figure 4.5(a). Since the cable network has four nodes, we have $d = 12$, and the deformed configuration is represented by $\boldsymbol{\varphi} \in \mathbb{R}^{12}$. The indices of elements of $\boldsymbol{\varphi}$, as well as the member indices, are defined as shown in Figure 4.5(a). Then the sets of degrees of freedom corresponding to the prescribed displacements and the prescribed external forces, respectively, are given by

$$\Gamma_{\mathrm{D}} = \{1, 2, 3, 10, 11, 12\}, \quad \Gamma_{\mathrm{N}} = \{4, 5, 6, 7, 8, 9\}.$$

The member length is obtained from the position vectors of the two nodes connected to the member under consideration, e.g., in the case of the member 1 we obtain

$$l_1(\boldsymbol{\varphi}) = \left\| \begin{bmatrix} \varphi_4 - \varphi_1 \\ \varphi_5 - \varphi_2 \\ \varphi_6 - \varphi_3 \end{bmatrix} \right\| - \underline{l}_1. \tag{4.39}$$

Hence, if we define $B_1 \in \mathbb{R}^{3\times 12}$ by

$$B_1 = \left[-I_3 | I_3 | O | O \right]$$

with $O \in \mathbb{R}^{3\times 3}$, then relation (4.39) is written in the form of (4.38). Similarly, for the other members, the matrices defined by

$$B_2 = \left[O | -I_3 | I_3 | O \right],$$

$$B_3 = \left[O | O | -I_3 | I_3 \right],$$

lead to the compatibility relations in the form of (4.38). ▯

Remark 4.2.4. As we can see in (4.38) and also in Example 4.1, the vector $B_i \boldsymbol{x} \in \mathbb{R}^3$ represents the member configuration at the reference state; it is the three-dimensional vector that is parallel with the configuration of the ith member at the undeformed state and satisfies $\|B_i \boldsymbol{x}\| = \underline{l}_i$. In other words, $B_i \boldsymbol{x}/\underline{l}_i$ is the *direction cosine* of the ith member at the reference configuration.

Similarly, the vector $B_i \boldsymbol{\varphi} \in \mathbb{R}^3$ corresponds to the member configuration at the deformed state, as illustrated in Figure 4.5(c). It then follows from (4.36) that $B_i \boldsymbol{u}$ represents the relative displacement of the two nodes. Analogous to continuum mechanics, $B_i \boldsymbol{\varphi}$ corresponds to the deformation gradient, and $B_i \boldsymbol{u}$ the displacement gradient. Thus the matrix B_i plays the role of the *gradient* operator in continuum mechanics. ■

Remark 4.2.5. The matrices B_i $(i = 1, \dots, m)$ are related to the incidence matrix used in the framework of graph theory[7] as follows. Consider a corresponding relation between the notion of structural elements and that of graph elements, given as

$$\begin{aligned} &\text{member} && \leftrightarrow && \text{edge}, \\ &\text{node} && \leftrightarrow && \text{vertex}. \end{aligned}$$

We assign a direction with each edge in an arbitrary way.[8] In other words, each edge is regarded as an ordered pair $(v_l, v_{l'})$ of vertices. We denote by $\mathcal{V}$ and $\mathcal{E}$ the sets of vertices and edges, respectively. Note that $|\mathcal{E}| = m$ is the number of members, and $|\mathcal{V}| = d/3$ is the number of nodes of the cable network. Then a pair $\mathcal{G} = (\mathcal{V}, \mathcal{E})$ corresponds to a *directed graph*, or for short, a *digraph*; see Figure 4.5(b) for an illustrative example. If $e = (v_l, v_{l'})$ is an edge of $\mathcal{G}$, then the start vertex v_l is called the *tail*, and the end vertex $v_{l'}$ is called the *head* of e.

For the digraph $\mathcal{G}$, the matrix $C = (C_{li}) \in \mathbb{R}^{(d/3) \times m}$ defined by

$$C_{li} = \begin{cases} 1 & \text{if the vertex } l \text{ is the tail of the edge } i, \\ -1 & \text{if the vertex } l \text{ is the head of the edge } i, \\ 0 & \text{otherwise}, \end{cases}$$

is called the *incidence matrix* of $\mathcal{G}$. Let $\tilde{C}_i \in \mathbb{R}^{d/3}$ denote the ith column vector of C, i.e.,

$$C = \begin{bmatrix} \tilde{C}_1 & \cdots & \tilde{C}_m \end{bmatrix}.$$

Then the matrices B_i's in (4.38) are written as

$$B_i = -\tilde{C}_i^{\mathrm{T}} \otimes I_3, \quad i = 1, \dots, m, \tag{4.40}$$

where for matrices P and Q we denote by $P \otimes Q$ their Kronecker product.[9]

---

[7] See, e.g., Diestel [112] and Jungnickel [199] for fundamentals of the graph theory.

[8] The term *arc* is also used instead of edge to distinguish between the directed and undirected cases.

[9] The Kronecker product $A \otimes B$ of $A = (A_{ij}) \in \mathbb{R}^{m_1 \times n_1}$ and $B \in \mathbb{R}^{m_2 \times n_2}$ is the $m_1 m_2 \times n_1 n_2$ matrix defined by

$$A \otimes B = \begin{bmatrix} A_{11}B & A_{12}B & \cdots & A_{1n_1}B \\ \vdots & \vdots & \ddots & \vdots \\ A_{m_1 1}B & A_{m_1 2}B & \cdots & A_{m_1 n_1}B \end{bmatrix}.$$

In the case of Example 4.1, an example of the digraph induced from the cable network in Figure 4.5(a) is illustrated in Figure 4.5(b), where the vertex set and the edge set are given by

$$\mathcal{V} = \{v_1, v_2, v_3, v_4\}, \quad \mathcal{E} = \{e_1, e_2, e_3\}.$$

Each element of $\mathcal{E}$ is defined as the ordered pair of vertices such as

$$e_1 = (v_1, v_2), \quad e_2 = (v_2, v_3), \quad e_3 = (v_3, v_4).$$

Then the incidence matrix C of $\mathcal{G} = (\mathcal{V}, \mathcal{E})$ is written as

$$C = \begin{bmatrix} 1 & 0 & 0 \\ -1 & 1 & 0 \\ 0 & -1 & 1 \\ 0 & 0 & -1 \end{bmatrix}.$$

Accordingly, B_i's defined by (4.40) with this matrix C coincide with the ones investigated in Example 4.1. ■

It should be clear in (4.38) that the relation between c_i and $\boldsymbol{\varphi}$ is valid even for the case subjected to the large deformation and the large strain. In the case of the large strain, the constitutive law in (4.1) should be interpreted as the one between the *Biot strain* and the *Biot stress*.[10]

By substituting (4.36) and (4.38) into problem (4.28), we obtain a specific formulation of the minimization problem of the potential energy as

$$\left.\begin{aligned} (\mathrm{PE}) : \min_{\boldsymbol{\varphi}, \boldsymbol{c}} \ & \sum_{i=1}^{m} \hat{w}^i_{\mathrm{cab}}(c_i) - \sum_{j \in \Gamma_{\mathrm{N}}} \underline{p}_j (\varphi_j - x_j) \\ \mathrm{s.\,t.} \ & c_i = \|B_i \boldsymbol{\varphi}\| - \underline{l}_i, \quad i = 1, \dots, m, \\ & \varphi_j = \underline{\varphi}_j, \quad j \in \Gamma_{\mathrm{D}}, \end{aligned}\right\} \tag{4.41}$$

where $\boldsymbol{c} \in \mathbb{R}^m$ and $\boldsymbol{\varphi} \in \mathbb{R}^d$ are the variables. According to Proposition 4.2.1, the conic optimization reformulation of the problem above is written as

$$\left.\begin{aligned} (\overline{\mathrm{PE}}) : \min_{\boldsymbol{\varphi}, \boldsymbol{c}_{\mathrm{e}}} \ & \sum_{i=1}^{m} \frac{1}{2} k_{\mathrm{e}i} c_{\mathrm{e}i}^2 - \sum_{j \in \Gamma_{\mathrm{N}}} \underline{p}_j (\varphi_j - x_j) \\ \mathrm{s.\,t.} \ & c_{\mathrm{e}i} \geq \|B_i \boldsymbol{\varphi}\| - \underline{l}_i, \quad i = 1, \dots, m, \\ & \varphi_j = \underline{\varphi}_j, \quad j \in \Gamma_{\mathrm{D}}. \end{aligned}\right\} \tag{4.42}$$

Here, the variables to be optimized are the elastic parts of the elongations, say, $\boldsymbol{c}_{\mathrm{e}} \in \mathbb{R}^m$, and the position vectors of the nodes, say, $\boldsymbol{\varphi} \in \mathbb{R}^d$.

---

[10] The Biot strain in continuum mechanics is defined by $(\boldsymbol{F}^{\mathrm{T}}\boldsymbol{F})^{1/2} - \boldsymbol{I}$ with the deformation gradient $\boldsymbol{F}$; see, e.g., Holzapfel [184]. As seen in Remark 4.2.4, $B_i\boldsymbol{\varphi}$ corresponds to the deformation gradient in the situation under consideration, and (4.38) can read $c_i = [(B_i\boldsymbol{\varphi})^{\mathrm{T}}(B_i\boldsymbol{\varphi})]^{1/2} - [(B_i\boldsymbol{x})^{\mathrm{T}}(B_i\boldsymbol{x})]^{1/2}$.

It should be clear that problem (4.42) is an SOCP problem. To see this, we first claim that the inequality constraint,

$$c_{ei} \geq \|B_i \boldsymbol{\varphi}\| - \underline{l}_i,$$

is written in the form of the second-order cone inequality in the four-dimensional space, say,

$$\begin{bmatrix} c_{ei} + \underline{l}_i \\ B_i \boldsymbol{\varphi} \end{bmatrix} \in \mathbb{L}^4,$$

where $\mathbb{L}^4 = \{(z_0, \boldsymbol{z}_1) \in \mathbb{R} \times \mathbb{R}^3 \mid z_0 \geq \|\boldsymbol{z}_1\|\}$. Moreover, to deal with the nonlinear terms in the objective function, we introduce additional variables, y_i $(i = 1, \dots, m)$, and convert the minimization of $(1/2)k_{ei}c_{ei}^2$ to the minimization of y_i under constraint $y_i \geq (1/2)k_{ei}c_{ei}^2$. Apply Fact 1.5.2 (b) to this convex quadratic inequality to see that

$$y_i \geq \frac{1}{2}k_{ei}c_{ei}^2 \quad \Leftrightarrow \quad (y_i/2k_{ei}) + 1 \geq \left\| \begin{bmatrix} (y_i/2k_{ei}) - 1 \\ c_{ei} \end{bmatrix} \right\|,$$

which is the second-order cone inequality in the three-dimensional space. Consequently, $(\overline{\mathrm{PE}})$ in (4.42) is equivalently reduced to

$$\left.\begin{aligned} &\min_{\boldsymbol{\varphi}, \boldsymbol{c}_{\mathrm{e}}, \boldsymbol{y}} \sum_{i=1}^{m} y_i - \sum_{j \in \Gamma_{\mathrm{N}}} \underline{p}_j(\varphi_j - x_j) \\ &\ \mathrm{s.t.} \quad (y_i/2k_{ei}) + 1 \geq \left\| \begin{bmatrix} (y_i/2k_{ei}) - 1 \\ c_{ei} \end{bmatrix} \right\|, \quad i = 1, \dots, m, \\ &\qquad\quad c_{ei} + \underline{l}_i \geq \|B_i \boldsymbol{\varphi}\|, \quad i = 1, \dots, m, \\ &\qquad\quad \varphi_j = \underline{\varphi}_j, \quad j \in \Gamma_{\mathrm{D}}, \end{aligned}\right\} \tag{4.43}$$

which is explicitly in the form of SOCP.

4.2.3 Principle for linear strain

Within the geometrically linear theory, the member elongation is considered to be a linear function of $\boldsymbol{u}$ by neglecting the nonlinear terms in the compatibility relation in (4.37). The elongation, defined in terms of the *linear strain* measure, is then written in the form of

$$\hat{c}_i(\boldsymbol{u}) = \boldsymbol{\beta}_i^{\mathrm{T}} \boldsymbol{u}, \tag{4.44}$$

where $\boldsymbol{\beta}_i \in \mathbb{R}^d$ is a constant vector. More precisely, as seen in Remark 4.2.4, the vector $B_i \boldsymbol{x}$ coincides with the undeformed member configuration, and $B_i \boldsymbol{x}/\underline{l}_i$ corresponds to the direction cosines. Hence, the vector $\boldsymbol{\beta}_i$, which is the gradient vector of $l_i(\boldsymbol{x} + \boldsymbol{u})$ with respect to $\boldsymbol{u}$, is explicitly given by

$$\boldsymbol{\beta}_i = -\frac{1}{\underline{l}_i}\tilde{C}_i \otimes (B_i \boldsymbol{x}) = \frac{1}{\underline{l}_i} B_i^{\mathrm{T}} B_i \boldsymbol{x}, \tag{4.45}$$

where $\tilde{C}_i$ was introduced in Remark 4.2.5.

Consequently, substituting (4.44) into problem (4.28) results in the specific form of minimization problem of the potential energy within the geometrically linear theory as

$$\left.\begin{aligned}(\text{PE}) : \min_{\boldsymbol{u},\boldsymbol{c}} \ & \sum_{i=1}^{m} \hat{w}_{\text{cab}}^{i}(c_i) - \sum_{j\in\Gamma_{\text{N}}} \underline{p}_j u_j \\ \text{s.t.}\ & c_i = \boldsymbol{\beta}_i^{\text{T}}\boldsymbol{u}, \quad i = 1,\dots,m, \\ & u_j = \underline{u}_j, \quad j \in \Gamma_{\text{D}}.\end{aligned}\right\} \tag{4.46}$$

In contrast to problem (4.41) for the geometrically nonlinear case, problem (4.46) above does not involve nonlinear constraints. However, $\hat{w}_{\text{cab}}^{i}$ in the objective function is defined by the disjunction in (4.4), which is a major difficulty arising from the nonsmoothness property of cables.

Applying Proposition 4.2.1 yields, for (PE) in (4.46), a tractable reformulation

$$\left.\begin{aligned}(\overline{\text{PE}}) : \min_{\boldsymbol{u},\boldsymbol{c}_{\text{e}}} \ & \sum_{i=1}^{m} \frac{1}{2} k_{\text{e}i} c_{\text{e}i}^2 - \sum_{j\in\Gamma_{\text{N}}} \underline{p}_j u_j \\ \text{s.t.}\ & c_{\text{e}i} \geq \boldsymbol{\beta}_i^{\text{T}}\boldsymbol{u}, \quad i = 1,\dots,m, \\ & u_j = \underline{u}_j, \quad j \in \Gamma_{\text{D}}.\end{aligned}\right\} \tag{4.47}$$

This is a QP problem[11] and hence is solved more effectively compared with (4.46).

4.2.4 Principle for the Green–Lagrange strain

We again consider the geometrically nonlinear theory. Unlike section 4.2.2, however, we assume that the strain is small, while the rotation is still considered large. The *Green–Lagrange strain* is then used as a measure of strain, where the compatibility relation in (4.37) is approximated by a quadratic function. Rewrite (4.37) as

$$(c_i + l_i(\boldsymbol{x}))^2 = l_i(\boldsymbol{\varphi})^2,$$

where c_i^2 can be neglected by assuming the small strain. Then, by using $l_i(\boldsymbol{\varphi}) = \|B_i\boldsymbol{\varphi}\|$ and $l_i(\boldsymbol{x}) = \underline{l}_i$, we obtain

$$\begin{aligned} c_i &= \frac{l_i(\boldsymbol{\varphi})^2 - l_i(\boldsymbol{x})^2}{2l_i(\boldsymbol{x})} \\ &= \frac{1}{2\underline{l}_i}\left[(B_i\boldsymbol{\varphi})^{\text{T}}(B_i\boldsymbol{\varphi}) - \underline{l}_i^2\right], \end{aligned} \tag{4.48}$$

[11]By QP we mean the convex quadratic program; see (1.31) in section 1.6.2.

which is the member elongation c_i with the Green–Lagrange strain measure.[12]

As a consequence, the minimization problem of the potential energy, in terms of the Green–Lagrange strain, is formulated from (4.28) as

$$\left.\begin{aligned} (\mathrm{PE}) : \min_{\boldsymbol{\varphi}, \boldsymbol{c}} \ & \sum_{i=1}^{m} \hat{w}^{i}_{\mathrm{cab}}(c_i) - \sum_{j \in \varGamma_{\mathrm{N}}} \underline{p}_j (\varphi_j - x_j) \\ \mathrm{s.\,t.} \ & c_i = \frac{1}{2\underline{l}_i} \left[(B_i \boldsymbol{\varphi})^{\mathrm{T}} (B_i \boldsymbol{\varphi}) - \underline{l}_i^2 \right], \quad i = 1, \dots, m, \\ & \varphi_j = \underline{\varphi}_j, \quad j \in \varGamma_{\mathrm{D}}. \end{aligned}\right\} \tag{4.49}$$

The convex optimization reformulation, corresponding to (4.29), is then obtained as

$$\left.\begin{aligned} (\overline{\mathrm{PE}}) : \min_{\boldsymbol{\varphi}, \boldsymbol{c}_{\mathrm{e}}} \ & \sum_{i=1}^{m} \frac{1}{2} k_{\mathrm{e}i} c_{\mathrm{e}i}^2 - \sum_{j \in \varGamma_{\mathrm{N}}} \underline{p}_j (\varphi_j - x_j) \\ \mathrm{s.\,t.} \ & c_{\mathrm{e}i} \geq \frac{1}{2\underline{l}_i} \left[(B_i \boldsymbol{\varphi})^{\mathrm{T}} (B_i \boldsymbol{\varphi}) - \underline{l}_i^2 \right], \quad i = 1, \dots, m, \\ & \varphi_j = \underline{\varphi}_j, \quad j \in \varGamma_{\mathrm{D}}, \end{aligned}\right\} \tag{4.50}$$

where $\boldsymbol{c}_{\mathrm{e}} \in \mathbb{R}^m$ and $\boldsymbol{\varphi} \in \mathbb{R}^d$ are the variables to be optimized.

It is noted that $(\overline{\mathrm{PE}})$ in (4.50) on this occasion is a QCQP (quadratically constrained quadratic programming) problem. This problem can further be reduced to a form of conic optimization, but such a reduction is not unique. For example, from Fact 1.3.10, the inequality constraints of (4.50) are reduced to

$$c_{\mathrm{e}i} \geq \frac{1}{2\underline{l}_i} \left[(B_i \boldsymbol{\varphi})^{\mathrm{T}} (B_i \boldsymbol{\varphi}) - \underline{l}_i^2 \right] \quad \Leftrightarrow \quad \left[\begin{array}{c|c} 2\underline{l}_i c_{\mathrm{e}i} + \underline{l}_i^2 & (B_i \boldsymbol{\varphi})^{\mathrm{T}} \\ \hline B_i \boldsymbol{\varphi} & I_3 \end{array} \right] \succeq O.$$

Moreover, section 1.6.2 showed that the quadratic terms in the objective function can be rewritten as a positive-semidefinite constraint (see, also Fact 1.5.2 (a)).

[12] Substitution of (4.36) into (4.48) yields

$$c_i = \frac{1}{2\underline{l}_i} \left([B_i(\boldsymbol{x} + \boldsymbol{u})]^{\mathrm{T}} [B_i(\boldsymbol{x} + \boldsymbol{u})] - \underline{l}_i^2 \right) = \frac{1}{2\underline{l}_i} \left[2(B_i \boldsymbol{x})^{\mathrm{T}} (B_i \boldsymbol{u}) + (B_i \boldsymbol{u})^{\mathrm{T}} (B_i \boldsymbol{u}) \right].$$

Therefore, by assuming that the displacement gradient is of small order, say, $\|B_i \boldsymbol{u}\| \ll 1$, we can see that the elongation in the Green–Lagrange strain measure coincides with the elongation in the linear strain measure, i.e., $\hat{c}_i$ defined in (4.44) with (4.45).

Consequently, $(\overline{\mathrm{PE}})$ in (4.50) is equivalently rewritten as

$$\left.\begin{array}{ll} \min\limits_{\boldsymbol{\varphi}, \boldsymbol{c}_{\mathrm{e}}, \boldsymbol{y}} & \sum\limits_{i=1}^{m} y_i - \sum\limits_{j \in \Gamma_{\mathrm{N}}} \underline{p}_j (\varphi_j - x_j) \\ \mathrm{s.\,t.} & \begin{bmatrix} y_i & c_{\mathrm{e}i} \\ c_{\mathrm{e}i} & 2/k_{\mathrm{e}i} \end{bmatrix} \succeq O, \quad i = 1, \dots, m, \\ & \left[\begin{array}{c|c} 2\underline{l}_i c_{\mathrm{e}i} + \underline{l}_i^2 & (B_i \boldsymbol{\varphi})^{\mathrm{T}} \\ \hline B_i \boldsymbol{\varphi} & I_3 \end{array}\right] \succeq O, \quad i = 1, \dots, m, \\ & \varphi_j = \underline{\varphi}_j, \quad j \in \Gamma_{\mathrm{D}}, \end{array}\right\} \tag{4.51}$$

which is an SDP problem.

Alternatively, it follows from Fact 1.5.2 (b) that the inequality constraints in (4.50) are rewritten as

$$\begin{aligned} & c_{\mathrm{e}i} \geq \frac{1}{2\underline{l}_i} \left[(B_i \boldsymbol{\varphi})^{\mathrm{T}} (B_i \boldsymbol{\varphi}) - \underline{l}_i^2 \right] \\ \Leftrightarrow \quad & \frac{2c_{\mathrm{e}i}}{\underline{l}_i} + 1 \geq \frac{1}{\underline{l}_i^2} (B_i \boldsymbol{\varphi})^{\mathrm{T}} (B_i \boldsymbol{\varphi}) \\ \Leftrightarrow \quad & (2c_{\mathrm{e}i}/\underline{l}_i + 1) + 1 \geq \left\| \begin{bmatrix} (2c_{\mathrm{e}i}/\underline{l}_i + 1) - 1 \\ 2B_i \boldsymbol{\varphi} \end{bmatrix} \right\| . \end{aligned}$$

Moreover, the quadratic terms in the objective function are dealt with in terms of second-order cone inequalities by applying the technique used in section 4.2.2 for reducing (4.42) to (4.43). As a consequence, $(\overline{\mathrm{PE}})$ in (4.50) is reduced to

$$\left.\begin{array}{ll} \min\limits_{\boldsymbol{\varphi}, \boldsymbol{c}_{\mathrm{e}}, \boldsymbol{v}} & \sum\limits_{i=1}^{m} y_i - \sum\limits_{j \in \Gamma_{\mathrm{N}}} \underline{p}_j (\varphi_j - x_j) \\ \mathrm{s.\,t.} & (y_i/2k_{\mathrm{e}i}) + 1 \geq \left\| \begin{bmatrix} (y_i/2k_{\mathrm{e}i}) - 1 \\ c_{\mathrm{e}i} \end{bmatrix} \right\|, \quad i = 1, \dots, m, \\ & (c_{\mathrm{e}i}/\underline{l}_i) + 1 \geq \left\| \begin{bmatrix} c_{\mathrm{e}i}/\underline{l}_i \\ B_i \boldsymbol{\varphi} \end{bmatrix} \right\|, \quad i = 1, \dots, m, \\ & \varphi_j = \underline{\varphi}_j, \quad j \in \Gamma_{\mathrm{D}}, \end{array}\right\} \tag{4.52}$$

which is an SOCP problem.

4.3 More on Cable Networks: Nonlinear Material Law

In section 4.1 and section 4.2, we supposed that while each cable member behaves linearly elastic in the taut state, it has zero stiffness in the slack state.

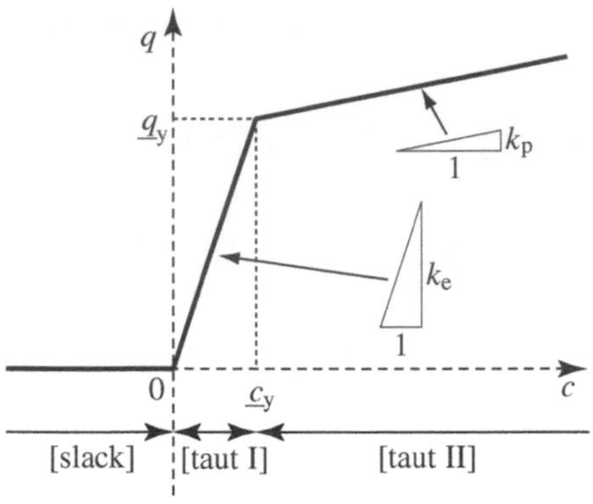

FIGURE 4.6: A piecewise-linear constitutive law of a cable member.

Thus the nonlinearity (or, more precisely, the nonsmoothness property) caused by slackening was discussed. As an extension, this section discusses the case in which the response function is also nonlinear in the taut state. In other words, we suppose that a cable member behaves as a nonlinear elastic material in a taut state.

Specifically, we deal with the *piecewise-linear law* for the taut state in section 4.3.1, which serves as a crude model for elastoplasticity when the monotonic loading is assumed. In section 4.3.2 we consider the *piecewise-cubic law*, which enables us to represent the stiffness reduction in the range of small axial force.

4.3.1 Piecewise-linear law

Consider a *piecewise-linear* constitutive law illustrated in Figure 4.6. The taut state is divided into two phases,[13] where the elastic moduli are $k_{\rm e} > 0$ and $k_{\rm p} > 0$. The elongation at the transition point of these phases is specified as $\underline{c}_{\rm y}$. Then we define $\underline{q}_{\rm y}$ by

$$\underline{q}_{\rm y} = k_{\rm e}\underline{c}_{\rm y}. \tag{4.53}$$

[13]A piecewise-linear law consisting of two phases is sometimes called a *bi-linear* law in literature in mechanics.

The response function, which describes the constitutive relation between the elongation and the axial force, is then written as[14]

$$\hat{q}_{\rm cab}(c) = \begin{cases} k_{\rm p}(c - \underline{c}_{\rm y}) + \underline{q}_{\rm y} & \text{if } c \geq \underline{c}_{\rm y}, \\ k_{\rm e}c & \text{if } c \in [0, \underline{c}_{\rm y}[, \\ 0 & \text{if } c < 0, \end{cases} \tag{4.54}$$

where the disjunction here means

$$\begin{cases} c \geq \underline{c}_{\rm y} & : [\text{taut (II)}], \\ c \in [0, \underline{c}_{\rm y}[& : [\text{taut (I)}], \\ c < 0 & : [\text{slack}]. \end{cases}$$

Accordingly, the stored energy function is obtained as

$$\begin{aligned} \hat{w}_{\rm cab}(c) &:= \int_0^c \hat{q}_{\rm cab}(c){\rm d}c \\ &= \begin{cases} \dfrac{1}{2}k_{\rm p}(c - \underline{c}_{\rm y})^2 + \underline{q}_{\rm y}(c - \underline{c}_{\rm y}) + \dfrac{1}{2}\underline{q}_{\rm y}\underline{c}_{\rm y}^2 & \text{if } c \geq \underline{c}_{\rm y}, \\ \dfrac{1}{2}k_{\rm e}c^2 & \text{if } c \in [0, \underline{c}_{\rm y}[, \\ 0 & \text{if } c < 0. \end{cases} \end{aligned} \tag{4.55}$$

Remark 4.3.1. The piecewise-linear constitutive law in (4.54) is often used for elastoplastic analysis. In that case, $k_{\rm e}\underline{c}_{\rm y}$ corresponds to the *yield* axial force, and $k_{\rm p}$ is the *tangent modulus* after yielding. The disjunction of the two taut phases come from the distinction of *elastic* and *plastic* responses.

If each member is monotonically loaded, i.e., in the occasion that the *elastic unloading* does not take place, then the axial force depends solely upon the elongation. In such a case the axial force is given by (4.54), and the equilibrium state of the cable network can be found by solving the minimization problem of the total potential energy as shown below.

In general, certainly, the elastoplastic response is path dependent, and hence the principle of potential energy cannot be written in terms of total deformations; it should be formulated terms of infinitesimal increments of kinematic variables. ■

In a manner similar to section 4.1.4, we introduce the additive decomposition of the elongation c as

$$c = c_{\rm p} + c_{\rm e} + c_{\rm s}, \tag{4.56}$$

where $c_{\rm s}$ is the slackening part, and $c_{\rm e}$ and $c_{\rm p}$ are the "elastic" and "plastic" parts, respectively.[15] In the slack state, say, when $c_{\rm s} < 0$, c in (4.56) does

[14] It follows from definition (4.53) of $\underline{q}_{\rm y}$ that $\hat{q}_{\rm cab}$ is a continuous (but nonsmooth) function.

[15] As mentioned in Remark 4.3.1, use of the constitutive law in (4.54) in the elastoplastic analysis requires the assumption of monotonic loading. However, we sometimes use the terms *elastic* and *plastic* to make an intuitive understanding of equations from the viewpoint of mechanical engineering.

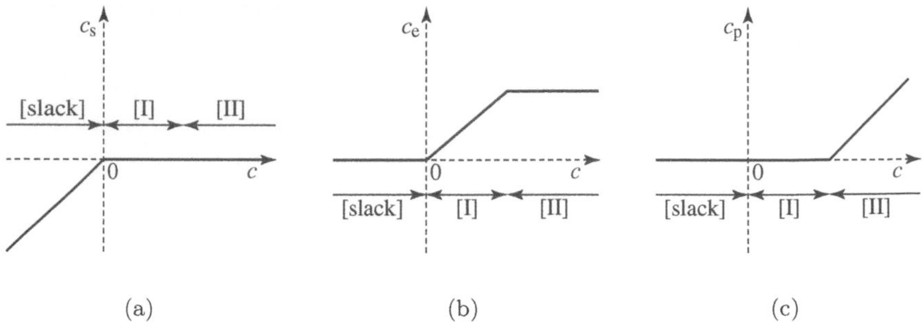

FIGURE 4.7: The decomposition of the elongation c in (4.56) and (4.57).

not make a non-zero axial force, which implies $c_p = c_e = 0$. Alternatively, when $c_p + c_e > 0$, the cable is taut state, and hence $c_s = 0$. Thus the complementarity condition holds between $-c_s$ and $c_e + c_p$. Moreover, $c_e < \underline{c}_y$ means that the cable is "elastic," say, $c_p = 0$. Conversely, $c_p > 0$ means that the cable is "plastic," and the yielding condition, $c_e = \underline{c}_y$, holds. Thus the complementarity condition also holds between $\underline{c}_y - c_e$ and c_p. Thus, in the decomposition (4.56), each part is determined as

$$c_p = \begin{cases} c - \underline{c}_y \\ 0 \\ 0 \end{cases} \quad c_e = \begin{cases} \underline{c}_y \\ c \\ 0 \end{cases} \quad c_s = \begin{cases} 0 & \text{if } c \geq \underline{c}_y, \\ 0 & \text{if } c \in [0, \underline{c}_y[, \\ c & \text{if } c < 0. \end{cases} \tag{4.57}$$

This relation is illustrated in Figure 4.7.

Regarding the constitutive law, $q = \hat{q}_{cab}(c)$ with (4.54), the axial force is also decomposed additively as

$$q = q_p + q_e, \tag{4.58}$$

where q_p and q_e are related to the elastic and plastic elongations as

$$q_p = k_p c_p, \quad q_e = k_e c_e. \tag{4.59}$$

Then the disjunction (4.57), in conjunction with (4.59), is reduced to

$$q_e \geq 0, \quad c_s \leq 0, \quad q_e c_s = 0, \tag{4.60}$$

$$q_p \geq 0, \quad c_e - \underline{c}_y \leq 0, \quad q_p(c_e - \underline{c}_y) = 0. \tag{4.61}$$

Condition (4.60), together with (4.58) and (4.59), corresponds to the distinction between the slack and the "elastic" taut states and hence is in the same form as (4.23a)–(4.23c) in section 4.1.4. In contrast, the complementarity condition in (4.61) represents the distinction between the "elastic" and "plastic" states and is a particular property introduced in this section. Consequently, the constitutive law in (4.54) is reduced to (4.58)–(4.61).

To formulating ($\overline{\text{PE}}$), i.e., the conic optimization form of the minimum principle of the potential energy, we further reduce (4.58)–(4.61) to a variational form, in a manner similar to section 4.1.3.

If the elongation c is decomposed as (4.56) with (4.57), then the stored energy in (4.55) is simply written as $\frac{1}{2}k_{\mathrm{p}}c_{\mathrm{p}}^2 + \underline{q}_{\mathrm{y}}c_{\mathrm{p}} + \frac{1}{2}k_{\mathrm{e}}c_{\mathrm{e}}^2$. We may conjecture that this function is to be minimized over some inequalities representing (4.57), which is the assertion stated in the following.

Proposition 4.3.2. *For $\hat{w}_{\mathrm{cab}}$ defined by* (4.55), *we have that*

$$\hat{w}_{\mathrm{cab}}(c) = \min_{c_{\mathrm{p}},c_{\mathrm{e}}} \left\{ \frac{1}{2}k_{\mathrm{p}}c_{\mathrm{p}}^2 + \underline{q}_{\mathrm{y}}c_{\mathrm{p}} + \frac{1}{2}k_{\mathrm{e}}c_{\mathrm{e}}^2 \mid c_{\mathrm{p}} + c_{\mathrm{e}} \geq c,\ c_{\mathrm{e}} \leq \underline{c}_{\mathrm{y}},\ c_{\mathrm{p}} \geq 0 \right\}. \tag{4.62}$$

Proof. We rewrite (4.58)–(4.61) as a complementarity problem. First, observe that $q_{\mathrm{e}} \geq 0$ in (4.60) is equivalent to $q \geq 0$. Moreover, because c_{s} is eliminated by using (4.56), we see that (4.60) is equivalently rewritten as

$$q \geq 0, \quad c - c_{\mathrm{p}} - c_{\mathrm{e}} \leq 0, \quad q(c - c_{\mathrm{p}} - c_{\mathrm{c}}) = 0.$$

We next derive an alternative form of (4.61). The condition $q_{\mathrm{p}} \geq 0$ in (4.61) is equivalent to $c_{\mathrm{p}} \geq 0$, because of (4.59). Moreover, we introduce a new variable q_{r} by

$$q_{\mathrm{r}} = \underline{q}_{\mathrm{y}} - q_{\mathrm{e}},$$

which represents the residual of the "yielding" condition. Then $c_{\mathrm{e}} - \underline{c}_{\mathrm{y}} \leq 0$ in (4.61) is equivalently rewritten as $q_{\mathrm{r}} \geq 0$, again because of (4.59). Thus (4.61) is alternatively written as

$$q_{\mathrm{r}} \geq 0, \quad c_{\mathrm{p}} \geq 0, \quad q_{\mathrm{r}}c_{\mathrm{p}} = 0.$$

Consequently, conditions (4.58)–(4.61), together with (4.56), are equivalently rewritten as

$$q \geq 0, \quad c_{\mathrm{p}} + c_{\mathrm{e}} - c \geq 0, \quad q(c_{\mathrm{p}} + c_{\mathrm{e}} - c) = 0, \tag{4.63a}$$
$$q = k_{\mathrm{p}}c_{\mathrm{p}} + (\underline{q}_{\mathrm{y}} - q_{\mathrm{r}}), \tag{4.63b}$$
$$q_{\mathrm{r}} \geq 0, \quad c_{\mathrm{p}} \geq 0 \quad q_{\mathrm{r}}c_{\mathrm{p}} = 0, \tag{4.63c}$$
$$q = q_{\mathrm{p}} + k_{\mathrm{e}}c_{\mathrm{e}}, \tag{4.63d}$$
$$q_{\mathrm{p}} \geq 0, \quad c_{\mathrm{e}} - \underline{c}_{\mathrm{y}} \leq 0, \quad q_{\mathrm{p}}(c_{\mathrm{e}} - \underline{c}_{\mathrm{y}}) = 0. \tag{4.63e}$$

Note that (4.63) includes some redundant conditions.

Finally, we show that condition (4.63) coincides with the optimality condition of the problem on the right-hand side of (4.62). Since this is a QP problem, we use the Lagrangian to derive the KKT conditions. The Lagrangian of this problem is defined by

$$L(c_{\mathrm{p}}, c_{\mathrm{e}}; \boldsymbol{z}^*) = \begin{cases} \frac{1}{2}k_{\mathrm{p}}c_{\mathrm{p}}^2 + \underline{q}_{\mathrm{y}}c_{\mathrm{p}} + \frac{1}{2}k_{\mathrm{e}}c_{\mathrm{e}}^2 & \\ \quad -z_1^*(c_{\mathrm{p}} + c_{\mathrm{e}} - c) & \\ \quad -z_2^*(\underline{c}_{\mathrm{y}} - c_{\mathrm{e}}) - z_3^*c_{\mathrm{p}} & \text{if } \boldsymbol{z}^* \geq \boldsymbol{0}, \\ -\infty & \text{otherwise,} \end{cases} \tag{4.64}$$

where $\boldsymbol{z}^* = (z_1^*, z_2^*, z_3^*)$ is the Lagrangian multiplier vector. Then, by rewriting the variables in (4.63) as

$$q = z_1^*, \quad q_{\mathrm{p}} = z_2^*, \quad q_{\mathrm{r}} = z_3^*,$$

it is easy to see that (4.63) coincides with the saddle-point condition for L in (4.64). Therefore, $(c_{\mathrm{p}}, c_{\mathrm{e}})$ is optimal for the problem in (4.62) if and only if there exists $(q, q_{\mathrm{p}}, q_{\mathrm{r}})$ satisfying (4.63), but this condition is further equivalent to (4.56) and (4.58)–(4.61). Accordingly, the assertion of this proposition is immediately obtained from definition (4.55) of $\hat{w}_{\mathrm{cab}}$. □

Recall that the minimization problem of the total potential energy for cable networks is formulated as (PE) in (4.41), within the large strain setting considered throughout section 4.2.2. By having the discussion same as in Proposition 4.2.1 (see section 4.2.1), Proposition 4.3.2 leads to $(\overline{\mathrm{PE}})$ for the piecewise-linear constitutive law. Specifically, a convex optimization formulation for (PE) is given by

$$\left.\begin{array}{lll} (\overline{\mathrm{PE}}): & \min\limits_{\boldsymbol{\varphi}, \boldsymbol{c}_{\mathrm{p}}, \boldsymbol{c}_{\mathrm{e}}} & \displaystyle\sum_{i=1}^{m} \left(\frac{1}{2} k_{\mathrm{p}i} c_{\mathrm{p}i}^2 + \underline{q}_{\mathrm{y}i} c_{\mathrm{p}i} + \frac{1}{2} k_{\mathrm{e}i} c_{\mathrm{e}i}^2 \right) - \sum_{j \in \Gamma_{\mathrm{N}}} \underline{p}_j (\varphi_j - x_j) \\ & \mathrm{s.\,t.} & c_{\mathrm{p}i} + c_{\mathrm{e}i} \geq \|B_i \boldsymbol{\varphi}\| - \underline{l}_i, \quad i = 1, \dots, m, \\ & & c_{\mathrm{e}i} \leq \underline{c}_{\mathrm{y}i}, \ c_{\mathrm{p}i} \geq 0, \quad i = 1, \dots, m, \\ & & \varphi_j = \underline{\varphi}_j, \quad j \in \Gamma_{\mathrm{D}}. \end{array}\right\} \quad (4.65)$$

Problem (4.65) for the piecewise-linear material law is compared with problem (4.42) for the linear one. Note that the objective function of (4.65) is a convex quadratic function. Therefore, in a manner similar to the reduction of (4.42) to (4.43), problem (4.65) can be reduced to an explicit form of the SOCP.

4.3.2 Piecewise-quadratic law

Consider a cable network with large self-weight, such as that used for a civil engineering structure. When the axial force of a cable member is relatively small, its elongation stiffness in the actual situation is often smaller than the value determined from the elastic modulus of the material. Such an apparent reduction of the stiffness may be understood as a consequence of the following two particular properties of cable members:

- A cable member is often realized as a stranded cable in the real world. While the axial force is small, the twist of fibers is still loose.
- Due to the self-weight, a cable member sags while the axial force is small.

Because of these properties, the elongation as a structural element can increase without introducing the actual strain, which yields the apparent reduction of the total elongation stiffness.

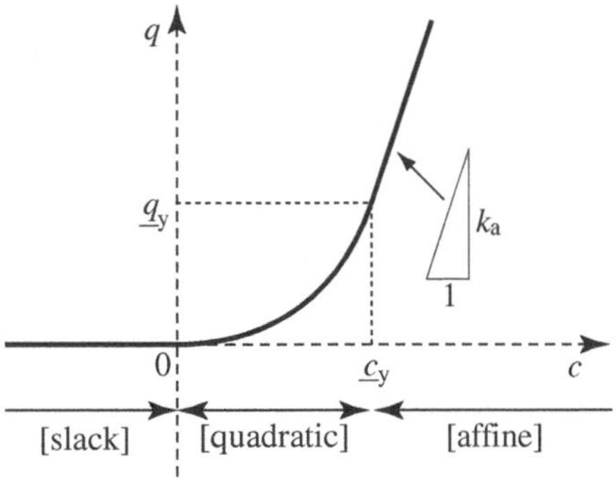

FIGURE 4.8: A piecewise-quadratic constitutive law.

To represent such a phenomenon, consider a constitutive relation depicted in Figure 4.8. The taut state is divided into two phases: the ones with and without an apparent stiffness reduction. When the axial force is sufficiently large, the member elongation stiffness coincides with that determined from the material property and hence is assumed to be an affine (or a linear) function. We also assume that the axial force is expressed by a (convex) quadratic function of the elongation for a small range. Thus the constitutive relation is given by

$$\hat{q}_{\mathrm{cab}}(c) = \begin{cases} k_{\mathrm{a}}(c - \underline{c}_{\mathrm{y}}) + \underline{q}_{\mathrm{y}} & \text{if } c \geq \underline{c}_{\mathrm{y}}, \\ k_{\mathrm{q}} c^2 & \text{if } c \in [0, \underline{c}_{\mathrm{y}}[, \\ 0 & \text{if } c < 0 \end{cases} \tag{4.66}$$

with constant k_{q}, $k_{\mathrm{a}} > 0$, and we put

$$\underline{q}_{\mathrm{y}} = k_{\mathrm{q}} \underline{c}_{\mathrm{y}}^2.$$

The disjunction in (4.66) corresponds to the one shown in Figure 4.8 as

$$\begin{cases} c \geq \underline{c}_{\mathrm{y}} & : \text{[affine]}, \\ c \in [0, \underline{c}_{\mathrm{y}}[& : \text{[quadratic]}, \\ c > 0 & : \text{[slack]}. \end{cases}$$

The stored energy function is obtained as

$$\hat{w}_{\mathrm{cab}}(c) = \begin{cases} \dfrac{1}{2} k_{\mathrm{a}}(c - \underline{c}_{\mathrm{y}})^2 + \underline{q}_{\mathrm{y}}(c - \underline{c}_{\mathrm{y}}) + \dfrac{1}{3} k_{\mathrm{q}} \underline{c}_{\mathrm{y}}^3 & \text{if } c \geq \underline{c}_{\mathrm{y}}, \\ \dfrac{1}{3} k_{\mathrm{q}} c^3 & \text{if } c \in [0, \underline{c}_{\mathrm{y}}[, \\ 0 & \text{if } c < 0. \end{cases} \tag{4.67}$$

As in section 4.3.1, we introduce the decomposition of the elongation as

$$c = c_{\mathrm{a}} + c_{\mathrm{q}} + c_{\mathrm{s}}, \tag{4.68}$$

where c_{a} and c_{q} are the affinely and quadratically responding parts, while c_{s} is the slackening part. Then, similar to Proposition 4.3.2, we can rewrite $\hat{w}_{\mathrm{cab}}$ in (4.67) as the optimal value of an optimization problem as follows:

$$\hat{w}_{\mathrm{cab}}(c) = \min_{c_{\mathrm{a}}, c_{\mathrm{q}}} \Big\{ \frac{1}{2} k_{\mathrm{a}} c_{\mathrm{a}}^2 + \underline{q}_{\mathrm{y}} c_{\mathrm{a}} + \frac{1}{3} k_{\mathrm{q}} c_{\mathrm{q}}^3 \mid c_{\mathrm{a}} + c_{\mathrm{q}} \geq c, \\ 0 \leq c_{\mathrm{q}} \leq \underline{c}_{\mathrm{y}},\ c_{\mathrm{a}} \geq 0 \Big\}. \tag{4.69}$$

In a manner similar to Proposition 4.2.1, substitution of (4.69) into (PE) in (4.28) results in the optimization

$$\left.\begin{array}{lll} (\overline{\mathrm{PE}}): & \displaystyle\min_{\varphi, c_{\mathrm{a}}, c_{\mathrm{q}}} & \displaystyle\sum_{i=1}^{m} \left(\frac{1}{2} k_{\mathrm{a}i} c_{\mathrm{a}i}^2 + \underline{q}_{\mathrm{y}i} c_{\mathrm{a}i} + \frac{1}{3} k_{\mathrm{q}i} c_{\mathrm{q}i}^3 \right) - \sum_{j \in \Gamma_{\mathrm{N}}} \underline{p}_j (\varphi_j - x_j) \\ & \mathrm{s.\,t.} & c_{\mathrm{a}i} + c_{\mathrm{q}i} \geq \|B_i \varphi\| - \underline{l}_i, \quad i = 1, \dots, m, \\ & & 0 \leq c_{\mathrm{q}i} \leq \underline{c}_{\mathrm{y}i},\ c_{\mathrm{a}i} \geq 0, \quad i = 1, \dots, m, \\ & & \varphi_j = \underline{\varphi}_j, \quad j \in \Gamma_{\mathrm{D}}, \end{array}\right\} \tag{4.70}$$

which is a convex optimization problem. Although $\frac{1}{3} k_{\mathrm{q}i} c_{\mathrm{q}i}^3$ appearing in the objective function is *not* a convex function, it is convex *over* $\{c_{\mathrm{q}i} \in \mathbb{R} \mid c_{\mathrm{q}i} \geq 0\}$. Therefore, $(\overline{\mathrm{PE}})$ in (4.70) is a convex optimization problem. Indeed, $(\overline{\mathrm{PE}})$ can be reduced to an SOCP problem as follows. Among the nonlinear terms in the objective function, the convex quadratic term, say, $\frac{1}{2} k_{\mathrm{a}i} c_{\mathrm{a}i}^2$, can be treated in the manner same as reduction of (4.42) to (4.43). Specifically, by introducing an additional variable $y_{\mathrm{a}i}$, minimizing this quadratic function is converted to minimizing $y_{\mathrm{a}i}$ under the inequality constraint,

$$y_{\mathrm{a}i} \geq \frac{1}{2} k_{\mathrm{a}i} c_{\mathrm{a}i}^2,$$

which can be transformed to a second-order cone constraint (see Fact 1.5.2 (b)). The remarkable point in the current problem is that the objective function under consideration involves a cubic term $\frac{1}{3} k_{\mathrm{q}i} c_{\mathrm{q}i}^3$. This term is transferred to the constraint, in conjunction with a constraint on $c_{\mathrm{q}i}$, as

$$y_{\mathrm{q}i} \geq \frac{1}{3} k_{\mathrm{q}i} c_{\mathrm{q}i}^3, \quad c_{\mathrm{q}i} \geq 0, \tag{4.71}$$

and $y_{\mathrm{q}i}$ is to be minimized. Introduce a new variable s_i, which is to become equal to $c_{\mathrm{q}i}^2$ at the optimal solution, to see that (4.71) is equivalent to

$$y_{\mathrm{q}i} c_{\mathrm{q}i} \geq \frac{1}{3} k_{\mathrm{q}i} s_i^2, \quad s_i \geq c_{\mathrm{q}i}^2, \quad c_{\mathrm{q}i} \geq 0. \tag{4.72}$$

Applying Fact 1.5.2 (b) to the two nonlinear inequalities in (4.72) yields

$$y_{\mathrm{q}i} + \frac{3c_{\mathrm{q}i}}{4k_{\mathrm{q}i}} \geq \left\| \begin{bmatrix} y_{\mathrm{q}i} - \dfrac{3c_{\mathrm{q}i}}{4k_{\mathrm{q}i}} \\ s_i \end{bmatrix} \right\|, \quad s_i + \frac{1}{4} \geq \left\| \begin{bmatrix} s_i - \dfrac{1}{4} \\ c_{\mathrm{q}i} \end{bmatrix} \right\|.$$

Accordingly, $(\overline{\mathrm{PE}})$ in (4.70) is equivalently rewritten as

$$\left.\begin{array}{ll} \min & \displaystyle\sum_{i=1}^{m} \left(y_{\mathrm{q}i} + \underline{q}_{\mathrm{y}i} c_{\mathrm{a}i} + y_{\mathrm{a}i} \right) - \sum_{j \in \Gamma_{\mathrm{N}}} \underline{p}_j (\varphi_j - x_j) \\ \mathrm{s.\,t.} & y_{\mathrm{q}i} + \dfrac{3c_{\mathrm{q}i}}{4k_{\mathrm{q}i}} \geq \left\| \begin{bmatrix} y_{\mathrm{q}i} - \dfrac{3c_{\mathrm{q}i}}{4k_{\mathrm{q}i}} \\ s_i \end{bmatrix} \right\|, \quad s_i + \dfrac{1}{4} \geq \left\| \begin{bmatrix} s_i - \dfrac{1}{4} \\ c_{\mathrm{q}i} \end{bmatrix} \right\| \quad (\forall i), \\ & \dfrac{y_{\mathrm{a}i}}{2k_{\mathrm{a}i}} + 1 \geq \left\| \begin{bmatrix} \dfrac{y_{\mathrm{a}i}}{2k_{\mathrm{a}i}} - 1 \\ c_{\mathrm{a}i} \end{bmatrix} \right\| \quad (\forall i), \\ & c_{\mathrm{q}i} + c_{\mathrm{a}i} \geq \|B_i \boldsymbol{\varphi}\| - \underline{l}_i \quad (\forall i), \\ & \underline{c}_{\mathrm{y}i} - c_{\mathrm{q}i} \geq 0, \; c_{\mathrm{a}i} \geq 0 \quad (\forall i), \\ & \varphi_j = \underline{\varphi}_j, \quad j \in \Gamma_{\mathrm{D}}, \end{array} \right\} \quad (4.73)$$

where $\boldsymbol{\varphi}$, $\boldsymbol{c}_{\mathrm{a}}$, $\boldsymbol{c}_{\mathrm{q}}$, $\boldsymbol{y}_{\mathrm{a}}$, $\boldsymbol{y}_{\mathrm{q}}$, and $\boldsymbol{s}$ are the variables for optimization. Problem (4.73) is clearly an SOCP problem.

Remark 4.3.3. Although only a piecewise-linear law was investigated in section 4.3.1 and a piecewise-quadratic one in section 4.3.2, an SOCP optimization formulation as seen in (4.73) can be enjoyed by more general nonlinear constitutive law for cable members. More precisely, if $\hat{q}_{\mathrm{cab}}$ is monotone and represented as a piecewise-polynomial function, then the corresponding problem $(\overline{\mathrm{PE}})$ can be rewritten as an SOCP problem; see Kanno–Ohsaki–Ito [208, section 4] for details. Notice here that the monotonicity of $\hat{q}_{\mathrm{cab}}$ implies that $\hat{w}_{\mathrm{cab}}$ is convex; see section 2.1.2. ■

4.4 Notes

Most of the discussions made in this chapter is taken from Kanno–Ohsaki–Ito [208], but the presentation was adjusted to the context of the book. The QCQP and SOCP formulations presented in section 4.2.4 for the Green–Lagrange strain are original. An extension of the result in section 4.2.2 to the case in the presence of distributed external loads is from Kanno–Ohsaki [206]; see Cannarozzi [69, 70] for related topics. The SOCP formulation can be extended to the frictionless contact problem of cable networks under considerably mild assumptions; see Kanno–Ohsaki [207].

Since cable is regarded as a one-dimensional structural element in the absence of the distributed load, it serves as one of the simplest subjects in nonsmooth mechanics. A variational inequality approach is in Panagiotopoulos [353, 356], while a numerical solution based on the relaxed strain energy, which was originally considered for membranes by Pipkin [369], is in Atai–Mioduchowski [22], Atai–Steigmann [23], and Haseganu–Steigmann [164]. A dynamic analysis of cables was performed by Feng–Tu [129] based on the complementarity system (or the differential inclusion).

There is a vast literature on numerical methods for equilibrium analysis, as well as design, of cable networks; see the survey by Knudson [231] for developments. The two major categories of those methods are:

- the Newton–Raphson method based on the tangent stiffness matrix
- the dynamic relaxation method

In the Newton–Raphson approach, which owes much to Argyris–Scharpf [21], the nonlinear equilibrium equation is solved incrementally, where unbalanced forces are repeatedly applied to the cable network until the equilibrium is obtained; see also Birnstiel [52], Freire–Negrão–Lopes [135], Fried [137], Kwan [248], and Pevrot–Goulois [363]. The dynamic relaxation method, which was originally considered by Otter [347, 348] for the analysis of prestressed concrete nuclear vessels, was applied to the static analysis of cable networks by Day–Bunce [104] and Lewis–Jones–Rushton [264]; see also Barnes [32] and Lewis [262]. As a method other than these two major ones, the nonlinear programming approach to the analysis of cable networks is in Contro–Maier–Zavelani [95], Coyette–Guisset [98], and Stefanou–Moossavi–Bishop–Koliopoulos [415].

In section 4.3 the material nonlinearity was treated. Elastoplastic analysis is not so common in cable networks, and only a few articles deal with this topic explicitly; see, e.g., Chisaliţa [83], Contro–Maier–Zavelani [95], Greenberg [154], Jonatowski–Birnstiel [198], and Murray–Willems [328].

Chapter 5

Duality in Cable Networks: Principles of Complementary Energy

Orientation Continuing our discussion of cable networks, in this chapter we deal with the variational principles for cable networks. Specifically, we consider the *dual* energy principle, called the *principle of complementary energy*, formulated in terms of static variables, such as stresses, axial forces, and nodal reactions. In contrast, it should be clear that the principle of potential energy is formulated in terms of kinematic variables, such as strains and displacements.

Static and kinematic variables are said to be dual when they are work-conjugate to each other. Furthermore, the minimization problem of the complementary energy (denoted (CE)) is considered to be dual to the minimization problem of the potential energy (denoted (PE)), from the mechanical point of view, because (PE) and (CE) provide the kinematic and static variables at the equilibrium state, respectively. In this chapter, we shall see that the duality theory of convex optimization, introduced in Chapter 2, can indeed serve as a mathematical tool for studying the duality of energy principles in mechanics. Thus the theory of mechanics is closely related to the theory of optimization.

The recipe of our investigation is as follows. In Chapter 4 we showed that (PE) for cable networks can be reformulated as a convex optimization problem subjected to conic constraints, which is denoted by $(\overline{\mathrm{PE}})$. We now reformulate $(\overline{\mathrm{PE}})$ in the form of

$$(\mathrm{P_F}): \quad \inf_{x}\{f(x) + g(\Lambda x) \mid x \in \mathbb{V}\}, \qquad \text{(see (2.28))}$$

which plays the role of a primal problem within the framework of the Fenchel duality theory. Here, $f : \mathbb{V} \to \mathbb{R} \cup \{+\infty\}$ and $g : \mathbb{Y} \to \mathbb{R} \cup \{+\infty\}$ are closed proper convex functions, and $\Lambda : \mathbb{V} \to \mathbb{Y}$ is a linear mapping. From the result in section 2.2, the dual to $(\mathrm{P_F})$ is known to be

$$(\mathrm{P_F^*}): \quad \sup_{z^*}\{-f^*(\Lambda^* z^*) - g^*(-z^*) \mid z^* \in \mathbb{Y}^*\}. \qquad \text{(see (2.31))}$$

By determining the explicit form of $(\mathrm{P_F^*})$, we obtain (CE).

It is to be shown rigorously that the optimal solution of the obtained Fenchel dual problem truly corresponds to the static variables at the equilibrium state. Two key tools for this are the strong duality (in Proposition 2.2.9) and optimality condition (in Proposition 2.2.5) in the framework of the Fenchel duality theory.

5.1 Duality in Cable Networks (1): Large Strain

We start the investigation on dual energy principles with the case of cable networks subjected to large deformation. The corresponding problem for minimizing the potential energy was studied in section 4.2.2, which is formulated as

$$\left.\begin{aligned} \text{(PE)} : \min_{\boldsymbol{\varphi}, \boldsymbol{c}} \ & \sum_{i=1}^{m} \hat{w}_{\mathrm{cab}}^{i}(c_i) - \sum_{j \in \Gamma_{\mathrm{N}}} \underline{p}_j (\varphi_j - x_j) \\ \text{s.t.} \ & c_i = \|B_i \boldsymbol{\varphi}\| - \underline{l}_i, \quad i = 1, \dots, m, \\ & \varphi_j = \underline{\varphi}_j, \quad j \in \Gamma_{\mathrm{D}}. \end{aligned}\right\} \qquad \text{(see (4.41))}$$

The major result of section 4.2.2 is that (PE) is equivalently rewritten as $(\overline{\text{PE}})$ in (4.42), say,

$$\left.\begin{aligned} (\overline{\text{PE}}) : \min_{\boldsymbol{\varphi}, \boldsymbol{c}_{\mathrm{e}}} \ & \sum_{i=1}^{m} \frac{1}{2} k_{\mathrm{e}i} c_{\mathrm{e}i}^2 - \sum_{j \in \Gamma_{\mathrm{N}}} \underline{p}_j (\varphi_j - x_j) \\ \text{s.t.} \ & c_{\mathrm{e}i} \geq \|B_i \boldsymbol{\varphi}\| - \underline{l}_i, \quad i = 1, \dots, m, \\ & \varphi_j = \underline{\varphi}_j, \quad j \in \Gamma_{\mathrm{D}}. \end{aligned}\right\} \qquad \text{(see (4.42))}$$

The relation between (PE) and $(\overline{\text{PE}})$ is stated in Proposition 4.2.1. Note that $(\overline{\text{PE}})$ is a convex optimization problem, because its objective function is a convex quadratic function, and its constraints are second-order cone constraints.

The goal of analysis in this section is to obtain the dual problem of (PE), which corresponds to the minimization problem of the complementary energy, denoted by (CE). To this end, we first derive the problem, denoted $(\overline{\text{CE}})$, which is the Fenchel dual of $(\overline{\text{PE}})$. The strong duality, as well as the optimality condition, is established between $(\overline{\text{PE}})$ and $(\overline{\text{CE}})$. We then formulate (CE) as a problem equivalent to $(\overline{\text{CE}})$.

5.1.1 Embedding to Fenchel form

Firstly, we show that $(\overline{\text{PE}})$ under consideration, say, (4.42), can be embedded into the form of $(\mathrm{P_F})$ in (2.28), which serves as the primal problem within the Fenchel duality theory introduced in section 2.2.5.

Define the primal variable x and related spaces by

$$x = \begin{bmatrix} \boldsymbol{\varphi} \\ \boldsymbol{c}_{\mathrm{e}} \end{bmatrix}, \quad \mathbb{V} = \mathbb{R}^d \times \mathbb{R}^m, \quad \mathbb{Y} = \mathbb{R}^m \times \mathbb{R}^{3m}, \tag{5.1}$$

where $x \in \mathbb{V}$. An element of $\mathbb{Y}$ is written as

$$z = \begin{bmatrix} \boldsymbol{z}_1 \\ \boldsymbol{z}_2 \end{bmatrix} \in \mathbb{Y},$$

where

$$\boldsymbol{z}_1 = (z_{1i} \in \mathbb{R} \mid i = 1, \dots, m), \quad \boldsymbol{z}_2 = (\boldsymbol{z}_{2i} \in \mathbb{R}^3 \mid i = 1, \dots, m).$$

Moreover, define $f : \mathbb{V} \to \mathbb{R} \cup \{+\infty\}$ and $g : \mathbb{Y} \to \mathbb{R} \cup \{+\infty\}$ by

$$f(x) = \begin{cases} \sum_{i=1}^{m} \frac{1}{2} k_{\mathrm{e}i} c_{\mathrm{e}i}^2 - \sum_{j \in \Gamma_{\mathrm{N}}} \underline{p}_j (\varphi_j - x_j) & \text{if } \varphi_j = \underline{\varphi}_j \ (j \in \Gamma_{\mathrm{D}}), \\ +\infty & \text{otherwise}, \end{cases} \tag{5.2}$$

$$g(z) = \begin{cases} 0 & \text{if } z_{1i} + \underline{l}_i \geq \|\boldsymbol{z}_{2i}\| \ (i = 1, \dots, m), \\ +\infty & \text{otherwise}, \end{cases} \tag{5.3}$$

which are easily seen to be proper convex functions. The action of the linear mapping $\Lambda : \mathbb{V} \to \mathbb{Y}$ is defined by

$$\Lambda x = \begin{bmatrix} \boldsymbol{c}_{\mathrm{e}} \\ -B_1 \boldsymbol{\varphi} \\ \vdots \\ -B_m \boldsymbol{\varphi} \end{bmatrix}. \tag{5.4}$$

Note that, from (5.1), the mapping Λ in (5.4) is simply regarded as a matrix $\Lambda \in \mathbb{R}^{(m+3m)\times(d+m)}$ defined by

$$\Lambda = \left[\begin{array}{c|c} O & I_m \\ \hline -B_1 & \\ \vdots & O \\ -B_m & \end{array}\right]. \tag{5.5}$$

With the setting above, the situation in the Fenchel duality is retrieved: ($\overline{\mathrm{PE}}$) in (4.42) is expressed as ($\mathrm{P_F}$) in (2.28). In what follows, we derive an explicit formulation of the dual problem ($\mathrm{P_F^*}$) in (2.31).

5.1.2 Dual problem

For the Fenchel dual problem, the variables spaces are given by

$$\mathbb{V}^* = \mathbb{R}^d \times \mathbb{R}^m, \quad \mathbb{Y}^* = \mathbb{R}^m \times \mathbb{R}^{3m}, \tag{5.6}$$

where $\mathbb{V}^* = \mathbb{V}$ and $\mathbb{Y}^* = \mathbb{Y}$ in (5.1), because they are finite-dimensional vector spaces. We write elements of $\mathbb{V}^*$ and $\mathbb{Y}^*$ as

$$x^* = \begin{bmatrix} \boldsymbol{\varphi}^* \\ \boldsymbol{c}_{\mathrm{e}}^* \end{bmatrix} \in \mathbb{V}^*, \quad z^* = \begin{bmatrix} \boldsymbol{z}_1^* \\ \boldsymbol{z}_2^* \end{bmatrix} \in \mathbb{Y}^*,$$

where $\boldsymbol{z}_1^* = (z_{1i}^* \in \mathbb{R} \mid i = 1, \dots, m)$ and $\boldsymbol{z}_2^* = (\boldsymbol{z}_{2i}^* \in \mathbb{R}^3 \mid i = 1, \dots, m)$.

We next derive the explicit forms of the conjugates of f, g, and Λ defined by (5.2), (5.3), and (5.4). Among them, derivation of g^* is essentially based on the self-duality of the second-order cone, which is a distinguishing character of our study posed within the context of conic optimization.

Proposition 5.1.1. *For g defined by (5.3), its conjugate function $g^* : \mathbb{Y}^* \to \mathbb{R} \cup \{+\infty\}$ is given by*

$$g^*(z^*) = \begin{cases} -\sum_{i=1}^{m} \underline{l}_i z_{1i}^* & \text{if } -z_{1i}^* \geq \| - \boldsymbol{z}_{2i}^* \| \quad (i = 1, \dots, m), \\ +\infty & \text{otherwise.} \end{cases}$$

Proof. Application of the definition of the conjugate function (see (2.13)) to g in (5.3) yields

$$\begin{aligned} g^*(z^*) &= \sup_{z} \{ \langle z^*, z \rangle - g(z) \mid z \} \\ &= \sup_{z} \{ \langle z^*, z \rangle \mid z \in \operatorname{dom} g \} \\ &= \sum_{i=1}^{m} \sup_{z_{1i}, \boldsymbol{z}_{2i}} \{ z_{1i}^* z_{1i} + \langle \boldsymbol{z}_{2i}^*, \boldsymbol{z}_{2i} \rangle \mid z_{1i} + \underline{l}_i \geq \| \boldsymbol{z}_{2i} \| \} \\ &= -\sum_{i=1}^{m} \inf_{z_{1i}, \boldsymbol{z}_{2i}} \left\{ \begin{bmatrix} -z_{1i}^* \\ -\boldsymbol{z}_{2i}^* \end{bmatrix} \cdot \begin{bmatrix} z_{1i} + \underline{l}_i \\ \boldsymbol{z}_{2i} \end{bmatrix} \mid z_{1i} + \underline{l}_i \geq \| \boldsymbol{z}_{2i} \| \right\} - \sum_{i=1}^{m} \underline{l}_i z_{1i}^*. \end{aligned}$$

Then the conclusion is obtained by applying Fact 1.4.4 (ii) to calculate the infimum. □

The explicit forms of f^* and Λ^* are derived through quite simple calculations as follows.

Proposition 5.1.2. *For f and Λ defined by (5.2) and (5.4), respectively, the following statements hold:*

(i) *The conjugate function $f^* : \mathbb{V}^* \to \mathbb{R} \cup \{+\infty\}$ is given by*

$$f^*(x^*) = \begin{cases} \sum_{i=1}^{m} \frac{1}{2} \frac{(c_{\mathrm{e}i}^*)^2}{k_{\mathrm{e}i}} + \sum_{j \in \Gamma_{\mathrm{D}}} \underline{\varphi}_j \varphi_j^* - \sum_{j \in \Gamma_{\mathrm{N}}} \underline{p}_j x_j & \text{if } \varphi_j^* + \underline{p}_j = 0 \ (j \in \Gamma_{\mathrm{N}}), \\ +\infty & \text{otherwise.} \end{cases}$$

(ii) *The action of the adjoint operator $\Lambda^* : \mathbb{Y}^* \to \mathbb{V}^*$ of Λ is defined by*

$$\Lambda^* z^* = \begin{bmatrix} -\sum_{i=1}^{m} B_i^{\mathrm{T}} \boldsymbol{z}_{2i}^* \\ \boldsymbol{z}_1^* \end{bmatrix}.$$

Proof. Write

$$\hat{w}_i(c_{ei}) = \frac{1}{2} k_{ei} c_{ei}^2 \tag{5.7}$$

for simplicity. To see (i), we begin by applying the conjugate transformation (5.8) to definition (5.2) of f as

$$\begin{aligned} f^*(x^*) &= \sup\{\langle x^*, x\rangle - f(x) \mid x \in \mathbb{V}\} \\ &= \sup_{\boldsymbol{\varphi}, \boldsymbol{c}_{\mathrm{e}}} \left\{ \begin{bmatrix} \boldsymbol{\varphi}^* \\ \boldsymbol{c}_{\mathrm{e}}^* \end{bmatrix} \cdot \begin{bmatrix} \boldsymbol{\varphi} \\ \boldsymbol{c}_{\mathrm{e}} \end{bmatrix} - \sum_{i=1}^{m} \hat{w}_i(c_{ei}) + \sum_{j \in \Gamma_{\mathrm{N}}} \underline{p}_j(\varphi_j - x_j) \mid \begin{bmatrix} \boldsymbol{\varphi} \\ \boldsymbol{c}_{\mathrm{e}} \end{bmatrix} \in \operatorname{dom} f \right\} \\ &= \sup_{\boldsymbol{c}_{\mathrm{e}}} \{\langle \boldsymbol{c}_{\mathrm{e}}^*, \boldsymbol{c}_{\mathrm{e}}\rangle - \sum_{i=1}^{m} \hat{w}_i(c_{ei}) \mid \boldsymbol{c}_{\mathrm{e}} \in \mathbb{R}^m\} \\ &\quad + \sup_{\boldsymbol{\varphi}} \left\{ \langle \boldsymbol{\varphi}^*, \boldsymbol{\varphi}\rangle + \sum_{j \in \Gamma_{\mathrm{N}}} \underline{p}_j \varphi_j \mid \varphi_j = \underline{\varphi}_j \ (j \in \Gamma_{\mathrm{D}}) \right\} - \sum_{j \in \Gamma_{\mathrm{N}}} \underline{p}_j x_j . \end{aligned} \tag{5.8}$$

The first term of (5.8) is further reduced to

$$\begin{aligned} &\sup_{\boldsymbol{c}_{\mathrm{e}}} \{\langle \boldsymbol{c}_{\mathrm{e}}^*, \boldsymbol{c}_{\mathrm{e}}\rangle - \sum_{i=1}^{m} \hat{w}_i(c_{ei}) \mid \boldsymbol{c}_{\mathrm{e}} \in \mathbb{R}^m\} \\ &\qquad = \sum_{i=1}^{m} \sup_{c_{ei}} \{\langle c_{ei}^*, c_{ei}\rangle - \hat{w}_i(c_{ei}) \mid c_{ei} \in \mathbb{R}\} \\ &\qquad = \sum_{i=1}^{m} \hat{w}_i^*(c_{ei}^*), \end{aligned} \tag{5.9}$$

where $\hat{w}_i^*$ is the conjugate function of $\hat{w}_i$. From definition (5.7) of $\hat{w}_i$, we immediately obtain

$$\hat{w}_i^*(c_{ei}^*) = \frac{1}{2} \frac{(c_{ei}^*)^2}{k_{ei}}.$$

The second term of (5.8) can be calculated as

$$\begin{aligned} &\sup_{\boldsymbol{\varphi}} \left\{ \langle \boldsymbol{\varphi}^*, \boldsymbol{\varphi}\rangle + \sum_{j \in \Gamma_{\mathrm{N}}} \underline{p}_j \varphi_j \mid \varphi_j = \underline{\varphi}_j \ (j \in \Gamma_{\mathrm{D}}) \right\} \\ &= \sup_{\varphi_j \ (j \in \Gamma_{\mathrm{N}})} \left\{ \sum_{j \in \Gamma_{\mathrm{D}}} \varphi_j^* \underline{\varphi}_j + \sum_{j \in \Gamma_{\mathrm{N}}} (\varphi_j^* + \underline{p}_j) \varphi_j \mid \varphi_j \in \mathbb{R} \ (j \in \Gamma_{\mathrm{N}}) \right\} \\ &= \sum_{j \in \Gamma_{\mathrm{D}}} \varphi_j^* \underline{\varphi}_j + \sum_{j \in \Gamma_{\mathrm{N}}} \sup_{\varphi_j} \{(\varphi_j^* + \underline{p}_j) \varphi_j \mid \varphi_j \in \mathbb{R}\} \\ &= \begin{cases} \displaystyle\sum_{j \in \Gamma_{\mathrm{D}}} \varphi_j^* \underline{\varphi}_j & \text{if } \varphi_j^* + \underline{p}_j = 0 \ (j \in \Gamma_{\mathrm{N}}), \\ +\infty & \text{otherwise.} \end{cases} \end{aligned} \tag{5.10}$$

Substituting (5.9) and (5.10) into (5.8) results in assertion (i).

To see (ii), recall observation in (5.4) and (5.5), which asserts that Λ is regarded as a conventional matrix. Thus the adjoint of Λ is identified with the transpose of Λ in the sense of a matrix, say,

$$\Lambda^* = \Lambda^{\mathrm{T}} = \left[\begin{array}{c|c} O & -B_1^{\mathrm{T}} \cdots -B_m^{\mathrm{T}} \\ \hline I_m & O \end{array}\right],$$

from which we immediately obtain assertion (ii). □

For writing an explicit form of the Fenchel dual problem, given abstractly in (2.31), we see from Proposition 5.1.2 that

$$g^*(-z^*) = \begin{cases} \sum_{i=1}^{m} \underline{l}_i z_{1i}^* & \text{if } z_{1i}^* \geq \|\boldsymbol{z}_{2i}^*\| \ (i = 1, \dots, m), \\ +\infty & \text{otherwise}. \end{cases} \tag{5.11}$$

Concerning $f^*(\Lambda^* z^*)$, we introduce new variables $\boldsymbol{t} \in \mathbb{R}^d$ by

$$\boldsymbol{t} = \sum_{i=1}^{m} B_i^{\mathrm{T}} \boldsymbol{z}_{2i}^*, \tag{5.12}$$

with which the result of Proposition 5.1.1 (ii) is simplified as

$$\Lambda^* z^* = \begin{bmatrix} -\boldsymbol{t} \\ \boldsymbol{z}_1^* \end{bmatrix}.$$

Substituting this into the result of Proposition 5.1.1 (ii) yields

$$f^*(\Lambda^* z^*) = \begin{cases} \sum_{i=1}^{m} \frac{1}{2} \frac{(z_{1i}^*)^2}{k_{\mathrm{e}i}} - \sum_{j \in \Gamma_{\mathrm{D}}} \underline{\varphi}_j t_j - \sum_{j \in \Gamma_{\mathrm{N}}} \underline{p}_j x_j & \text{if } t_j = \underline{p}_j \ (j \in \Gamma_{\mathrm{N}}), \\ +\infty & \text{otherwise}, \end{cases} \tag{5.13}$$

where (5.12) should be satisfied. By rewriting the dual variables as[1]

$$q_i = z_{1i}^*, \quad \boldsymbol{v}_i = \boldsymbol{z}_{2i}^*, \tag{5.14}$$

[1]The mechanical interpretation of these dual variables shall be formally given through the investigation in section 5.1.3. Indeed, we shall see that q_i and $\boldsymbol{v}_i$ correspond to the axial force and the internal force vector of the ith member, respectively.

and by using (5.11) and (5.13), $(\mathrm{P}_{\mathrm{F}}^{*})$ in (2.31) is explicitly written as

$$\left.\begin{aligned}
(\overline{\mathrm{CE}}) : \max_{\boldsymbol{q},\boldsymbol{v},\boldsymbol{t}} \ & -\sum_{i=1}^{m} \frac{1}{2}\frac{q_i^2}{k_{\mathrm{e}i}} - \sum_{i=1}^{m} \underline{l}_i q_i + \sum_{j\in\Gamma_{\mathrm{D}}} \underline{\varphi}_j t_j + \sum_{j\in\Gamma_{\mathrm{N}}} \underline{p}_j x_j \\
\mathrm{s.t.} \ & \sum_{i=1}^{m} B_i^{\mathrm{T}} \boldsymbol{v}_i = \boldsymbol{t}, \\
& q_i \geq \|\boldsymbol{v}_i\|, \quad i = 1,\dots,m, \\
& t_j = \underline{p}_j, \quad j \in \Gamma_{\mathrm{N}}.
\end{aligned}\right\} \qquad (5.15)$$

In the objective function of this problem, the term $\frac{1}{2}(q_i^2/k_{\mathrm{e}i})$ corresponds to the complementary stored energy function of a truss member, in accordance with the fact that the objective function of $(\overline{\mathrm{PE}})$ involves the stored energy function $\frac{1}{2}k_{\mathrm{e}i}c_{\mathrm{e}i}^2$. As investigated in detail in section 4.1.4, a truss member serves as a reference elastic body of a cable member. Thence, the conic optimization formulation $(\overline{\mathrm{PE}})$ for the minimization of the potential energy of a cable network is obtained by replacing the stored energy $\hat{w}_{\mathrm{cab}}^i$ in (PE) with $\frac{1}{2}k_{\mathrm{e}i}c_{\mathrm{e}i}^2$ for the reference elastic body and replacing the compatibility relation in (PE) with an inequality constraint. Similarly, $(\overline{\mathrm{CE}})$ involves the complementary stored energy $\frac{1}{2}(q_i^2/k_{\mathrm{e}i})$ as well as the inequality constraints $q_i \geq \|\boldsymbol{v}_i\|$. Thus a truss member serves as a reference elastic body also in $(\overline{\mathrm{CE}})$.

5.1.3 Duality and optimality

We now have the primal-dual pair of conic optimization problems: $(\overline{\mathrm{PE}})$ in (4.42) and $(\overline{\mathrm{CE}})$ in (5.15). By applying Proposition 2.2.9, strong duality is established between $(\overline{\mathrm{PE}})$ and $(\overline{\mathrm{CE}})$ as follows.

Proposition 5.1.3. *Suppose that there exists a vector* $\boldsymbol{v} = (\boldsymbol{v}_i \in \mathbb{R}^3 \mid i = 1,\dots,m)$ *satisfying*

$$\left(\sum_{i=1}^{m} B_i^{\mathrm{T}} \boldsymbol{v}_i\right)_j = \underline{p}_j, \quad j \in \Gamma_{\mathrm{N}}.$$

Then, $(\overline{\mathrm{PE}})$ *and* $(\overline{\mathrm{CE}})$ *each has an optimal solution, and* $\min(\overline{\mathrm{PE}}) = \max(\overline{\mathrm{CE}})$.

Proof. The assertion is proven by showing that $(\overline{\mathrm{PE}})$ and $(\overline{\mathrm{CE}})$ satisfy the assumptions required in Proposition 2.2.12, because $(\overline{\mathrm{PE}})$ and $(\overline{\mathrm{CE}})$ correspond to $(\mathrm{P}_{\mathrm{F}})$ and $(\mathrm{P}_{\mathrm{F}}^{*})$, respectively.

From the definitions of f and g (see (5.2) and (5.3)), we immediately obtain

$$\mathrm{ri}(\mathrm{dom}\, f) = \{(\boldsymbol{\varphi}, \boldsymbol{c}_{\mathrm{e}}) \mid \varphi_j = \underline{\varphi}_j \ (j \in \Gamma_{\mathrm{D}})\},$$

$$\mathrm{ri}(\mathrm{dom}\, g) = \{(\boldsymbol{z}_1, \boldsymbol{z}_2) \mid z_{1i} + \underline{l}_i > \|\boldsymbol{z}_{2i}\| \ (i = 1,\dots,m)\}.$$

From the definition of Λ in (5.4), assumption (i) required in Proposition 2.2.12, say,

$$\exists x \in \mathrm{ri}(\mathrm{dom}\, f): \quad \Lambda x \in \mathrm{ri}(\mathrm{dom}\, g), \tag{5.16}$$

can be interpreted in our situation that there exists $(\boldsymbol{\varphi}, \boldsymbol{c}_{\mathrm{e}})$ satisfying

$$\varphi_j = \underline{\varphi}_j, \quad j \in \Gamma_{\mathrm{D}}, \tag{5.17}$$

$$c_{\mathrm{e}i} + \underline{l}_i > \|B_i \boldsymbol{\varphi}\|, \quad i = 1, \dots, m. \tag{5.18}$$

It is easy to see for any $\boldsymbol{\varphi}$ satisfying (5.17) that there always exists $\boldsymbol{c}_{\mathrm{e}}$ satisfying (5.18), because c_{ei} is not bounded from above. Thus (5.16) is always satisfied.

Finally, consider assumption (ii) of Proposition 2.2.12, say,

$$\exists z^* \in \mathrm{ri}(\mathrm{dom}\, g^*): \quad -\Lambda^* z^* \in \mathrm{ri}(\mathrm{dom}\, f^*). \tag{5.19}$$

Recall the explicit forms of f^* and g^* (see Proposition 5.1.1 and Proposition 5.1.2 (i)) to obtain

$$\mathrm{ri}(\mathrm{dom}\, f^*) = \{(\boldsymbol{\varphi}^*, \boldsymbol{c}_{\mathrm{e}}^*) \mid \varphi_j^* + \underline{p}_j = 0 \ (j \in \Gamma_{\mathrm{N}})\},$$
$$\mathrm{ri}(\mathrm{dom}\, g^*) = \{(-\boldsymbol{z}_1^*, -\boldsymbol{z}_2^*) \mid z_{1i}^* > \|\boldsymbol{z}_{2i}^*\| \ (i = 1, \dots, m)\}.$$

From the action of Λ^* (see Proposition 5.1.2 (ii)), (5.19) requires that there exists $(\boldsymbol{q}, \boldsymbol{v})$ satisfying

$$\Big(\sum_{i=1}^m B_i^{\mathrm{T}} \boldsymbol{v}_i\Big)_j = \underline{p}_j, \quad j \in \Gamma_{\mathrm{N}}, \tag{5.20}$$

$$q_i > \|\boldsymbol{v}_i\|, \quad i = 1, \dots, m. \tag{5.21}$$

Suppose that (5.20) has a solution. Then, for any $\boldsymbol{v}$ satisfying (5.20), we can always choose $\boldsymbol{q}$ satisfying (5.21), which concludes the proof. □

We next derive the optimality conditions for ($\overline{\mathrm{PE}}$) and ($\overline{\mathrm{CE}}$) by applying Proposition 2.2.9.

Proposition 5.1.4. *Under the hypotheses in Proposition 5.1.3, the following statements hold:*

(i) $(\bar{\boldsymbol{\varphi}}, \bar{\boldsymbol{c}}_{\mathrm{e}})$ *and* $(\bar{\boldsymbol{q}}, \bar{\boldsymbol{v}}, \bar{\boldsymbol{t}})$ *are optimal for* ($\overline{\mathrm{PE}}$) *and* ($\overline{\mathrm{CE}}$), *respectively, if and only if they satisfy*

$$\bar{q}_i = k_{\mathrm{e}i} \bar{c}_{\mathrm{e}i}, \quad i = 1, \dots, m, \tag{5.22a}$$

$$\sum_{i=1}^m B_i^{\mathrm{T}} \bar{\boldsymbol{v}}_i = \bar{\boldsymbol{t}}, \tag{5.22b}$$

$$\bar{\varphi}_j = \underline{\varphi}_j, \quad j \in \Gamma_{\mathrm{D}}; \qquad \bar{t}_j = \underline{p}_j, \quad j \in \Gamma_{\mathrm{N}}, \tag{5.22c}$$

$$\bar{c}_{\mathrm{e}i} + \underline{l}_i \geq \|-B_i \bar{\boldsymbol{\varphi}}\|, \ \bar{q}_i \geq \|\bar{\boldsymbol{v}}_i\|, \quad i = 1, \dots, m, \tag{5.22d}$$

$$\begin{bmatrix} \bar{c}_{\mathrm{e}i} + \underline{l}_i \\ -B_i \bar{\boldsymbol{\varphi}} \end{bmatrix} \cdot \begin{bmatrix} \bar{q}_i \\ \bar{\boldsymbol{v}}_i \end{bmatrix} = 0, \quad i = 1, \dots, m. \tag{5.22e}$$

(ii) *Optimal solutions* $(\bar{\boldsymbol{\varphi}}, \bar{\boldsymbol{c}}_{\mathrm{e}})$ *and* $(\bar{\boldsymbol{q}}, \bar{\boldsymbol{v}}, \bar{\boldsymbol{t}})$ *of* $(\overline{\mathrm{PE}})$ *and* $(\overline{\mathrm{CE}})$ *satisfy*

$$\bar{c}_{\mathrm{e}i} - \hat{c}_i(\bar{\boldsymbol{\varphi}}) \geq 0, \quad \bar{q}_i \geq 0, \quad (\bar{c}_{\mathrm{e}i} - \hat{c}_i(\bar{\boldsymbol{\varphi}}))\bar{q}_i = 0 \tag{5.23}$$

for each $i = 1, \dots, m$, *where* $\hat{c}_i(\bar{\boldsymbol{\varphi}}) = \|B_i \bar{\boldsymbol{\varphi}}\| - \underline{l}_i$.

Proof. To see the assertion (i), we apply the expression of the optimality condition in Proposition 2.2.9 (b) for $(\mathrm{P_F})$ and $(\mathrm{P_F^*})$ to the pair $(\overline{\mathrm{PE}})$ and $(\overline{\mathrm{CE}})$.

For the first equality in Proposition 2.2.9 (b), recall that f, f^*, and Λ^* are given by (5.2) and Proposition 5.1.2, where the variables are defined by (5.1), (5.12), and (5.14). Then we see that $f(x) + f^*(\Lambda^* z^*)$ is finite if and only if (5.22b) and (5.22c) are satisfied. If that is the case we obtain

$$f(x) = \sum_{i=1}^{m} \frac{1}{2} k_{\mathrm{e}i} c_{\mathrm{e}i}^2 - \sum_{j \in \Gamma_{\mathrm{N}}} \underline{p}_j (\varphi_j - x_j),$$

$$f^*(\Lambda^* z^*) = \sum_{i=1}^{m} \frac{1}{2} \frac{q_i^2}{k_{\mathrm{e}i}} - \sum_{j \in \Gamma_{\mathrm{D}}} \underline{\varphi}_j t_j - \sum_{j \in \Gamma_{\mathrm{N}}} \underline{p}_j x_j.$$

Simultaneously, under (5.22b) and (5.22c), we have

$$\begin{aligned}\langle \Lambda^* z^*, x \rangle &= \begin{bmatrix} -\sum_{i=1}^{m} B_i^{\mathrm{T}} \boldsymbol{v}_i \\ \boldsymbol{q} \end{bmatrix} \cdot \begin{bmatrix} \boldsymbol{\varphi} \\ \boldsymbol{c}_{\mathrm{e}} \end{bmatrix} = \begin{bmatrix} -\boldsymbol{t} \\ \boldsymbol{q} \end{bmatrix} \cdot \begin{bmatrix} \boldsymbol{\varphi} \\ \boldsymbol{c}_{\mathrm{e}} \end{bmatrix} \\ &= \sum_{i=1}^{m} q_i c_{\mathrm{e}i} + \sum_{j \in \Gamma_{\mathrm{D}}} t_j \underline{\varphi}_j - \sum_{j \in \Gamma_{\mathrm{N}}} \underline{p}_j \varphi_j.\end{aligned}$$

Consequently, $f(x) + f^*(\Lambda^* z^*) = \langle \Lambda^* z^*, x \rangle$ is equivalent to

$$\sum_{i=1}^{m} \frac{1}{2} k_{\mathrm{e}i} c_{\mathrm{e}i}^2 + \sum_{i=1}^{m} \frac{1}{2} \frac{q_i^2}{k_{\mathrm{e}i}} = \sum_{i=1}^{m} q_i c_{\mathrm{e}i}.$$

Rewrite this equation as

$$\sum_{i=1}^{m} \left(\frac{1}{2} k_{\mathrm{e}i} c_{\mathrm{e}i}^2 + \sum_{i=1}^{m} \frac{1}{2} \frac{q_i^2}{k_{\mathrm{e}i}} - q_i c_{\mathrm{e}i} \right) = \frac{1}{2k_{\mathrm{e}i}} \sum_{i=1}^{m} (k_{\mathrm{e}i} c_{\mathrm{e}i} - q_i)^2 = 0$$

to obtain (5.22a).

For considering the second equality in Proposition 2.2.9 (b), recall that g, g^*, and Λ are given in (5.3), (5.4), and Proposition 5.1.1. Then it is immediately evident that $g(\Lambda x) + g(-z^*)$ is finite if and only if (5.22d) is satisfied. If that is the case we obtain

$$g(\Lambda x) + g(-z^*) = \sum_{i=1}^{m} \underline{l}_i q_i.$$

On the other hand, the right-hand side of the goal equation in Proposition 2.2.9 (b) is reduced to

$$\langle -z^*, \Lambda x\rangle = \begin{bmatrix} -\boldsymbol{q} \\ -\boldsymbol{v}_1 \\ \vdots \\ -\boldsymbol{v}_m \end{bmatrix} \cdot \begin{bmatrix} \boldsymbol{c}_{\rm e} \\ -B_1\boldsymbol{\varphi} \\ \vdots \\ -B_m\boldsymbol{\varphi} \end{bmatrix} = -\sum_{i=1}^{m} \left(\begin{bmatrix} q_i \\ \boldsymbol{v}_i \end{bmatrix} \cdot \begin{bmatrix} c_{{\rm e}i} \\ -B_i\boldsymbol{\varphi} \end{bmatrix} \right).$$

Thus the equality $g(\Lambda x) + g(-z^*) = \langle -z^*, \Lambda x\rangle$ under consideration is equivalent to (5.22e).

Concerning assertion (ii), observe that $(\bar{\boldsymbol{\varphi}}, \bar{\boldsymbol{c}}_{\rm e})$ satisfies (4.31) in Proposition 4.2.1, because it is optimal for $(\overline{\rm PE})$. From (5.22a), we see that $\bar{c}_{{\rm e}i} > 0$ ($\bar{c}_{{\rm e}i} = 0$, resp.) implies $\bar{q}_i > 0$ ($\bar{q}_i = 0$, resp.). In conjunction with (4.31), we have

$$\bar{c}_{{\rm e}i} - \hat{c}_i(\bar{\boldsymbol{\varphi}}) \in \begin{cases} \{0\} & \text{if } \hat{c}_i(\bar{\boldsymbol{u}}) \geq 0, \\ [0, +\infty[& \text{otherwise}, \end{cases}$$

$$\bar{q}_i \in \begin{cases} [0, +\infty[& \text{if } \hat{c}_i(\bar{\boldsymbol{u}}) \geq 0, \\ \{0\} & \text{otherwise}, \end{cases}$$

and hence (5.23) is satisfied. □

To further explore the implication of the optimality condition in (5.22), we next investigate the property of the optimal solution of $(\overline{\rm CE})$.

Proposition 5.1.5. *If $(\bar{\boldsymbol{q}}, \bar{\boldsymbol{v}}, \bar{\boldsymbol{t}})$ is optimal for $(\overline{\rm CE})$, then it satisfies*

$$\bar{q}_i = \|\bar{\boldsymbol{v}}_i\|, \quad i = 1, \dots, m. \tag{5.24}$$

Proof. Observe that $(\overline{\rm CE})$ is a maximization problem, and the variable q_i appears as the term $-\underline{l}_i q_i$ in the objective function. Moreover, q_i is subjected solely to the constraint

$$q_i \geq \|\boldsymbol{v}_i\|. \tag{5.25}$$

Therefore, q_i is maximized at the optimal solution of $(\overline{\rm CE})$, as far as (5.25) is satisfied, which concludes the proof. □

Since (5.22a) corresponds to the constitutive law, the variable q_i can be interpreted as the axial force of the ith member. Proposition 5.1.5 implies that, at the optimal solution, the modulus of $\boldsymbol{v}_i$ becomes equal to the axial force. For perceiving the mechanical meaning of the vector $\boldsymbol{v}_i$, its direction is determined as follows.

Proposition 5.1.6. *Suppose that $(\bar{\boldsymbol{\varphi}}, \bar{\boldsymbol{c}}_{\rm e})$ and $(\bar{\boldsymbol{q}}, \bar{\boldsymbol{v}}, \bar{\boldsymbol{t}})$ are optimal for $(\overline{\rm PE})$ and $(\overline{\rm CE})$, respectively. Then they satisfy*

$$\bar{\boldsymbol{v}}_i = \bar{q}_i \frac{B_i\bar{\boldsymbol{\varphi}}}{\|B_i\bar{\boldsymbol{\varphi}}\|}, \quad i = 1, \dots, m,$$

where we define $\boldsymbol{0}/0 = \boldsymbol{0}$.

Proof. This assertion follows from Proposition 5.1.5 and the property of the complementarity condition over the second-order cone investigated in section 1.4.3 (see Table 1.4 for a summary; also (1.23)–(1.25c)).

Among the optimality conditions established in Proposition 5.1.4, we see that (5.22d) and (5.22e) coincide with the complementarity condition over the second-order cone. It follows from the results in section 1.4.3 that, when (5.22d) holds, condition (5.22e) is satisfied if and only if

$$\begin{aligned}
&\bar{c}_{ei} + \underline{l}_i > \|-B_i\bar{\boldsymbol{\varphi}}\| && \Rightarrow \quad \bar{q}_i = \|\bar{\boldsymbol{v}}_i\| = 0, \\
&\bar{c}_{ei} + \underline{l}_i = \|-B_i\bar{\boldsymbol{\varphi}}\|, \ \bar{q}_i = \|\bar{\boldsymbol{v}}_i\| && \Rightarrow \quad \|-B_i\bar{\boldsymbol{\varphi}}\|\bar{\boldsymbol{v}}_i + \|\bar{\boldsymbol{v}}_i\|(-B_i\bar{\boldsymbol{\varphi}}) = \mathbf{0}
\end{aligned}$$

hold. In the former case, the assertion of the proposition is satisfied because $\bar{\boldsymbol{v}}_i = \mathbf{0}$. Alternatively, in the latter case, applying Proposition 5.1.5 yields

$$\|B_i\bar{\boldsymbol{\varphi}}\|\bar{\boldsymbol{v}}_i = \bar{q}_i(B_i\bar{\boldsymbol{\varphi}}),$$

which concludes the proof. □

The mechanical meaning of $\bar{\boldsymbol{v}}_i$ is now clearly explained as follows. Proposition 5.1.6 implies that, at the optimal solutions of $(\overline{\text{PE}})$ and $(\overline{\text{CE}})$, the vector $\bar{\boldsymbol{v}}_i$ becomes parallel with $B_i\bar{\boldsymbol{\varphi}}$. This means that $\bar{\boldsymbol{v}}_i$ is parallel with member i at the deformed configuration, because so is $B_i\bar{\boldsymbol{\varphi}}$ (see Remark 4.2.4). Consequently, $\bar{\boldsymbol{v}}_i$ is interpreted as the internal force vector at the deformed state. Thus, for the the kinematic variables, say, $B_1\boldsymbol{\varphi}, \dots, B_m\boldsymbol{\varphi}$, and the static variables, say, $\boldsymbol{v}_1, \dots, \boldsymbol{v}_m$, the compatibility of the directions is guaranteed at the optimal solutions of $(\overline{\text{PE}})$ and $(\overline{\text{CE}})$. This is an imperative in establishing the variational principles in large deformation theory, and it should be emphasized that this result is successfully obtained through the analysis within the frameworks of conic optimization.

5.1.4 Principle of complementary energy

In light of the analysis in section 5.1.1 through section 5.1.3, we can now establish the principle of complementary energy for cable networks subjected to large deformation.

Since the principle of complementary energy is usually stated in the form of a minimum principle, we transform $(\overline{\text{CE}})$ in (5.15) to a minimization problem by changing the sign of the objective function. Moreover, by substituting the result of Proposition 5.1.5 to $(\overline{\mathrm{CE}})$, we obtain[2]

$$\left.\begin{aligned}
(\mathrm{CE}):\ \min_{\boldsymbol{q},\boldsymbol{v},\boldsymbol{t}}\ & \sum_{i=1}^{m}\frac{1}{2}\frac{q_i^2}{k_{\mathrm{e}i}}+\sum_{i=1}^{m}\underline{l}_i\|\boldsymbol{v}_i\|-\sum_{j\in\Gamma_{\mathrm{D}}}\underline{\varphi}_j t_j-\sum_{j\in\Gamma_{\mathrm{N}}}\underline{p}_j x_j\\
\mathrm{s.\,t.}\ & \sum_{i=1}^{m}B_i^{\mathrm{T}}\boldsymbol{v}_i=\boldsymbol{t},\\
& q_i=\|\boldsymbol{v}_i\|,\quad i=1,\dots,m,\\
& t_j=\underline{p}_j,\quad j\in\Gamma_{\mathrm{N}}.
\end{aligned}\right\}\qquad(5.26)$$

In the following, we show that the optimal solution of (CE) indeed corresponds to the static variables in the equilibrium state. To see this, we demonstrate the solution property of the two energy minimization problems, say, (PE) in (4.41) and (CE) in (5.26), by using the optimality conditions for their convex formulations, say, $(\overline{\mathrm{PE}})$ in (4.42) and $(\overline{\mathrm{CE}})$ in (5.15).

Proposition 5.1.7. *Under the hypotheses in Proposition 5.1.3,* $(\bar{\boldsymbol{\varphi}},\bar{\boldsymbol{c}})$ *and* $(\bar{\boldsymbol{q}},\bar{\boldsymbol{v}},\bar{\boldsymbol{t}})$ *are optimal solutions of* (PE) *and* (CE), *respectively, if and only if they correspond to the kinematic and static variables*[3] *at the equilibrium state of the cable network. Moreover,* $\min(\mathrm{PE})=-\min(\mathrm{CE})$.

Proof. The proof is essentially based on the optimality conditions for $(\overline{\mathrm{PE}})$ and $(\overline{\mathrm{CE}})$ given in Proposition 5.1.4, say, (5.22).

The relation between optimal solutions of $(\overline{\mathrm{PE}})$ and (PE) is stated in Proposition 4.2.1. We thus see that, instead of (5.22a) and (5.22d), the optimal solution of (PE) satisfies $\bar{q}_i=\hat{q}^i_{\mathrm{cab}}(\bar{c}_i)$ and $\bar{c}_i=\hat{c}_i(\bar{\boldsymbol{\varphi}})$. On the other hand, it is clear from the construction procedure of (CE) that $(\overline{\mathrm{PE}})$ and (CE) share the same optimal solution. Therefore, the optimal solution of (CE) also satisfies the condition stated in Proposition 5.1.6. Conversely, this condition implies (5.22d) and (5.22e), because Proposition 5.1.6 is a consequence of (5.22d) and (5.22e). Consequently, necessary and sufficient conditions for optimality of (PE) and (CE) are

[2]Since any feasible solution of (5.26) satisfies $q_i\geq 0$, the term $q_i^2/(2k_{\mathrm{e}i})$ in the objective function can be replaced with $\hat{w}^{i*}_{\mathrm{cab}}(q_i)$ defined by (see (4.9))

$$\hat{w}^{i*}_{\mathrm{cab}}(q_i)=\begin{cases}q_i^2/(2k_{\mathrm{e}i}) & \text{if } q_i\geq 0,\\ +\infty & \text{if } q_i<0.\end{cases}$$

The duality between (PE) and (CE) may be captured more clearly with this $\hat{w}^{i*}_{\mathrm{cab}}(q_i)$, because $\hat{w}^{i*}_{\mathrm{cab}}$ is the conjugate function of $\hat{w}^{i}_{\mathrm{cab}}$ in (PE).

[3]As we discussed in section 4.2.1 and section 5.1.3, $\boldsymbol{\varphi}$ and $\boldsymbol{c}$ are the nodal location vector and the member elongation vector in the Biot strain measure, respectively, while q_i, $\boldsymbol{v}_i$, and $\boldsymbol{t}$ are the member axial force, the internal force vector in the Biot stress measure, and the nodal force vector, respectively.

written as

$$\bar{q}_i = \hat{q}^i_{\rm cab}(\bar{c}_i), \quad i = 1, \dots, m, \tag{5.27}$$

$$\bar{c}_i = \|B_i \bar{\boldsymbol{\varphi}}\| - \underline{l}_i, \quad i = 1, \dots, m, \tag{5.28}$$

$$\bar{\boldsymbol{v}}_i = \bar{q}_i \frac{B_i \bar{\boldsymbol{\varphi}}}{\|B_i \bar{\boldsymbol{\varphi}}\|}, \quad i = 1, \dots, m, \tag{5.29}$$

$$\sum_{i=1}^{m} B_i^{\rm T} \bar{\boldsymbol{v}}_i = \bar{\boldsymbol{t}}, \tag{5.30}$$

$$\bar{\varphi}_j = \underline{\varphi}_j, \quad j \in \Gamma_{\rm D}; \qquad \bar{t}_j = \underline{p}_j, \quad j \in \Gamma_{\rm N}, \tag{5.31}$$

where we define $\mathbf{0}/0 = \mathbf{0}$ in (5.29).

We now clearly see that the system of (5.27)–(5.31) forms the governing equations for the equilibrium of a cable network. Condition (5.27) is the constitutive law of "no-compression" cable members, (5.28) is the compatibility relation of the kinematic variables, (5.29) gives the compatibility of the static variables at the deformed configuration, (5.30) is the force-balance equation, and the boundary conditions are stated in (5.31). This means that finding the equilibrium state of a cable network is equivalent to solving (5.27)–(5.31) and thence solving either (PE) or (CE). □

Proposition 5.1.7 also confirms that the *minimum* principle of potential energy holds for a cable network, because any solution to (5.27)–(5.31) is shown to be a minimizer of (PE). This is a distinguishing property of structures consisting of cables only, while only the *stationarity* principle can be stated for general structures subjected to large deformation (see also Remark 4.2.2).

Concerning (CE) in (5.26), it is noted that only the static variables are involved, while, in general, the equilibrium equation at the deformed configuration, say, (5.30), is written in conjunction with the compatibility relation between the deformation and the static variables, as seen in (5.29). In contrast, such a constraint is not involved in (CE) but is shown to be satisfied at the optimal solution. Thus it is rather surprising that (CE) provides an equilibrium solution correctly although only the equilibrium equation between the internal and external forces is involved as a constraint.

The objective function of (CE) consists of the following four factors:

- $\dfrac{1}{2}\dfrac{\|\boldsymbol{v}_i\|^2}{k_{ei}}$: the complementary stored energy
- $-\sum_{j \in \Gamma_{\rm D}} \underline{\varphi}_j t_j$: the complementary work done by the prescribed displacements
- $-\sum_{j \in \Gamma_{\rm N}} \underline{p}_j x_j$: a constant term depending on the reference configuration
- $\underline{l}_i \|\boldsymbol{v}_i\|$: a characteristic term for the large rotation theory

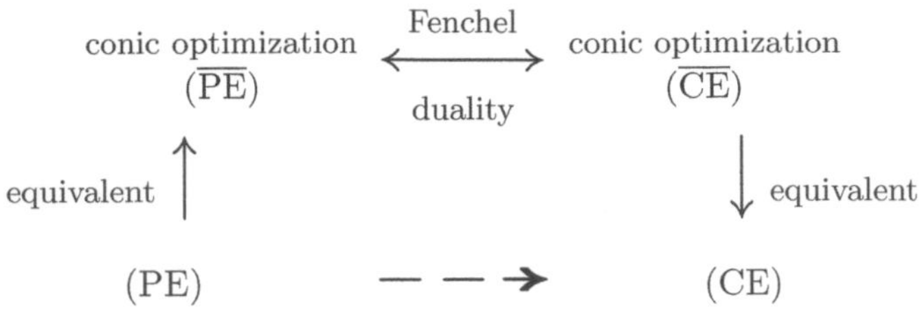

FIGURE 5.1: Relation among variational principles and their conic optimization formulations. (PE): the principle of potential energy; $(\overline{\mathrm{PE}})$: the conic reformulation of (PE); $(\overline{\mathrm{CE}})$: the conic formulation dual to $(\overline{\mathrm{PE}})$; (CE): the principle of complementary energy.

A mechanical interpretation of the term $\underline{l}_i \|\boldsymbol{v}_i\|$ is given in Remark 5.1.11.

Remark 5.1.8. We arrive at the formulation (CE), not directly from (PE) but through the pair $(\overline{\mathrm{PE}})$ and $(\overline{\mathrm{CE}})$. The procedure is summarized in Figure 5.1. This "circumvention" enables us to enjoy some preferable properties of the duality in conic optimization; the duality gap between $(\overline{\mathrm{PE}})$ and $(\overline{\mathrm{CE}})$ vanishes, the necessary and sufficient condition for optimality is provided by the duality theory, and $(\overline{\mathrm{PE}})$ and $(\overline{\mathrm{CE}})$ do not share the same variables. In Part IV we apply the scheme in Figure 5.1 to various problems in nonsmooth mechanics. In Remark 5.1.9, it is examined that a dualization procedure is "directly" applied to (PE) to demonstrate the advantages of (CE) above. ■

Remark 5.1.9. Recall (PE) in (4.41). The Lagrangian $L^{(\mathrm{PE})} : \mathbb{R}^{d+m} \times \mathbb{R}^{m+|\Gamma_{\mathrm{D}}|} \to \mathbb{R}$ of (PE) is defined by

$$
\begin{aligned}
L^{(\mathrm{PE})}(\boldsymbol{\varphi}, \boldsymbol{c}, \boldsymbol{q}, \boldsymbol{t}_{\mathrm{D}}) = \sum_{i=1}^{m} \hat{w}^i_{\mathrm{cab}}(c_i) - \sum_{j \in \Gamma_{\mathrm{N}}} \underline{p}_j (\varphi_j - x_j) \\
- \sum_{i=1}^{m} q_i (c_i - \|B_i \boldsymbol{\varphi}\| + \underline{l}_i) - \sum_{j \in \Gamma_{\mathrm{D}}} t_{\mathrm{D}j} (\varphi_j - \underline{\varphi}_j),
\end{aligned}
\tag{5.32}
$$

where $\boldsymbol{q} \in \mathbb{R}^m$ and $\boldsymbol{t}_{\mathrm{D}} \in \mathbb{R}^{|\Gamma_{\mathrm{D}}|}$ are the Lagrange multipliers. The stationarity condition of $L^{(\mathrm{PE})}$ is written as

$$
q_i = \hat{q}^i_{\mathrm{cab}}(c_i), \quad i = 1, \dots, m, \tag{5.33a}
$$

$$
\sum_{i=1}^{m} B_i^{\mathrm{T}} \left(\frac{B_i \boldsymbol{\varphi}}{\|B_i \boldsymbol{\varphi}\|} q_i \right) = \boldsymbol{t}, \tag{5.33b}
$$

$$
\varphi_j = \underline{\varphi}_j, \quad j \in \Gamma_{\mathrm{D}}; \qquad t_j = \underline{p}_j, \quad j \in \Gamma_{\mathrm{N}}, \tag{5.33c}
$$

$$
c_i + \underline{l}_i = \|B_i \boldsymbol{\varphi}\|, \quad i = 1, \dots, m, \tag{5.33d}
$$

where we define $\mathbf{0}/0 = \mathbf{0}$ in (5.33b). Thus the force-balance equation, (5.33b), involves the unknown deformation $\boldsymbol{\varphi}$, which is as usual in the geometrically nonlinear theory.

Condition (5.33) should be compared with (5.22), i.e., the optimality condition for ($\overline{\text{PE}}$) (and also for ($\overline{\text{CE}}$)). Note that (5.22) corresponds to the stationarity condition of the function

$$L^{(\overline{\text{PE}})}(\boldsymbol{\varphi}, \boldsymbol{c}_{\text{e}}, \boldsymbol{q}, \boldsymbol{v}, \boldsymbol{t}_{\text{D}}) = \begin{cases} \displaystyle\sum_{i=1}^{m} \frac{1}{2} k_{\text{e}i} c_{\text{e}i}^2 - \sum_{j \in \Gamma_{\text{N}}} \underline{p}_j (\varphi_j - x_j) \\ \displaystyle\quad - \sum_{i=1}^{m} \begin{bmatrix} q_i \\ \boldsymbol{v}_i \end{bmatrix} \cdot \begin{bmatrix} c_{\text{e}i} + \underline{l}_i \\ B_i \boldsymbol{\varphi} \end{bmatrix} - \sum_{j \in \Gamma_{\text{D}}} t_{\text{D}j} (\varphi_j - \underline{\varphi}_j) & \text{if } q_i \geq \|\boldsymbol{v}_i\| \ (i = 1, \dots, m), \\ +\infty & \text{otherwise}, \end{cases}$$

which serves as the Lagrangian for ($\overline{\text{PE}}$).

There are some significant differences between these two systems. Among (5.22a)–(5.22e), only (5.22d) and (5.22e) are nonlinear conditions, which form the complementarity conditions over second-order cones. All nonlinear properties coming from the geometrical nonlinearity are "hidden" in (5.22d) and (5.22e), and hence the force-balance equation (5.22b) is linear. The hidden nonlinearity is the compatibility relation between the deformation $\boldsymbol{\varphi}$ and the internal force vector $\boldsymbol{v}_i$. Without using the second-order cone complementarity, this compatibility condition should be written explicitly, which certainly involves both the kinematic variables $\boldsymbol{\varphi}$ and the static ones $\boldsymbol{v}_i$ as seen in (5.29) (note that (5.33b) is a consequence of substitution of (5.29) into (5.30)). Thus, in (5.22) the coupling of the kinematic and static variables is avoided by introducing the second-order cones complementarity. In contrast, it is not possible, or at least not straightforward, to derive a principle involving only static variables from (5.33).

It is not necessarily possible to reformulate the energy principle as a convex optimization problem over cones. In the case of trusses, for example, the minimization problem of the potential energy, i.e., the problem (4.35), is essentially nonconvex. Therefore, the coupling of the kinematic and static variables in the force-balance equation cannot be avoided. What enables us to formulate ($\overline{\text{PE}}$) is the no-compression property of cables. ■

Remark 5.1.10. As commented in Remark 5.1.9, the minimization problem of the potential energy for geometrically nonlinear trusses is essentially nonconvex, and it generally has a stationary point that is not minimal. Therefore, we can state not the *minimum* principle, but any *stationary* point to be considered for equilibrium analysis. Accordingly, the complementary energy principle, in a form similar to (CE) in (5.26) for cable networks, cannot be formulated; more precisely, the minimum principle is no longer true, and moreover, it is impossible (as far as the author's knowledge) to formulate a variational principle involving only the static variables.

Within the geometrically nonlinear framework, the existence of a variational principle in the *truly complementary form*, i.e., the one involves only the static variables, has been a long-time challenging problem in applied mechanics. This dates back most likely to the article by Hellinger [171] in 1914, as pointed out by Koiter [236]. As clearly stated by Koiter [236], in many papers (see, e.g., Levinson [258]) the existence of the inversion of the constitutive law, written in terms of the first Piola–Kirchhoff stress and the displacement gradient, is implicitly assumed, although it is

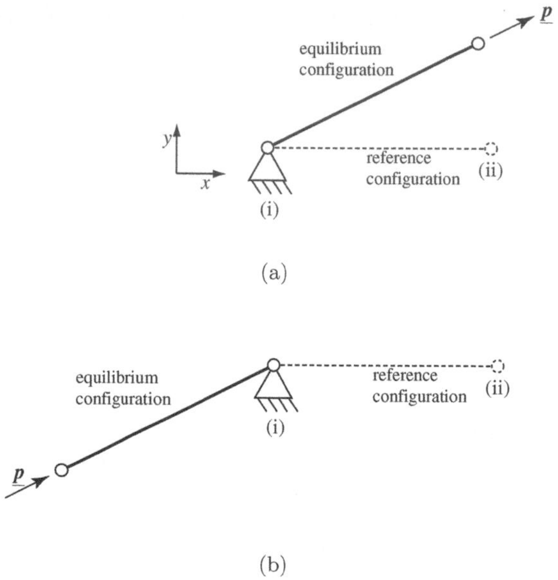

FIGURE 5.2: A counter-example of uniqueness of the equilibrium configuration of trusses. The equilibrium states in (a) and (b) share the same internal force vector $\boldsymbol{v}$.

really a crucial point in this problem; see also Bertóti [50], Campos–Oden [68], Fraeijs de Veubeke [134], Gao [143], Valid [443], and Zubov [473].

Regarding trusses, the stationarity principle involving the unknown axial forces and unknown rotations is from Libov [267] and Jennings [196]. Mikkola [310] aimed to establish a variational principle for trusses in terms of the internal force vectors[4]; however, as a serious drawback, Mikkola's principle inevitably includes indefinite signs in the "energy function" (hence, it is not a single-valued function).

Such an indefiniteness can be understood by using a simple example shown in Figure 5.2. Consider a one-member truss, where node (i) is fixed and node (ii) is free. When the external load $\underline{\boldsymbol{p}}$ is applied at node (i), then the equilibrium configuration of the member is parallel with $\underline{\boldsymbol{p}}$ as illustrated in Figure 5.2(a), where the member has a tensile axial force, say, $q_i = \|\underline{\boldsymbol{p}}\|$. This equilibrium state is also attained if the member is a cable. The remarkable point is that the truss member has the other equilibrium state shown in Figure 5.2(b), where the member configuration is in parallel with $\underline{\boldsymbol{p}}$ but now the axial force is in compression, i.e., $q_i = -\|\underline{\boldsymbol{p}}\|$. Thus, in both case (a) and case (b), we see that the internal force vector is $\boldsymbol{v} = \underline{\boldsymbol{p}}$. This means that the deformed configuration of a truss is not determined uniquely if only

[4]Recall that we denote by $q_i \in \mathbb{R}$ the axial force and by $\boldsymbol{v}_i \in \mathbb{R}^3$ the internal force vector. The internal force vector involves information of the direction of the ith member at the deformed state.

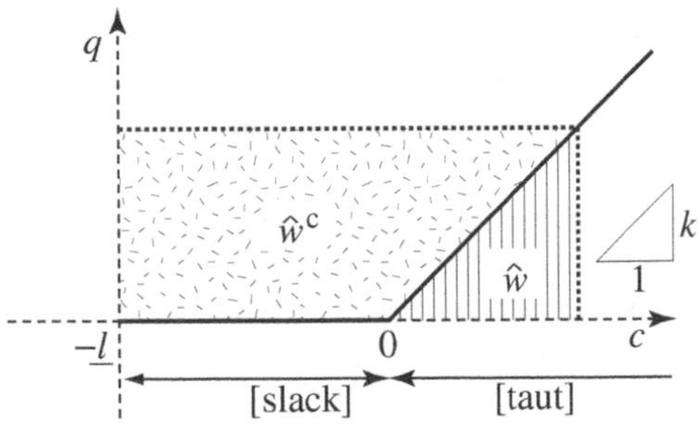

FIGURE 5.3: An interpretation of the complementary stored energy of a cable member undergoing large deformation.

the internal force vectors of members are given. In the case of a cable network, in contrast, the equilibrium state (b) cannot exist (because a cable cannot transmit a compression force), and hence the deformed configuration is determined uniquely only from the specified internal force vectors. ■

Remark 5.1.11. From a mechanical point of view, the objective function of (CE) (in (5.26)) may be interpreted as follows. Let $\boldsymbol{x} = \mathbf{0}$ for simplicity, which means that in the reference configuration all the nodes gather together at the origin, and subsequently the cable network is expanded to attain the equilibrium configuration. In such a situation the last term of the objective function vanishes. The term $-\sum_{j \in \Gamma_{\rm D}} \underline{\varphi}_j t_j$ is regarded as the conventional complementary work done by the prescribed displacements,

The complementary stored energy of the ith member, denoted $\hat{w}_i^{\rm c}(q_i)$, is evaluated as the complementary work done during the loading process from $q_i = 0$ to $q_i = q_i$. For the terms $\frac{1}{2}(\|\boldsymbol{v}_i\|^2/k_{{\rm e}i}) + \underline{l}_i\|\boldsymbol{v}_i\|$, we recall that at the reference configuration the two nodes of the ith member coincide, because both are at the origin. Thus the complementary stored energy (and the stored energy also) should be evaluated in the course of expansion of the cable from $c_i = -\underline{l}_i$ to $c_i = c_i$, as shown in Figure 5.3. The corresponding complementary work stored in the member is then written as $\hat{w}_i^{\rm c}(q_i) = \frac{1}{2}(\|\boldsymbol{v}_i\|^2/k_{{\rm e}i}) + \underline{l}_i\|\boldsymbol{v}_i\|$, while the stored energy is $\hat{w}_i(c_i) = \frac{1}{2}k_{{\rm e}i}c_i^2$. Thus the complementary stored energy depends on the reference configuration in the geometrically nonlinear theory, which explains the role of the additional term $\underline{l}_i\|\boldsymbol{v}_i\|$. In contrast, the stored energy is independent of the reference configuration, and is the same as that in the geometrically linear theory. See Kanno–Ohsaki [204, section 6] for more discussion. ■

5.1.5 Existence and uniqueness of solution

The hypothesis of Proposition 5.1.3, say,

$$\exists \boldsymbol{v}_1, \dots, \boldsymbol{v}_m : \quad \Big(\sum_{i=1}^{m} B_i^{\rm T} \boldsymbol{v}_i \Big)_j = \underline{p}_j \ (\forall j \in \Gamma_{\rm N}), \tag{5.34}$$

guarantees that ($\underline{\text{PE}}$) and ($\underline{\text{CE}}$) have optimal solutions, and so do (PE) and (CE). Consequently, (5.34) guarantees the existence of equilibrium states of the cable network under consideration. We here investigate a sufficient condition that (5.34) is satisfied for any $\underline{\boldsymbol{p}}$.

To see this, we make use of the digraph $\mathcal{G} = (\mathcal{V}, \mathcal{E})$ associated with the cable network, which was introduced in Remark 4.2.5 of section 4.2.2. The vertex set $\mathcal{V}$ is defined as the set of nodes of the cable network, the edge set $\mathcal{E}$ is the set of cable members, and the direction of each edge is given arbitrarily. Then the matrices $B_1, \dots, B_m$ in (5.34) are written as (4.40), where $\tilde{C}_i$ is the ith column vector of the incidence matrix C of $\mathcal{G}$.

Proposition 5.1.12. *Suppose that $\mathcal{G}(\mathcal{V}, \mathcal{E})$ is connected and includes no loops or parallel edges. If some nodes of the cable network are fixed so that the rigid-body motion is not admissible, then* (5.34) *is satisfied for any* $\underline{\boldsymbol{p}}$.

Proof. For each member i, $\boldsymbol{v}_i$ is the internal force vector in $\mathbb{R}^3$, which is written as

$$\boldsymbol{v}_i = (v_{i\alpha} \mid \alpha = 1, 2, 3).$$

Here, $\alpha = 1, 2$, and 3 correspond to x-, y-, and z-directions as seen, for example, in Figure 4.5(a). Accordingly, the index set of the degrees of freedom, Γ_{N}, is partitioned as $\bigcup_{1 \le \alpha \le 3} \Gamma_{\mathrm{N}}^{\alpha}$. For example, Γ_{N}^{1} consists of the degrees of freedom along the x-direction, among those included in Γ_{N}. We denote by $l(j)$ the index of the node associated with the jth degree of freedom.[5] Then the system of equations in (5.34) is rewritten with respect to each direction as

$$-\sum_{i=1}^{m} C_{l(j)i} v_{i\alpha} = \underline{p}_j \ (\forall j \in \Gamma_{\mathrm{N}}^{\alpha}), \quad \alpha = 1, 2, 3, \tag{5.35}$$

where $C = (C_{l(j)i}) \in \mathbb{R}^{d/3 \times m}$, and $d/3$ is the number of nodes.

Define $\hat{C}^{\alpha} \in \mathbb{R}^{|\Gamma_{\mathrm{N}}^{\alpha}| \times m}$ by

$$\hat{C}^{\alpha} = (C_{l(j)i} \mid j \in \Gamma_{\mathrm{N}}^{\alpha},\ i = 1, \dots, m), \quad \alpha = 1, 2, 3$$

for simplicity. Then (5.35) has a solution for any $\underline{\boldsymbol{p}}$ if and only if $\operatorname{rank} \hat{C}^{\alpha} = |\Gamma_{\mathrm{N}}^{\alpha}|$ $(\alpha = 1, 2, 3)$. Since the rigid-body motion is inadmissible, for each $\alpha = 1, 2, 3$ there exists at least one $\check{l}_{\alpha}$ such that $\check{l}_{\alpha} \notin \Gamma_{\mathrm{N}}^{\alpha}$. By using such $\check{l}_{\alpha}$, define $\check{C}^{\alpha}$ by

$$\check{C}^{\alpha} = (C_{li} \mid l \in \{1, \dots, d/3\} \setminus \{\check{l}_{\alpha}\},\ i = 1, \dots, m), \quad \alpha = 1, 2, 3.$$

Then $\check{C}^{\alpha} \in \mathbb{R}^{(d/3-1) \times m}$ is called the *truncated incidence matrix* and is known to satisfy $\operatorname{rank} \check{C}^{\alpha} = d/3 - 1$ $(\alpha = 1, 2, 3)$; see [89, Chap. 19]. Since any row vector of $\hat{C}^{\alpha}$ is contained in $\check{C}^{\alpha}$, $\hat{C}^{\alpha}$ has full row rank, i.e., $\operatorname{rank} \hat{C}^{\alpha} = |\Gamma_{\mathrm{N}}^{\alpha}|$ $(\alpha = 1, 2, 3)$, which concludes the proof. □

[5]The reader should not confuse $l(j)$ used in this proof with $\underline{l}_i$ in the other places; $l(j)$ is the index of component of matrix, while $\underline{l}_i$ is the initial length of the ith member.

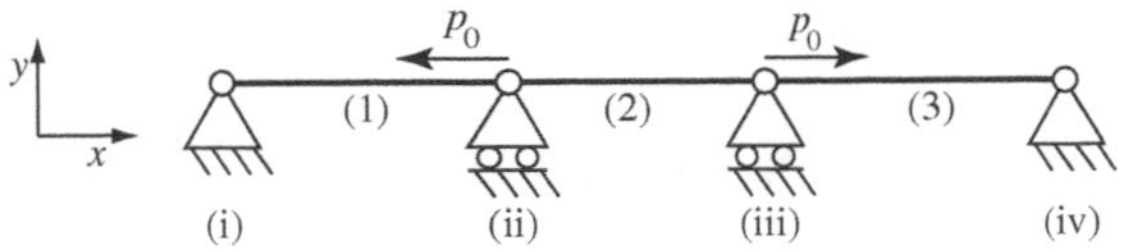

FIGURE 5.4: An example of an external load with which the cable network has a non-unique equilibrium configuration.

Practically, a cable network in the real world satisfies the hypothesis postulated in Proposition 5.1.12. For example, suppose that the graph $\mathcal{G}$ induced by a cable network does not have a loop or a parallel edge and that only one node is fixed in the three-dimensional space. This is the case of $\hat{C}^{\alpha} = \check{C}^{\alpha}$ ($\alpha = 1, 2, 3$) in the proof of Proposition 5.1.12, and thence such a cable network has an equilibrium state for any external load.

Assume the existence of solution to (CE). Since the objective function of (CE) is strictly convex, its optimal solution exists uniquely. Thus, under the hypothesis of Proposition 5.1.12, the set of static variables in the equilibrium state is determined uniquely for any external load. In this sense a cable network does not experience bifurcation phenomena.

In contrast, the optimal solution of (PE) is not necessarily unique. In (5.22a), $\bar{q}_i$ exists uniquely (from the uniqueness of the solution to (CE)), and hence $\bar{c}_{ei}$ is determined uniquely. However, if $\bar{c}_{ei} = 0$, then $\bar{q}_i = 0$ and $\bar{\boldsymbol{v}}_i = \mathbf{0}$, and hence the configuration $\bar{\varphi}$ satisfying (5.22d) and (5.22e) is not necessarily unique.

A simple example of the non-uniqueness of equilibrium configuration is illustrated in Figure 5.4, where we consider a planar cable network. Nodes (i) and (iv) are fixed, while (ii) and (iii) are free in the x-direction. The external forces of p_0 are applied to nodes (ii) and (iii) in the opposite directions. At the equilibrium state, member 2 is taut while members 1 and 3 become slack. The set of axial forces is determined uniquely; $(\bar{q}_1, \bar{q}_2, \bar{q}_3) = (0, p_0, 0)$. However, member 2 in the taut state can move freely as long as members 1 and 3 are slack. Thus the locations of nodes (ii) and (iii) are not determined uniquely.

5.2 Duality in Cable Networks (2): Linear Strain

We next establish a dual energy principle of cable networks under the assumption of small deformation. The corresponding primal principles were investigated in section 4.2.3, where the minimization problem of the potential energy has been formulated as

$$\left.\begin{aligned}(\mathrm{PE}) : \min_{\boldsymbol{u},\boldsymbol{c}} & \sum_{i=1}^{m} \hat{w}_{\mathrm{cab}}^{i}(c_i) - \sum_{j\in\Gamma_{\mathrm{N}}} \underline{p}_j u_j \\ \text{s. t. } & c_i = \boldsymbol{\beta}_i^{\mathrm{T}}\boldsymbol{u}, \quad i = 1,\dots,m, \\ & u_j = \underline{u}_j, \quad j \in \Gamma_{\mathrm{D}}. \end{aligned}\right\} \qquad \text{(see (4.46))}$$

It has been shown that (4.46) is equivalently rewritten as

$$\left.\begin{aligned}(\overline{\mathrm{PE}}) : \min_{\boldsymbol{u},\boldsymbol{c}_{\mathrm{e}}} & \sum_{i=1}^{m} \frac{1}{2} k_{\mathrm{e}i} c_{\mathrm{e}i}^2 - \sum_{j\in\Gamma_{\mathrm{N}}} \underline{p}_j u_j \\ \text{s. t. } & c_{\mathrm{e}i} \geq \boldsymbol{\beta}_i^{\mathrm{T}}\boldsymbol{u}, \quad i = 1,\dots,m, \\ & u_j = \underline{u}_j, \quad j \in \Gamma_{\mathrm{D}}, \end{aligned}\right\} \qquad \text{(see (4.47))}$$

which is a QP problem.

We now proceed following the same recipe as that given in section 5.1.

5.2.1 Embedding to Fenchel form

($\overline{\mathrm{PE}}$) in (4.47) is reformulated in the form of the primal problem of the Fenchel duality, i.e., ($\mathrm{P_F}$) in (2.28), as follows.

Define the primal variable, together with the primal spaces, by

$$x = \begin{bmatrix} \boldsymbol{u} \\ \boldsymbol{c}_{\mathrm{e}} \end{bmatrix}, \quad \mathbb{V} = \mathbb{R}^d \times \mathbb{R}^m, \quad \mathbb{Y} = \mathbb{R}^m, \tag{5.36}$$

where $x \in \mathbb{V}$. We write an element of $\mathbb{Y}$ as

$$z = \boldsymbol{z}, \quad \boldsymbol{z} = (z_i \in \mathbb{R} \mid i = 1,\dots,m).$$

Define the two convex functions $f : \mathbb{V} \to \mathbb{R} \cup \{+\infty\}$ and $g : \mathbb{Y} \to \mathbb{R} \cup \{+\infty\}$ by

$$f(x) = \begin{cases} \displaystyle\sum_{i=1}^{m} \frac{1}{2} k_{\mathrm{e}i} c_{\mathrm{e}i}^2 - \sum_{j\in\Gamma_{\mathrm{N}}} \underline{p}_j u_j & \text{if } u_j = \underline{u}_j \ (j \in \Gamma_{\mathrm{D}}), \\ +\infty & \text{otherwise,} \end{cases} \tag{5.37}$$

$$g(z) = \begin{cases} 0 & \text{if } \boldsymbol{z} \geq \boldsymbol{0}, \\ +\infty & \text{otherwise.} \end{cases} \tag{5.38}$$

The action of the linear mapping $\Lambda : \mathbb{V} \to \mathbb{Y}$ is defined by

$$\Lambda x = \begin{bmatrix} c_{\mathrm{e}1} - \boldsymbol{\beta}_1^{\mathrm{T}}\boldsymbol{u} \\ \vdots \\ c_{\mathrm{e}m} - \boldsymbol{\beta}_m^{\mathrm{T}}\boldsymbol{u} \end{bmatrix}. \tag{5.39}$$

From (5.36) and (5.39), Λ is identified with the $m \times (d+m)$ matrix defined by

$$\Lambda = \left[\begin{array}{c|c} -\boldsymbol{\beta}_1^{\mathrm{T}} & \\ \vdots & I_m \\ -\boldsymbol{\beta}_m^{\mathrm{T}} & \end{array}\right]. \tag{5.40}$$

Thus $(\overline{\mathrm{PE}})$ in (4.47) is represented as $(\mathrm{P_F})$ in (2.28).

5.2.2 Dual problem

From (5.36), the spaces for the Fenchel dual problem are given by

$$\mathbb{V}^* = \mathbb{R}^d \times \mathbb{R}^m, \quad \mathbb{Y}^* = \mathbb{R}^m,$$

where $\mathbb{V}^* = \mathbb{V}$ and $\mathbb{Y}^* = \mathbb{Y}$. We write elements of $\mathbb{V}^*$ and $\mathbb{Y}^*$ as

$$x^* = \begin{bmatrix} \boldsymbol{u}^* \\ \boldsymbol{c}_{\mathrm{e}}^* \end{bmatrix} \in \mathbb{V}^*, \quad z^* = \boldsymbol{z}^* \in \mathbb{Y}^*.$$

For f and Λ defined by (5.39), their conjugates, denoted by $f^* : \mathbb{V}^* \to \mathbb{R} \cup \{+\infty\}$ and $\Lambda^* : \mathbb{Y}^* \to \mathbb{V}^*$, are obtained in a manner similar to Proposition 5.1.2 as

$$f^*(x^*) = \begin{cases} \displaystyle\sum_{i=1}^m \frac{1}{2} \frac{(c_{\mathrm{e}i}^*)^2}{k_{\mathrm{e}i}} + \sum_{j \in \Gamma_{\mathrm{D}}} \underline{u}_j u_j^* & \text{if } u_j^* + \underline{p}_j = 0 \ (j \in \Gamma_{\mathrm{N}}), \\ +\infty & \text{otherwise,} \end{cases} \tag{5.41}$$

$$\Lambda^* z^* = \begin{bmatrix} -\displaystyle\sum_{i=1}^m z_i^* \boldsymbol{\beta}_i \\ \boldsymbol{z}^* \end{bmatrix}. \tag{5.42}$$

The conjugate function of g reflects the self-duality of the nonnegative orthant as follows.

Proposition 5.2.1. *For g defined by (5.38), its conjugate function $g^* : \mathbb{V}^* \to \mathbb{R} \cup \{+\infty\}$ is given by*

$$g^*(z^*) = \begin{cases} 0 & \text{if } -\boldsymbol{z}^* \geq \boldsymbol{0}, \\ +\infty & \text{otherwise.} \end{cases} \tag{5.43}$$

Proof. Application of the definition of the conjugate function (see (2.13)) to g under consideration yields

$$\begin{aligned} g^*(z^*) &= \sup_z \{\langle z^*, z \rangle \mid z \in \operatorname{dom} g\} \\ &= \sum_{i=1}^m \sup_{z_i} \{z_i^* z_i \mid z_i \geq 0\}, \end{aligned}$$

which concludes the proof. □

Now we proceed to derive the explicit form of the Fenchel dual problem, say, $(\mathrm{P}_{\mathrm{F}}^*)$ in (2.31), by using (5.41), (5.42) and (5.43). Rewrite the dual variables as[6]

$$q_i = z_i^* .$$

For simplicity, we introduce the additional variables $\boldsymbol{t} \in \mathbb{R}^d$ by

$$\boldsymbol{t} = \sum_{i=1}^{m} z_i^* \boldsymbol{\beta}_i$$

to see that (5.42) reads

$$\Lambda^* z^* = \begin{bmatrix} -\sum_{i=1}^m q_i \boldsymbol{\beta}_i \\ \boldsymbol{q} \end{bmatrix} = \begin{bmatrix} -\boldsymbol{t} \\ \boldsymbol{q} \end{bmatrix}.$$

Thus, from f^* and g^* in (5.41) and (5.43), the Fenchel dual problem (2.31) is explicitly formulated as

$$\left.\begin{aligned} (\overline{\mathrm{CE}}) : \max_{\boldsymbol{q},\boldsymbol{t}} \ & -\sum_{i=1}^{m} \frac{1}{2} \frac{q_i^2}{k_{ei}} + \sum_{j \in \Gamma_{\mathrm{D}}} \underline{u}_j t_j \\ \text{s. t.} \ & \sum_{i=1}^{m} q_i \boldsymbol{\beta}_i = \boldsymbol{t}, \\ & \boldsymbol{q} \geq \boldsymbol{0}, \\ & t_j = \underline{p}_j, \quad j \in \Gamma_{\mathrm{N}}. \end{aligned}\right\} \tag{5.44}$$

5.2.3 Duality and optimality

For the primal-dual pair, $(\overline{\mathrm{PE}})$ in (4.47) and $(\overline{\mathrm{CE}})$ in (5.44), for the geometrically linear problem of cable network we can establish the strong duality in a manner similar to discussions in section 5.1.3 for the geometrically nonlinear theory. These results are given without proofs, because they can be shown similarly to the corresponding results in section 5.1.3.

Proposition 5.2.2. *Suppose that there exists a vector $\boldsymbol{q} \in \mathbb{R}^m$ satisfying*

$$\begin{aligned} &\Big(\sum_{i=1}^{m} q_i \boldsymbol{\beta}_i\Big)_j = \underline{p}_j, \quad j \in \Gamma_{\mathrm{N}}, \\ &\boldsymbol{q} \geq \boldsymbol{0}. \end{aligned}$$

Then, each of $(\overline{\mathrm{PE}})$ and $(\overline{\mathrm{CE}})$ has an optimal solution, and $\min(\overline{\mathrm{PE}}) = \max(\overline{\mathrm{CE}})$.

[6]In section 5.2.3, it shall be clearly shown from the optimality condition that q_i corresponds to the axial force of the ith member.

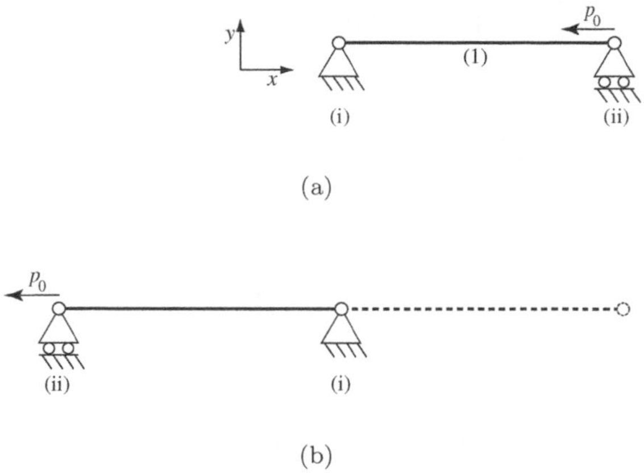

FIGURE 5.5: An example of loading condition to explain the existence condition of an equilibrium solution. (a) The cable network does not possess the equilibrium state within small deformation; (b) The equilibrium state exists if we allow large deformation.

Remark 5.2.3. The hypothesis postulated in Proposition 5.2.2 is much more restrictive compared with the one in Proposition 5.1.3 in the case of large deformation. Indeed, as observed in section 5.1.5, for a broad class of cable networks, it is guaranteed that an equilibrium solution exists for any external load if we allow large deformation. However, when we assume small deformation, the existence of an equilibrium state highly depends upon the loading condition. Roughly speaking, the hypothesis in Proposition 5.1.3 is the existence of a solution to a system of equations, while the one in Proposition 5.2.2 is the existence of a solution to a system of inequalities; the latter one is more restrictive than the former one.

As a simple example, consider a planar cable network (consisting of one member) illustrated in Figure 5.5(a). The external force of p_0 is applied at node (ii) in the negative direction of the x-axis. Since the external force leads to a compressive force for the cable member, the elongation stiffness of the member vanishes, and hence no equilibrium solution exists within small deformation.[7] In contrast, if we allow large deformation, then node (ii) can move drastically to transform the external force to a tensile force as shown in Figure 5.5(b). Thus, in the framework of the geometrically nonlinear theory, this cable network possesses an equilibrium state for any external force. ■

The optimality conditions for ($\overline{\text{PE}}$) and ($\overline{\text{CE}}$) are given below.

[7]Needless to say, for an external load in the positive direction of the x-axis, the cable network always possesses an equilibrium state within the geometrically linear theory.

Proposition 5.2.4. *Under the hypotheses in Proposition 5.2.2,* $(\bar{\boldsymbol{u}}, \bar{\boldsymbol{c}}_{\rm e})$ *and* $(\bar{\boldsymbol{q}}, \bar{\boldsymbol{t}})$ *are optimal for* $(\overline{\rm PE})$ *and* $(\overline{\rm CE})$*, respectively, if and only if they satisfy*

$$\bar{q}_i = k_{{\rm e}i} c_{{\rm e}i}, \quad i = 1, \dots, m, \tag{5.45a}$$

$$\sum_{i=1}^{m} \bar{q}_i \boldsymbol{\beta}_i = \bar{\boldsymbol{t}}, \tag{5.45b}$$

$$\bar{u}_j = \underline{u}_j, \quad j \in \Gamma_{\rm D}; \qquad \bar{t}_j = \underline{p}_j, \quad j \in \Gamma_{\rm N}, \tag{5.45c}$$

$$\bar{c}_{{\rm e}i} \geq \boldsymbol{\beta}_i^{\rm T} \bar{\boldsymbol{u}}, \ \bar{q}_i \geq 0, \quad i = 1, \dots, m, \tag{5.45d}$$

$$(\bar{c}_{{\rm e}i} - \boldsymbol{\beta}_i^{\rm T} \bar{\boldsymbol{u}}) \bar{q}_i = 0, \quad i = 1, \dots, m. \tag{5.45e}$$

Since (5.45a) is interpreted as a constitutive law and (5.45b) is the force-balance equation, we can see that q_i corresponds to the axial force of the ith member.

Remark 5.2.5. Optimality condition (5.45) forms the LCP (linear complementarity problem). ■

5.2.4 Principle of complementary energy

We can transform $(\overline{\rm CE})$ in (5.44) from maximization to minimization by changing the sign of the objective function. Notice here that minimizing $\frac{1}{2}(q_i^2/k_{{\rm e}i})$ under the constraint $q_i \geq 0$ is equivalent to minimizing

$$\hat{w}_{\rm cab}^{i*}(q_i) = \begin{cases} \dfrac{1}{2} \dfrac{q_i^2}{k_{{\rm e}i}} & \text{if } q_i \geq 0, \\ +\infty & \text{otherwise} \end{cases}$$

with respect to $q_i \in \mathbb{R}$. The function $\hat{w}_{\rm cab}^{i*}$ corresponds to the conjugate function of $\hat{w}_{\rm cab}^{i}$; see (4.9) in section 4.1.2. Consequently, $(\overline{\rm CE})$ can be reformulated without changing the optimal solution as

$$\left. \begin{aligned} ({\rm CE}) : \min_{\boldsymbol{q}, \boldsymbol{t}} \ & \sum_{i=1}^{m} \hat{w}_{\rm cab}^{i*}(q_i) - \sum_{j \in \Gamma_{\rm D}} \underline{u}_j t_j \\ \text{s. t.} \ & \sum_{i=1}^{m} q_i \boldsymbol{\beta}_i = \boldsymbol{t}, \\ & t_j = \underline{p}_j, \quad j \in \Gamma_{\rm N}. \end{aligned} \right\} \tag{5.46}$$

Note that, by definition, $\hat{w}_{\rm cab}^{i*}$ is the complementary stored energy function of the ith cable member. It is clear that $\underline{u}_j t_j$ $(j \in \Gamma_{\rm D})$ corresponds to the complementary external work done by prescribed displacement, where t_j is the nodal reaction force. Therefore, the objective function of (CE) in (5.46) represents the complementary energy of the cable network. The constraints, meanwhile, consist of the static equation of the equilibrium (where q_i is the internal axial force of the ith member) and the boundary conditions concerning the external loads. Thus (CE) involves static variables only, is a minimization problem of the complementary energy over the force-balance equation, and provides us with the equilibrium solution at its optimal solution.

Remark 5.2.6. Although (CE) in (5.46) is equivalent to $(\overline{\mathrm{CE}})$ in (5.44) in the sense that they share the same optimal solution, $(\overline{\mathrm{CE}})$ is more tractable both from theoretical and numerical points of view, because $(\overline{\mathrm{CE}})$ is formulated as a QP problem. ■

5.3 Duality in Cable Networks (3): Green–Lagrange Strain

We here consider a cable network undergoing large rotation and small strain. The corresponding primal energy principle is formulated as

$$\left.\begin{aligned}
(\mathrm{PE}) : \min_{\boldsymbol{\varphi}, \boldsymbol{c}_{\mathrm{e}}} & \sum_{i=1}^{m} \hat{w}^{i}_{\mathrm{cab}}(c_i) - \sum_{j \in \Gamma_{\mathrm{N}}} \underline{\boldsymbol{p}}_j^{} (\boldsymbol{\varphi}_j - \boldsymbol{x}_j) \\
\mathrm{s.\,t.}\ & c_i = \frac{1}{2\underline{l}_i} \left[(B_i \boldsymbol{\varphi})^{\mathrm{T}} (B_i \boldsymbol{\varphi}) - \underline{l}_i^2 \right], \quad i = 1, \dots, m, \\
& \boldsymbol{\varphi}_j = \underline{\boldsymbol{\varphi}}_j, \quad j \in \Gamma_{\mathrm{D}},
\end{aligned}\right\} \qquad \text{(see (4.49))}$$

as studied in section 4.2.4. The convex optimization reformulation is then given by

$$\left.\begin{aligned}
(\overline{\mathrm{PE}}) : \min_{\boldsymbol{\varphi}, \boldsymbol{c}_{\mathrm{e}}} & \sum_{i=1}^{m} \frac{1}{2} k_{\mathrm{e}i} c_{\mathrm{e}i}^2 - \sum_{j \in \Gamma_{\mathrm{N}}} \underline{\boldsymbol{p}}_j^{} (\boldsymbol{\varphi}_j - \boldsymbol{x}_j) \\
\mathrm{s.\,t.}\ & c_{\mathrm{e}i} \geq \frac{1}{2\underline{l}_i} \left[(B_i \boldsymbol{\varphi})^{\mathrm{T}} (B_i \boldsymbol{\varphi}) - \underline{l}_i^2 \right], \quad i = 1, \dots, m, \\
& \boldsymbol{\varphi}_j = \underline{\boldsymbol{\varphi}}_j, \quad j \in \Gamma_{\mathrm{D}}.
\end{aligned}\right\} \qquad \text{(see (4.50))}$$

We again proceed following the same recipe as that given in section 5.1 to derive $(\overline{\mathrm{CE}})$ and (CE) (see also Figure 5.1).

5.3.1 Embedding to Fenchel form

We first rewrite $(\overline{\mathrm{PE}})$ in (4.50) in the form of $(\mathrm{P_F})$ in (2.28), which serves as the primal problem in the framework of the Fenchel duality. It follows from Fact 1.3.10 that the inequality constraints of $(\overline{\mathrm{PE}})$ are reduced to

$$c_{\mathrm{e}i} \geq \frac{1}{2\underline{l}_i} \left[(B_i \boldsymbol{\varphi})^{\mathrm{T}} (B_i \boldsymbol{\varphi}) - \underline{l}_i^2 \right] \quad \Leftrightarrow \quad \left[\begin{array}{c|c} c_{\mathrm{e}i} + (\underline{l}_i/2) & -(1/2)(B_i \boldsymbol{\varphi})^{\mathrm{T}} \\ \hline -(1/2) B_i \boldsymbol{\varphi} & (\underline{l}_i/2) I_3 \end{array} \right] \succeq O.$$

Therefore, $(\overline{\mathrm{PE}})$ can be rewritten as

$$
\left.\begin{aligned}
(\overline{\mathrm{PE}}) : \min_{\boldsymbol{\varphi}, \boldsymbol{c}_{\mathrm{e}}} & \sum_{i=1}^{m} \frac{1}{2} k_{\mathrm{e}i} c_{\mathrm{e}i}^2 - \sum_{j \in \Gamma_{\mathrm{N}}} \underline{p}_j (\varphi_j - x_j) \\
\text{s. t.} & \begin{bmatrix} c_{\mathrm{e}i} + (\underline{l}_i/2) & -(1/2)(B_i \boldsymbol{\varphi})^{\mathrm{T}} \\ -(1/2) B_i \boldsymbol{\varphi} & (\underline{l}_i/2) I_3 \end{bmatrix} \succeq O, \quad i = 1, \dots, m, \\
& \varphi_j = \underline{\varphi}_j, \quad j \in \Gamma_{\mathrm{D}}.
\end{aligned}\right\} \tag{5.47}
$$

Note that the inequality constraints of this problem are the positive-semidefinite constraints. Since with the formulation in (5.47) we can enjoy the self-duality property of the positive-semidefinite cone (specifically in Proposition 5.3.1), in what follows we treat (5.47) instead of (4.50).

Associated with the form of $(\mathrm{P_F})$, define primal variables by[8]

$$x = (\boldsymbol{\varphi}, \boldsymbol{c}_{\mathrm{e}}) \tag{5.48}$$

with the spaces

$$\mathbb{V} = \mathbb{R}^d \times \mathbb{R}^m, \quad \mathbb{Y} = \prod_{i=1}^{m} \mathbb{S}^4, \tag{5.49}$$

where $x \in \mathbb{V}$. We write an element of $\mathbb{Y}$ as

$$z = (Z_i \in \mathbb{S}^4 \mid i = 1, \dots, m), \quad Z_i = \begin{bmatrix} z_{i11} & \boldsymbol{z}_{i21}^{\mathrm{T}} \\ \boldsymbol{z}_{i21} & Z_{i22} \end{bmatrix}. \tag{5.50}$$

Define $f : \mathbb{V} \to \mathbb{R} \cup \{+\infty\}$ and $g : \mathbb{Y} \to \mathbb{R} \cup \{+\infty\}$ by

$$f(x) = \begin{cases} \displaystyle\sum_{i=1}^{m} \frac{1}{2} k_{\mathrm{e}i} c_{\mathrm{e}i}^2 - \sum_{j \in \Gamma_{\mathrm{N}}} \underline{p}_j (\varphi_j - x_j) & \text{if } \varphi_j = \underline{\varphi}_j \ (j \in \Gamma_{\mathrm{D}}), \\ +\infty & \text{otherwise}, \end{cases} \tag{5.51}$$

$$g(z) = \begin{cases} 0 & \text{if } \begin{bmatrix} z_{i11} + (l_i/2) & \boldsymbol{z}_{i21}^{\mathrm{T}} \\ \boldsymbol{z}_{i21} & Z_{i22} + (\underline{l}_i/2) I_3 \end{bmatrix} \succeq O \ (i = 1, \dots, m), \\ +\infty & \text{otherwise}. \end{cases} \tag{5.52}$$

Moreover, define the action of the linear mapping $\Lambda : \mathbb{V} \to \mathbb{Y}$ on x by

$$\Lambda x = (\Lambda_i x \mid i = 1, \dots, m) \tag{5.53}$$

with

$$\Lambda_i x = \begin{bmatrix} c_{\mathrm{e}i} & -(1/2)(B_i \boldsymbol{\varphi})^{\mathrm{T}} \\ -(1/2) B_i \boldsymbol{\varphi} & O \end{bmatrix}. \tag{5.54}$$

With the setting of (5.48)–(5.54), $(\overline{\mathrm{PE}})$ in (5.47) (and hence (4.50) also) is identical to $(\mathrm{P_F})$ in (2.28).

[8]The reader should not confuse x in (5.48) with $\boldsymbol{x}$ in (5.47); $\boldsymbol{x} \in \mathbb{R}^d$ is the given position vector of the nodes in the reference configuration of the cable network, while x is the variable vector of the optimization problem in the standard form of the Fenchel duality theory.

5.3.2 Dual problem

The spaces for the dual problem are

$$\mathbb{V}^* = \mathbb{R}^d \times \mathbb{R}^m, \quad \mathbb{Y}^* = \prod_{i=1}^{m} \mathbb{S}^4, \tag{5.55}$$

where $\mathbb{V}^* = \mathbb{V}$ and $\mathbb{Y}^* = \mathbb{Y}$ in (5.49). We write the elements of $\mathbb{V}^*$ and $\mathbb{Y}^*$ as

$$x^* = (\boldsymbol{\varphi}^*, \boldsymbol{c}_{\mathrm{e}}^*), \quad z^* = (Z_1^*, \dots, Z_m^*), \tag{5.56}$$

where

$$Z_i^* = \begin{bmatrix} z_{i11}^* & (\boldsymbol{z}_{i21}^*)^{\mathrm{T}} \\ \boldsymbol{z}_{i21}^* & Z_{i22}^* \end{bmatrix} \in \mathbb{S}^4.$$

We next determine the conjugate functions explicitly. For f in (5.51), f^* was obtained in Proposition 5.1.2 as

$$f^*(x^*) = \begin{cases} \displaystyle\sum_{i=1}^{m} \frac{1}{2} \frac{(c_{\mathrm{e}i}^*)^2}{k_{\mathrm{e}i}} + \sum_{j \in \Gamma_{\mathrm{D}}} \underline{\varphi}_j \varphi_j^* - \sum_{j \in \Gamma_{\mathrm{N}}} \underline{p}_j x_j & \text{if } \varphi_j^* + \underline{p}_j = 0 \ (j \in \Gamma_{\mathrm{N}}), \\ +\infty & \text{otherwise.} \end{cases} \tag{5.57}$$

For deriving g^*, we enjoy the self-duality of the positive-semidefinite cone as follows.

Proposition 5.3.1. *For g defined by* (5.52), *we have*

$$g^*(z^*) = \begin{cases} \displaystyle -\sum_{i=1}^{m} \frac{\underline{l}_i}{2} (z_{i11}^* + \operatorname{tr} Z_{i22}^*) & \text{if } -\begin{bmatrix} z_{i11}^* & (\boldsymbol{z}_{i21}^*)^{\mathrm{T}} \\ \boldsymbol{z}_{i21}^* & Z_{i22}^* \end{bmatrix} \succeq O \ (i = 1, \dots, m), \\ +\infty & \text{otherwise.} \end{cases}$$

Proof. Write

$$\hat{Z}_i(Z_i) = \begin{bmatrix} z_{i11} + (\underline{l}_i/2) & \boldsymbol{z}_{i21}^{\mathrm{T}} \\ \boldsymbol{z}_{i21} & Z_{i22} \end{bmatrix}$$

to simplify the notation. Then $Z \in \operatorname{dom} g$ if and only if $\hat{Z}_i(Z_i) \succeq O$

$(i = 1, \dots, m)$. We thus have

$$\begin{aligned} g^*(z^*) &= \sup_{z}\{\langle z^*, z\rangle \mid z \in \operatorname{dom} g\} \\ &= \sum_{i=1}^{m} \sup_{Z_i}\{Z_i^* \bullet Z_i \mid \hat{Z}_i(Z_i) \succeq O\} \\ &= \sum_{i=1}^{m} \sup_{Z_i}\{\hat{Z}_i^*(Z_i) \bullet \hat{Z}_i - (\underline{l}_i/2)(z_{i11}^* + \operatorname{tr} Z_{i22}^*) \mid \hat{Z}_i \succeq O\} \\ &= -\sum_{i=1}^{m} \inf_{Z_i}\{(-Z_i^*) \bullet \hat{Z}_i(Z_i) \mid \hat{Z}_i(Z_i) \succeq O\} \\ &\quad - \sum_{i=1}^{m} \frac{\underline{l}_i}{2}(z_{i11}^* + \operatorname{tr} Z_{i22}^*), \end{aligned}$$

and applying Fact 1.3.17 (ii) concludes the proof. □

From the definition of Λ, say, (5.53) and (5.54), we obtain

$$\begin{aligned} \langle z^*, \Lambda x\rangle &= \sum_{i=1}^{m} \begin{bmatrix} z_{i11}^* & (\boldsymbol{z}_{i21}^*)^{\mathrm{T}} \\ \boldsymbol{z}_{i21}^* & Z_{i22}^* \end{bmatrix} \bullet \begin{bmatrix} c_{ei} & -(1/2)(B_i\boldsymbol{\varphi})^{\mathrm{T}} \\ -(1/2)B_i\boldsymbol{\varphi} & O \end{bmatrix} \\ &= \sum_{i=1}^{m} \left[z_{i11}^* c_{ei} - (B_i^{\mathrm{T}}\boldsymbol{z}_{i21}^*)^{\mathrm{T}}\boldsymbol{\varphi} \right], \end{aligned}$$

which is equal to $\langle \Lambda^* z^*, x\rangle$. Therefore, $\Lambda^* z^* \in \mathbb{V}^*$ is given by

$$\Lambda^* z^* = \left(-\sum_{i=1}^{m} (B_i^{\mathrm{T}}\boldsymbol{z}_{i21}^*), z_{i11}^*, \dots, z_{m11}^* \right). \tag{5.58}$$

Now we are in a position to state the Fenchel dual problem explicitly; for $(\mathrm{P}_{\mathrm{F}}^*)$ in (2.31), f^*, g^*, and Λ^* are given by (5.57), Proposition 5.3.1, and (5.58). Rewrite the dual variables as[9]

$$\begin{bmatrix} q_i & \boldsymbol{\pi}_i^{\mathrm{T}} \\ \boldsymbol{\pi}_i & \tau_i^{\varphi} \end{bmatrix} = \begin{bmatrix} z_{i11}^* & (\boldsymbol{z}_{i21}^*)^{\mathrm{T}} \\ \boldsymbol{z}_{i21}^* & Z_{i22}^* \end{bmatrix}, \tag{5.59}$$

and introduce new variables, $\boldsymbol{t} \in \mathbb{R}^d$, by (5.12) to see that (5.58) is reduced to

$$\Lambda^* z^* = (\boldsymbol{t}, \boldsymbol{q}).$$

[9]It shall be shown in section 5.3.3 that q_i, $\boldsymbol{\pi}_i$, and τ_i^{φ} correspond to the axial forces in terms of the second Piola–Kirchhoff stress measure, the first Piola–Kirchhoff stress measure, and the Kirchhoff stress measure, respectively.

We thus write the problem dual to $(\overline{\mathrm{PE}})$ as

$$\left.\begin{aligned}
(\overline{\mathrm{CE}}):\ &\max_{\boldsymbol{q},\boldsymbol{\pi}_i,\tau_i^{\varphi},\boldsymbol{t}} -\sum_{i=1}^{m}\frac{1}{2}\frac{q_i^2}{k_{\mathrm{e}i}}-\sum_{i=1}^{m}\frac{\underline{l}_i}{2}(q_i+\operatorname{tr}\tau_i^{\varphi})+\sum_{j\in\Gamma_{\mathrm{D}}}\underline{\varphi}_j t_j+\sum_{j\in\Gamma_{\mathrm{N}}}\underline{p}_j x_j\\
\text{s. t.}\quad &\sum_{i=1}^{m}B_i^{\mathrm{T}}\boldsymbol{\pi}_i=\boldsymbol{t},\\
&\begin{bmatrix} q_i & \boldsymbol{\pi}_i^{\mathrm{T}}\\ \boldsymbol{\pi}_i & \tau_i^{\varphi}\end{bmatrix}\succeq O,\quad i=1,\dots,m,\\
&t_j=\underline{p}_j,\quad j\in\Gamma_{\mathrm{N}}.
\end{aligned}\right\}\tag{5.60}$$

5.3.3 Duality and optimality

A primal-dual pair, $(\overline{\mathrm{PE}})$ in (5.47) (or equivalently (4.50)) and $(\overline{\mathrm{CE}})$ in (5.60), is set parallel with the pair of standard forms, $(\mathrm{P_F})$ in (2.28) and $(\mathrm{P_F^*})$ in (2.31). In this section, through in-depth study of the optimality conditions for $(\overline{\mathrm{PE}})$ and $(\overline{\mathrm{CE}})$, we confirm that their optimal solutions correspond to the kinematic and static variables in the equilibrium. To begin, similar to section 5.1.3, we state the strong duality between $(\overline{\mathrm{PE}})$ and $(\overline{\mathrm{CE}})$.

Proposition 5.3.2. *Suppose that there exists a vector $\boldsymbol{\pi}=(\boldsymbol{\pi}_i\in\mathbb{R}^3\mid i=1,\dots,m)$ satisfying*

$$\Bigl(\sum_{i=1}^{m}B_i^{\mathrm{T}}\boldsymbol{\pi}_i\Bigr)_j=\underline{p}_j,\quad j\in\Gamma_{\mathrm{N}}.$$

Then, $(\overline{\mathrm{PE}})$ and $(\overline{\mathrm{CE}})$ each has an optimal solution, and $\min(\overline{\mathrm{PE}})=\max(\overline{\mathrm{CE}})$.

Proof. It suffices to show that the assumptions required in Proposition 2.2.9 are satisfied by the pair of $(\overline{\mathrm{PE}})$ and $(\overline{\mathrm{CE}})$ under consideration.

Assumption (i) in Proposition 2.2.12 is reduced in this case to the condition that there exists $(\boldsymbol{\varphi},\boldsymbol{c}_{\mathrm{e}})$ satisfying

$$\begin{bmatrix} c_{\mathrm{e}i}+(\underline{l}_i/2) & -(1/2)(B_i\boldsymbol{\varphi})^{\mathrm{T}}\\ -(1/2)B_i\boldsymbol{\varphi} & (\underline{l}_i/2)I_3\end{bmatrix}\succ O,\quad i=1,\dots,m,\tag{5.61}$$

$$\varphi_j=\underline{\varphi}_j,\quad j\in\Gamma_{\mathrm{D}},\tag{5.62}$$

while assumption (ii) is reduced to the condition that there exists $(\boldsymbol{q},\boldsymbol{\pi}_1,\dots,\boldsymbol{\pi}_m,\tau_1^{\varphi},\dots,\tau_m^{\varphi},\boldsymbol{t})$ satisfying

$$\Bigl(\sum_{i=1}^{m}B_i^{\mathrm{T}}\boldsymbol{\pi}_i\Bigr)_j=\underline{p}_j,\quad j\in\Gamma_{\mathrm{N}},\tag{5.63}$$

$$\begin{bmatrix} q_i & \boldsymbol{\pi}_i^{\mathrm{T}}\\ \boldsymbol{\pi}_i & \tau_i^{\varphi}\end{bmatrix}\succ O,\quad i=1,\dots,m.\tag{5.64}$$

We here omit the details of the reductions above, which are similar to Proposition 5.1.3. For the primal problem, choose a vector $\boldsymbol{\varphi}$ satisfying (5.62). Then there always exists a vector $\boldsymbol{c}_{\rm e}$ satisfying (5.61), because $c_{{\rm e}i}$ is not bounded from above. For the dual problem, suppose that (5.63) has a solution. Then, for any solution $\boldsymbol{\pi}$, we can choose a q_i with an arbitrary large value and a τ_i^{φ} with arbitrary large eigenvalues, so that (5.63) holds. Thus, the system of (5.61)–(5.64) becomes feasible if (5.63) has a solution. □

In what follows, instead of $(\boldsymbol{q}, \boldsymbol{\pi}_1, \dots, \boldsymbol{\pi}_m, \tau_1^{\varphi}, \dots, \tau_m^{\varphi}, \boldsymbol{t})$, we often write $(\boldsymbol{q}, \boldsymbol{\pi}_i, \tau_i^{\varphi}, \boldsymbol{t})$ for simple notation.[10]

We next establish the optimality conditions for $(\overline{\rm PE})$ and $(\overline{\rm CE})$.

Proposition 5.3.3. *Under the hypotheses in Proposition 5.3.2, the following statements hold:*

(i) $(\bar{\boldsymbol{\varphi}}, \bar{\boldsymbol{c}}_{\rm e})$ *and* $(\bar{\boldsymbol{q}}, \bar{\boldsymbol{\pi}}_i, \bar{\tau}_i^{\varphi}, \bar{\boldsymbol{t}})$ *are optimal for* $(\overline{\rm PE})$ *and* $(\overline{\rm CE})$*, respectively, if and only if they satisfy*

$$\bar{q}_i = k_{{\rm e}i}\bar{c}_{{\rm e}i}, \quad i = 1, \dots, m, \tag{5.65a}$$

$$\sum_{i=1}^{m} B_i^{\rm T}\bar{\boldsymbol{\pi}}_i = \bar{\boldsymbol{t}}, \tag{5.65b}$$

$$\bar{\varphi}_j = \underline{\varphi}_j, \quad j \in \Gamma_{\rm D}; \qquad \bar{t}_j = \underline{p}_j, \quad j \in \Gamma_{\rm N}, \tag{5.65c}$$

$$\begin{bmatrix} \bar{c}_{{\rm e}i} + (\underline{l}_i/2) & -(B_i\bar{\boldsymbol{\varphi}})^{\rm T}/2 \\ -B_i\bar{\boldsymbol{\varphi}}/2 & (\underline{l}_i/2)I_3 \end{bmatrix} \succeq O, \quad \begin{bmatrix} \bar{q}_i & \bar{\boldsymbol{\pi}}_i^{\rm T} \\ \bar{\boldsymbol{\pi}}_i & \bar{\tau}_i^{\varphi} \end{bmatrix} \succeq O, \quad i = 1, \dots, m, \tag{5.65d}$$

$$\begin{bmatrix} \bar{c}_{{\rm e}i} + (\underline{l}_i/2) & -(B_i\bar{\boldsymbol{\varphi}})^{\rm T}/2 \\ -B_i\bar{\boldsymbol{\varphi}}/2 & (\underline{l}_i/2)I_3 \end{bmatrix} \bullet \begin{bmatrix} \bar{q}_i & \bar{\boldsymbol{\pi}}_i^{\rm T} \\ \bar{\boldsymbol{\pi}}_i & \bar{\tau}_i^{\varphi} \end{bmatrix} = 0, \quad i = 1, \dots, m. \tag{5.65e}$$

(ii) *Optimal solutions* $(\bar{\boldsymbol{\varphi}}, \bar{\boldsymbol{c}}_{\rm e})$ *and* $(\bar{\boldsymbol{q}}, \bar{\boldsymbol{\pi}}_i, \tau_i^{\varphi}, \bar{\boldsymbol{t}})$ *of* $(\overline{\rm PE})$ *and* $(\overline{\rm CE})$ *satisfy*

$$\bar{c}_{{\rm e}i} - \hat{c}_i(\bar{\boldsymbol{\varphi}}) \geq 0, \quad \bar{q}_i \geq 0, \quad (\bar{c}_{{\rm e}i} - \hat{c}_i(\bar{\boldsymbol{\varphi}}))\bar{q}_i = 0$$

for each $i = 1, \dots, m$*, where* $\hat{c}_i(\bar{\boldsymbol{\varphi}}) = [(B_i\bar{\boldsymbol{\varphi}})^{\rm T}(B_i\bar{\boldsymbol{\varphi}}) - \underline{l}_i^2]/(2\underline{l}_i)$.

Proof. In a manner similar to Proposition 5.1.4 (i), assertion (i) is obtained by applying Proposition 2.2.9; the calculations are omitted. Assertion (ii) then follows from Proposition 4.2.1, which is also similar to Proposition 5.1.4 (ii). □

As a consequence of Proposition 5.3.3, we see in (5.65a) that the dual variable q_i corresponds to the axial force of the ith member, because (5.65a) is the constitutive law that relates the elongation $c_{{\rm e}i}$ to the axial force. For clarifying the physical meanings of the other variables, the solution property of $(\overline{\rm CE})$ is investigated below.

[10] It should be clear that $\tau_i^{\varphi} \in \mathbb{S}^3$ $(i = 1, \dots, m)$.

Proposition 5.3.4. *If* $(\bar{\boldsymbol{q}}, \bar{\boldsymbol{\pi}}_i, \bar{\tau}_i^{\varphi}, \bar{\boldsymbol{t}})$ *is optimal for* $(\overline{\text{CE}})$, *then it satisfies*

$$\bar{\tau}_i^{\varphi} = \begin{cases} \bar{\boldsymbol{\pi}}_i \bar{\boldsymbol{\pi}}_i^{\mathrm{T}} / \bar{q}_i & \text{if } \bar{q}_i \neq 0, \\ O & \text{if } \bar{q}_i = 0, \end{cases} \tag{5.66}$$

for each $i = 1, \dots, m$.

Proof. Suppose that $\bar{q}_i \neq 0$. In this case, it follows from Fact 1.3.10 that the positive-semidefinite constraint in $(\overline{\text{CE}})$ can be rewritten as

$$\begin{bmatrix} q_i & \boldsymbol{\pi}_i^{\mathrm{T}} \\ \boldsymbol{\pi}_i & \tau_i^{\varphi} \end{bmatrix} \succeq O \quad \Leftrightarrow \quad \tau_i^{\varphi} \succeq \frac{\boldsymbol{\pi}_i \boldsymbol{\pi}_i^{\mathrm{T}}}{q_i}, \; q_i > 0 \tag{5.67}$$

without changing the optimal solution. Observe that, in $(\overline{\text{CE}})$, the variable τ_i^{φ} is subjected solely to (5.67). In the objective function, τ_i^{φ} is included only in the term $-\underline{l}_i \operatorname{tr} \tau_i^{\varphi}/2$, which is to be maximized. Since $\underline{l}_i > 0$, the optimal solution should satisfy

$$\bar{\tau}_i^{\varphi} \in \arg \min_{\tau_i^{\varphi} \in \mathbb{S}^3} \{ \operatorname{tr} \tau_i^{\varphi} \mid \tau_i^{\varphi} \succeq \bar{\boldsymbol{\pi}}_i \bar{\boldsymbol{\pi}}_i^{\mathrm{T}} / \bar{q}_i \}. \tag{5.68}$$

Noting that $\operatorname{tr} \tau_i^{\varphi}$ is the sum of eigenvalues of τ_i^{φ}, we see that the minimum in (5.68) is attained at $\bar{\tau}_i^{\varphi} = \bar{\boldsymbol{\pi}}_i \bar{\boldsymbol{\pi}}_i^{\mathrm{T}} / \bar{q}_i$ (see Proposition 9.4.7 for a rigorous proof).

Alternatively, suppose that $\bar{q}_i = 0$. Then the constraint of $(\overline{\text{CE}})$ is reduced to

$$\begin{bmatrix} 0 & \boldsymbol{\pi}_i^{\mathrm{T}} \\ \boldsymbol{\pi}_i & \tau_i^{\varphi} \end{bmatrix} \succeq O,$$

but this requires $\boldsymbol{\pi}_i = \mathbf{0}$. Hence, τ_i^{φ} is subjected to

$$\begin{bmatrix} 0 & \mathbf{0}^{\mathrm{T}} \\ \mathbf{0} & \tau_i^{\varphi} \end{bmatrix} \succeq O. \tag{5.69}$$

Minimizing $\operatorname{tr} \tau_i^{\varphi}$ under (5.69) leads to $\bar{\tau}_i^{\varphi} = O$. □

By combining Proposition 5.3.3 and Proposition 5.3.4, we obtain the following property of optimal solutions, which acts as a bridge between the primal and the dual variables.

Proposition 5.3.5. *Suppose that* $(\bar{\varphi}, \bar{\boldsymbol{c}}_{\mathrm{e}})$ *and* $(\bar{\boldsymbol{q}}, \bar{\boldsymbol{\pi}}_i, \bar{\tau}_i^{\varphi}, \bar{\boldsymbol{t}})$ *are optimal for* $(\overline{\text{PE}})$ *and* $(\overline{\text{CE}})$, *respectively. Then they satisfy*

$$\bar{\boldsymbol{\pi}}_i = \bar{q}_i \frac{B_i \bar{\varphi}}{\underline{l}_i}, \quad i = 1, \dots, m.$$

Proof. If $\bar{q}_i = 0$, then it follows from Proposition 5.3.4 and its proof that $\bar{\tau}_i^{\varphi} = O$ and $\bar{\boldsymbol{\pi}}_i = \mathbf{0}$, and hence nothing is to be proved. Therefore, we suppose that $\bar{q}_i \neq 0$.

Define δ_i by

$$\delta_i = \begin{bmatrix} \bar{c}_{ei} + (\underline{l}_i/2) & -(B_i\bar{\boldsymbol{\varphi}})^{\mathrm{T}}/2 \\ -B_i\bar{\boldsymbol{\varphi}}/2 & (\underline{l}_i/2)I_3 \end{bmatrix} \bullet \begin{bmatrix} \bar{q}_i & \bar{\boldsymbol{\pi}}_i^{\mathrm{T}} \\ \bar{\boldsymbol{\pi}}_i & \bar{\tau}_i^{\varphi} \end{bmatrix}. \tag{5.70}$$

It follows from Proposition 5.3.3 that the optimal solution satisfies (5.65). Among (5.65a)–(5.65e), (5.65d) implies $\delta_i \geq 0$ $(i = 1, \dots, m)$, because of the self-duality of the positive-semidefinite cone (see Fact 1.3.15). Hence, (5.65e) holds if and only if

$$\delta_i = 0, \quad i = 1, \dots, m, \tag{5.71}$$

where (5.70) is reduced to

$$\delta_i = \bar{q}_i\left(\bar{c}_{ei} + \frac{\underline{l}_i}{2}\right) - \bar{\boldsymbol{\pi}}_i^{\mathrm{T}} B_i\bar{\boldsymbol{\varphi}} + \frac{\underline{l}_i}{2}\operatorname{tr}\bar{\tau}_i^{\varphi} \tag{5.72}$$

after simple calculations.

From the hypothesis $\bar{q} > 0$ and Proposition 5.3.3 (ii), an optimal solution $(\bar{\boldsymbol{\varphi}}, \bar{\boldsymbol{c}}_{\mathrm{e}})$ satisfies $\bar{c}_{ei} - \hat{c}_i(\bar{\boldsymbol{\varphi}}) = 0$, i.e.,

$$\bar{c}_{ei} = \frac{1}{2\underline{l}_i}[(B_i\bar{\boldsymbol{\varphi}})^{\mathrm{T}}(B_i\bar{\boldsymbol{\varphi}}) - \underline{l}_i^2]. \tag{5.73}$$

On the other hand, from Proposition 5.3.4 we have

$$\operatorname{tr}\bar{\tau}_i^{\varphi} = \frac{\bar{\boldsymbol{\pi}}_i^{\mathrm{T}}\bar{\boldsymbol{\pi}}_i}{\bar{q}_i}. \tag{5.74}$$

Substitution of (5.73) and (5.74) into (5.72) yields

$$\begin{aligned} \delta_i &= \frac{\bar{q}_i}{2\underline{l}_i}(B_i\bar{\boldsymbol{\varphi}})^{\mathrm{T}}(B_i\bar{\boldsymbol{\varphi}}) - \bar{\boldsymbol{\pi}}_i^{\mathrm{T}} B_i\bar{\boldsymbol{\varphi}} + \frac{\underline{l}_i}{2}\frac{\bar{\boldsymbol{\pi}}_i^{\mathrm{T}}\bar{\boldsymbol{\pi}}_i}{\bar{q}_i} \\ &= \frac{\underline{l}_i}{2\bar{q}_i}\left[\frac{\bar{q}_i^2}{\underline{l}_i^2}(B_i\bar{\boldsymbol{\varphi}})^{\mathrm{T}}(B_i\bar{\boldsymbol{\varphi}}) - 2\bar{\boldsymbol{\pi}}_i^{\mathrm{T}}\left(\frac{\bar{q}_i}{\underline{l}_i}B_i\bar{\boldsymbol{\varphi}}\right)\frac{\bar{\boldsymbol{\pi}}_i^{\mathrm{T}}\bar{\boldsymbol{\pi}}_i}{\bar{q}_i}\right] \\ &= \frac{\underline{l}_i}{2\bar{q}_i}[\bar{\boldsymbol{\pi}}_i - (\bar{q}_i/\underline{l}_i)B_i\bar{\boldsymbol{\varphi}}]^{\mathrm{T}}[\bar{\boldsymbol{\pi}}_i - (\bar{q}_i/\underline{l}_i)B_i\bar{\boldsymbol{\varphi}}]. \end{aligned}$$

Consequently, (5.71) implies $\bar{\boldsymbol{\pi}}_i = (\bar{q}_i/\underline{l}_i)B_i\bar{\boldsymbol{\varphi}}$. □

Recall that, as studied in Remark 4.2.4, vector $B_i\bar{\boldsymbol{\varphi}} \in \mathbb{R}^3$ is regarded as a deformation gradient, in the sense that it has the same direction and modulus as the ith member at the deformed configuration. Therefore, Proposition 5.3.5 means that $\bar{\boldsymbol{\pi}}_i$ is parallel with the deformed configuration of the ith member. Moreover, if the strain is small enough (which is the case when the Green–Lagrange strain is valid), then the variation of the member length is negligibly small, say, $\|B_i\bar{\boldsymbol{\varphi}}\| \simeq \underline{l}_i$. Consequently, Proposition 5.3.5 implies $\|\bar{\boldsymbol{\pi}}_i\| \simeq \bar{q}_i$. Thus $\bar{\boldsymbol{\pi}}_i$ has the same direction as the deformed member configuration and almost the same modulus as the axial force,[11] and hence $\bar{\boldsymbol{\pi}}_i$ is interpreted

[11]This condition is to be compared with Proposition 5.1.6 established for the case of finite strain, in which $\|\bar{\boldsymbol{v}}_i\| = \bar{q}_i$ holds, and thence we have seen that $\bar{\boldsymbol{v}}_i$ is regarded as the internal force vector in the Biot stress measure.

as the internal force vector at the deformed state. Thus the compatibility between the primal (kinematic) variables and dual (static) variables at the optimal solutions is ensured by the complementarity condition (5.65e) over the positive-semidefinite cones in (5.65d).

In the terminology of continuum mechanics, the dual variables are interpreted as the internal forces corresponding to the following stresses:

- q_i: the second Piola–Kirchhoff stress
- $\boldsymbol{\pi}_i$: the first Piola–Kirchhoff stress
- τ_i^{φ}: the Kirchhoff stress

It is known in continuum mechanics that, for the second Piola–Kirchhoff stress tensor $\boldsymbol{S}$ and deformation gradient $\boldsymbol{F}$, the first Piola–Kirchhoff stress tensor and the Kirchhoff stress tensor $\boldsymbol{\tau}^{\varphi}$ are given by $\boldsymbol{\Pi} = \boldsymbol{F}\boldsymbol{S}$ and $\boldsymbol{\tau}^{\varphi} = \boldsymbol{\Pi}\boldsymbol{F}$. In the current situation, $c_{\mathrm{e}i}$ is in the Green–Lagrange strain measure, and (5.65a) implies that $c_{\mathrm{e}i}$ and q_i are work-conjugate. Hence, q_i corresponds to the second Piola–Kirchhoff stress. From the discussion above, the result in Proposition 5.3.5 asserts that $\boldsymbol{\pi}_i$ corresponds to the first Piola–Kirchhoff stress. Substituting this into the result in Proposition 5.3.4, we obtain $\tau_i^{\varphi} = \boldsymbol{\pi}_i(B_i\boldsymbol{\varphi}/\underline{l}_i)^{\mathrm{T}}$, which means that τ_i^{φ} corresponds to the Kirchhoff stress.

5.3.4 Principle of complementary energy

We are now in a position to state the minimum principle of complementary energy in terms of the first and second Piola–Kirchhoff stresses and the Kirchhoff stress. By changing the sign of the objective function, $(\overline{\mathrm{CE}})$ in (5.60) is converted to a minimization problem. Moreover, by substituting the result of Proposition 5.3.4, we obtain

$$
\left.
\begin{aligned}
(\mathrm{CE}): \quad & \min_{\boldsymbol{q},\boldsymbol{\pi}_i,\tau_i^{\varphi},\boldsymbol{t}} \ \sum_{i=1}^{m} \frac{1}{2}\frac{q_i^2}{k_{\mathrm{e}i}} + \sum_{i=1}^{m} \frac{\underline{l}_i}{2}(q_i + \operatorname{tr}\tau_i^{\varphi}) \\
& \qquad - \sum_{j\in\Gamma_{\mathrm{D}}} \underline{\varphi}_j t_j - \sum_{j\in\Gamma_{\mathrm{N}}} \underline{p}_j x_j \\
\mathrm{s.t.} \quad & \sum_{i=1}^{m} B_i^{\mathrm{T}}\boldsymbol{\pi}_i = \boldsymbol{t}, \\
& q_i\tau_i^{\varphi} = \boldsymbol{\pi}_i\boldsymbol{\pi}_i^{\mathrm{T}}, \ q_i \ge 0, \quad i = 1,\dots,m, \\
& t_j = \underline{p}_j, \quad j \in \Gamma_{\mathrm{N}}.
\end{aligned}
\right\}
\tag{5.75}
$$

For the pair of energy minimization problems, i.e., (PE) in (4.49) and (CE) in (5.75),[12] the optimality conditions are derived below by using the results es-

[12] In a manner similar to (5.26), the term $q_i^2/(2k_{\mathrm{e}i})$ in the objective function of (5.26) can be replaced with $\hat{w}_{\mathrm{cab}}^{i*}(q_i)$ in (4.9), because (5.26) includes the constraint $q_i \ge 0$. The duality between (PE) and (CE) may be captured more clearly with this $\hat{w}_{\mathrm{cab}}^{i*}(q_i)$, because $\hat{w}_{\mathrm{cab}}^{i}$ in (PE) is the conjugate function of $\hat{w}_{\mathrm{cab}}^{i*}$.

tablished for $(\overline{\mathrm{PE}})$ and $(\overline{\mathrm{CE}})$ to show that their optimal solutions corresponds to the equilibrium states.

Proposition 5.3.6. *Under the hypotheses in Proposition 5.3.2,* $(\bar{\boldsymbol{\varphi}}, \bar{\boldsymbol{c}})$ *and* $(\bar{\boldsymbol{q}}, \bar{\boldsymbol{\pi}}_i, \bar{\tau}_i^{\varphi}, \bar{\boldsymbol{t}})$ *are optimal solutions of* (PE) *and* (CE)*, respectively, if and only if they corresponds to the kinematic and static variables*[13] *at the equilibrium state of the cable network. Moreover,* $\min(\mathrm{PE}) = -\min(\mathrm{CE})$.

Proof. The assertion is proved by writing the optimality conditions for (PE) and (CE) in a manner similar to Proposition 5.1.7. We employ the optimality conditions for $(\overline{\mathrm{PE}})$ and $(\overline{\mathrm{CE}})$ (in Proposition 5.3.3) and the relation between $(\overline{\mathrm{PE}})$ and (PE) (in Proposition 4.2.1). Moreover, it is clear from its construction procedure that (CE) has the same optimality condition as that for $(\overline{\mathrm{CE}})$. Thus the optimality conditions for $(\overline{\mathrm{PE}})$ and $(\overline{\mathrm{CE}})$ can be transfered to those for (PE) and (CE). Particularly, the condition stated in Proposition 5.3.5 is also satisfied at the optimal solutions of $(\overline{\mathrm{PE}})$ and (CE). As a consequence, necessary and sufficient conditions for optimality of (PE) and (CE) are written as

$$\bar{q}_i = \hat{q}^i_{\mathrm{cab}}(\bar{c}_i), \quad i = 1, \dots, m, \tag{5.76}$$

$$\bar{c}_i = \frac{1}{2\underline{l}_i}\left[(B_i\bar{\boldsymbol{\varphi}})^{\mathrm{T}}(B_i\bar{\boldsymbol{\varphi}}) - \underline{l}_i^2\right], \quad i = 1, \dots, m, \tag{5.77}$$

$$\bar{\boldsymbol{\pi}}_i = \bar{q}_i \frac{B_i\bar{\boldsymbol{\varphi}}}{\underline{l}_i}, \quad i = 1, \dots, m, \tag{5.78}$$

$$\bar{\tau}_i^{\varphi} = \frac{\bar{\boldsymbol{\pi}}_i\bar{\boldsymbol{\pi}}_i^{\mathrm{T}}}{\bar{q}_i}, \quad i = 1, \dots, m, \tag{5.79}$$

$$\sum_{i=1}^{m} B_i^{\mathrm{T}}\bar{\boldsymbol{\pi}}_i = \bar{\boldsymbol{t}}, \tag{5.80}$$

$$\bar{\varphi}_j = \underline{\varphi}_j, \quad j \in \Gamma_{\mathrm{D}}; \qquad \bar{t}_j = \underline{p}_j, \quad j \in \Gamma_{\mathrm{N}}, \tag{5.81}$$

where we define $O/0 = O$ in (5.79).

We thus see that solving (PE) and (CE) is equivalent to solving (5.76)–(5.81), where the latter one corresponds to the the system of governing equations for the equilibrium of a cable network. Indeed, (5.76) is the constitutive law of "no-compression" cable members, (5.77) is the compatibility relation of the static variables, while (5.79) states the compatibility of the static variables. The compatibility between the kinematic and static variables is given in (5.78), the force-balance equation is in (5.80), and the boundary conditions are in (5.81). Thus the optimal solutions of (PE) and (CE) truly correspond to the equilibrium states. □

[13] As stated in section 4.2.1, $\boldsymbol{\varphi}$ and c_i are the nodal location vector and the elongation of the ith member in the Green–Lagrange strain measure, respectively. Moreover, as discussed in section 5.3.3, q_i, $\boldsymbol{\pi}_i$, and τ_i^{φ} are the internal force vectors in the second Piola–Kirchhoff stress, first Piola–Kirchhoff stress, and the Kirchhoff stress, respectively, while $\boldsymbol{t}$ is the nodal force vector.

The objective function of (CE) corresponds to the complementary energy function for cable networks undergoing large rotation. Indeed, this function consists of the following four factors:

- $\dfrac{1}{2}\dfrac{\|\boldsymbol{v}_i\|^2}{k_{ei}}$: the complementary stored energy
- $-\sum_{j\in\Gamma_{\rm D}} \underline{\varphi}_j t_j$: the complementary work by the prescribed displacements.
- $-\sum_{j\in\Gamma_{\rm N}} \underline{p}_j x_j$: a constant term depending on the reference configuration
- $\underline{l}_i(q_i + \operatorname{tr}\tau_i^{\varphi})/2$: the complementary work due to the large rotation

The term $\underline{l}_i(q_i + \operatorname{tr}\tau_i^{\varphi})/2$ is characteristic of the large rotation theory (see Remark 5.1.11 for the case of large strain).

5.4 Notes

The minimum principles of complementary energy for cable networks were investigated in this chapter. The case for large strain, presented in section 5.1, is in Kanno–Ohsaki [204]. The analysis and results in section 5.3 for the Green–Lagrange strain are original.

The complementary energy principle in the geometrically nonlinear theory is an age-old question in applied and computational mechanics; see Remark 5.1.10, Gao [143], and Koiter [236]. See also Bertóti [50], Campos–Oden [68], Fraeijs de Veubeke [134], Gao [142, 143, 144], Guo [156], Labisch [249], Reissner [386], Valid [443], and Zubov [473].

An extension to a cable network subjected to distributed load can be found in Kanno–Ohsaki [206].

The existence and uniqueness of static equilibrium solutions of cable networks were studied by Atai–Steigmann [23], Kanno–Ohsaki [204], and Volokh–Vilnay [448].

Part III

Numerical Methods

Chapter 6

Algorithms for Conic Optimization

Orientation This chapter treats numerical methods for the solution of conic optimization problems as well as some related problems. Among others, our attention should focus on two approaches: the primal-dual interior-point methods; and the reformulation (with/without smoothing) methods. When the problem under consideration includes conic constraints associated with a self-dual cone, these two methods can completely enjoy the self-duality property of the cone.

This chapter assumes familiarity with the conventional Newton method for nonlinear equations.

6.1 Primal-Dual Interior-Point Method

This section sketches the typical primal-dual interior-point methods for conic optimization problems. Following a short overview in section 6.1.1 of the interior-point methods, the primal-dual interior-point methods for LP (linear programming) and SDP (semidefinite programming) are introduced in section 6.1.2 and section 6.1.3, respectively.

6.1.1 Outline of interior-point methods

The interior-point method solves, in general, a nonlinear optimization problem by generating iterates that strictly satisfy the inequality constraints included in the problem. In this sense, the interior-point method is originated from the *SUMT* (sequential unconstrained minimization technique) proposed by Fiacco–McCormick [131] in the 1960s, which is usually called the *barrier method* today.

A subclass of the interior-point methods, known as the primal-dual interior-point methods, was proposed as solution methods for LP problems in the late 1980s [238]. By the early 1990s, the primal-dual interior-point method, rather than the simplex method, was recognized as the most efficient practical approach to LP problems. From a theoretical aspect, the efficiency of the primal-dual interior-point method is ensured because it is a polynomial-time algorithm, which means that the number of arithmetic operations before the solution is obtained is bounded by a polynomial of the problem size. In the primal-dual interior-point method, the "curve" defined in the feasible set of the LP problem, called the *central path*, plays a crucial role; roughly speaking, the solution of the LP is found by tracing the central path numerically. The central path is defined as the solution set of a family of modified KKT conditions but can also be characterized as the set of solutions of the family of barrier problems [307]. Thus the primal-dual interior-point method is related to the classical barrier method, though a significant distinction is that the primal-dual interior-point method updates both the primal and dual variables at each iteration, while the classical barrier method treats the primal and dual variables in different ways.

Following the success in LP problems, the primal-dual interior-point method has been extended to nonlinear optimization problems primarily by two ways. The first one is based on the self-concordant barrier function proposed by Nesterov–Nemirovski [333], which provided the polynomial-time algorithms for QP (quadratic programming) problems [239, 315], SDP problems [9, 173, 243, 335], and SOCP (second-order cone programming) problems [317, 438]. Thus, roughly speaking, a vast class of convex optimization problems is solved efficiently by the primal-dual interior-point methods. For the second way, the primal-dual interior-point methods for NLP (nonlinear programming) problems (which are not convex in general) have been developed; see a survey by Forsgren–Gill–Wright [132]. The algorithms in this category can be compared with general nonlinear programming approaches, e.g., SQP (sequential quadratic programming) methods [153]; they have no polynomial-time property and in general converge to a local optimal solution of the nonlinear programming problem.

For solving a conic optimization problem,[1] the algorithms in the former category, which can fully enjoy the favorable property of the problem, have many more advantages than those in the latter category. Therefore, the methods specifically designed for the conic optimization, which fall into the former category, are the focus of this chapter. From the viewpoint of practical computation, many efficient software packages based on the primal-dual interior-point methods are available today. Hence, only the outlines of these algorithms are sketched, and no convergence analysis is given in the present volume; see, e.g., [340, 460] for comprehensive analysis of the interior-point methods.

6.1.2 Interior-point method for linear programming

This section introduces the interior-point methods for LP problems. Algorithms for solving QP problems can be designed in a similar manner; see Vanderbei [445] and Ye [460].

[1] Almost all problems considered in this book can be reduced either to LP, SOCP, or SDP; An exception is the frictional contact problem with Coulomb friction treated in Chapter 10.

6.1.2.1 Optimality condition

We consider the LP problem in the standard form,[2] i.e.,

$$(\mathrm{P_{LP}}) : \left.\begin{array}{l} \min\limits_{\boldsymbol{x}} \boldsymbol{c}^{\mathrm{T}}\boldsymbol{x} \\ \text{s.t. } A\boldsymbol{x} = \boldsymbol{b}, \\ \qquad \boldsymbol{x} \geq \boldsymbol{0}, \end{array}\right\} \tag{6.1}$$

where $\boldsymbol{x} \in \mathbb{R}^n$ is the variable vector, $\boldsymbol{b} \in \mathbb{R}^m$ and $\boldsymbol{c} \in \mathbb{R}^n$ are constant vectors, and $A \in \mathbb{R}^{m \times n}$ is a constant matrix. Note that we may assume without loss of generality that $\operatorname{rank} A = m$. We say that $\boldsymbol{x}$ is a feasible solution of $(\mathrm{P_{LP}})$ if it satisfies all the constraints in (6.1) .

The problem dual to (6.1) is given by

$$(\mathrm{D_{LP}}) : \left.\begin{array}{l} \max\limits_{\boldsymbol{y},\boldsymbol{s}} \boldsymbol{b}^{\mathrm{T}}\boldsymbol{y} \\ \text{s.t. } A^{\mathrm{T}}\boldsymbol{y} + \boldsymbol{s} = \boldsymbol{c}, \\ \qquad \boldsymbol{s} \geq \boldsymbol{0}, \end{array}\right\} \tag{6.2}$$

where $\boldsymbol{y} \in \mathbb{R}^m$ and $\boldsymbol{s} \in \mathbb{R}^n$ are the variable vectors. Like the case of $(\mathrm{P_{LP}})$, we say that $(\boldsymbol{y}, \boldsymbol{s})$ is a feasible solution of $(\mathrm{D_{LP}})$ if $\boldsymbol{y}$ and $\boldsymbol{s}$ satisfy all the constraints in (6.2).

In the following, for a vector $\boldsymbol{x} \in \mathbb{R}^n$ we write $\boldsymbol{x} > \boldsymbol{0}$ if $x_i > 0$ $(i = 1, \dots, n)$. Define $\mathcal{F}$, $\mathcal{F}^\circ \subset \mathbb{R}^n \times \mathbb{R}^m \times \mathbb{R}^n$ by

$$\mathcal{F} = \{(\boldsymbol{x}, \boldsymbol{y}, \boldsymbol{s}) \mid A\boldsymbol{x} = \boldsymbol{b},\ A^{\mathrm{T}}\boldsymbol{y} + \boldsymbol{s} = \boldsymbol{c},\ \boldsymbol{x} \geq \boldsymbol{0},\ \boldsymbol{s} \geq \boldsymbol{0}\},$$
$$\mathcal{F}^\circ = \{(\boldsymbol{x}, \boldsymbol{y}, \boldsymbol{s}) \mid A\boldsymbol{x} = \boldsymbol{b},\ A^{\mathrm{T}}\boldsymbol{y} + \boldsymbol{s} = \boldsymbol{c},\ \boldsymbol{x} > \boldsymbol{0},\ \boldsymbol{s} > \boldsymbol{0}\},$$

where $\mathcal{F}^\circ = \operatorname{ri} \mathcal{F}$. Note that $\mathcal{F}$ is the set of all the feasible solutions of $(\mathrm{P_{LP}})$ and $(\mathrm{D_{LP}})$ and hence is called the *feasible set*. We call $\mathcal{F}^\circ$ the *strictly feasible set*, and we say that $(\boldsymbol{x}, \boldsymbol{y}, \boldsymbol{s})$ is a strictly feasible solution if $(\boldsymbol{x}, \boldsymbol{y}, \boldsymbol{s}) \in \mathcal{F}^\circ$.

Suppose that $\mathcal{F} \neq \emptyset$. Then $(\mathrm{P_{LP}})$ and $(\mathrm{D_{LP}})$ have optimal solutions, and their optimal values coincide. Moreover, the optimal solutions of $(\mathrm{P_{LP}})$ and $(\mathrm{D_{LP}})$ are characterized by the KKT conditions,[3] stated as

$$A\boldsymbol{x} = \boldsymbol{b}, \tag{6.3a}$$
$$A^{\mathrm{T}}\boldsymbol{y} + \boldsymbol{s} = \boldsymbol{c}, \tag{6.3b}$$
$$x_i s_i = 0, \quad i = 1, \dots, n, \tag{6.3c}$$
$$\boldsymbol{x} \geq \boldsymbol{0}, \quad \boldsymbol{s} \geq \boldsymbol{0}. \tag{6.3d}$$

Note that (6.3a) and (6.3b) are linear equations, which can be handled easily. The nonlinear equations (6.3c) as well as the linear inequalities (6.3d) require particular treatments in numerical solutions.

[2]See Definition 1.6.1 in section 1.6.1, together with Remark 1.6.2.

[3]Which can be derived in a manner similar to section 2.2.7.

For considering primal-dual interior-point methods, we rewrite (6.3) as

$$\begin{bmatrix} A\boldsymbol{x} \\ A^{\mathrm{T}}\boldsymbol{y}+\boldsymbol{s} \\ X\boldsymbol{s} \end{bmatrix} = \begin{bmatrix} \boldsymbol{b} \\ \boldsymbol{c} \\ \boldsymbol{0} \end{bmatrix}, \tag{6.4a}$$

$$(\boldsymbol{x},\boldsymbol{s}) \geq \boldsymbol{0}, \tag{6.4b}$$

where

$$X = \mathrm{diag}(\boldsymbol{x}).$$

Roughly speaking, a primal-dual interior-point method solves (6.4) numerically with respect to $(\boldsymbol{x},\boldsymbol{y},\boldsymbol{s})$, to the optimal solutions of ($\mathrm{P_{LP}}$) and ($\mathrm{D_{LP}}$) simultaneously.

Each iteration of an interior-point method is characterized by the search direction and the step size. The central path, introduced below, plays a crucial role in determining the search direction.

6.1.2.2 Central path and path-following method

To understand the importance of the notion of central path, we first consider a crude procedure in which the conventional Newton method is employed to solve the KKT conditions (6.4).

At the current point $(\boldsymbol{x}^k,\boldsymbol{y}^k,\boldsymbol{s}^k)$, the Newton direction for (6.4a) is defined as the solution of the following Newton equations:

$$\begin{bmatrix} A & O & O \\ O & A^{\mathrm{T}} & I \\ S^k & O & X^k \end{bmatrix} \begin{bmatrix} \Delta\boldsymbol{x} \\ \Delta\boldsymbol{y} \\ \Delta\boldsymbol{s} \end{bmatrix} = - \begin{bmatrix} \boldsymbol{r}_b^k \\ \boldsymbol{r}_c^k \\ X^k\boldsymbol{s}^k \end{bmatrix}, \tag{6.5}$$

where $X^k = \mathrm{diag}(\boldsymbol{x}^k)$, $S^k = \mathrm{diag}(\boldsymbol{s}^k)$, and

$$\boldsymbol{r}_b^k = A\boldsymbol{x}^k - \boldsymbol{b}, \tag{6.6}$$

$$\boldsymbol{r}_c^k = A^{\mathrm{T}}\boldsymbol{y}^k + \boldsymbol{s}^k - \boldsymbol{c}. \tag{6.7}$$

The linear equations (6.5) are solved to obtain the search direction, called the *affine scaling direction*, with which the new point is defined as

$$\begin{bmatrix} \boldsymbol{x}^{k+1} \\ \boldsymbol{y}^{k+1} \\ \boldsymbol{s}^{k+1} \end{bmatrix} = \begin{bmatrix} \boldsymbol{x}^{k} \\ \boldsymbol{y}^{k} \\ \boldsymbol{s}^{k} \end{bmatrix} + \alpha_k \begin{bmatrix} \Delta\boldsymbol{x}^{k} \\ \Delta\boldsymbol{y}^{k} \\ \Delta\boldsymbol{s}^{k} \end{bmatrix}. \tag{6.8}$$

Here, $\alpha_k > 0$ is called the *step size.*

Suppose that a strictly feasible initial solution is given, i.e., $(\boldsymbol{x}^0,\boldsymbol{y}^0,\boldsymbol{s}^0) \in \mathcal{F}^{\circ}$. Consider the iteration defined by (6.8), in which $\alpha_k \in (0,1]$ is chosen so that $(\boldsymbol{x}^{k+1},\boldsymbol{s}^{k+1}) > \boldsymbol{0}$ is satisfied. Repeating this procedure, we can (approximately) solve (6.3) [316]. However, such a method using (6.5) is less effective practically speaking, because often only a very small step size, say, $\alpha \ll 1$, is accepted.

Roughly speaking, this happens when the current solution $(\boldsymbol{x}^k, \boldsymbol{y}^k, \boldsymbol{s}^k)$ is located "near" the boundary of $\mathcal{F}$. One way to avoid going toward the boundary is to consider a curve lying at the "center" of $\mathcal{F}$ that leads to the solution of (6.3). Then we can arrive at the solution by tracing that curve without approaching the boundary of $\mathcal{F}$ at the intermediate iterations, which allows large step sizes. Such a curve serving as a "guide" to the solution is called the *central path*.

Let $\nu > 0$ be a constant scalar, and consider

$$A\boldsymbol{x} = \boldsymbol{b}, \tag{6.9a}$$

$$A^{\mathrm{T}}\boldsymbol{y} + \boldsymbol{s} = \boldsymbol{c}, \tag{6.9b}$$

$$x_i s_i = \nu, \quad i = 1, \dots, n, \tag{6.9c}$$

$$\boldsymbol{x} > \boldsymbol{0}, \quad \boldsymbol{s} > \boldsymbol{0}. \tag{6.9d}$$

We call (6.9) the *modified KKT conditions*. Indeed, by putting $\nu = 0$ in (6.9c) and replacing "$>$" in (6.9d) with "$\geq$", we recover the KKT conditions in (6.3). It is known that the solution to (6.9) exists uniquely (if $\mathcal{F}^{\circ} \neq \emptyset$).

A trajectory of the solutions $(\boldsymbol{x}, \boldsymbol{y}, \boldsymbol{s})$ of (6.9) with respect to the parameter $\nu > 0$ is called the central path. More precisely, the central path, denoted $\mathcal{C}$, is the subset of $\mathbb{R}^{2n+m}$ defined by

$$\mathcal{C} = \{(\boldsymbol{x}, \boldsymbol{y}, \boldsymbol{s}) \mid \boldsymbol{g}(\boldsymbol{x}, \boldsymbol{y}, \boldsymbol{s}; \nu) = \boldsymbol{0},\ (\boldsymbol{x}, \boldsymbol{s}) > \boldsymbol{0},\ \nu > 0\}, \tag{6.10}$$

where $\boldsymbol{g}(\cdot; \nu) : \mathbb{R}^{2n+m} \to \mathbb{R}^{2n+m}$ is defined by

$$\boldsymbol{g}(\boldsymbol{x}, \boldsymbol{y}, \boldsymbol{s}; \nu) = \begin{bmatrix} A\boldsymbol{x} \\ A^{\mathrm{T}}\boldsymbol{y} + \boldsymbol{s} \\ X\boldsymbol{s} \end{bmatrix} - \begin{bmatrix} \boldsymbol{b} \\ \boldsymbol{c} \\ \nu\boldsymbol{1} \end{bmatrix} \tag{6.11}$$

with $\boldsymbol{1} = (1, 1, \dots, 1)^{\mathrm{T}}$. Then the central path $\mathcal{C}$ is traced numerically by decreasing $\nu \searrow 0$ to find the solution of (6.3).

For defining the search direction specifically, we use the *duality measure* defined by

$$\rho_k = \frac{(\boldsymbol{x}^k)^{\mathrm{T}}\boldsymbol{s}^k}{n} \tag{6.12}$$

at the current solution $(\boldsymbol{x}^k, \boldsymbol{y}^k, \boldsymbol{s}^k)$. Choose the reduction factor $\sigma_k \in]0, 1]$ to define ν for the next iteration by

$$\nu_{k+1} = \sigma_k \rho_k. \tag{6.13}$$

With this ν_{k+1}, the search direction is obtained as the Newton direction for $\boldsymbol{g}(\boldsymbol{x}, \boldsymbol{y}, \boldsymbol{s}; \nu) = \boldsymbol{0}$. Referring to (6.11), the Newton equations are written as

$$\begin{bmatrix} A & O & O \\ O & A^{\mathrm{T}} & I \\ S^k & O & X^k \end{bmatrix} \begin{bmatrix} \Delta\boldsymbol{x} \\ \Delta\boldsymbol{y} \\ \Delta\boldsymbol{s} \end{bmatrix} = - \begin{bmatrix} \boldsymbol{r}_b^k \\ \boldsymbol{r}_c^k \\ X^k\boldsymbol{s}^k - \nu_{k+1}\boldsymbol{1} \end{bmatrix}, \tag{6.14}$$

where $\boldsymbol{r}_b^k$ and $\boldsymbol{r}_c^k$ are defined by (6.6) and (6.7). Particularly, when $\sigma_k = 1$, the solution of (6.14) is called the *centering direction*. By using the search direction given by (6.14), we obtain an algorithm, known as the *path-following primal-dual interior-point method*, described formally below.

Algorithm 6.1.1.

Step 0: Let $(\boldsymbol{x}^0, \boldsymbol{y}^0, \boldsymbol{s}^0)$ satisfying $(\boldsymbol{x}^0, \boldsymbol{s}^0) > \boldsymbol{0}$ be an initial solution.

Step 1: If

$$\rho_k < \epsilon, \quad \|A\boldsymbol{x}^k - \boldsymbol{b}\| < \epsilon, \quad \|A^{\mathrm{T}}\boldsymbol{y}^k + \boldsymbol{s}^k - \boldsymbol{c}\| < \epsilon,$$

then stop. Otherwise, choose $\sigma_k \in]0, 1]$, and compute ν_{k+1} by (6.12) and (6.13).

Step 2: Solve (6.14) to find the solution $(\Delta\boldsymbol{x}^k, \Delta\boldsymbol{y}^k, \Delta\boldsymbol{s}^k)$.

Step 3: Update the solution according to (6.8) by choosing α_k so that $(\boldsymbol{x}^{k+1}, \boldsymbol{s}^{k+1}) > \boldsymbol{0}$. Let $k \leftarrow k + 1$, and go to Step 1.

In practical algorithms, often different step lengths are used for updating the primal variables $\boldsymbol{x}^k$ and the dual variables $(\boldsymbol{y}^k, \boldsymbol{s}^k)$. Moreover, the predictor-corrector method in Mehrotra [308] drastically enhances the efficiency of practical implementation. See, e.g., Nocedal–Wright [340, Chap. 14] for these issues.

The relation between the interior-point method and the classical barrier method can be described as follows. We can see that (6.9), defining the central path, is obtained as the KKT conditions for the following problem:

$$\left.\begin{aligned} &\min_{\boldsymbol{x}} \ \boldsymbol{c}^{\mathrm{T}}\boldsymbol{x} - \nu \sum_{i=1}^{n} \ln x_i \\ &\text{s.t.} \ \ A\boldsymbol{x} = \boldsymbol{b}. \end{aligned}\right\} \tag{6.15}$$

The function $-\nu \ln x_i$ is a logarithmic-barrier function for the constraint $x_i \geq 0$, and hence problem (6.15) serves as a barrier problem for $(\mathrm{P_{LP}})$ in (6.1). Since the objective function of (6.15) is strictly convex, it has a unique optimal solution. Moreover, one can verify that a necessary and sufficient condition for (6.15) coincides with (6.9). Therefore, the central path $\mathcal{C}$ in (6.10) is also regarded as a trajectory of optimal solutions of a family of barrier problems (6.15). Simultaneously, modified KKT conditions in (6.9) also coincide with the KKT conditions for the problem

$$\left.\begin{aligned} &\max_{\boldsymbol{y},\boldsymbol{s}} \ \boldsymbol{b}^{\mathrm{T}}\boldsymbol{y} + \nu \sum_{i=1}^{n} \log s_i \\ &\text{s.t.} \ \ A^{\mathrm{T}}\boldsymbol{y} + \boldsymbol{s} = \boldsymbol{c}, \end{aligned}\right\}$$

which is a barrier problem for $(\mathrm{D_{LP}})$ in (6.2).

6.1.3 Interior-point method for semidefinite programming

The primal-dual interior-point methods for LP can be extended, with a slight modification, to SDP because the set of symmetric positive-semidefinite matrices enjoys the self-dual property like the nonnegative orthant.

6.1.3.1 Optimality condition

We consider the SDP problem in the standard form and its dual,[4] which are given by

$$\left.\begin{aligned}(\mathrm{P_{SDP}}):\ &\min_{X\in\mathbb{S}^n}\ C\bullet X\\ &\text{s.t.}\ \ A_i\bullet X=b_i,\quad i=1,\dots,m,\\ &\qquad\ X\succeq O;\end{aligned}\right\}\tag{6.16}$$

$$\left.\begin{aligned}(\mathrm{D_{SDP}}):\ &\max_{\boldsymbol{y},S}\ \sum_{i=1}^m b_iy_i\\ &\text{s.t.}\ \ \sum_{i=1}^m y_iA_i+S=C,\\ &\qquad\ S\succeq O.\end{aligned}\right\}\tag{6.17}$$

Here, $X\in\mathbb{S}^n$ is considered a variable matrix of the *primal* problem (6.16), $\boldsymbol{y}=(y_i)\in\mathbb{R}^m$ and $S\in\mathbb{S}^n$ are the variables of the *dual* problem (6.17), $A_1,\dots,A_m\in\mathbb{S}^n$ and $C\in\mathbb{S}^n$ are constant matrices, and $\boldsymbol{b}=(b_i)\in\mathbb{R}^m$ is a constant vector. We may assume without loss of generality that the matrices $A_1,\dots,A_m$ are linearly independent, i.e., that $\sum_{i=1}^m\eta_iA_i=O$ implies $\eta_i=0$ $(i=1,\dots,m)$.

Concerning the feasibility for $(\mathrm{P_{SDP}})$ and $(\mathrm{D_{SDP}})$, define $\mathcal{F}$, $\mathcal{F}^\circ\subset\mathbb{S}^n\times\mathbb{R}^m\times\mathbb{S}^n$ by

$$\mathcal{F}=\left\{(X,\boldsymbol{y},S)\ \middle|\ \begin{aligned}&A_i\bullet X=b_i\ (i=1,\dots,m),\\ &\sum_{i=1}^m y_iA_i+S=C,\ X\succeq O,\ S\succeq O\end{aligned}\right\},$$

$$\mathcal{F}^\circ=\left\{(X,\boldsymbol{y},S)\ \middle|\ \begin{aligned}&A_i\bullet X=b_i\ (i=1,\dots,m),\\ &\sum_{i=1}^m y_iA_i+S=C,\ X\succ O,\ S\succ O\end{aligned}\right\}.$$

Note that $\mathcal{F}^\circ=\operatorname{ri}\mathcal{F}$. Like the case of LP, $\mathcal{F}$ and $\mathcal{F}^\circ$ are called the feasible set and the strictly feasible set, respectively. Suppose that $\mathcal{F}^\circ\neq\emptyset$, then $(\mathrm{P_{SDP}})$ and $(\mathrm{D_{SDP}})$ have the optimal solutions. As shown in Proposition 2.3.3 (see section 2.3.2), X and $(\boldsymbol{y},S)$ are optimal for $(\mathrm{P_{SDP}})$ and $(\mathrm{D_{SDP}})$, respectively,

[4] See Definition 1.6.4 in section 1.6.2.

if and only if they satisfy the optimality conditions

$$A_i \bullet X = b_i, \quad i = 1, \dots, m, \tag{6.18a}$$

$$\sum_{i=1}^{m} y_i A_i + S = C, \tag{6.18b}$$

$$X \bullet S = 0, \tag{6.18c}$$

$$X \succeq O, \quad S \succeq O. \tag{6.18d}$$

In (6.18), we see from Fact 1.3.18 (see also Fact 1.3.20) that (6.18c) can be replaced with

$$XS = O. \tag{6.19}$$

This system of nonlinear equations can basically be attacked by the Newton method, but a slight modification is required. To see this, write the Newton equations for (6.18a), (6.18b), and (6.19) as

$$A_i \bullet \Delta\check{X} = -r_{bi}^{\;k}, \quad i = 1, \dots, m, \tag{6.20}$$

$$\sum_{i=1}^{m} \Delta\check{y}_i A_i + \Delta\check{S} = -R_C^k, \tag{6.21}$$

$$X^k \Delta\check{S} + \Delta\check{X} S^k = -X^k S^k, \tag{6.22}$$

where $\Delta\check{X}$, $\Delta\check{\boldsymbol{y}}$, and $\Delta\check{S}$ are the variables, and

$$r_{bi}^{\;k} = A_i \bullet X^k - b_i, \quad i = 1, \dots, m, \tag{6.23}$$

$$R_C^k = \sum_{i=1}^{m} y_i^k A_i + S^k - C. \tag{6.24}$$

The source of difficulty is that, at the solution of (6.20)–(6.22), $\Delta\check{X}$ and $\Delta\check{S}$ are not symmetric matrices in general (see, e.g., Helmberg [172, section 7]). In the next section we introduce some approaches for managing this difficulty.

6.1.3.2 Central path

In a manner similar to (6.15) for LP, we consider a barrier problem for ($\mathrm{P_{SDP}}$) in (6.16). For the constraint $X \succeq O$, we consider a logarithmic-barrier function given as

$$-\nu \sum_{i=1}^{n} \ln \lambda_i(X) = -\nu \ln\Big(\prod_{i=1}^{n} \lambda_i(X)\Big) = -\nu \ln(\det X),$$

where $\lambda_i(X)$ is the ith eigenvalue of X, and $\nu > 0$ is a constant. With this barrier function, a barrier problem for ($\mathrm{P_{SDP}}$) is obtained as

$$\left.\begin{array}{ll} \displaystyle\min_{X \in \mathbb{S}^n} & C \bullet X - \nu \ln(\det X) \\ \text{s.t.} & A_i \bullet X = b_i, \quad i = 1, \dots, m. \end{array}\right\} \tag{6.25}$$

Note that the objective function of (6.25) is strictly convex. A necessary and sufficient condition for optimality can be written as

$$A_i \bullet X = b_i, \quad i = 1, \dots, m, \tag{6.26a}$$

$$\sum_{i=1}^{m} y_i A_i + S = C, \tag{6.26b}$$

$$XS = \nu I, \tag{6.26c}$$

$$X \succ O, \quad S \succ O. \tag{6.26d}$$

The central path for SDP is defined as a trajectory of the solutions $(X, \boldsymbol{y}, S)$ of (6.26) with respect to the parameter $\nu > 0$. Indeed, by putting $\nu = 0$ in (6.26c), and by replacing "$\succ$" in (6.26d) with "$\succeq$", we recover the optimality condition in (6.18), because (6.19) is equivalent to (6.18c) when (6.18d) is satisfied. Therefore, by tracing the central path by decreasing $\nu \searrow 0$, we can attain the optimal solutions of ($\mathrm{P_{SDP}}$) and ($\mathrm{D_{SDP}}$). It is known that the solution of (6.26) for a given $\nu > 0$ exists uniquely. Alternatively, (6.26) is also regarded as the optimality condition for the problem

$$\left.\begin{aligned} \max_{\boldsymbol{y}, S} \ & \sum_{i=1}^{m} b_i y_i + \nu \ln(\det S) \\ \text{s.t.} \ & \sum_{i=1}^{m} y_i A_i + S = C, \end{aligned}\right\} \tag{6.27}$$

which serves as a barrier problem for ($\mathrm{D_{SDP}}$).

Let $\mathbb{S}^n_{\text{skew}}$ be the set of all $n \times n$ skew-symmetric matrices. We denote by $\tilde{\mathcal{L}}$ a maximal (i.e., $n(n-1)/2$-dimensional) monotone linear subspace[5] of $\mathbb{S}^n_{\text{skew}} \times \mathbb{S}^n_{\text{skew}}$. In a manner similar to the case of LP, we define ν_{k+1} by (6.13), where the duality measure for SDP is defined by

$$\rho_k = \frac{X^k \bullet S^k}{n}$$

instead of (6.12). At the current solution $(X^k, \boldsymbol{y}^k, S^k)$, the solution $(\Delta X, \Delta \boldsymbol{y}, \Delta S) \in \mathbb{S}^n \times \mathbb{R}^m \times \mathbb{S}^n$ to the following system of equation is known as the search direction in the KSH family [243]:

$$A_i \bullet \Delta X = -r_{b_i}^k, \quad i = 1, \dots, m, \tag{6.28a}$$

$$\sum_{i=1}^{m} \Delta y_i A_i + \Delta S = -R_C^k, \tag{6.28b}$$

$$\exists (\Delta\tilde{X}, \Delta\tilde{S}) \in \tilde{\mathcal{L}}: \quad (\Delta X + \Delta\tilde{X}) S^k + X^k (\Delta S + \Delta\tilde{S}) = -H^k, \tag{6.28c}$$

[5]We say that a linear subspace $\tilde{\mathcal{L}}$ of $\mathbb{S}^n_{\text{skew}} \times \mathbb{S}^n_{\text{skew}}$ is *monotone* if $\tilde{X} \bullet \tilde{S} \geq 0$ for any $(\tilde{X}, \tilde{S}) \in \tilde{\mathcal{L}}$.

where $\boldsymbol{r}_b^k = (r_{bi}^k) \in \mathbb{R}^m$ and $R_C^k \in \mathbb{S}^n$ are given in (6.23) and (6.24), and $H^k \in \mathbb{S}^n$ is defined by

$$H^k = X^k S^k - \nu_{k+1} I. \tag{6.29}$$

If $X^k, S^k \succ O$, then it is known that (6.28) has a unique solution $(\Delta X, \Delta \boldsymbol{y}, \Delta S) \in \mathbb{S}^n \times \mathbb{R}^m \times \mathbb{S}^n$; see Theorem 4.2 in Kojima–Shindoh–Hara [243].

As a particular case of the KSH family, suppose that we choose $\tilde{\mathcal{L}}$ as

$$\tilde{\mathcal{L}} := \mathbb{S}^n_{\mathrm{skew}} \times \{O\}.$$

Then (6.28c) is reduced to

$$\begin{aligned} &(\Delta X + \Delta \tilde{X}) S^k + X^k \Delta S = -H^k, \\ &\qquad \text{where } (\Delta X, \Delta S, \Delta \tilde{X}) \in \mathbb{S}^n \times \mathbb{S}^n \times \mathbb{S}^n_{\mathrm{skew}}. \end{aligned} \tag{6.30}$$

The search direction obtained as the solution of (6.28a), (6.28b), and (6.30) is known as the HRVW/KSH/M search direction.[6]

Alternatively, if we choose $\tilde{\mathcal{L}}$ as

$$\tilde{\mathcal{L}} := \{O\} \times \mathbb{S}^n_{\mathrm{skew}},$$

then (6.28c) is reduced to

$$\begin{aligned} &\Delta X\, S^k + X^k (\Delta S + \Delta \tilde{S}) = -H^k, \\ &\qquad \text{where } (\Delta X, \Delta S, \Delta \tilde{S}) \in \mathbb{S}^n \times \mathbb{S}^n \times \mathbb{S}^n_{\mathrm{skew}}. \end{aligned} \tag{6.31}$$

The search direction defined as the solution of (6.28a), (6.28b), and (6.31) is called the dual HRVW/KSH/M search direction.

Define $W \in \mathbb{S}^n_+$ by

$$\begin{aligned} W &= X^{1/2} (X^{1/2} S X^{1/2})^{-1/2} X^{1/2} \\ &= S^{-1/2} (S^{1/2} X S^{1/2})^{1/2} S^{-1/2}, \end{aligned} \tag{6.32}$$

where the superscript k is omitted for simplicity. Note that the definition (6.32) of W is motivated by the relation

$$W^{-1/2} X^k W^{-1/2} = W^{1/2} S^k W^{1/2},$$

which means that $W^{1/2}$ scales X^k and S^k to the same matrix. Choose $\tilde{\mathcal{L}}$ as[7]

$$\tilde{\mathcal{L}} := \{(W^{1/2} U W^{1/2}, W^{-1/2} U W^{-1/2}) \mid U \in \mathbb{S}^n_{\mathrm{skew}}\}. \tag{6.33}$$

[6]Which is due to Helmberg–Rendl–Vanderbei–Wolkowicz [173], Kojima–Shindoh–Hara [243], and Monteiro [314].

[7]For any $V \in \mathbb{R}^{n \times n}$ that satisfies $VV^{\mathrm{T}} = W$, one can verify that $\{(VUV^{\mathrm{T}}, V^{-1}UV^{-1}) \mid U \in \mathbb{S}^n_{\mathrm{skew}}\}$ does not depend on the choice of V. We here simply put $V = W^{1/2}$.

Then the solution of (6.28) with (6.33) coincides with the NT direction proposed by Nesterov–Todd [334]; see also [242].

Besides the HRVW/KSH/M and NT directions, various search directions have been proposed by many authors; see Todd [433] for a survey on the search directions.

Once a search direction is computed, then the iteration is performed in a manner similar to Algorithm 6.1.1; see Todd [434] and Wolkowicz–Saigal–Vandenberghe [454] for more details. Regarding the practical implementation, see, e.g., Borchers–Young [58], Sturm [424, 426], Tütüncü–Toh–Todd [439], and Yamashita–Fujisawa–Kojima [459].

6.2 Reformulation and Smoothing Method

We begin by considering solution methods for the mixed complementarity problem[8]

$$\left.\begin{array}{l} \boldsymbol{f}(\boldsymbol{x}, \boldsymbol{y}, \boldsymbol{s}) = \boldsymbol{0}, \\ x_i \geq 0, \;\; s_i \geq 0, \;\; x_i s_i = 0 \quad (i = 1, \ldots, n), \end{array}\right\} \tag{6.34}$$

where $\boldsymbol{x}, \boldsymbol{s} \in \mathbb{R}^n$ and $\boldsymbol{y} \in \mathbb{R}^m$ are the variables, and $\boldsymbol{f} : \mathbb{R}^n \times \mathbb{R}^m \times \mathbb{R}^n \to \mathbb{R}^{n+m}$. The problem (6.34) includes, for example, LP as a particular case: indeed, if we define $\boldsymbol{f}$ by

$$\boldsymbol{f}(\boldsymbol{x}, \boldsymbol{y}, \boldsymbol{s}) = \begin{bmatrix} A^{\mathrm{T}}\boldsymbol{y} + \boldsymbol{s} - \boldsymbol{c} \\ A\boldsymbol{x} - \boldsymbol{b} \end{bmatrix},$$

then (6.34) is reduced to (6.3), which is the system of the KKT conditions for LP. In section 6.2.3 the results on (6.34) are generalized to the complementarity conditions over nonlinear convex cones.

6.2.1 Reformulation method

The essential idea of the reformulation methods is to reformulate a complementarity problem as a system of (nonsmooth) nonlinear equations. Then the system is solved by using a nonsmooth Newton method.

We begin by considering a complementarity condition

$$x_i \geq 0, \quad s_i \geq 0, \quad x_i s_i = 0. \tag{6.35}$$

The source of difficulty for huddling (6.35) is that it involves inequality constraints. To deal with the complementarity problems in the framework of

[8]See Definition 1.2.2 in section 1.2.2 for mixed complementarity problem.

numerical solutions for nonlinear equations, restate (6.35) as a single equation as follows.

A function $\psi : \mathbb{R}^2 \to \mathbb{R}$ is said to be the *complementarity function* if it satisfies

$$\psi(x_i, s_i) = 0 \quad \Leftrightarrow \quad (6.35) \qquad (\forall x_i, s_i \in \mathbb{R}). \tag{6.36}$$

An extremely simple example of complementarity function is the min-function defined by

$$\psi_{\min}(x_i, s_i) := \min\{x_i, s_i\}. \tag{6.37}$$

Note that $\psi_{\min}$ is alternatively represented as

$$\begin{aligned} \psi_{\min}(x_i, s_i) &= x_i - \max\{x_i - s_i, 0\} \\ &= x_i - \mathrm{p}_{\mathbb{R}_+}(x_i - s_i), \end{aligned} \tag{6.38}$$

where $\mathrm{p}_{\mathbb{R}_+} : \mathbb{R} \to \mathbb{R}$ is the projection on $\mathbb{R}_+$ defined by

$$\mathrm{p}_{\mathbb{R}_+}(x_i - s_i) = \arg\min_{\check{z}}\{|\check{z} - (x_i - s_i)| \mid \check{z} \in \mathbb{R}\}. \tag{6.39}$$

We easily see that (6.35) is equivalently rewritten as

$$\psi_{\min}(x_i, s_i) = 0. \tag{6.40}$$

Thus the condition including inequalities is reduced to a single nonlinear equation. Note that $\psi_{\min}$ is not differentiable (in the sense of Fréchet derivative) at any point (x_i, s_i) satisfying $x_i = s_i$.

Another important complementarity function is the Fischer–Burmeister function defined by

$$\psi_{\mathrm{FB}}(x_i, s_i) := \sqrt{x_i^2 + s_i^2} - (x_i + s_i). \tag{6.41}$$

Like $\psi_{\min}$, ψ_{FB} is a nonsmooth function: ψ_{FB} is not differentiable at $(x_i, s_i) = (0, 0)$. Note that ψ_{FB}^2 is a continuously differentiable function. Besides $\psi_{\min}$ and ψ_{FB}, various complementarity functions have been proposed; see Facchinei–Pang [125, Chap. 1], and also Chen–Chen–Kanzow [75], Kanzow–Yamashita–Fukushima [218], and Mangasarian [292].

Using a complementarity function, the complementarity problem (6.34) can be restated as a system of nonlinear equations. Define $\boldsymbol{h} : \mathbb{R}^n \times \mathbb{R}^m \times \mathbb{R}^n \to \mathbb{R}^{n+m} \times \mathbb{R}^n$ by

$$\boldsymbol{h}(\boldsymbol{x}, \boldsymbol{y}, \boldsymbol{s}) = \begin{bmatrix} \boldsymbol{f}(\boldsymbol{x}, \boldsymbol{y}, \boldsymbol{s}) \\ \psi(x_1, s_1) \\ \vdots \\ \psi(x_n, s_n) \end{bmatrix} \tag{6.42}$$

to see from (6.36) that (6.34) is equivalent to

$$\boldsymbol{h}(\boldsymbol{x}, \boldsymbol{y}, \boldsymbol{s}) = \boldsymbol{0}. \tag{6.43}$$

For a nonsmooth complementarity function, such as $\psi_{\min}$ or ψ_{FB}, (6.43) results in the system of nonsmooth nonlinear equations. The nonsmoothness property seems to be a drawback on one hand, because the conventional Newton methods are likely to fail for solving (6.43). However, there are some reasons a nonsmooth equation reformulation is considered to be more preferable to a reformulation with a smooth complementarity function. Indeed, it is known that with a smooth complementarity function the Jacobian of $\boldsymbol{h}$ in (6.42) becomes singular at any degenerate solution of (6.34), which shows that we cannot expect to develop practical methods with the locally fast property based on smooth complementarity functions; see Proposition 9.1.1 in Facchinei–Pang [125]. Thus, using a nonsmooth complementarity function, overcoming the difficulty arising from the nonsmoothness property, is much more preferable to the use of a smooth complementarity function.

The nonsmooth Newton method for complementarity problems basically solves the nonlinear equation reformulation (6.43) with a nonsmooth complementarity function. In what follows we write

$$\boldsymbol{z} = (\boldsymbol{x}, \boldsymbol{y}, \boldsymbol{s})$$

for simplicity. A typical approach to designing a globally convergent algorithm is to consider the *merit function* defined by

$$\theta(\boldsymbol{z}) := \frac{1}{2}\boldsymbol{h}(\boldsymbol{z})^{\mathrm{T}}\boldsymbol{h}(\boldsymbol{z}). \tag{6.44}$$

For a given $\boldsymbol{z}$, $\boldsymbol{h}(\boldsymbol{z})$ is said to be B-differentiable at $\boldsymbol{z}$, if it is Lipshitz continuous in a neighborhood of $\boldsymbol{z}$ and directionally differentiable at $\boldsymbol{z}$. We say that $\boldsymbol{h}(\boldsymbol{z})$ is B-differentiable if it is B-differentiable at any point $\boldsymbol{z}$. For example, $\psi_{\min}$ is B-differentiable. Therefore, if $\boldsymbol{f}$ involved in the original complementarity problem (6.34) is continuously differentiable, then $\boldsymbol{h}$ defined by (6.42) becomes B-differentiable with $\psi := \psi_{\min}$. In such a situation, we may consider the following algorithm, known as the B-differentiable Newton method (or the semismooth Newton method, more generally), to solve (6.42) based on minimizing the merit function θ.

Algorithm 6.2.1.

Step 0: Let $\boldsymbol{z}^0$ be a given initial solution.

Step 1: If $\theta(\boldsymbol{z}^k) < \epsilon$, then stop. Otherwise, solve

$$\boldsymbol{h}'(\boldsymbol{z}^k; \Delta\boldsymbol{z}^k) = -\boldsymbol{h}(\boldsymbol{z}^k) \tag{6.45}$$

to obtain the search direction $\Delta\boldsymbol{z}^k$.[9]

[9]Recall that $\boldsymbol{h}'(\boldsymbol{z}^k; \Delta\boldsymbol{z}^k)$ denotes the directional derivative of $\boldsymbol{h}$ at $\boldsymbol{z}^k$ with respect to the direction $\Delta\boldsymbol{z}^k$; see section 2.1.4.

Step 2: Choose the step length α_k such that $\theta(\boldsymbol{z}^k+\alpha_k\Delta\boldsymbol{z}^k)$ accommodates a sufficient decrease from the current value $\theta(\boldsymbol{z}^k)$.

Step 3: Update the solution by $\boldsymbol{z}^{k+1} = \boldsymbol{z}^k + \alpha_k\Delta\boldsymbol{z}^k$. Let $k \leftarrow k+1$, and go to Step 1.

If $\boldsymbol{h}$ is differentiable (in the Fréchet sense) at $\boldsymbol{z}^k$, then the directional derivative $\boldsymbol{h}'(\boldsymbol{z}^k;\Delta\boldsymbol{z}^k)$ is reduced to $\nabla\boldsymbol{h}(\boldsymbol{z}^k)\Delta\boldsymbol{z}^k$, and hence (6.45) coincides with the conventional Newton equation. Otherwise, (6.45) is a nonlinear equation. Thus Algorithm 6.2.1 is regarded as a generalization of the Newton method for smooth nonlinear equations. Since Algorithm 6.2.1 is essentially based on the Newton method, it does not necessarily converge to a solution of the complementarity problem (6.34) for an arbitrary $\boldsymbol{f}$; to ensure the global convergence to a solution, it is necessary to assume, for example, that $\boldsymbol{f}$ is monotone[10]; see Facchinei–Pang [125] and Harker–Pang [161].

6.2.2 Smoothing method

For a nonsmooth (continuous) function $g : \mathbb{R}^n \to \mathbb{R}$, we consider a family of functions $g_\nu : \mathbb{R}^n \to \mathbb{R}$ with a parameter $\nu > 0$ satisfying the following property:

- For any fixed $\nu > 0$, g_ν is continuously differentiable.
- $\lim_{\nu \searrow 0} g_\nu(\boldsymbol{x}) = g(\boldsymbol{x})$ for any $\boldsymbol{x} \in \mathbb{R}^n$.

Such a g_ν is called the *smoothing function* of g. See Definition 11.8.1 in [125] for a more comprehensive way of defining the smoothing function.

For example, define $\psi_{\min\nu} : \mathbb{R}^2 \to \mathbb{R}$ by

$$\psi_{\min\nu}(x_i, s_i) = \sqrt{(x_i - s_i)^2 + 4\nu} - (x_i + s_i), \tag{6.46}$$

where $\nu > 0$ is the smoothing parameter. Then $\psi_{\min\nu}$ serves as a smoothing function of $\psi_{\min}$. As another example, we may consider

$$\psi_{\mathrm{FB}\nu}(x_i, s_i) = \sqrt{x_i^2 + s_i^2 + 2\nu} - (x_i + s_i), \tag{6.47}$$

which, in turn, serves as a smoothing function of ψ_{FB}. In the smoothing methods using $\psi_{\mathrm{FB}\nu}$, $\boldsymbol{h}$ in (6.42) is to be replaced with

$$\boldsymbol{h}_\nu(\boldsymbol{x}, \boldsymbol{y}, \boldsymbol{s}) = \begin{bmatrix} \boldsymbol{f}(\boldsymbol{x}, \boldsymbol{y}, \boldsymbol{s}) \\ \psi_{\mathrm{FB}\nu}(x_1, s_1) \\ \vdots \\ \psi_{\mathrm{FB}\nu}(x_n, s_n) \end{bmatrix}. \tag{6.48}$$

[10]The monotonicity of $\boldsymbol{f}$ and ψ imply the convexity of θ (see Proposition 2.1.4). Therefore, solving the complementarity problem (6.34) is reduced to finding the minimal point of the convex function $\theta(\boldsymbol{z})$, which can be solved quite easily in general.

The search direction $\Delta \boldsymbol{z}^k$ is then defined as the Newton direction of $\boldsymbol{h}_\nu$ in (6.48) as

$$\nabla \boldsymbol{h}_\nu(\boldsymbol{z})\, \Delta \boldsymbol{z}^k = -\boldsymbol{h}_\nu(\boldsymbol{z}^k), \tag{6.49}$$

where $\boldsymbol{z} = (\boldsymbol{x}, \boldsymbol{y}, \boldsymbol{s})$ for short. Accordingly, the merit function, denoted θ_ν, is naturally defined by

$$\theta_\nu(\boldsymbol{z}) = \frac{1}{2}\boldsymbol{h}_\nu(\boldsymbol{z})^{\mathrm{T}}\boldsymbol{h}_\nu(\boldsymbol{z}), \tag{6.50}$$

which serves as a smoothing function of θ in (6.44). The step length α_k is determined by employing the line search based on the merit function θ_ν in (6.50). Thus the smoothing methods solves the minimization of θ_ν by gradually decreasing $\nu > 0$, so that a solution of the original complementarity problem (6.34) is obtained by taking the limit $\nu \searrow 0$.

6.2.3 Extensions to conic complementarity problems

We are now in a position to extend the reformulation and smoothing methods introduced in section 6.2.1 and section 6.2.2 to the complementarity problems over $\mathbb{L}^n$ (the second-order cone) or $\mathbb{S}^n_+$ (the positive-semidefinite cone). Recall that the second-order cone in $\mathbb{R}^n$ is defined by

$$\mathbb{L}^n = \{(x_0, \boldsymbol{x}_1) \mid x_0 \geq \|\boldsymbol{x}_1\|\}.$$

The complementarity condition over the second-order cone is then defined by

$$\boldsymbol{x} \in \mathbb{L}^n, \quad \boldsymbol{s} \in \mathbb{L}^n, \quad \boldsymbol{x}^{\mathrm{T}}\boldsymbol{s} = 0, \tag{6.51}$$

which is the principal subject dealt with in the first-half of this section.

In a manner similar to (6.39), define the projection $\mathbf{p}_{\mathbb{L}^n} : \mathbb{R}^n \to \mathbb{R}^n$ on $\mathbb{L}^n$ by

$$\mathbf{p}_{\mathbb{L}^n}(\boldsymbol{z}) = \arg\min_{\check{\boldsymbol{z}}}\{\|\check{\boldsymbol{z}} - \boldsymbol{z}\| \mid \check{\boldsymbol{z}} \in \mathbb{L}^n\}. \tag{6.52}$$

Note that $\mathbf{p}_{\mathbb{L}^n}(\boldsymbol{z})$ can be written explicitly in terms of the Jordan frame[11] of $\boldsymbol{z}$ with respect to $\mathbb{L}^n$ by

$$\mathbf{p}_{\mathbb{L}^n}(\boldsymbol{z}) = \max\{\lambda_{(1)}, 0\}\boldsymbol{\phi}^{(1)} + \max\{\lambda_{(2)}, 0\}\boldsymbol{\phi}^{(2)}, \tag{6.53}$$

where $\lambda_{(1)}$ and $\lambda_{(2)}$ are the spectral values, and $\{\boldsymbol{\phi}^{(1)}, \boldsymbol{\phi}^{(2)}\}$ is a Jordan frame of $\boldsymbol{z}$ with respect to $\mathbb{L}^n$. Analogous to the definition of $\psi_{\min}$ in (6.38), define $\boldsymbol{\psi}^{\mathbb{L}^n}_{\min} : \mathbb{R}^n \times \mathbb{R}^n \to \mathbb{R}^n$ by

$$\boldsymbol{\psi}^{\mathbb{L}^n}_{\min}(\boldsymbol{x}, \boldsymbol{s}) = \boldsymbol{x} - \mathbf{p}_{\mathbb{L}^n}(\boldsymbol{x} - \boldsymbol{s}). \tag{6.54}$$

[11] See section 1.4.3 for the Jordan frame, as well as the spectral factorization, associated with the second-order cone $\mathbb{L}^n$.

Then $\psi_{\min}^{\mathbb{L}^n}$ satisfies[12]

$$\psi_{\min}^{\mathbb{L}^n}(\boldsymbol{x}, \boldsymbol{s}) = \boldsymbol{0} \quad \Leftrightarrow \quad (6.51). \tag{6.55}$$

Because of the property in (6.55), $\psi_{\min}^{\mathbb{L}^n}$ in (6.54) is called a second-order cone complementarity function.

A class of smoothing functions of $\psi_{\min}^{\mathbb{L}^n}$ was presented by Fukushima–Luo–Tseng [140]. For example, define for $\nu > 0$ the mapping $\mathbf{p}_{\mathbb{L}^n \nu} : \mathbb{R}^n \times \mathbb{R}^n \to \mathbb{R}^n$ by

$$\mathbf{p}_{\mathbb{L}^n \nu}(\boldsymbol{z}) = \sum_{i=1}^{2} \frac{\sqrt{(\lambda_{(i)})^2 + 4\nu^2} + \lambda_{(i)}}{2} \boldsymbol{\phi}^{(i)},$$

which serves as a smoothing function of $\mathbf{p}_{\mathbb{L}^n}$. By using this $\mathbf{p}_{\mathbb{L}^n \nu}$, a smoothing function of $\psi_{\min}^{\mathbb{L}^n}$ is obtained as

$$\psi_{\min \nu}^{\mathbb{L}^n}(\boldsymbol{x}, \boldsymbol{s}) = \boldsymbol{x} - \mathbf{p}_{\mathbb{L}^n \nu}(\boldsymbol{x} - \boldsymbol{s}).$$

Besides the min-function, it is also possible to extend the Fischer–Burmeister function, defined by (6.41), for the complementarity condition over the second-order cone. Recall that the Jordan product, denoted $\boldsymbol{x} \circ \boldsymbol{s}$, is defined by (1.14). It is shown that, for any $\boldsymbol{x} \in \mathbb{R}^n$, there exists a unique vector in $\mathbb{L}^n$, denoted $\boldsymbol{x}^{1/2}$, such that $\boldsymbol{x}^{1/2} \circ \boldsymbol{x}^{1/2} = \boldsymbol{x}$. Then, we see that

$$\psi_{\mathrm{FB}}^{\mathbb{L}^n}(\boldsymbol{x}, \boldsymbol{s}) = (\boldsymbol{x} \circ \boldsymbol{x} + \boldsymbol{s} \circ \boldsymbol{s})^{1/2} - (\boldsymbol{x} + \boldsymbol{s}) \tag{6.56}$$

is well defined as a function $\psi_{\mathrm{FB}}^{\mathbb{L}^n} : \mathbb{R}^n \times \mathbb{R}^n \to \mathbb{R}^n$. One can show that $\psi_{\mathrm{FB}}^{\mathbb{L}^n}$ is a second-order cone complementarity function; i.e., it satisfies[13] "$\psi_{\mathrm{FB}}^{\mathbb{L}^n}(\boldsymbol{x}, \boldsymbol{s}) = \boldsymbol{0} \Leftrightarrow$ (6.51)."

The Fischer–Burmeister function is also extended for the complementarity condition over the positive-semidefinite cone. Recall that the square root of a positive-semidefinite matrix is defined by Definition 1.3.4. Hence, the mapping $\psi_{\mathrm{FB}}^{\mathbb{S}^n} : \mathbb{S}^n \times \mathbb{S}^n \to \mathbb{S}^n$

$$\psi_{\mathrm{FB}}^{\mathbb{S}^n}(X, S) = (X^2 + S^2)^{1/2} - (X + S) \tag{6.57}$$

is well defined for any X, $S \in \mathbb{S}^n$. Then it can be shown that $\psi_{\mathrm{FB}}^{\mathbb{S}^n}$ satisfies[14]

$$\psi_{\mathrm{FB}}^{\mathbb{S}^n}(X, S) = O \quad \Leftrightarrow \quad X \succeq O,\ S \succeq O,\ X \bullet S = 0. \tag{6.58}$$

Thus, from (6.58) we can restate the complementarity condition over the positive-semidefinite cone as nonlinear equations. For example, the KKT

[12] See Fukushima–Luo–Tseng [140] for a proof of (6.55).

[13] See Fukushima–Luo–Tseng [140] for a proof.

[14] See [216, 437] for proofs of (6.58).

conditions, (6.18), for SDP are rewritten as the following system of nonlinear equations:

$$A_i \bullet X - b_i = 0, \quad i = 1, \dots, m, \tag{6.59a}$$

$$\sum_{i=1}^{m} y_i A_i + S - C = O, \tag{6.59b}$$

$$\psi_{\mathrm{FB}}^{\mathbb{S}^n}(X, S) = O. \tag{6.59c}$$

Like the smoothing function (6.47) for ψ_{FB}, define $\psi_{\mathrm{FB}\nu}^{\mathbb{S}^n}$ for $\nu > 0$ by

$$\psi_{\mathrm{FB}\nu}^{\mathbb{S}^n}(X, S) = (X^2 + S^2 + 2\nu I)^{1/2} - (X + S).$$

It is of interest to note that $\psi_{\mathrm{FB}\nu}^{\mathbb{S}^n}$ satisfies[15]

$$\psi_{\mathrm{FB}\nu}^{\mathbb{S}^n}(X, S) = O \quad \Leftrightarrow \quad X \succ O, \ S \succ O, \ XS = \nu I$$

for any X, $S \in \mathbb{S}^n$. Therefore, for example, the central path of SDP, defined by (6.26), can be equivalently written as a system of nonlinear equations consisting of (6.26a), (6.26b), and $\psi_{\mathrm{FB}\nu}^{\mathbb{S}^n}(X, S) = O$.

6.3 Notes

1. The primal-dual interior-point methods for LP, QP, SOCP, and SDP are the polynomial-time algorithm, in general; see section 1.7 for the bibliographical notes. For algorithmic aspects, refer to [9, 173, 243, 314, 334, 335, 433] for SDP, [317, 334, 335, 438] for SOCP.

On the other hand, the primal-dual interior-point methods for general nonlinear programming problems are not polynomial-time algorithms, although these two categories of algorithms are known under the same name. See Forsgren–Gill–Wright [132] for a survey on the algorithms falling into the latter category. The interior-point method used for solving the elastoplastic problems by Krabbenhøft–Lyamin–Sloan–Wriggers [246] is the algorithm for general nonlinear programming problems.

The interior-point methods for solving nonlinear complementarity problems have been proposed by many authors; see, e.g., Kojima–Noma–Yoshise [241], Potra–Ye [375], Tseng [436], Vanderbei–Shanno [446], Wang–Monteiro–Pang [451], and Zhao–Li [469]. The one from Wang–Monteiro–Pang [451] was used for solving the frictional contact problems by Christensen–Klarbring–Pang–Strömberg [87].

[15] See Kanzow–Nagel [216, 217] for a proof; also, see Sun–Sun [427].

2. The study of the nonsmooth Newton methods for solving complementarity problems began in the early 1990s; see Facchinei–Pang [125, section 9.6] for a comprehensive survey.

Making use of θ in (6.44), defined with $\psi_{\min}$, is initiated by Pang [357] for globalizing the convergence of a nonsmooth Newton method, in which the so-called B-differentiable Newton method is proposed. A similar nonsmooth Newton method was independently proposed by Alart–Curnier [6] for solving the frictional contact problems. For more nonsmooth methods, including semismooth Newton methods, one may refer to De Luca–Facchinei–Kanzow [113, 114], Pieraccini–Gasparo–Pasquali [365], andd Qi–Sun [378].

Applications of the B-differentiable Newton method to frictional/frictionless contact problems are due to Björkman [54] and Christensen–Klarbring–Pang–Strömberg [87]. A semismooth Newton method for solving elastoplastic frictional contact problems can be found in Christensen [85] and Christensen–Pang [88].

3. Concerning the smoothing methods for the complementarity problems, one may refer to Chen–Chen [74], Kanzow [214], Li–Fukushima [266], Qi–Liao [377], Qi–Sun [379], and Zhang–Jiang–Wang [468]. Numerical approaches based on smoothing methods to frictional contact problems are in Kanno–Martins–Pinto da Costa [202], Leung–Chen–Chen [257], Zhang–He–Li [464], and Zhang–He–Li–Wriggers [465].

4. We did not treat an alternative approach based on the pivoting algorithm, namely, the simplex method for the linear programming problems and the Lemke method for the linear complementarity problems; see Cottle–Pang–Stone [96, section 4.4] for the Lemke method. The software package known as the PATH for the nonlinear complementarity problems is due to Dirkse–Ferris [119], which solves linear complementarity subproblems by using the Lemke method to generate a path for a new iteration. The PATH solver is used for solving the complementarity problems arising from the elastoplastic problems by Tin-Loi–Xia [431, 432], and the frictional contact problems by Pinto da Costa–Martins–Figueiredo–Júdice [368].

5. Developing the nonsmooth/smoothing Newton methods for the complementarity problems subjected to self-dual cone constraints has attracted recent interest. For the ones solving second-order cone complementarity problems, see Chen–Chen–Teng [76], Chen–Tseng [79], Fukushima–Luo–Tseng [140], Hayashi–Yamashita–Fukushima [170], Kanzow–Ferenczi–Fukushima [215], and Pan–Chen [352]; for semidefinite complementarity problems, see Chen–Tseng [78], Sun–Sun [427], and Tseng [437].

In Kanno–Martins–Pinto da Costa [202] the frictional contact problems were solved by using the smoothing Newton method for the second-order cone complementarity problems, which was proposed by Hayashi–Yamashita–Fukushima [170] based on $\boldsymbol{\psi}^{\mathbb{L}^n}_{\min \nu}$.

Chapter 7

Numerical Analysis of Cable Networks

Orientation We discuss here the numerical analysis of cable networks, the theory of which was dealt with in Chapters 4 and 5 based on the methodology presented in this book. This chapter shows the advantage of this methodology in numerical computation over the conventional methods in computational mechanics. Once a problem in nonsmooth mechanics is reformulated as a conic optimization problem, we can solve that problem efficiently by using the algorithm designed for convex optimization introduced in Chapter 6. In continuation of the preceding three chapters, we here focus on the polynomial-time primal-dual interior-point method for solving the equilibrium analysis, as well as some related problems, of cable networks.

In section 7.1, we present the equilibrium analysis of cable network undergoing large deformation using the primal-dual interior-point method for SOCP (second-order cone programming) problems. As an extension, in section 7.2 we investigate the equilibrium analysis of a cable network with sliding joints, which are not fixed on cables and hence can move along cables freely. Section 7.3 discusses the form-finding problem of a cable network, which is a design problem of a cable network with the specified axial forces. We shall see that all the problems dealt with in this chapter can be solved by using the primal-dual interior-point method for SOCP.

7.1 Cable Networks with Pin-Joints

In this section we consider the numerical analysis of cable networks within the framework of the geometrically nonlinear theory by using the modern algorithm for convex optimization.

We begin by recalling the results established in Chapter 4. Consider a cable network placed in the physical three-dimensional space. We denote by m and d the number of members and the number of the degrees of freedom of displacements, respectively. It has been shown that the minimization problem of the total potential energy can be formulated as (see (4.41))

$$\left.\begin{aligned} \text{(PE)} : \min_{\boldsymbol{\varphi}, \boldsymbol{c}} \ & \sum_{i=1}^{m} \hat{w}_{\text{cab}}^{i}(c_i) - \sum_{j \in \Gamma_{\text{N}}} \underline{p}_j(\varphi_j - x_j) \\ \text{s.t.} \ & c_i = \|B_i \boldsymbol{\varphi}\| - \underline{l}_i, \quad i = 1, \dots, m, \\ & \varphi_j = \underline{\varphi}_j, \quad j \in \Gamma_{\text{D}}. \end{aligned}\right\} \tag{7.1}$$

Here, $\boldsymbol{x} = (x_j) \in \mathbb{R}^d$ denotes the given position vector of the nodes in the reference configuration, $\boldsymbol{\varphi} = (\varphi_j) \in \mathbb{R}^d$ is the unknown position vector of the nodes in the deformed configuration, c_i is the unknown elongation of the ith member, $\underline{l}_i$ is the given initial unstressed member length, and $B_i \in \mathbb{R}^{3 \times d}$ $(i = 1, \dots, m)$ are given matrices that represent the member connectivity of the cable network. As the boundary conditions, the deformation is prescribed as $\underline{\varphi}_j$ for each $j \in \Gamma_{\text{D}}$, while the external load is prescribed as $\underline{p}_j$ for each $j \in \Gamma_{\text{N}}$.

The stored energy function $\hat{w}_{\text{cab}}^{i}$ for the member i is given by

$$\hat{w}_{\text{cab}}^{i}(c_i) = \begin{cases} \dfrac{1}{2} k_{\text{e}i} c_i^2 & \text{if } c_i \geq 0, \\ 0 & \text{otherwise.} \end{cases} \tag{7.2}$$

Here, the constant $k_{\text{e}i}$ is written as

$$k_{\text{e}i} = \frac{E a_i}{\underline{l}_i}, \tag{7.3}$$

where a_i is the member cross-sectional area, and E is Young's modulus. The constitutive law of the ith member, which relates the axial force q_i to the elongation c_i, is then given by

$$q_i = \hat{q}_{\text{cab}}^{i}(c_i) := \begin{cases} k_{\text{e}i} c_i & \text{if } c_i \geq 0, \\ 0 & \text{otherwise.} \end{cases} \tag{7.4}$$

The major difficulty in the numerical analysis of a cable network stems from the fact that (7.2) (and (7.4) also) involves the transition point of the response, which corresponds to the transition between the taut and slack states of a cable.

In section 4.2.2, we showed that (PE) above can be rewritten equivalently as $(\overline{\text{PE}})$ in (4.42), i.e.,

$$\left.\begin{aligned} \min_{\boldsymbol{\varphi}, \boldsymbol{c}_{\text{e}}} \ & \sum_{i=1}^{m} \frac{1}{2} k_{\text{e}i} c_{\text{e}i}^2 - \sum_{j \in \Gamma_{\text{N}}} \underline{p}_j(\varphi_j - x_j) \\ \text{s.t.} \ & c_{\text{e}i} \geq \|B_i \boldsymbol{\varphi}\| - \underline{l}_i, \quad i = 1, \dots, m, \\ & \varphi_j = \underline{\varphi}_j, \quad j \in \Gamma_{\text{D}}. \end{aligned}\right\} \tag{7.5}$$

Note that the optimal solution of problem (7.5) does not depend on $\boldsymbol{x}$. Hence, in what follows we put $\boldsymbol{x} = \boldsymbol{0}$, which results in

$$\left.\begin{array}{ll} (\overline{\mathrm{PE}}) : \min\limits_{\boldsymbol{\varphi}, \boldsymbol{c}_{\mathrm{e}}} & \displaystyle\sum_{i=1}^{m} \frac{1}{2} k_{\mathrm{e}i} c_{\mathrm{e}i}^2 - \sum_{j \in \Gamma_{\mathrm{N}}} \underline{p}_j \varphi_j \\ \quad \mathrm{s.\,t.} & c_{\mathrm{e}i} \geq \|B_i \boldsymbol{\varphi}\| - \underline{l}_i, \quad i = 1, \dots, m, \\ & \varphi_j = \underline{\varphi}_j, \quad j \in \Gamma_{\mathrm{D}}. \end{array}\right\} \qquad (7.6)$$

Thus the equilibrium configuration of the cable network can be obtained as the optimal solution of problem (7.6).[1] Since $(\overline{\mathrm{PE}})$ above is a conic optimization problem, it is much more tractable compared with (PE) in (7.1), both from the numerical and theoretical points of view. Note that $(\overline{\mathrm{PE}})$ is essentially an SOCP problem because the minimization of the convex quadratic function can be converted to a minimization of a linear function under the second-order cone constraints; see (4.43).

In section 5.1, the problem dual to $(\overline{\mathrm{PE}})$ was derived as (see (5.15))

$$\left.\begin{array}{ll} (\overline{\mathrm{CE}}) : \max\limits_{\boldsymbol{q}, \boldsymbol{v}, \boldsymbol{t}} & \displaystyle -\sum_{i=1}^{m} \frac{1}{2} \frac{q_i^2}{k_{\mathrm{e}i}} - \sum_{i=1}^{m} \underline{l}_i q_i + \sum_{j \in \Gamma_{\mathrm{D}}} \underline{\varphi}_j t_j \\ \quad \mathrm{s.\,t.} & \displaystyle\sum_{i=1}^{m} B_i^{\mathrm{T}} \boldsymbol{v}_i = \boldsymbol{t}, \\ & q_i \geq \|\boldsymbol{v}_i\|, \quad i = 1, \dots, m, \\ & t_j = \underline{p}_j, \quad j \in \Gamma_{\mathrm{N}}, \end{array}\right\} \qquad (7.7)$$

where $\boldsymbol{x} = \boldsymbol{0}$ was used. Here, q_i is the axial force of the ith member, $\boldsymbol{v}_i \in \mathbb{R}^3$ is the internal force vector, and $\boldsymbol{t} \in \mathbb{R}^d$ is the vector of nodal forces. Like $(\overline{\mathrm{PE}})$, we can reformulate $(\overline{\mathrm{CE}})$ to the standard form of SOCP.

Compared with treating (PE), it is much more reasonable to solve the pair $(\overline{\mathrm{PE}})$ and $(\overline{\mathrm{CE}})$ from the viewpoint of modern optimization. This methodology is quite different from the conventional approaches in computational mechanics, such as the Newton–Raphson method based on the tangent stiffness matrix, which is essentially derived from (PE). However, this new methodology relying on convex optimization is much more promising compared with the conventional approaches; the pair $(\overline{\mathrm{PE}})$ and $(\overline{\mathrm{CE}})$ can be solved by using a polynomial-time algorithm, i.e., the primal-dual interior-point method introduced in section 6.1. Thus the convergence to the solutions of $(\overline{\mathrm{PE}})$ and $(\overline{\mathrm{CE}})$ is theoretically guaranteed. In contrast, a conventional approach based on the tangent stiffness usually requires some sort of iteration procedure (e.g., the trial-and-error procedure) due to the nonsmoothness property of the constitutive law in (7.4). In general, the convergence of such an iteration procedure is not guaranteed. The primal-dual interior-point method certainly involves *no* iteration procedure for dealing with the nonsmoothness property in (7.4).

[1]See Proposition 4.2.1 in section 4.2.1.

As observed above, the reformulation of (PE) to the conic optimization form ($\overline{\text{PE}}$) provides us with not only the theoretical results explored in Chapter 5 but also a robust and efficient numerical solution. In what follows, we show through numerical examples that ($\overline{\text{PE}}$) can be solved efficiently by using the primal-dual interior-point method.

As mentioned in section 6.1, the search direction of the primal-dual interior-point method is derived from the modified optimality condition; see section 6.1.2.2 for the case of LP. It has been shown in Proposition 5.1.4 that the optimality conditions for ($\overline{\text{PE}}$) in (7.6) and ($\overline{\text{CE}}$) in (7.7) are given by (see (5.22))

$$q_i = k_{\text{e}i} c_{\text{e}i}, \quad i = 1, \dots, m, \tag{7.8a}$$

$$\sum_{i=1}^{m} B_i^{\text{T}} \boldsymbol{v}_i = \boldsymbol{t}, \tag{7.8b}$$

$$\varphi_j = \underline{\varphi}_j, \quad j \in \Gamma_{\text{D}}; \qquad t_j = \underline{p}_j, \quad j \in \Gamma_{\text{N}}, \tag{7.8c}$$

$$c_{\text{e}i} + \underline{l}_i \geq \|-B_i \boldsymbol{\varphi}\|, \; q_i \geq \|\boldsymbol{v}_i\|, \quad i = 1, \dots, m, \tag{7.8d}$$

$$\begin{bmatrix} c_{\text{e}i} + \underline{l}_i \\ -B_i \boldsymbol{\varphi} \end{bmatrix} \cdot \begin{bmatrix} q_i \\ \boldsymbol{v}_i \end{bmatrix} = 0, \quad i = 1, \dots, m. \tag{7.8e}$$

When (7.8d) is satisfied, it follows from Fact 1.4.6 that (7.8e) is equivalently rewritten as

$$\begin{bmatrix} c_{\text{e}i} + \underline{l}_i & -(B_i \boldsymbol{\varphi})^{\text{T}} \\ -B_i \boldsymbol{\varphi} & (c_{\text{e}i} + \underline{l}_i) I_3 \end{bmatrix} \begin{bmatrix} q_i \\ \boldsymbol{v}_i \end{bmatrix} = \begin{bmatrix} 0 \\ \boldsymbol{0} \end{bmatrix}. \tag{7.9}$$

Note that the left-hand side of (7.9) is the Jordan product of the two vectors $\begin{bmatrix} c_{\text{e}i} + \underline{l}_i \\ -B_i \boldsymbol{\varphi} \end{bmatrix}$ and $\begin{bmatrix} q_i \\ \boldsymbol{v}_i \end{bmatrix}$ defined in the Euclidean Jordan algebra associated with the second-order cone $\mathbb{L}^4$; see section 1.4.3. By introducing a parameter $\nu > 0$, the modified optimality condition is then given by

$$q_i = k_{\text{e}i} c_{\text{e}i}, \quad i = 1, \dots, m, \tag{7.10a}$$

$$\sum_{i=1}^{m} B_i^{\text{T}} \boldsymbol{v}_i = \boldsymbol{t}, \tag{7.10b}$$

$$\varphi_j = \underline{\varphi}_j, \quad j \in \Gamma_{\text{D}}; \qquad t_j = \underline{p}_j, \quad j \in \Gamma_{\text{N}}, \tag{7.10c}$$

$$\begin{bmatrix} c_{\text{e}i} + \underline{l}_i & -(B_i \boldsymbol{\varphi})^{\text{T}} \\ -B_i \boldsymbol{\varphi} & (c_{\text{e}i} + \underline{l}_i) I_3 \end{bmatrix} \begin{bmatrix} q_i \\ \boldsymbol{v}_i \end{bmatrix} = \nu \begin{bmatrix} 1 \\ \boldsymbol{0} \end{bmatrix}, \quad i = 1, \dots, m, \tag{7.10d}$$

$$c_{\text{e}i} + \underline{l}_i > \|-B_i \boldsymbol{\varphi}\|, \; q_i > \|\boldsymbol{v}_i\|. \quad i = 1, \dots, m. \tag{7.10e}$$

Here, $\begin{bmatrix} 1 \\ \boldsymbol{0} \end{bmatrix} \in \mathbb{R}^4$ on the right-hand side of (7.10d) is the identity element of the Euclidean Jordan algebra.

The search direction $(\Delta\boldsymbol{\varphi}, \Delta\boldsymbol{c}_{\mathrm{e}}, \Delta\boldsymbol{q}, \Delta\boldsymbol{v}, \Delta\boldsymbol{t})$ of the primal-dual interior-point method is defined as the solution of the Newton equations for (7.10a)–(7.10d). In a manner similar to Algorithm 6.1.1 for LP, we start from an initial solution satisfying (7.10e). For a fixed $\nu > 0$, we solve the Newton equations for (7.10a)–(7.10d) to obtain the search direction. The solution is updated as

$$\begin{bmatrix} \boldsymbol{\varphi}^{k+1} \\ \boldsymbol{c}_{\mathrm{e}}^{k+1} \\ \boldsymbol{q}^{k+1} \\ \boldsymbol{v}^{k+1} \\ \boldsymbol{t}^{k+1} \end{bmatrix} = \begin{bmatrix} \boldsymbol{\varphi}^{k} \\ \boldsymbol{c}_{\mathrm{e}}^{k} \\ \boldsymbol{q}^{k} \\ \boldsymbol{v}^{k} \\ \boldsymbol{t}^{k} \end{bmatrix} + \alpha_k \begin{bmatrix} \Delta\boldsymbol{\varphi} \\ \Delta\boldsymbol{c}_{\mathrm{e}} \\ \Delta\boldsymbol{q} \\ \Delta\boldsymbol{v} \\ \Delta\boldsymbol{t} \end{bmatrix},$$

where the step length $\alpha_k > 0$ is chosen so that the inequalities

$$c_{\mathrm{e}i}^{k+1} + \underline{l}_i > \|-B_i\boldsymbol{\varphi}^{k+1}\|, \; q_i^{k+1} > \|\boldsymbol{v}_i^{k+1}\|, \quad i = 1, \dots, m$$

are satisfied. Then we decrease $\nu > 0$ in an appropriate way, and repeat the procedure above.

This is a rough sketch of the primal-dual interior-point method directly applied to the pair ($\overline{\mathrm{PE}}$) and ($\overline{\mathrm{CE}}$). Alternatively, there exist several well-developed software packages based on the primal-dual interior-point method, and we can use such a package to solve the pair ($\overline{\mathrm{PE}}$) and ($\overline{\mathrm{CE}}$). As shown in section 4.2.2 (see (4.43)), ($\overline{\mathrm{PE}}$) can be rewritten in the standard form of SOCP as

$$\left.\begin{aligned} \min_{\boldsymbol{\varphi}, \boldsymbol{c}_{\mathrm{e}}, \boldsymbol{y}} \quad & \sum_{i=1}^{m} y_i - \sum_{j \in \Gamma_{\mathrm{N}}} \underline{p}_j \varphi_j \\ \mathrm{s.\,t.} \quad & (y_i/2k_{\mathrm{e}i}) + 1 \geq \left\| \begin{bmatrix} (y_i/2k_{\mathrm{e}i}) - 1 \\ c_{\mathrm{e}i} \end{bmatrix} \right\|, \quad i = 1, \dots, m, \\ & c_{\mathrm{e}i} + \underline{l}_i \geq \|B_i \boldsymbol{\varphi}\|, \quad i = 1, \dots, m, \\ & \varphi_j = \underline{\varphi}_j, \quad j \in \Gamma_{\mathrm{D}}, \end{aligned}\right\} \tag{7.11}$$

where y_i $(i = 1, \dots, m)$ are the extra variables. Several packages of the primal-dual interior-point method, e.g., SDPT3 [435] and SeDuMi [424], can solve the SOCP problem (7.11) directly. Since a second-order cone constraint can be expressed as a positive-semidefinite constraint,[2] problem (7.11) can further be reduced to an SDP problem. Then the implementation of the primal-dual interior-point method for SDP, e.g., SDPA [459] and CSDP [57], is available for solving the resulting SDP problem.

In the following numerical examples, we solve problem (7.11) by using SeDuMi Ver. 1.05 [424]. Computation has been carried out on a Core2 Duo (2.26 GHz with 4 GB memory) with MATLAB® Ver. 7.9 [305].

[2] See Fact 1.5.1.

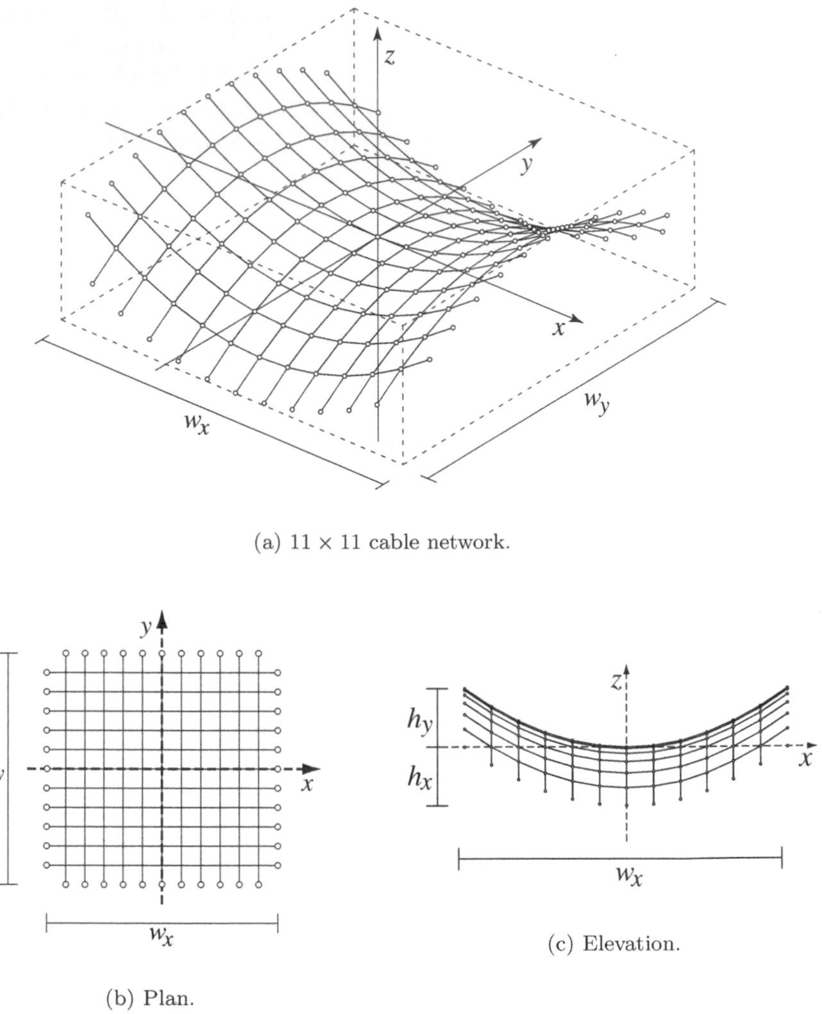

(a) 11 × 11 cable network.

(b) Plan.

(c) Elevation.

FIGURE 7.1: A hyperbolic paraboloid-like cable network. The nodal locations are used for defining the initial member lengths.

Example 7.1

Consider the cable network illustrated in Figure 7.1. For defining the boundary conditions as well as the initial member lengths, we first identify the reference configuration. Suppose that the projection of the cable network onto the x-y plane forms a square grid as shown in Figure 7.1(b). The z-coordinate of each node is then defined by

$$z = \rho(x^2 - y^2), \tag{7.12}$$

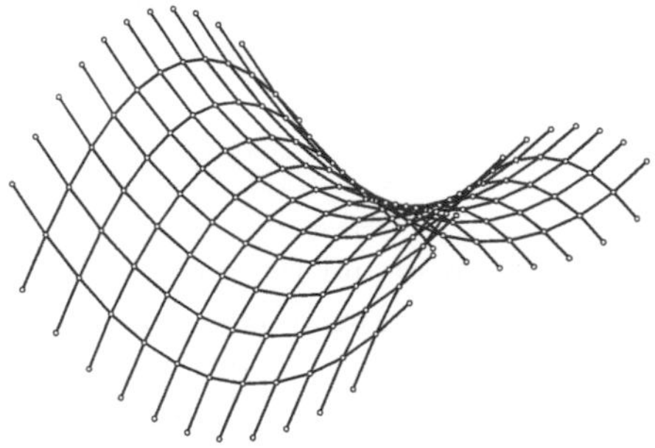

(a) Reference configuration.

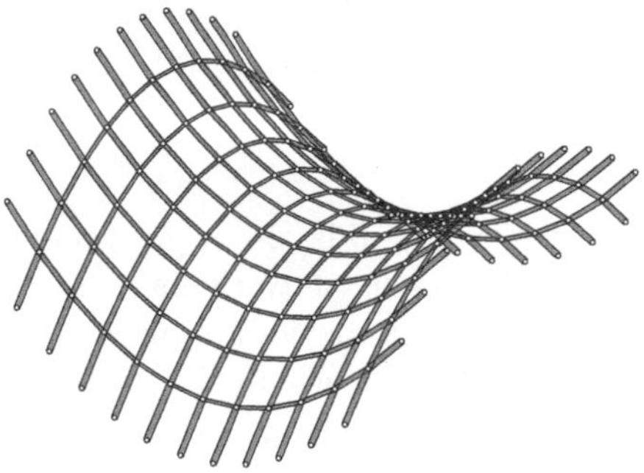

(b) Equilibrium configuration.

FIGURE 7.2: Self-equilibrium configuration of the pin-jointed cable network in Figure 7.1.

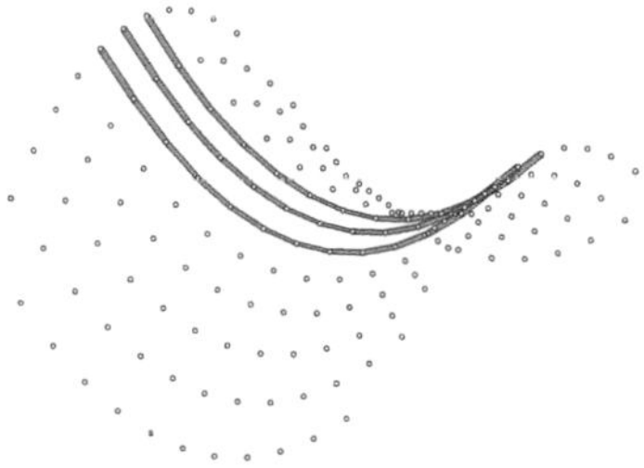

(a) $\underline{p}_j = 3.50 \times 10^3$ kN.

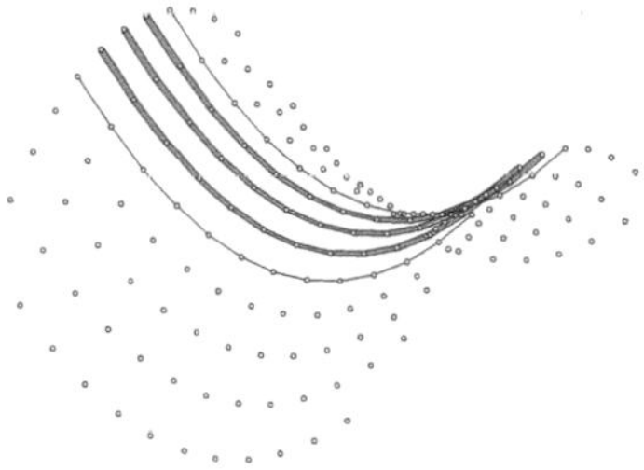

(b) $\underline{p}_j = 3.75 \times 10^3$ kN.

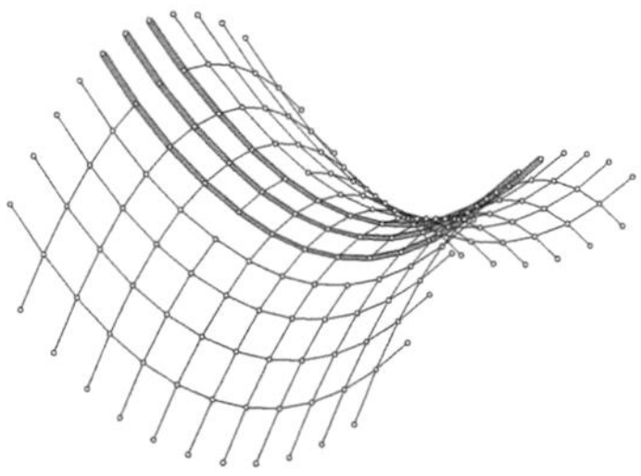

(c) $\underline{p}_j = 4.00 \times 10^3$ kN.

FIGURE 7.3: Equilibrium analysis of the pin-jointed cable network in Figure 7.1 subjected to the external loads.

where $\rho > 0$ is a parameter. We denote by $\check{l}_i$ the member length of the ith member in this reference configuration. The initial unstressed member length $\underline{l}_i$ is defined by

$$\underline{l}_i = \alpha \check{l}_i, \quad i = 1, \dots, m, \tag{7.13}$$

where $\alpha > 0$ is also a parameter.

The cable network consists of $m = 264$ members and 165 nodes. The nodes on the boundaries are pin-supported; i.e., the locations of the nodes indicated by circles in Figure 7.1(b) are fixed. The $11 \times 11 = 121$ internal nodes are free nodes, and hence the number of degrees of freedom of displacements is $d - |\Gamma_{\mathrm{D}}| = 363$.

We set $w_x = w_y = 12\,\mathrm{m}$, $h_x = h_y = 3.6\,\mathrm{m}$, and $\rho = 0.1$ in (7.12). The corresponding reference configuration is shown in Figure 7.2(a) The cross-sectional area of each member is $a_i = 1000.0\ \mathrm{mm}^2$, and Young's modulus is $E = 200.0$ GPa. The initial member lengths are defined by (7.13) with $\alpha = 0.95$. No external loads are considered to be applied to the internal nodes, i.e., $\underline{p}_j = 0\ (j \in \Gamma_{\mathrm{N}})$. The equilibrium configuration found by solving $(\overline{\mathrm{PE}})$ is shown in Figure 7.2(b), where the width of each member is proportional to its axial force at the equilibrium state.

We next consider situations in which many members are slack in the presence of the external loads. The initial member lengths are defined by (7.13) with $\alpha = 1.005$. At an equilibrium configuration[3] without external loads, all the members are slack. In this sense, this situation is rather unrealistic from a practical point of view, especially for use in civil engineering structures. We apply the downward vertical loads of $\underline{p}_j = 3.5 \times 10^3\,\mathrm{kN}$ to the center nine nodes. The optimal solution of $(\overline{\mathrm{PE}})$ is shown in Figure 7.3(a), where only taut members are illustrated. Although many members are slack, the primal-dual interior-point method can find the optimal solution without any difficulty. The computational results are listed in Table 7.1. The larger the external loads, the larger the number of taut members, as observed in Figures 7.3(a)–(c). □

Example 7.2

We next consider the cable network illustrated in Figure 7.5(a). In a manner similar to Example 7.1, the cable network consists of $m = 264$ members with $11 \times 11 = 121$ internal free nodes. The projections of the boundary nodes onto the x-y plane are on the edges of a square, where $w_x = w_y = 12\,\mathrm{m}$ in Figure 7.5(a). The z-coordinate of each boundary node is then given by

$$z = \rho |x + y|, \tag{7.14}$$

[3]Which is not unique.

TABLE 7.1: Computational results of the hyperbolic paraboloid-like cable network in Figure 7.3.

$\underline{p}_j$ ($\times 10^3$ kN)	CPU (s)	Iter.	Opt. ($\times 10^3$ kN $\cdot$ m)
0	0.3	18	0.0000
3.50	0.3	20	22.3623
3.75	0.5	26	24.9082
4.00	0.4	20	27.5643

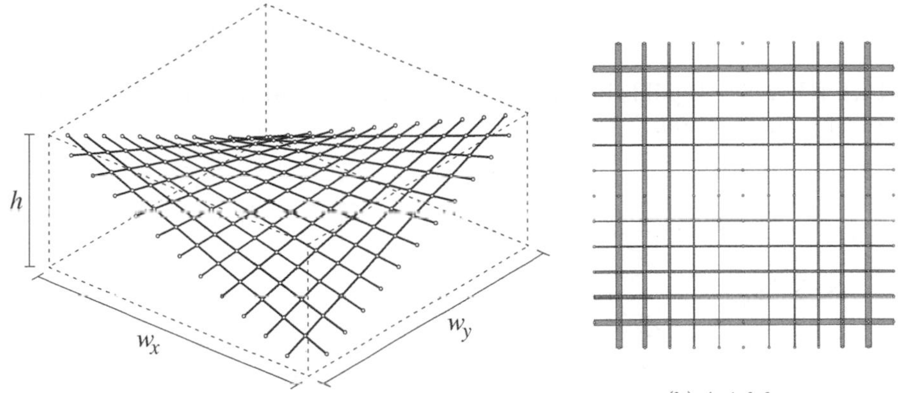

(a) Equilibrium configuration.

(b) Axial forces.

FIGURE 7.4: A cable network with straight boundaries.

where $\rho = 0.45$ and $h = 5.4\,\mathrm{m}$. The initial unstressed length of each member is defined by

$$\underline{l}_i = \alpha \times 1.0\,\mathrm{m}, \quad i = 1, \dots, m \tag{7.15}$$

with $\alpha = 1.0$.

Figure 7.4(a) shows the equilibrium configuration without external loads. The axial forces at this equilibrium state is depicted in Figure 7.4(b), where the width of each member is proportional to its axial force, and the slack members are not illustrated. It is observed that the elongations of the $2 \times 12 = 24$ crosswise members in Figure 7.4(b) are equal to zero; i.e., these members are at the transition points between the taut and slack states.

At the center node of this cable network, we apply the downward vertical load of 10.0×10^3 kN. The equilibrium configuration obtained by solving $(\overline{\mathrm{PE}})$ is shown in Figure 7.5, where all the members are taut. We next consider the variation of initial member lengths by changing the parameter α in (7.15). Figure 7.6 and Figure 7.7 show the equilibrium configurations for $\alpha = 1.05$ and

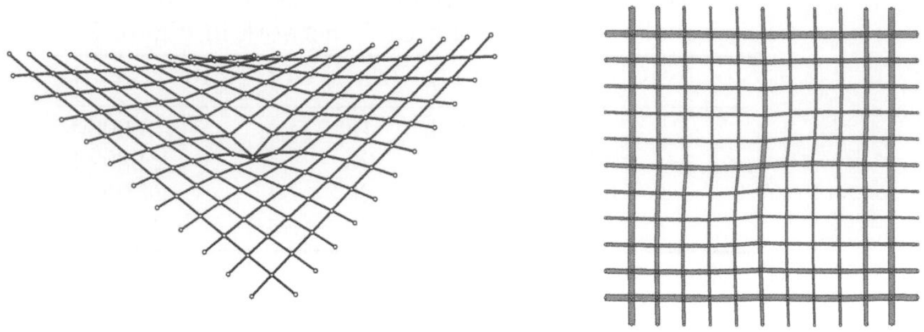

(a) Equilibrium configuration.

(b) Axial forces.

FIGURE 7.5: Equilibrium analysis of the pin-jointed cable network in Figure 7.4 with $\alpha = 1.00$.

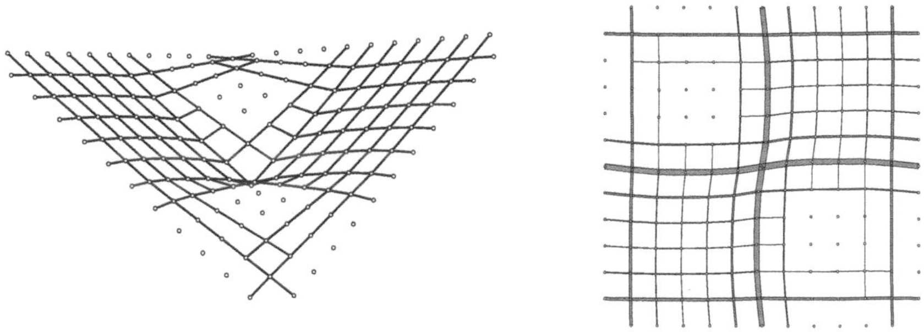

(a) Equilibrium configuration.

(b) Axial forces.

FIGURE 7.6: Equilibrium analysis of the pin-jointed cable network in Figure 7.4 with $\alpha = 1.05$.

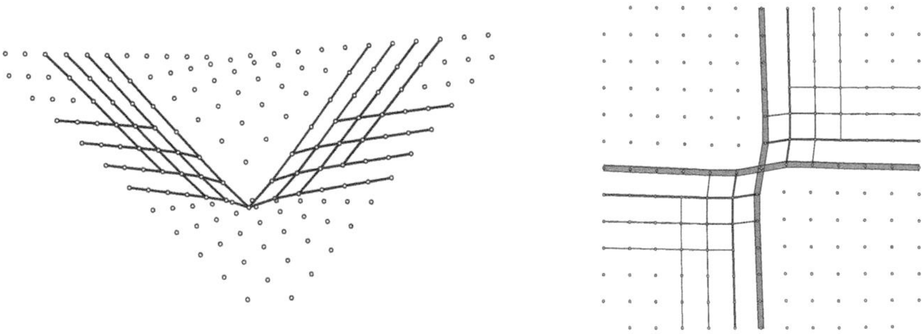

(a) Equilibrium configuration.

(b) Axial forces.

FIGURE 7.7: Equilibrium analysis of the pin-jointed cable network in Figure 7.4 with $\alpha = 1.10$.

$\alpha = 1.1$, respectively, where only the taut members are illustrated. Although a large number of members are slack at the equilibrium state in Figure 7.7, the primal-dual interior-point method can solve $(\overline{\text{PE}})$ without any difficulty. $\square$

7.2 Cable Networks with Sliding Joints

The methodology presented in section 7.1 for the equilibrium analysis of pin-jointed cable networks can be extended to the analysis of cable networks with sliding joints.

Sliding joint is a model of physical joint used to construct cable networks in civil engineering structures, such as a stadium roof. In usual construction process, cables are first tightened to the boundary supports, and internal joints are not completely fastened with cables. Then the initial tensions (so-called prestresses) are introduced to the cables using jacks. After the static equilibrium expected in structural design is reached, these internal joints are fixed to cables to generate pin-joints. Thus, during the initial tensioning process, an internal joint can slide on the cables to which the joint is connected. As a simplified model of such an internal joint, we define a sliding joint as an internal joint that can slide on the cables without friction. In this section, we show that the equilibrium configuration of a cable network with sliding joints can also be found by solving an SOCP problem.

In what follows, a single cable connecting two pin-joints is simply referred to as a *cable*. If some sliding joints are located on a cable, a portion of the cable between two adjacent sliding joints (or the one between a sliding joint and an adjacent pin-joint) is called a *member*. Figure 7.8(a) illustrates a cable divided into three members by the two sliding joints (depicted by "$\circ$"). In Figure 7.8(b) there exist two cables between the fixed supports (depicted by "$\triangle$"), and each cable consists of two members. Since we suppose that the external forces are applied only to joints, the configuration of each member at the equilibrium state forms a straight line.

For a cable network with sliding joints, let m^{c} and m denote the number of cables and the number of members, respectively. For each $r = 1, \dots, m^{\mathrm{c}}$, define $\mathcal{I}_r$ as the set of indices of the members that belong to the rth cable. Thus the set $\{1, \dots, m\}$ is partitioned as $\mathcal{I}_1 \cup \cdots \cup \mathcal{I}_{m^{\mathrm{c}}} = \{1, \dots, m\}$.

For the member i, let $l_i(\varphi)$ denote the member length at the deformed configuration φ. We denote by $\underline{l}_r$ the initial length of the cable r. Then the

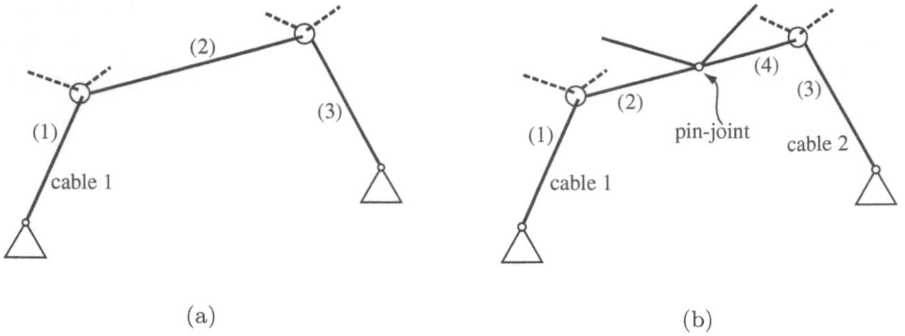

FIGURE 7.8: Sliding joints. (a) two sliding joints are located on the cable 1, i.e., $\mathcal{I}_1 = \{1, 2, 3\}$; (b) two sliding joints and one pin-joint are located on the cables 1 and 2, i.e., $\mathcal{I}_1 = \{1, 2\}$ and $\mathcal{I}_2 = \{3, 4\}$.

elongation of the rth cable, denoted c_r, is written as

$$c_r = \sum_{i \in \mathcal{I}_r} l_i(\boldsymbol{\varphi}) - \underline{l}_r \tag{7.16}$$

$$= \sum_{i \in \mathcal{I}_r} \|B_i \boldsymbol{\varphi}\| - \underline{l}_r, \tag{7.17}$$

where (4.38) has been used.

The sliding joints can freely slide on the cables, before the equilibrium configuration is attained. Therefore, the members belonging to $\mathcal{I}_r$ share a same axial force, denoted q_r. The constitutive law that relates q_r to c_r is given as

$$q_r = \begin{cases} k_{\mathrm{e}r} c_r & \text{if } c_r \geq 0, \\ 0 & \text{otherwise} \end{cases} \tag{7.18}$$

with the constant $k_{\mathrm{e}r}$ defined in a manner similar to (7.3). The stored energy function is then obtained by integrating (7.19) as

$$\hat{w}^r_{\mathrm{cab}}(c_r) = \begin{cases} \dfrac{1}{2} k_{\mathrm{e}r} c_r^2 & \text{if } c_r \geq 0, \\ 0 & \text{otherwise.} \end{cases} \tag{7.19}$$

Regarding the boundary conditions, in a manner similar to that described in section 7.1, we suppose that the nodal force is prescribed as $\underline{p}_j$ for the degree of freedom j belonging to $\mathit{\Gamma}_{\mathrm{N}}$, while the location of the node is prescribed as $\underline{\varphi}_j$ for $j \in \mathit{\Gamma}_{\mathrm{D}}$. By using (7.17) and (7.19), the minimization problem of the potential energy is formulated as[4]

$$\left.\begin{aligned} (\mathrm{PE}) : \min_{\boldsymbol{\varphi}, \boldsymbol{c}} \ & \sum_{r=1}^{m^{\mathrm{c}}} \hat{w}^{r}_{\mathrm{cab}}(c_r) - \sum_{j \in \Gamma_{\mathrm{N}}} \underline{p}_j \varphi_j \\ \mathrm{s.\,t.} \ & c_r = \sum_{i \in \mathcal{I}_r} \|B_i \boldsymbol{\varphi}\| - \underline{l}_r, \quad r = 1, \dots, m^{\mathrm{c}}, \\ & \varphi_j = \underline{\varphi}_j, \quad j \in \Gamma_{\mathrm{D}}. \end{aligned}\right\} \tag{7.20}$$

According to Proposition 4.1.1, $\hat{w}^{r}_{\mathrm{cab}}$ can be rewritten as

$$\hat{w}^{r}_{\mathrm{cab}}(c_r) = \min_{c_{\mathrm{e}r}} \left\{ \frac{1}{2} k_{\mathrm{e}r} c_{\mathrm{e}r}^2 \mid c_{\mathrm{e}r} \geq c_r \right\}, \tag{7.21}$$

where $c_{\mathrm{e}r}$ optimal for (7.21) corresponds to the elastic part of the elongation of the rth cable. Substituting eqrefn.eq.slide.hat.w.alt yields for problem (7.20) the following equivalent reformulation:

$$\left.\begin{aligned} (\overline{\mathrm{PE}}) : \min_{\boldsymbol{\varphi}, \boldsymbol{c}_{\mathrm{e}}} \ & \sum_{r=1}^{m^{\mathrm{c}}} \frac{1}{2} k_{\mathrm{e}r} c_{\mathrm{e}r}^2 - \sum_{j \in \Gamma_{\mathrm{N}}} \underline{p}_j \varphi_j \\ \mathrm{s.\,t.} \ & c_{\mathrm{e}r} \geq \sum_{i \in \mathcal{I}_r} \|B_i \boldsymbol{\varphi}\| - \underline{l}_r, \quad r = 1, \dots, m^{\mathrm{c}}, \\ & \varphi_j = \underline{\varphi}_j, \quad j \in \Gamma_{\mathrm{D}}, \end{aligned}\right\} \tag{7.22}$$

which is in the form of conic optimization. The precise relation between the optimal solutions of (PE) and $(\overline{\mathrm{PE}})$ above can be established in a manner similar to Proposition 4.2.1.

Like the case of a pin-jointed cable network, $(\overline{\mathrm{PE}})$ in (7.22) is an SOCP problem. To see this more explicitly, we make use of the member length l_i in (7.16) as a variable to be optimized. Then each second-order cone constraint in $(\overline{\mathrm{PE}})$ is rewritten as

$$\begin{aligned} & c_{\mathrm{e}r} \geq \sum_{i \in \mathcal{I}_r} \|B_i \boldsymbol{\varphi}\| - \underline{l}_r \\ \Leftrightarrow \quad & c_{\mathrm{e}r} = \sum_{i \in \mathcal{I}_r} l_i - \underline{l}_r, \ l_i \geq \|B_i \boldsymbol{\varphi}\| \ (i \in \mathcal{I}_r) \end{aligned} \tag{7.23}$$

By using (7.23), we can eliminate $c_{\mathrm{e}r}$ in problem (7.22). Moreover, by introducing an extra variable $y_r \in \mathbb{R}$, the minimization of $(1/2) k_{\mathrm{e}r} c_{\mathrm{e}r}^2$ can be converted to the minimization of y_r under constraint $y_r \geq (1/2) k_{\mathrm{e}r} c_{\mathrm{e}r}^2$. There-

[4]Without loss of generality, we put $\boldsymbol{x} = \mathbf{0}$, because the optimal solution of (PE) does not depend on $\boldsymbol{x}$.

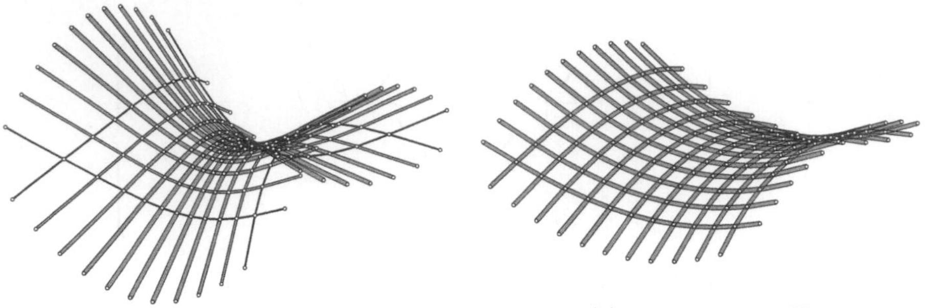

(a) $\rho = 0.1$, $\alpha = 0.95$.

(b) $\rho = 0.05$, $\alpha = 0.95$.

FIGURE 7.9: Equilibrium configurations of the cable networks with sliding joints. The reference configuration is defined as Figure 7.1.

fore, ($\overline{\text{PE}}$) can be rewritten equivalently as

$$\left.\begin{aligned} &\min_{\boldsymbol{\varphi},\boldsymbol{l},\boldsymbol{y}} \sum_{r=1}^{m^{\mathrm{c}}} y_r - \sum_{j\in\Gamma_{\mathrm{N}}} \underline{p}_j \varphi_j \\ &\text{s. t.}\ \ y_r \geq \frac{1}{2} k_{\mathrm{e}r} \Big(\sum_{i\in\mathcal{I}_r} l_i - \underline{l}_r \Big)^2, \quad r = 1,\dots,m^{\mathrm{c}}, \\ &\qquad l_i \geq \|B_i \boldsymbol{\varphi}\|, \quad i = 1,\dots,m, \\ &\qquad \varphi_j = \underline{\varphi}_j, \quad j \in \Gamma_{\mathrm{D}}, \end{aligned}\right\} \tag{7.24}$$

where $\boldsymbol{\varphi} \in \mathbb{R}^d$, $\boldsymbol{l} \in \mathbb{R}^m$, and $\boldsymbol{y} \in \mathbb{R}^{m^{\mathrm{c}}}$ are the variables to be optimized in this problem. Furthermore, each quadratic inequality constraint in this problem can be rewritten as a second-order cone constraint, i.e.,[5]

$$y_r \geq \frac{1}{2} k_{\mathrm{e}r} \Big(\sum_{i\in\mathcal{I}_r} l_i - \underline{l}_r \Big)^2 \quad \Leftrightarrow \quad (y_r/2k_{\mathrm{e}r}) + 1 \geq \left\| \begin{bmatrix} (y_r/2k_{\mathrm{e}r}) - 1 \\ \sum_{i\in\mathcal{I}_r} l_i - \underline{l}_r \end{bmatrix} \right\|. \tag{7.25}$$

Thus ($\overline{\text{PE}}$) is explicitly rewritten in the form of SOCP.

Example 7.3

In continuation of Example 7.1, consider the cable network illustrated in Figure 7.1. We now suppose that all the internal nodes are sliding joints. Therefore, the cable network consists of $m^{\mathrm{c}} = 22$ cables, each of which consists of $m/m^{\mathrm{c}} = 12$ members.

The initial length of each member is defined in a manner similar to Example 7.1, where $\rho = 0.95$ in (7.12) and $\alpha = 0.1$ in (7.13). No external load is

[5]See Fact 1.5.2 (b).

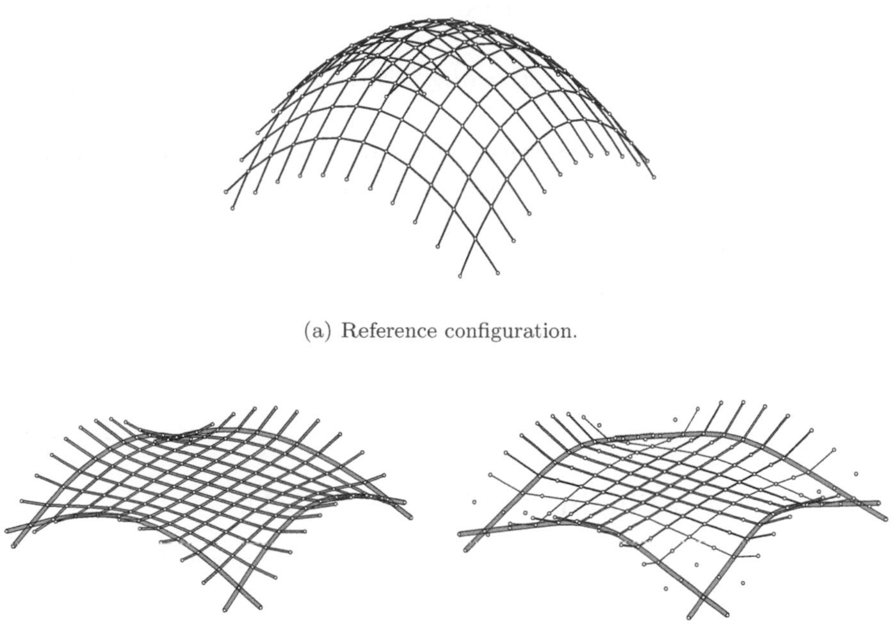

(a) Reference configuration.

(b) $\rho = 1.0$, $\alpha = 0.8$.

(c) $\rho = 1.0$, $\alpha = 0.85$.

FIGURE 7.10: Equilibrium analysis of a dome-like cable network with sliding joints.

applied to the internal nodes, i.e., $\underline{p}_j = 0$ ($j \in \Gamma_{\mathrm{N}}$). The equilibrium configuration found by solving ($\overline{\mathrm{PE}}$) in (7.22) is shown in Figure 7.9(a). It is observed that the sliding joints gather around the center of the cable network compared with the case of the pin-jointed cable network shown in Figure 7.2(b).

As a shallower example, we define the reference configuration by putting $\alpha = 0.05$ in (7.13). The equilibrium configuration of this example is illustrated in Figure 7.9(b), where the width of each member is proportional to its axial force. □

Example 7.4

We next consider the cable network shown in Figure 7.10(a). In a manner similar to Example 7.1, the reference configuration is defined such that the projection of the cable network onto the x-y plane forms a square grid. Then the z-coordinate of each node is given by

$$z = \rho(-x^2 - y^2), \tag{7.26}$$

as illustrated in Figure 7.10(a). We put $\rho = 1.0$ in this example. The initial length of each member is defined

$$\underline{l}_i = \alpha \check{l}_i, \quad i = 1, \dots, m, \tag{7.27}$$

where $\check{l}_i$ is the length of member i in the reference configuration defined above.

All the internal loads are assumed to be sliding joints. Hence, the cable network consists of $m^{\rm c} = 22$ cables, each of which is divided into 12 members. Figure 7.10(b) shows the equilibrium configuration for $\alpha = 0.8$, where all the members are taut. In contrast, four cables are slack at the equilibrium configuration for $\alpha = 0.85$, as shown in Figure 7.10(c). □

7.3 Form-Finding of Cable Networks

As a closely related topic of the analysis of cable networks, here we take up the *form-finding problem* of cable networks.

The form-finding problem has a different aspect from the other problems treated in this book. In the other parts of the book, we consider the problem for finding the equilibrium state of a given structure under the specified boundary conditions. In contrast, the form-finding problem determines the design of a structure and its equilibrium configuration simultaneously, when some other state variables, e.g., the stresses, at the equilibrium state are specified. Thus the form-finding problem can be regarded as a sort of the inverse problem. The configuration of a tension structure, such as a cable network or a membrane structure, depends on the stress distribution, and an arbitrary given curved surface in the three-dimensional space cannot be necessarily realized as a tension structure. Therefore, the form-finding problem plays an important role in the process of structural design of a tension structure.

In this section, we consider a form-finding problem of pin-jointed cable networks. The design variables to be determined are the initial unstressed length of each member as well as the equilibrium geometry of the cable network. We assume that the topology (i.e., the connectivity of the members) of the cable network, represented by the matrices B_i $(i = 1, \dots, m)$, and the elongation stiffness $k_{{\rm e}i}$ of each member are fixed. Furthermore, the boundary conditions, i.e., the prescribed deformations $\underline{\varphi}_j$ for $j \in \varGamma_{\rm D}$ and the prescribed nodal loads $\underline{p}_j$ for $j \in \varGamma_{\rm N}$, are also specified. Unlike the problem of equilibrium analysis, however, we specify the member axial forces $\underline{q}_i$ $(i = 1, \dots, m)$ at the (unknown) equilibrium state. Under this situation, we aim to find the initial member lengths $\underline{l}_i$ $(i = 1, \dots, m)$ and the coordinates of nodes φ at the equilibrium state. We show that this form-finding problem can be formulated as an SOCP problem.

7.3.1 Form-finding with specified axial forces

Consider a design process of a pin-jointed cable network. Let B_i $(i = 1,\dots,m)$, $k_{\mathrm{e}i}$ $(i = 1,\dots,m)$, $\varphi_j = \underline{\varphi}_j$ $(j \in \Gamma_{\mathrm{D}})$, and $\underline{p}_j$ $(j \in \Gamma_{\mathrm{N}})$ be given. Furthermore, suppose that the member axial forces $\underline{q}_i > 0$ $(i = 1,\dots,m)$ are specified. Then deciding on the design of the cable network corresponds to determining the initial unstressed member lengths $\underline{l}_i$ $(i = 1,\dots,m)$.

Instead of $\underline{l}_i$'s, we choose the member lengths at the equilibrium state, denoted ℓ_i $(i = 1,\dots,m)$, as the design variables. Once ℓ_i is obtained, we can compute the initial member length $\underline{l}_i$ from $k_{\mathrm{e}i}$, $\underline{q}_i$, and ℓ_i. The form-finding problem is thence formulated with the variables $\boldsymbol{\ell} = (\ell_i) \in \mathbb{R}^m$ and $\boldsymbol{\varphi} \in \mathbb{R}^d$.

Consider the following optimization problem:

$$\left.\begin{aligned} \min_{\boldsymbol{\ell},\boldsymbol{\varphi}} \ & \sum_{i=1}^{m} \underline{q}_i \ell_i - \sum_{j\in\Gamma_{\mathrm{N}}} \underline{p}_j \varphi_j, \\ \mathrm{s.t.} \ & \ell_i \geq \|B_i \boldsymbol{\varphi}\|, \quad i = 1,\dots,m, \\ & \varphi_j = \underline{\varphi}_j, \quad j \in \Gamma_{\mathrm{D}}, \end{aligned}\right\} \tag{7.28}$$

which is an SOCP problem. Note that this problem is obtained from $(\overline{\mathrm{PE}})$ in (7.6) by replacing $\frac{1}{2}k_{\mathrm{e}i}c_{\mathrm{e}i}^2$ and $c_{\mathrm{e}i} + \underline{l}_i$ with $\underline{q}_i\ell_i$ and ℓ_i, respectively. Thus problem (7.28) can be regarded as a variant of the minimization problem of the total potential energy. However, it should be clear that the member length ℓ_i is treated as a variable, while the member axial force $\underline{q}_i$ is a constant.

The optimality condition of problem (7.28) can be obtained in a manner similar to Proposition 5.1.4; $(\bar{\boldsymbol{\ell}}, \bar{\boldsymbol{\varphi}})$ is an optimal solution of problem (7.28) if and only if there exist $\bar{\boldsymbol{v}}_i \in \mathbb{R}^3$ $(i = 1,\dots,m)$ and $\bar{\boldsymbol{t}} \in \mathbb{R}^d$ satisfying

$$\sum_{i=1}^{m} B_i^{\mathrm{T}} \bar{\boldsymbol{v}}_i = \bar{\boldsymbol{t}}, \tag{7.29a}$$

$$\varphi_j = \underline{\varphi}_j, \quad j \in \Gamma_{\mathrm{D}}; \qquad \bar{t}_j = \underline{p}_j, \quad j \in \Gamma_{\mathrm{N}}, \tag{7.29b}$$

$$\bar{\ell}_i \geq \|-B_i\bar{\boldsymbol{\varphi}}\|, \ \underline{q}_i \geq \|\bar{\boldsymbol{v}}_i\|, \quad i = 1,\dots,m, \tag{7.29c}$$

$$\begin{bmatrix} \bar{\ell}_i \\ -B_i\bar{\boldsymbol{\varphi}} \end{bmatrix} \cdot \begin{bmatrix} \underline{q}_i \\ \bar{\boldsymbol{v}}_i \end{bmatrix} = 0, \quad i = 1,\dots,m. \tag{7.29d}$$

Here, (7.29a) is the force-balance equation in terms of the internal force vectors $\boldsymbol{v}_i$ $(i = 1,\dots,m)$ and the nodal force vector $\boldsymbol{t} \in \mathbb{R}^d$, while (7.29b) is the boundary condition. Conditions (7.29c) and (7.29d) form the complementarity conditions over second-order cones $\mathbb{L}^4$. By using the optimal $\bar{\ell}_i$, define $\underline{l}_i$ and $c_{\mathrm{e}i}$ by

$$\underline{l}_i = \bar{\ell}_i - c_{\mathrm{e}i}, \tag{7.30}$$

$$c_{\mathrm{e}i} = \frac{\underline{q}_i}{k_{\mathrm{e}i}} \tag{7.31}$$

for each $i = 1, \dots, m$. Substituting (7.30) and (7.31) into (7.29) yields

$$\underline{q}_i = k_{ei} c_{ei}, \quad i = 1, \dots, m, \tag{7.32a}$$

$$\sum_{i=1}^{m} B_i^{\mathrm{T}} \bar{\boldsymbol{v}}_i = \bar{\boldsymbol{t}}, \tag{7.32b}$$

$$\varphi_j = \underline{\varphi}_j, \quad j \in \Gamma_{\mathrm{D}}; \qquad \bar{t}_j = \underline{p}_j, \quad j \in \Gamma_{\mathrm{N}}, \tag{7.32c}$$

$$c_{ei} + \underline{l}_i \geq \|-B_i \bar{\boldsymbol{\varphi}}\|, \ \underline{q}_i \geq \|\bar{\boldsymbol{v}}_i\|, \quad i = 1, \dots, m, \tag{7.32d}$$

$$\begin{bmatrix} c_{ei} + \underline{l}_i \\ -B_i \bar{\boldsymbol{\varphi}} \end{bmatrix} \cdot \begin{bmatrix} \underline{q}_i \\ \bar{\boldsymbol{v}}_i \end{bmatrix} = 0, \quad i = 1, \dots, m. \tag{7.32e}$$

This condition coincides with (5.22) in Proposition 5.1.4, which has been shown in section 5.1.3 to be the optimality condition of ($\overline{\mathrm{PE}}$) and ($\overline{\mathrm{CE}}$) in (7.6) and (7.7). This means that $\bar{\boldsymbol{\varphi}}$ satisfying (7.32) corresponds to the equilibrium configuration of the cable network, the initial member lengths of which are specified as $\underline{l}_i$ ($i = 1, \dots, m$). Furthermore, at this equilibrium configuration, the axial force of each member coincides with $\underline{q}_i$.

In sum, the design of the cable network is determined by $\underline{l}_i$ ($i = 1, \dots, m$) in (7.30) by using $(\bar{\boldsymbol{\ell}}, \bar{\boldsymbol{\varphi}})$ optimal for problem (7.28), and the equilibrium configuration coincides with $\bar{\boldsymbol{\varphi}}$. Thus, the initial member lengths and the equilibrium configuration can be found simultaneously by solving problem (7.28); i.e., this problem serves as the form-finding problem. Since problem (7.28) is also an SOCP, we can solve it efficiently by using the primal-dual interior-point method in a manner similar to section 7.1.

7.3.2 Special cases

As a special case of the form-finding problem in (7.28), suppose that no external load is applied; i.e., $\underline{p}_j = 0$ ($j \in \Gamma_{\mathrm{N}}$). The corresponding equilibrium configuration of a cable network, in the presence of the member axial forces, is called the *self-equilibrium configuration*. Then the form-finding problem is to find ℓ_i ($i = 1, \dots, m$), together with $\boldsymbol{\varphi}$, such that the axial force of each member coincides with the specified value $\underline{q}_i$ at the (unknown) self-equilibrium configuration $\boldsymbol{\varphi}$.

In this case, problem (7.28) is reduced to

$$\left.\begin{aligned} \min_{\boldsymbol{\ell}, \boldsymbol{\varphi}} \ & \sum_{i=1}^{m} \underline{q}_i \ell_i \\ \text{s.t.} \ & \ell_i \geq \|B_i \boldsymbol{\varphi}\|, \quad i = 1, \dots, m, \\ & \varphi_j = \underline{\varphi}_j, \quad j \in \Gamma_{\mathrm{D}}. \end{aligned}\right\} \tag{7.33}$$

Since all the inequality constraints of problem (7.33) become active at the optimal solution, this problem is equivalently rewritten as

$$\left.\begin{aligned} &\min_{\boldsymbol{\varphi}} \sum_{i=1}^{m} \underline{q}_i \|B_i \boldsymbol{\varphi}\| \\ &\text{s.t.}\ \varphi_j = \underline{\varphi}_j, \quad j \in \Gamma_{\mathrm{D}}. \end{aligned}\right\} \qquad (7.34)$$

Since $\|B_i \boldsymbol{\varphi}\|$ represents the length of the ith member in the deformed configuration, this problem can be interpreted as the minimization of the weighted sum of member lengths.

If $\underline{q}_1 = \cdots = \underline{q}_m$ and $\underline{p}_j = 0$ $(j \in \Gamma_{\mathrm{N}})$, then problem (7.28) is further simplified as

$$\left.\begin{aligned} &\min_{\boldsymbol{\ell}, \boldsymbol{\varphi}} \sum_{i=1}^{m} \ell_i \\ &\text{s.t.}\ \ell_i \geq \|B_i \boldsymbol{\varphi}\|, \quad i = 1, \dots, m, \\ &\qquad \varphi_j = \underline{\varphi}_j, \quad j \in \Gamma_{\mathrm{D}}. \end{aligned}\right\} \qquad (7.35)$$

It is easy to see that this SOCP problem is equivalent to

$$\left.\begin{aligned} &\min_{\boldsymbol{\varphi}} \sum_{i=1}^{m} \|B_i \boldsymbol{\varphi}\| \\ &\text{s.t.}\ \varphi_j = \underline{\varphi}_j, \quad j \in \Gamma_{\mathrm{D}}, \end{aligned}\right\} \qquad (7.36)$$

which is interpreted as the minimization problem of the sum of the member lengths at the deformed state. Thus the self-equilibrium configuration with the uniform distribution of axial forces is identified with the cable network that has the minimal total length of members. This fact was originally shown by Schek [400, Theorem 2].

It is interesting to see the similarity between the form-finding problem (7.36) and the minimal-surface problem, which is to find the curved surface with the minimal surface area for the given boundary. The minimal surfaces are often used in the design process of membrane structures, because the minimal surface coincides with the geometry of an isotropic membrane that has the uniform stress distribution; see, e.g., Bletzinger–Wüchner–Daoud–Camprubí [56]. Problem (7.36) for a cable network can be interpreted as a variant of the minimal-surface problem, in the sense that the surface area for a membrane is replaced with the total length of members for a cable network.

Example 7.5

We here consider the form-finding problem of a pin-jointed cable network with parabolic boundaries.

The locations of boundary nodes, shown in Figure 7.11(a), are defined in the same manner as Example 7.1, where $w_x = w_y = 12.0\,\mathrm{m}$, and $h = 7.2\,\mathrm{m}$. These nodes are equally spaced with respect to the horizontal plane, and their

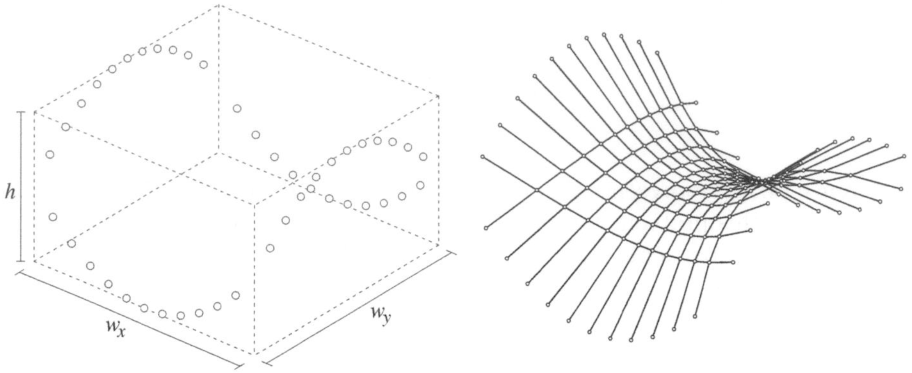

(a) Boundary nodes.

(b) Equilibrium configuration.

FIGURE 7.11: Form-finding of a hyperbolic paraboloid-like cable network.

z-coordinates are defined by a hyperbolic paraboloid in (7.12) with $\rho = 0.1$. The topology of the cable network is specified such that the members form quadrilaterals. No external forces are considered. We then aim to find the initial member lengths of the cable network, which has the uniform distribution of the axial forces at the equilibrium state.

Since we consider the situation of $\underline{q}_i$ $(i = 1, \dots, m)$ and $\underline{p}_j = 0$ $(j \in \Gamma_{\rm N})$, the form-finding problem (7.28) can be reduced to problem (7.35). The optimal solution of this problem is shown in Figure 7.11(b), at which all the members share the same axial force. ∎

Example 7.6

We next consider the case in which boundary nodes are on a paraboloid as shown in Figure 7.12(a). The projections of these nodes onto the x-y plane are equally placed on the edges of a square with $w_x = w_y = 12.0\,\mathrm{m}$. The z-coordinates are defined by (7.26) with $\rho = 0.1$ in Example 7.4.

By solving problem (7.35), we find the geometry depicted in Figure 7.12(b), which realizes the uniform distribution of the axial forces in the absence of the external loads.

Alternatively, suppose that the vertical external load $\underline{p}_j$ is applied to each node to the downward direction. We aim to find the initial member lengths, which realize the uniform distribution of the axial forces, with the magnitudes of $|\underline{p}_j / \underline{q}_i| = 0.05$ in the presence of the external loads. The solution obtained by solving problem (7.28) is shown in Figure 7.12(c), which is the equilibrium configuration of the obtained cable network subjected to the specified external loads. ∎

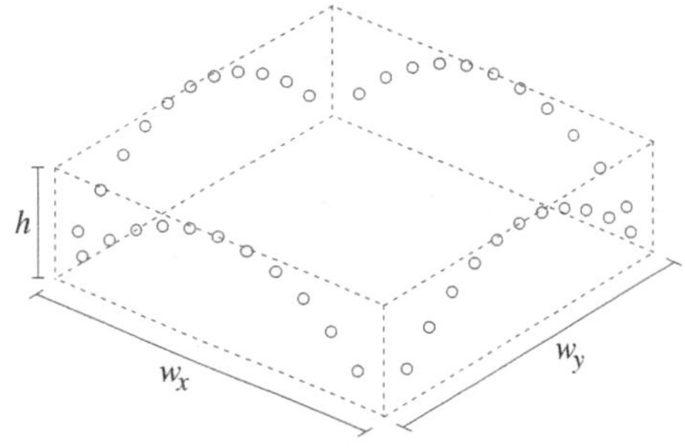

(a) Boundary nodes.

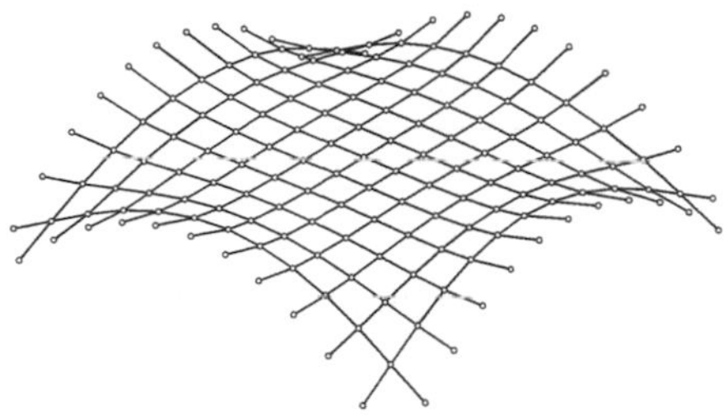

(b) Without external loads.

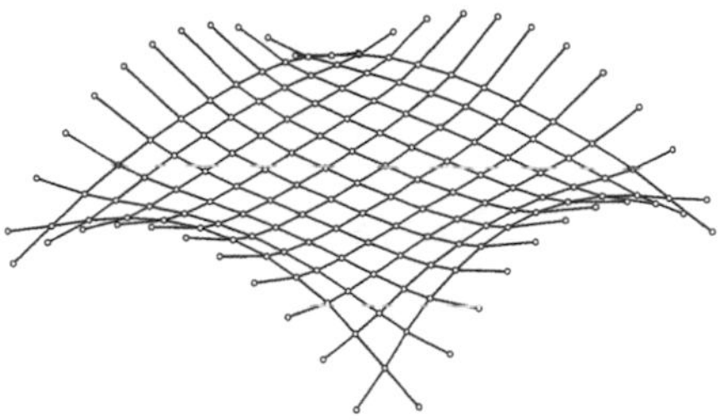

(c) With vertical external loads.

FIGURE 7.12: Form-finding of a dome-like cable network with parabolic boundaries.

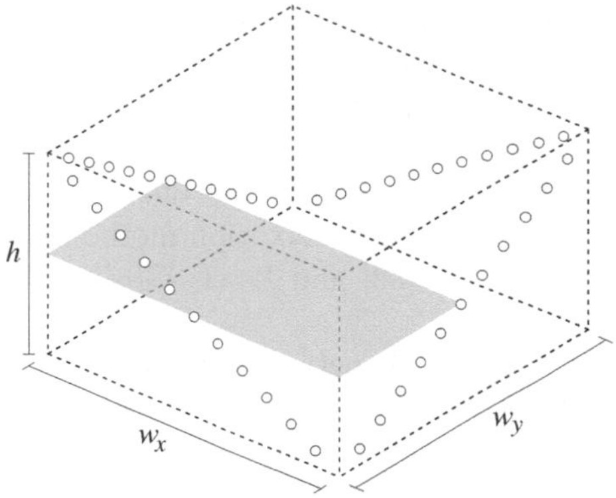

(a) Boundary nodes.

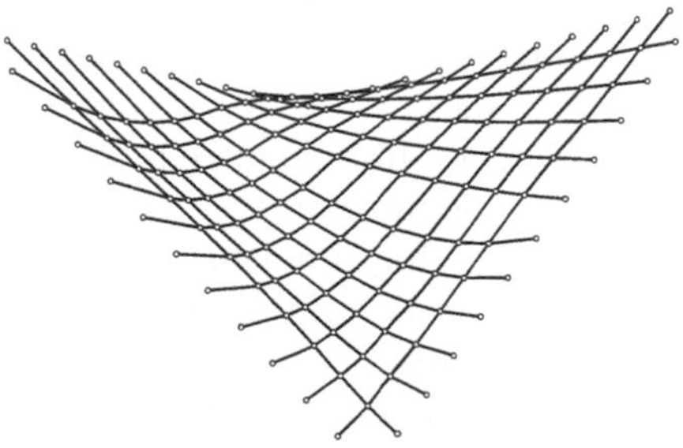

(b) Equilibrium configuration.

FIGURE 7.13: Form-finding of a cable network with straight edges.

Example 7.7
In this example we consider a cable network with straight boundaries, as shown in Figure 7.13(a). In a manner similar to Example 7.2, the locations of the boundary nodes are defined by (7.14) with $\rho = 0.6$, where $w_x = w_y = 12.0\,\mathrm{m}$, and $h = 7.2\,\mathrm{m}$ in Figure 7.13(a).

As the loading condition, we suppose that the uniform vertical downward loads are applied to the nodes in the front side shadowed in Figure 7.13(a). No external loads are applied to the nodes in the back side. In the presence of these external loads, we specify the uniformly distributed axial forces as $|\underline{p}_j/\underline{q}_i| = 0.15$.

The geometry of the cable network obtained by solving problem (7.28) is shown in Figure 7.13(b), which corresponds to the equilibrium configuration under the specified external loads. ▯

7.4 Notes

For a survey of numerical methods for the equilibrium analysis of cable networks, see section 4.4.

Concerning the form-finding problem of tension structures, the two major categories of numerical methods are as follows:

- the dynamic relaxation method
- the force-density method

More fundamental approaches based on the tangent stiffness matrix can be found in Argyris–Angelopoulos–Bichat [20].

The dynamic relaxation method is used both for the (static) equilibrium analysis and the form-finding problem of cable networks. For finding the static equilibrium configuration, the dynamic relaxation method performs the dynamic analysis by considering the fictitious masses, damping, and time. The static equilibrium state is then obtained as the long-time limit of the fictitious dynamic problem [262]. Extensions of this dynamic relaxation method to the form-finding problem can be found in Barnes [32] and Lewis–Gosling [265]; see also the review paper by Barnes [33].

As a convenient method for form-finding of cable networks, the force density method was originally proposed by Linkwitz–Schek [269] and Schek [400]. In the force-density method, we specify the ratio of the axial force to the member length (which is called the *force density*) for each member. Then the equilibrium configuration can be found by solving a system of linear equations.

Although the force density cannot be interpreted clearly from the physical point of view, the force-density method is very simple in the sense that only a system of linear equations is required to be solved. For this reason, the force-density method has been extended to various structures; see Maurin–Motro [306] and Pauletti–Pimenta [360] for membrane structures; Vassart–Motro [447] and Zhang–Ohsaki [467] for tensegrity structures; and Den–Jiang–Kwan [108] for cable–strut structures. For cable networks, Hernández-Montes–Jurado-Piña–Bayo [175] and Levy–Spillers [260] proposed practical form-finding methods, in which the force-density method is performed repeatedly.

The formulation studied in section 7.3 for the form-finding problem of cable networks is similar to the minimal surface method for membranes; see, e.g., Bletzinger–Ramm [55] and Bletzinger–Wüchner–Daoud–Camprubí [56] for the minimal-surface method. The surface area is minimized in the minimal-surface method, while the total length of members, with modification due to the loading and the non-uniform distribution of axial forces, is minimized in problem (7.28). An extension of problem (7.28) to the form-finding problem of cable domes can be found in Ohsaki–Kanno [344].

Part IV

Problems in Nonsmooth Mechanics

Chapter 8

Masonry Structures

Orientation In this chapter we take up static analyses of masonry structures consisting of perfectly no-tension material. The constitutive law of the no-tension model involves nonsmoothness properties because of the absence of tensile stiffnesses. The principal objective of this chapter is to reformulate the minimization problems of potential and complementary energies of masonry structures as the primal-dual pair of the conic optimization problems and then to investigate the duality between those two problems. We cope with the nonsmoothness property of the stiffness of no-tension material based upon the complementarity condition over the positive-semidefinite cone (see section 1.3), and the duality of the two energy minimizations is established using the Fenchel duality theory (see section 2.2.5).

8.1 Introduction

In the chapters in Part IV, we apply the abstract results on duality and complementarity over convex cones to various concrete problems arising from nonsmooth mechanics. With a view to applying the concept of duality and complementarity, we set the problem under consideration within the Fenchel duality methodology investigated in section 2.2.5, which was proven through the preceding parts of this book to be a powerful tool for analyzing various convex optimization problems subjected to conic constraints.

Recall that the primal problem, in the Fenchel duality theory, is defined by

$$(\mathrm{P_F}): \quad \inf_{x}\{f(x) + g(\Lambda x) \mid x \in \mathbb{V}\}, \qquad \text{(see (2.28))}$$

where $f : \mathbb{V} \to \mathbb{R} \cup \{+\infty\}$ and $g : \mathbb{Y} \to \mathbb{R} \cup \{+\infty\}$ are closed proper convex functions, and $\Lambda : \mathbb{V} \to \mathbb{Y}$ is a continuous linear mapping. The problem dual to $(\mathrm{P_F})$ is given by

$$(\mathrm{P_F^*}): \quad \sup_{z^*}\{-f^*(\Lambda^* z^*) - g^*(-z^*) \mid z^* \in \mathbb{Y}^*\}. \qquad \text{(see (2.31))}$$

Here, $\mathbb{V}^*$ and $\mathbb{Y}^*$ are the (topological) dual spaces of $\mathbb{V}$ and $\mathbb{Y}$, respectively. We show that various problems in nonsmooth mechanics can be formulated in the form of (P_F), and derive the explicit form of (P_F^*) for the problem under consideration; see Figure 5.1 for a summary of this methodology. Then we apply Proposition 2.2.9 (the strong duality) and Proposition 2.2.5 (the optimality condition) to represent a particular nonsmoothness property in the mechanical problem as a conic optimization problem. The duality and optimality for the problems in nonsmooth mechanics are thus obtained systematically within the framework of the Fenchel duality theory and the methodology of the conic optimization. Besides those theoretical results, we demonstrate that numerical solutions for such a problem in nonsmooth mechanics can be obtained efficiently by solving the pair (P_F) and (P_F^*) with the polynomial-time primal-dual interior-point method, an algorithm that also enjoys the favorable property of conic optimization problems. Thus through these applications we shed new light on nonsmooth mechanics from the perspective of modern optimization.

In this chapter we focus on the static analysis of a particular class of masonry structures. In general, by a masonry structure we mean a structure that consists of brittle materials, such as mortar, stones, bricks, or non-reinforced concrete. Besides varying widely in mechanical behavior, those materials share the common characteristic that they are extremely weak if we try to apply tensile loads but are reasonably elastic under compressive loads. A drastic idealization of such materials is known as a perfectly *no-tension material*,[1] by which we mean an isotropic non-dissipative material incapable of sustaining tensile stresses at all but otherwise like linear elastic.

Thus, no-tension material is a crude modeling of mechanical behavior. Many authors have given it much attention for several decades, from the viewpoint of both theoretical concerns [19, 63, 100, 106, 107, 118, 118, 145, 280, 350] and numerical solutions [7, 141, 276, 277, 279, 291, 297, 351, 397]. It is worth noting two major advantages of the no-tension material model. First, the no-tension model is an idealized simple model involving a few parameters while, in contrast, the actual mechanical elements constituting masonries usually require too large a number of material parameters. Thus the no-tension model is often less information-intensive than a more sophisticated model. Second, from a practical point of view, in several situations the no-tension model can approximate with reasonable accuracy the response of many important structures. For example, the no-tension model may be able to approximate responses of stone block structures with poor or no bending stiffnesses, mortar interfaces, and ancient masonries that are already cracked and thence unable to withstand any further tension loads. The crisis of such structures is mostly due to crack openings rather than reaching the limit of compressive strength [176, 177].

Since the tensile stress is assumed to be zero while the compressive stiffness

[1]A no-tension material is also called a *masonry-like material* by some authors, e.g., Anzellotti [19], Del Piero [106], and Lucchesi–Padovani–Pagni [276].

is as the linear elasticity, the nonsmoothness property apparently appears in the constitutive law of the no-tension material model. This nonsmoothness property precisely corresponds to the crack opening and closing phenomena. Furthermore, the nonsmoothness property also appears near the boundary of the admissible set of the stress tensor (and also that of the crack strain tensor as seen below), which can be recognized as follows. The lack of resistance to tensile stresses means that any principal stress cannot be positive. In other words, the symmetric stress tensor, denoted $\boldsymbol{\sigma}$, should be negative semidefinite. Thus the boundary of the admissible set of stress tensor is given by $\min_i\{\lambda_i(\boldsymbol{\sigma})\} = 0$, where $\lambda_i(\boldsymbol{\sigma})$ is the eigenvalue of $\boldsymbol{\sigma}$. Since the minimum eigenvalue, $\min_i\{\lambda_i(\boldsymbol{\sigma})\}$, of a symmetric tensor $\boldsymbol{\sigma}$ is a nonsmooth function, the boundary of the set of admissible stress tensor is nonsmooth.

In this chapter we consider these nonsmoothness properties of the no-tension masonry structures without resorting to any regularization process, by virtue of the concept of the complementarity conditions over positive-semidefinite cones. Following several words regarding notation used particularly in this chapter, section 8.2 explores the constitutive law of the no-tension model to reformulate the minimization problem of the potential energy in the form of a conic optimization. We establish the minimum principle of the complementary energy for masonry structures in section 8.3 through an in-depth investigation of the duality between those two energy principles. In section 8.4, for numerical computation we apply a spatial discretization to the presented minimization problem of the potential energy, and show several numerical examples.

8.1.1 Notation

By a *tensor* we mean a second-order tensor $\boldsymbol{\alpha} = (\alpha_{ij})$. Since we always consider the Cartesian coordinate system, we can ignore the distinction between the covariant and contravariant components of a tensor, and we always use subscripts to show indices. The set of all such tensors can be identified with the set of square matrices with the corresponding order. Therefore, we write $\boldsymbol{\alpha} \in \mathbb{R}^{n \times n}$ if $\boldsymbol{\alpha}$ is a second-order tensor in the n-dimensional space. Moreover, we write $\boldsymbol{\alpha} \in \mathbb{S}^n$ if $\boldsymbol{\alpha} \in \mathbb{R}^{n \times n}$ is symmetric. Accordingly, following Definition 1.3.12 for matrices, for two tensors $\boldsymbol{\alpha}$, $\boldsymbol{\beta} \in \mathbb{R}^{n \times n}$, we represent their scalar product by $\boldsymbol{\alpha} \bullet \boldsymbol{\beta} = \sum_{i=1}^{n} \sum_{j=1}^{n} \alpha_{ij}\beta_{ij}$. However, we also, and mostly, write $\boldsymbol{\alpha} : \boldsymbol{\beta}$ to represent the scalar product, following the conventional notation in tensor analysis. Similarly, the scalar product of two vectors $\boldsymbol{p}$, $\boldsymbol{q} \in \mathbb{R}^n$ is denoted by $\boldsymbol{p} \cdot \boldsymbol{q} = \boldsymbol{p}^{\mathrm{T}}\boldsymbol{q} = \sum_{i=1}^{n} p_i q_i$.

We denote by ∇ the gradient operator. Thus for a scalar field $f(\boldsymbol{x})$ and a vector field $\boldsymbol{p}(\boldsymbol{x})$ in the n-dimensional space, with a position vector $\boldsymbol{x}$, we have that

$$\nabla f(\boldsymbol{x}) = \left(\frac{\partial f}{\partial x_i}(\boldsymbol{x}) \right) \in \mathbb{R}^n,$$

$$\nabla \boldsymbol{p}(\boldsymbol{x}) = \left(\frac{\partial p_i}{\partial x_j}(\boldsymbol{x}) \right) \in \mathbb{R}^{n \times n}.$$

By "div" we mean the divergence operator. The divergence for a vector field $\boldsymbol{p}(\boldsymbol{x})$ and a tensor field $\boldsymbol{\beta}(\boldsymbol{x})$ reads

$$\operatorname{div} \boldsymbol{p}(\boldsymbol{x}) = \sum_{i=1}^{n} \frac{\partial p_i}{\partial x_i}(\boldsymbol{x}) \in \mathbb{R},$$

$$\operatorname{div} \boldsymbol{\beta}(\boldsymbol{x}) = \left(\sum_{j=1}^{n} \frac{\partial \beta_{ij}}{\partial x_j}(\boldsymbol{x}) \right) \in \mathbb{R}^n.$$

For a continuously differentiable function $g : \mathbb{R}^{n \times n} \to \mathbb{R}$, its derivative is denoted by

$$\frac{\partial g}{\partial \boldsymbol{\alpha}}(\boldsymbol{\alpha}) = \left(\frac{\partial g}{\partial \alpha_{ij}}(\boldsymbol{\alpha}) \right) \in \mathbb{R}^{n \times n}.$$

8.2 Principle of Potential Energy for Masonry Structures

The constitutive law of a perfectly no-tension model is defined, and is shown to be reduced to the complementarity condition over the positive-semidefinite cone. The minimization problem of the total potential energy for masonry structures is formulated, and then an equivalent convex optimization is derived.

8.2.1 Principle of potential energy

Consider a masonry-like deformable body in three-dimensional space. The body occupies the domain $\overline{\Omega} := \operatorname{cl} \Omega$, where $\Omega \subset \mathbb{R}^3$ is a given bounded and connected open set.[2] At each point $\boldsymbol{x} \in \overline{\Omega}$ we denote by $\boldsymbol{u}(\boldsymbol{x}) \in \mathbb{R}^3$ the displacement vector, so that a displacement of the whole body is expressed by the vector field $\boldsymbol{u} : \overline{\Omega} \to \mathbb{R}^3$, which is assumed to be sufficiently smooth.[3]

---

[2]More precisely, Ω is assumed to be a *Lipschitzian domain*; see [221, pp. 15–16] for its definition.

[3]We shall specify the smoothness property of $\boldsymbol{u}$ by employing the notion of Sobolev spaces in section 8.3.1; there, $\boldsymbol{u}$ is considered as a function defined on Ω.

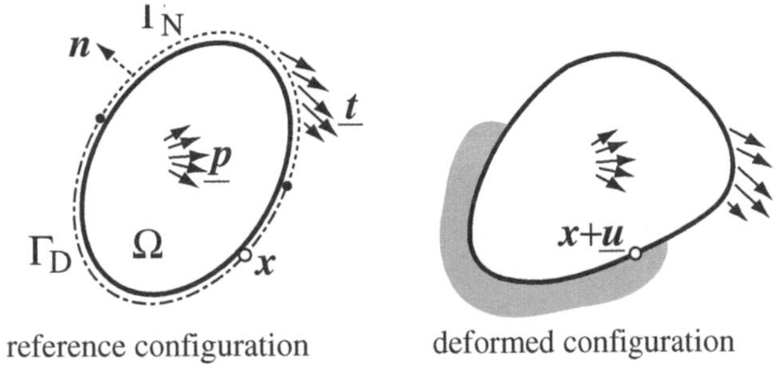

FIGURE 8.1: Reference and deformed configurations of a continuum media.

Assuming the small deformation, the compatibility relation between the linear strain tensor $\boldsymbol{\varepsilon}$ and the displacement vector $\boldsymbol{u}$ is given by

$$\boldsymbol{\varepsilon} = \hat{\boldsymbol{\varepsilon}}(\boldsymbol{u}) := \frac{1}{2}(\nabla \boldsymbol{u}^{\mathrm{T}} + \nabla \boldsymbol{u}). \tag{8.1}$$

By definition, $\boldsymbol{\varepsilon}$ is a 3×3 symmetric tensor.

By the no-tension material we mean a non-dissipative elastic material that cannot transmit tensile normal stresses at all, whereas under the compressive normal stresses it behaves as the conventional elastic material. More precisely, the total strain is decomposed to the elastic and inelastic parts so that the stress depends solely on the elastic strain (where we assume for simplicity the linearity and isotropy of this dependency), while the inelastic strain corresponds to the crack opening behavior.[4] The non-dissipation property means that the response of no-tension material is reversible and path independent. More intuitively, even if the material experienced the crack opening, no damage remains after the crack closes. Therefore, there exists a strain energy function,[5] denoted $\hat{w}_{\mathrm{mas}} : \mathbb{S}^3 \to \mathbb{R}$, with which the constitutive equation (stress–strain relation) of a no-tension material can be given as

$$\boldsymbol{\sigma} \in \partial \hat{w}_{\mathrm{mas}}(\boldsymbol{\varepsilon}), \tag{8.2}$$

where $\boldsymbol{\sigma}$ denotes the symmetric stress tensor corresponding to the linear strain $\boldsymbol{\varepsilon}$. We postpone a concrete definition of the function $\hat{w}_{\mathrm{mas}}$ to section 8.2.2.[6]

Let $\boldsymbol{n} \in \mathbb{R}^3$ denote the unit vector outward and normal to $\Gamma := \mathrm{bd}\, \Omega$, as shown in Figure 8.1 . Suppose that Γ is partitioned into two disjoint subsets,

[4]By the inelastic strain we mean the part of the total strain that does not contribute to yield stress at all. The inelastic strain should not be confused with the notion of *irreversible* (or, equivalently, *dissipative*) strain, which includes, e.g., the plastic strain.

[5]The strain energy function is also called the *stored energy function* [90, p. 141].

[6]We shall see that $\boldsymbol{\sigma}$ in (8.2) is determined uniquely for a given $\boldsymbol{\varepsilon}$; see the exposition given just before Proposition 8.2.6.

denoted $\Gamma_{\rm N}$ and Γ_D, so that $\Gamma_{\rm N} \cup \Gamma_{\rm D} = \Gamma$. Let $\underline{\boldsymbol{t}} : \Gamma_{\rm N} \to \mathbb{R}^3$ denote the traction per unit area in the reference configuration, specified at each point $\boldsymbol{x} \in \Gamma_{\rm N}$. In contrast, the displacement is prescribed at each point $\boldsymbol{x} \in \Gamma_{\rm D}$ by $\underline{\boldsymbol{u}} : \Gamma_{\rm D} \to \mathbb{R}^3$. We denote by $\underline{\boldsymbol{p}} : \Omega \to \mathbb{R}^3$ the applied body force per unit volume in the reference configuration. The applied loads are assumed to be sufficiently smooth[7] and conservative. Consequently, no energy dissipates during deformation, and hence the equilibrium state is independent of the loading history.

The functional W defined for $\boldsymbol{u}$ by

$$W(\boldsymbol{u}) = \int_{\Omega} \hat{w}_{\rm mas}(\hat{\boldsymbol{\varepsilon}}(\boldsymbol{u}(\boldsymbol{x})))\mathrm{d}\Omega$$

is called the *strain energy*. The *total potential energy*, denoted Π, is the functional defined by

$$\Pi(\boldsymbol{u}) = W(\boldsymbol{u}) - \int_{\Omega} \underline{\boldsymbol{p}}(\boldsymbol{x}) \cdot \boldsymbol{u}(\boldsymbol{x})\mathrm{d}\Omega - \int_{\Gamma_{\rm N}} \underline{\boldsymbol{t}}(\boldsymbol{x}) \cdot \boldsymbol{u}(\boldsymbol{x})\mathrm{d}\Gamma.$$

With the assumptions above, and by using (8.1), the minimization problem of the total potential energy is formulated as

$$\left.\begin{aligned} \text{(PE)} : \min_{\boldsymbol{u},\boldsymbol{\varepsilon}} \ & \int_{\Omega} \left(\hat{w}_{\rm mas}(\boldsymbol{\varepsilon}) - \underline{\boldsymbol{p}} \cdot \boldsymbol{u}\right) \mathrm{d}\Omega - \int_{\Gamma_{\rm N}} \underline{\boldsymbol{t}} \cdot \boldsymbol{u}\mathrm{d}\Gamma \\ \text{s. t. } \ & \boldsymbol{\varepsilon} = \frac{1}{2}(\nabla \boldsymbol{u}^{\rm T} + \nabla \boldsymbol{u}) \quad \text{in } \Omega, \\ & \boldsymbol{u} = \underline{\boldsymbol{u}} \quad \text{on } \Gamma_{\rm D}. \end{aligned}\right\} \qquad (8.3)$$

The principle of minimum total potential energy states that[8] any (smooth enough) optimal solution $(\bar{\boldsymbol{u}}, \bar{\boldsymbol{\varepsilon}})$ of (PE) in (8.3) corresponds to the displacement vector and the strain tensor in the equilibrium state.

The crucial difficulty in the analysis of masonry structures arises from the complication of the definition of $\hat{w}_{\rm mas}$ in (8.2). We next present an entire study of the constitutive law in section 8.2.2 and then reformulate (PE) into a tractable form, i.e., the form of conic optimization, in section 8.2.3.

8.2.2 Constitutive law

The no-tension material behaves like a conventional elastic material when it is subjected to compressive normal stresses, but it cannot withstand tensile

[7]Together with the comment on the functional space to which $\boldsymbol{u}$ belongs, we postpone specification of the smoothness properties of applied loads to section 8.3.1.

[8]See, e.g., [457, section 3.4.3] and [90, section 4.1] for the principle of minimum total potential energy. Although we here accept this principle as a postulate tentatively, it shall be justified rigorously in section 8.3.4 by showing that the necessary and sufficient condition for the optimality of (PE) coincides with the boundary value problem for the equilibrium of a masonry structure.

normal stresses. More specifically, for a given strain ε the response of no-tension material is founded on the following four physical hypotheses[9]:

(i) Any tensile stress cannot be sustained.

(ii) The strain ε can be additively decomposed to two parts—the elastic and inelastic strains.

(iii) The inelastic strain corresponds to the reversible crack-opening behavior, which can take place perpendicular to the plane of nonnegative principal stress.

(iv) The compressive stress is caused elastically solely by the elastic stress.

We next restate the hypotheses above formally. Let $\lambda_i(\boldsymbol{\sigma})$ $(i = 1, 2, 3)$ denote the eigenvalues of $\boldsymbol{\sigma}$, i.e., the principal stresses, where $\lambda_1(\boldsymbol{\sigma}) \geq \lambda_2(\boldsymbol{\sigma}) \geq \lambda_3(\boldsymbol{\sigma})$. Then hypothesis (i) is equivalently rewritten as

$$\lambda_i(\boldsymbol{\sigma}) \leq 0, \quad i = 1, 2, 3. \tag{8.4}$$

According to hypothesis (ii), we decompose $\boldsymbol{\varepsilon}$ as

$$\boldsymbol{\varepsilon} = \boldsymbol{\varepsilon}_{\mathrm{e}} + \boldsymbol{\varepsilon}_{\mathrm{c}}, \tag{8.5}$$

where $\boldsymbol{\varepsilon}_{\mathrm{e}}$ is the elastic part, and $\boldsymbol{\varepsilon}_{\mathrm{c}}$ is the inelastic part. We call $\boldsymbol{\varepsilon}_{\mathrm{c}}$ the *crack strain* in accordance with its physical meaning given by hypothesis (iii). The crack-opening behavior is intuitively interpreted as a separation of the plane of maximum stress into two parallel planes with positive gap.[10] Since overlap of two separated bodies is not accepted, the principal values of the crack strain, denoted $\omega_i(\boldsymbol{\varepsilon}_{\mathrm{c}})$, should be nonnegative, i.e.,

$$\omega_i(\boldsymbol{\varepsilon}_{\mathrm{c}}) \geq 0, \quad i = 1, 2, 3, \tag{8.6}$$

where we set $\omega_1(\boldsymbol{\varepsilon}_{\mathrm{c}}) \geq \omega_2(\boldsymbol{\varepsilon}_{\mathrm{c}}) \geq \omega_3(\boldsymbol{\varepsilon}_{\mathrm{c}})$. From the perpendicularity stated in hypothesis (iii), any positive principal value of the crack strain should be in the direction of the principal direction of the maximum stress, which is equal to 0 because of (8.4). Thus, we have that $\lambda_i(\boldsymbol{\sigma}) = 0$ in any direction, which can be represented as a linear combination of the eigenvectors corresponding to $\omega_i(\boldsymbol{\varepsilon}_{\mathrm{c}}) > 0$. Moreover, in any direction corresponds to negative principal stress the crack cannot open; i.e., we have that $\omega_i(\boldsymbol{\varepsilon}_{\mathrm{c}}) = 0$ in any direction, which can be represented as a linear combination of the eigenvectors corresponding to $\lambda_i(\boldsymbol{\sigma}) < 0$. Therefore, we see that[11]

$$\boldsymbol{\sigma} \text{ and } \boldsymbol{\varepsilon}_{\mathrm{c}} \text{ are coaxial} \tag{8.7}$$

[9]Which essentially follow Cuomo–Ventura [100].

[10]This rough sketch of the meaning of $\boldsymbol{\varepsilon}_{\mathrm{c}}$ should be interpreted in the sense of *smeared* way in the context of continuum mechanics.

[11]We mean by (8.7) that $\boldsymbol{\sigma}$ and $\boldsymbol{\varepsilon}_{\mathrm{c}}$ share a same system of eigenvectors.

and that

$$\omega_i(\boldsymbol{\varepsilon}_{\rm c}) > 0 \quad \Rightarrow \quad \lambda_i(\boldsymbol{\sigma}) = 0, \tag{8.8}$$

$$\lambda_i(\boldsymbol{\sigma}) < 0 \quad \Rightarrow \quad \omega_i(\boldsymbol{\varepsilon}_{\rm c}) = 0 \tag{8.9}$$

for each $i = 1, 2, 3$. Finally, hypothesis (iv) means that there exists a function $\hat{w}_{\rm ref} : \mathbb{S}^3 \to \mathbb{R}$, with which the response $\boldsymbol{\sigma}$ is expressed as

$$\boldsymbol{\sigma} \in \partial \hat{w}_{\rm ref}(\boldsymbol{\varepsilon}_{\rm e}). \tag{8.10}$$

Note that such a function $\hat{w}_{\rm ref}$ is not unique.

By regarding $\hat{w}_{\rm ref}$ in (8.10) as a strain energy function, we may introduce a fictitious elastic body whose stress–strain relation is given by

$$\boldsymbol{\sigma} \in \partial \hat{w}_{\rm ref}(\boldsymbol{\varepsilon}). \tag{8.11}$$

Such a fictitious body is called a *reference elastic body* for the no-tension model.[12] In other words, if all the cracks of the no-tension body are forced to close artificially, then the obtained stress response is regarded as that of the reference elastic body.

In what follows, we assume linear isotropy of the no-tension material for simplicity. Consequently, as a reference elastic body we may choose the conventional linear elastic material, the strain energy function of which is given by

$$\hat{w}_{\rm ref}(\boldsymbol{\varepsilon}) = \frac{1}{2}\lambda(\operatorname{tr}\boldsymbol{\varepsilon})^2 + \mu\boldsymbol{\varepsilon} : \boldsymbol{\varepsilon}. \tag{8.12}$$

Here, λ and μ are the Lamé moduli. Then the relation (8.10) of $\boldsymbol{\sigma}$ and $\boldsymbol{\varepsilon}_{\rm e}$ is written explicitly as

$$\begin{aligned}\boldsymbol{\sigma} = \hat{\boldsymbol{\sigma}}_{\rm ref}(\boldsymbol{\varepsilon}_{\rm e}) &:= \frac{\partial \hat{w}_{\rm ref}}{\partial \boldsymbol{\varepsilon}_{\rm e}}(\boldsymbol{\varepsilon}_{\rm e}) \\ &= (\lambda \operatorname{tr}\boldsymbol{\varepsilon}_{\rm e})\boldsymbol{I} + 2\mu\boldsymbol{\varepsilon}_{\rm e},\end{aligned} \tag{8.13}$$

where $\boldsymbol{I}$ denotes the unit tensor.

Remark 8.2.1. In (8.12) we assume the linear isotropy for simplicity, but this assumption is not crucial for the subsequent discussions; we need only to assume that $\hat{w}_{\rm ref}$ is strictly convex, as well as hypotheses (i)–(iv).

Note that the isotropy implies that $\hat{\boldsymbol{\sigma}}_{\rm ref}(\boldsymbol{\varepsilon}_{\rm e})$ and $\boldsymbol{\varepsilon}_{\rm e}$ are coaxial. It then follows from (8.5), (8.7), and (8.10) that $\boldsymbol{\varepsilon}$, $\boldsymbol{\varepsilon}_{\rm e}$, $\boldsymbol{\varepsilon}_{\rm c}$, and $\boldsymbol{\sigma}$ all share the same system of eigenvectors. Thus the assumption of isotropy provides us with the simplest case. In an anisotropic case, in contrast, $\boldsymbol{\varepsilon}$, $\boldsymbol{\varepsilon}_{\rm e}$, and $\boldsymbol{\sigma}$ are not coaxial in general to each other, although hypothesis (iii) still means that $\boldsymbol{\varepsilon}_{\rm c}$ and $\boldsymbol{\sigma}$ are coaxial. ■

---

[12] For a comparison, recall that a truss member serves as a reference elastic body of a cable member as seen in section 4.1.4.

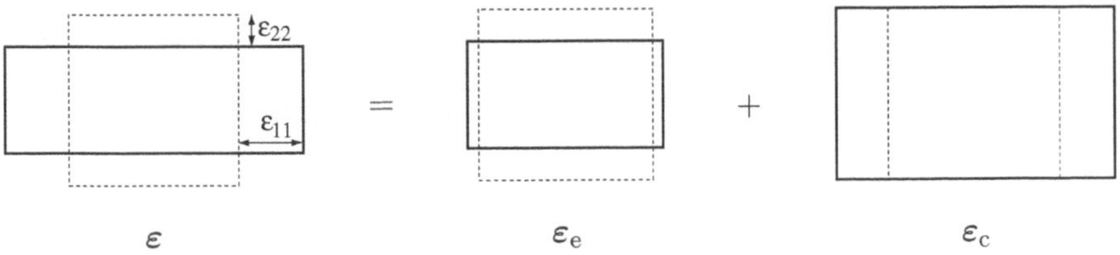

FIGURE 8.2: A schematic exposition of the constitutive law of no-tension material in the plane-stress case investigated in Example 8.1, where $\varepsilon_{11} > -\nu\varepsilon_{22}$ and $\varepsilon_{22} < 0$. '......': initial configuration; '—': deformed configuration.

Example 8.1
Consider the plane-stress case, which is characterized by $\sigma_{33} = \sigma_{23} = \sigma_{31} = 0$. Then the constitutive relation of the reference elastic body, given by (8.10) with (8.13), is reduced to

$$\hat{\boldsymbol{\sigma}}_{\mathrm{ref}}(\boldsymbol{\varepsilon}) = \frac{E}{1-\nu^2}\begin{bmatrix} \varepsilon_{11} + \nu\varepsilon_{22} & (1-\nu)\varepsilon_{12} \\ (1-\nu)\varepsilon_{12} & \nu\varepsilon_{11} + \varepsilon_{22} \end{bmatrix}, \tag{8.14}$$

where E and ν are *Young's modulus* and *Poisson's ratio* defined by

$$E = \frac{\mu(3\lambda + 2\mu)}{\lambda + \mu}, \quad \nu = \frac{\lambda}{2(\lambda + \mu)}.$$

Suppose that $\boldsymbol{\varepsilon}$ is given and that $\varepsilon_{12} = 0$ for simplicity. Substituting this into (8.14) results in

$$\hat{\boldsymbol{\sigma}}_{\mathrm{ref}}(\boldsymbol{\varepsilon}) = \frac{E}{1-\nu^2}\begin{bmatrix} \varepsilon_{11} + \nu\varepsilon_{22} & 0 \\ 0 & \nu\varepsilon_{11} + \varepsilon_{22} \end{bmatrix}. \tag{8.15}$$

Consider the situation in which $\varepsilon_{11} > -\nu\varepsilon_{22}$ and $\varepsilon_{22} < 0$, as illustrated in the left figure in Figure 8.2. We give physical observation of the crack and stress states of the no-tension body and verify that the physical phenomenon is represented correctly by (8.4)–(8.10) with (8.15).

Because of the positive (large) strain ε_{11}, we intuitively consider that a crack opens in the x_1-direction.[13] In contrast, it seems that the compressive stress exists in the x_2-direction due to $\varepsilon_{22} < 0$. Then the body is subjected to the uniaxial compression state, because in the x_1-direction no stress can be sustained due to crack opening. Therefore, the stress state is obtained as

$$\sigma_{11} = 0, \tag{8.16a}$$

$$\sigma_{22} = E\varepsilon_{22}. \tag{8.16b}$$

[13]The assertion "if $\varepsilon_{11} > 0$, then a crack opens in the x_1-direction" is not always true; because of the Poisson effect, the body is in the bi-axial compression state when $0 < \varepsilon_{11} < -\nu\varepsilon_{22}$, and in this case no crack opens.

In (8.16) we easily see that σ_{11} and σ_{22} correspond to the principal stresses, and hence (8.4) is satisfied.

Suppose that the reference elastic body undergoes the stress state in (8.16). The stress state in this situation corresponds to the elastic strain $\boldsymbol{\varepsilon}_{\rm e}$ of the no-tension body, according to (8.10). By using (8.15), we thus obtain $\boldsymbol{\varepsilon}_{\rm e}$ as

$$\varepsilon_{\rm e11} = -\nu\varepsilon_{22}, \tag{8.17a}$$

$$\varepsilon_{\rm e22} = \varepsilon_{22}, \tag{8.17b}$$

and $\varepsilon_{\rm e12} = 0$. It then follows from (8.5) that the crack strain $\boldsymbol{\varepsilon}_{\rm c}$ is determined as

$$\varepsilon_{\rm c11} = \varepsilon_{11} + \nu\varepsilon_{22}, \tag{8.18a}$$

$$\varepsilon_{\rm c22} = 0, \tag{8.18b}$$

and certainly $\varepsilon_{\rm c12} = 0$ (see Figure 8.2). Since $\varepsilon_{11} > -\nu\varepsilon_{22}$ is assumed, we see that $\boldsymbol{\varepsilon}_{\rm c}$ in (8.18) satisfies (8.6). Furthermore, from (8.16) and (8.18) we easily verify that (8.7), (8.8), and (8.9) are satisfied. Thus, $\boldsymbol{\sigma}$, $\boldsymbol{\varepsilon}_{\rm e}$, and $\boldsymbol{\varepsilon}_{\rm c}$ in (8.16)–(8.18) satisfy (8.4)–(8.10) with (8.15). □

Remark 8.2.2. As seen in (8.17) of Example 8.1, the elastic strain $\boldsymbol{\varepsilon}_{\rm e}$ is not necessarily negative semidefinite. In contrast, the crack strain $\boldsymbol{\varepsilon}_{\rm c}$ is always positive semidefinite as mentioned before. ■

Being oriented toward the perspective of complementarity over convex cones, a key observation stated in Proposition 8.2.3 below is that the relation between $\boldsymbol{\varepsilon}_{\rm c}$ and $-\boldsymbol{\sigma}$ in (8.4) and (8.6)–(8.9) is interpreted in parallel with the relation of two positive-semidefinite matrices satisfying the complementarity condition investigated in section 1.3.4; see Table 1.3 for a summary.

Proposition 8.2.3. *The constitutive law of no-tension material, defined by* (8.4)–(8.10) *and* (8.12), *is equivalently rewritten as*

$$\boldsymbol{\sigma} = (\lambda \operatorname{tr} \boldsymbol{\varepsilon}_{\rm e})\boldsymbol{I} + 2\mu\boldsymbol{\varepsilon}_{\rm e}, \tag{8.19a}$$

$$\boldsymbol{\varepsilon} = \boldsymbol{\varepsilon}_{\rm e} + \boldsymbol{\varepsilon}_{\rm c}, \tag{8.19b}$$

$$\boldsymbol{\varepsilon}_{\rm c} \succeq \boldsymbol{o}, \quad -\boldsymbol{\sigma} \succeq \boldsymbol{o}, \quad \boldsymbol{\varepsilon}_{\rm c} : (-\boldsymbol{\sigma}) = 0. \tag{8.19c}$$

Proof. The relation between $\boldsymbol{\varepsilon}_{\rm c}$ and $-\boldsymbol{\sigma}$ given in (8.4) and (8.6)–(8.9) is the same as the relation between X and S in Table 1.3, and then equivalent to (8.19c) because of Fact 1.3.20. It follows from (8.13) that (8.19a) is equivalent to (8.13) with (8.12). Condition (8.19b) is nothing but (8.5). □

As a direct consequence of this proposition, we see that the constitutive law in (8.2) is alternatively written as

$$\boldsymbol{\sigma} \in \partial\hat{w}_{\rm mas}(\boldsymbol{\varepsilon}) \quad \Leftrightarrow \quad \exists\boldsymbol{\varepsilon}_{\rm e}, \boldsymbol{\varepsilon}_{\rm c} \in \mathbb{S}^3 : (8.19), \tag{8.20}$$

where it is obvious from the mechanical point of view that $\hat{w}_{\rm mas}(\boldsymbol{o}) = 0$ should also be satisfied.

Remark 8.2.4. Condition (8.19c) in Proposition 8.2.3 means that $\boldsymbol{\sigma}$ compatible with $\boldsymbol{\varepsilon}_{\mathrm{c}}$ can be determined from an analogy of the maximum dissipation law in the plasticity theory. This law is originally stated that, among admissible stress tensors, the *true* stress tensor compatible with a given plastic strain rate is the one that maximizes the energy dissipation due to the plastic deformation. In the situation of the no-tension material, for a given $\boldsymbol{\varepsilon}_{\mathrm{c}} \succeq \boldsymbol{o}$ ($\boldsymbol{\varepsilon}_{\mathrm{c}} \neq \boldsymbol{o}$), condition (8.19c) is equivalently rewritten as

$$\boldsymbol{\sigma} \in \arg\max_{\check{\boldsymbol{\sigma}}}\{\boldsymbol{\varepsilon}_{\mathrm{c}} : \check{\boldsymbol{\sigma}} \mid \check{\boldsymbol{\sigma}} \preceq \boldsymbol{o}\}. \tag{8.21}$$

This means that the *true* stress tensor $\boldsymbol{\sigma}$ corresponding to $\boldsymbol{\varepsilon}_{\mathrm{c}}$ is the one that maximizes $\boldsymbol{\varepsilon}_{\mathrm{c}} : \check{\boldsymbol{\sigma}}$ among admissible stress tensors. It should be clear that the crack-opening phenomena are assumed to be non-dissipative as mentioned before; i.e., the work $\boldsymbol{\varepsilon}_{\mathrm{c}} : \check{\boldsymbol{\sigma}}$ done by the crack strain is not dissipative. Thus the maximum dissipation law holds for the no-tention material as an analogy.

The maximum dissipation law (8.21) implies that *normality rule*, which is known well in the plasticity theory, is satisfied by the no-tension material as an analogy. The normality rule states that if the crack strain $\boldsymbol{\varepsilon}_{\mathrm{c}} \succeq \boldsymbol{o}$ ($\boldsymbol{\varepsilon}_{\mathrm{c}} \neq \boldsymbol{o}$) takes place, then the following two properties are satisfied:

(a) The corresponding stress tensor $\boldsymbol{\sigma}$ is on the boundary of the admissible set of stress tensors; i.e., $\boldsymbol{\sigma} \in \mathrm{bd}\{\check{\boldsymbol{\sigma}} \mid \check{\boldsymbol{\sigma}} \preceq \boldsymbol{o}\}$.

(b) $\boldsymbol{\varepsilon}_{\mathrm{c}}$ is in the outward normal direction of $\mathrm{bd}\{\check{\boldsymbol{\sigma}} \mid \check{\boldsymbol{\sigma}} \preceq \boldsymbol{o}\}$ at $\boldsymbol{\sigma}$.

This normality rule is postulated for the no-tention material by some authors, e.g., Maier–Nappi [291], instead of hypothesis (iii). ■

8.2.3 Conic optimization formulation

As seen in Proposition 8.2.3, the constitutive law of a no-tension material is represented by (8.19). We next give an alternative representation of $\hat{w}_{\mathrm{mas}}$ based on the conic optimization formulation, which is established in the proposition below.

Proposition 8.2.5. *Suppose that $\boldsymbol{\varepsilon} \in \mathbb{S}^3$ is given. Consider the following optimization problem in the variable $\boldsymbol{\varepsilon}_{\mathrm{e}} \in \mathbb{S}^3$:*

$$\left.\begin{array}{ll} \min\limits_{\boldsymbol{\varepsilon}_{\mathrm{e}}} & \hat{w}_{\mathrm{ref}}(\boldsymbol{\varepsilon}_{\mathrm{e}}) \\ \mathrm{s.t.} & \boldsymbol{\varepsilon}_{\mathrm{e}} \preceq \boldsymbol{\varepsilon}, \end{array}\right\} \tag{8.22}$$

where $\hat{w}_{\mathrm{ref}}$ is defined by (8.12)*. Then $\boldsymbol{\varepsilon}_{\mathrm{e}} \in \mathbb{S}^3$ is optimal for* (8.22) *if and only if there exist $\boldsymbol{\sigma}$ and $\boldsymbol{\varepsilon}_{\mathrm{c}}$ satisfying* (8.19)*. Moreover, the optimal value of* (8.22) *is equal to $\hat{w}_{\mathrm{mas}}(\boldsymbol{\varepsilon})$, where $\hat{w}_{\mathrm{mas}}$ is defined by* (8.20)*.*

Proof. Problem (8.22) consists of the objective function bounded below and the convex feasible set with a nonempty interior. Hence, we apply Proposition 2.2.9 (b) to problem (8.22) to derive the optimality condition.

We first show that problem (8.22) is embedded into the form of ($\mathrm{P_F}$) in (2.28), which serves as the primal problem within the Fenchel duality theory. We write the primal variables and the associated spaces as

$$x = \boldsymbol{\varepsilon}_{\mathrm{e}}, \quad z = \boldsymbol{Z}, \quad \mathbb{V} = \mathbb{Y} = \mathbb{S}^3,$$

where $x \in \mathbb{V}$ and $z \in \mathbb{Y}$. Define $f : \mathbb{V} \to \mathbb{R}$, $g : \mathbb{Y} \to \mathbb{R} \cup \{+\infty\}$, and $\Lambda : \mathbb{V} \to \mathbb{Y}$ by

$$f(\boldsymbol{\varepsilon}_{\mathrm{e}}) = \frac{1}{2}\lambda(\mathrm{tr}\,\boldsymbol{\varepsilon}_{\mathrm{e}})^2 + \mu\boldsymbol{\varepsilon}_{\mathrm{e}} : \boldsymbol{\varepsilon}_{\mathrm{e}}, \tag{8.23}$$

$$g(\boldsymbol{Z}) = \begin{cases} 0 & \text{if } \boldsymbol{\varepsilon} - \boldsymbol{Z} \succeq \boldsymbol{o}, \\ +\infty & \text{otherwise}, \end{cases} \tag{8.24}$$

$$\Lambda\boldsymbol{\varepsilon}_{\mathrm{e}} = \boldsymbol{\varepsilon}_{\mathrm{e}}. \tag{8.25}$$

Then problem (8.22) is rewritten as $\inf_x\{f(x) + g(\Lambda x) \mid x \in \mathbb{V}\}$, which is in the form of ($\mathrm{P_F}$).

We write the dual variable and associated space used in Proposition 2.2.9 (b) as

$$z^* = \boldsymbol{Z}^*, \quad \mathbb{Y}^* = \mathbb{S}^3.$$

It follows from Proposition 2.1.12 and (8.25) that the first equation in Proposition 2.2.9 (b) can be reduced to

$$\bar{\boldsymbol{Z}}^* = \frac{\partial f}{\partial \boldsymbol{\varepsilon}_{\mathrm{e}}}(\bar{\boldsymbol{\varepsilon}}_{\mathrm{e}}) = (\lambda\,\mathrm{tr}\,\bar{\boldsymbol{\varepsilon}}_{\mathrm{e}})\boldsymbol{I} + 2\mu\bar{\boldsymbol{\varepsilon}}_{\mathrm{e}}, \tag{8.26}$$

because f in (8.23) is differentiable. Concerning the second equation in Proposition 2.2.9 (b), the conjugate function of g in (8.24) is obtained by using Fact 1.3.17 (ii) as

$$\begin{aligned} g^*(\boldsymbol{Z}^*) &= \sup_{\boldsymbol{Z} \in \mathbb{S}^3} \{\boldsymbol{Z}^* : \boldsymbol{Z} \mid \boldsymbol{\varepsilon} - \boldsymbol{Z} \succeq \boldsymbol{o}\} \\ &= -\inf_{\boldsymbol{Z} \in \mathbb{S}^3} \{\boldsymbol{Z}^* : (\boldsymbol{\varepsilon} - \boldsymbol{Z}) \mid \boldsymbol{\varepsilon} - \boldsymbol{Z} \succeq \boldsymbol{o}\} - \boldsymbol{Z}^* \bullet \boldsymbol{\varepsilon} \\ &= \begin{cases} -\boldsymbol{\varepsilon} : \boldsymbol{Z}^* & \text{if } \boldsymbol{Z}^* \succeq \boldsymbol{o}, \\ +\infty & \text{otherwise}. \end{cases} \end{aligned}$$

Hence, $g(\Lambda\bar{x}) + g^*(-\bar{z}^*) = g(\bar{\boldsymbol{\varepsilon}}_{\mathrm{e}}) + g^*(-\bar{\boldsymbol{Z}}^*)$ is finite if and only if

$$\boldsymbol{\varepsilon} - \bar{\boldsymbol{\varepsilon}}_{\mathrm{e}} \succeq \boldsymbol{o}, \quad -\bar{\boldsymbol{Z}}^* \succeq \boldsymbol{o}. \tag{8.27}$$

If this is the case the condition $g(\Lambda\bar{x}) + g^*(-\bar{z}^*) = \langle -\bar{z}^*, \Lambda\bar{x}\rangle$ is reduced to

$$-\boldsymbol{\varepsilon} : \bar{\boldsymbol{Z}}^* = -\bar{\boldsymbol{Z}}^* : \bar{\boldsymbol{\varepsilon}}_{\mathrm{e}}. \tag{8.28}$$

Consequently, $\bar{\boldsymbol{\varepsilon}}_{\mathrm{e}}$ is optimal for (8.22) if and only if there exists $\bar{\boldsymbol{Z}}^*$ satisfying (8.26), (8.27), and (8.28). These optimality conditions are reduced to (8.19) by putting

$$\bar{\boldsymbol{\varepsilon}}_{\mathrm{c}} := \boldsymbol{\varepsilon} - \bar{\boldsymbol{\varepsilon}}_{\mathrm{e}}, \quad \bar{\boldsymbol{\sigma}} := \bar{\boldsymbol{Z}}^*.$$

It is then immediate to see that the optimal value of (8.22) coincides with $\hat{w}_{\mathrm{mas}}(\boldsymbol{\varepsilon})$. □

Proposition 8.2.5 provides us with a variational form of the constitutive law of the no-tension material. It is emphasized that the objective function of (8.22) is the strain energy function of the linear isotropic material, which serves as a reference elastic body for the no-tension body.

The latter assertion in Proposition 8.2.5, i.e., the relation

$$\hat{w}_{\mathrm{mas}}(\boldsymbol{\varepsilon}) = \min_{\boldsymbol{\varepsilon}_{\mathrm{e}}}\{\hat{w}_{\mathrm{ref}}(\boldsymbol{\varepsilon}_{\mathrm{e}}) \mid \boldsymbol{\varepsilon}_{\mathrm{e}} \preceq \boldsymbol{\varepsilon}\}, \tag{8.29}$$

provides us with two interesting properties of $\hat{w}_{\mathrm{mas}}$. First, observe that the optimization problem on the right-hand side of (8.29) is a minimization of a strictly convex function under a convex constraint. Therefore, its optimal solution $\boldsymbol{\varepsilon}_{\mathrm{e}}$ exists uniquely. This means that, for a given $\boldsymbol{\varepsilon}$, the decomposition in (8.19b) is determined uniquely, and thence $\boldsymbol{\sigma}$ is also determined uniquely from (8.19a). For this reason there exists a function $\hat{\boldsymbol{\sigma}}_{\mathrm{mas}}$ satisfying

$$\boldsymbol{\sigma} = \hat{\boldsymbol{\sigma}}_{\mathrm{mas}}(\boldsymbol{\varepsilon}) \quad \Leftrightarrow \quad (8.19\mathrm{c}). \tag{8.30}$$

As the second consequence of (8.29), the convexity of $\hat{w}_{\mathrm{mas}}$ can be shown as follows.

Proposition 8.2.6. *The strain energy function $\hat{w}_{\mathrm{mas}}$ of no-tension material is convex.*

Proof. We make use of the fact established in Proposition 2.2.1. To this end, define $\Phi : \mathbb{S}^3 \times \mathbb{S}^3$ by

$$\Phi(\boldsymbol{\varepsilon}_{\mathrm{e}}, \boldsymbol{\varepsilon}) = \hat{w}_{\mathrm{ref}}(\boldsymbol{\varepsilon}_{\mathrm{e}}) + \delta_{\mathbb{S}^3_+}(\boldsymbol{\varepsilon} - \boldsymbol{\varepsilon}_{\mathrm{e}}), \tag{8.31}$$

where, in the situation of Proposition 2.2.1, Φ is introduced in (2.15), $F = \hat{w}_{\mathrm{ref}}$, $\mathbb{V} = \mathbb{Y} = \mathbb{S}^3$, $x = \boldsymbol{\varepsilon}_{\mathrm{e}}$, and $z = \boldsymbol{\varepsilon}$. Since $\hat{w}_{\mathrm{ref}}$ and $\delta_{\mathbb{S}^3_+}$ are closed proper convex functions, so is Φ. Thus Proposition 2.2.1 implies that θ defined by (2.16) is convex. Turning around the present situation, Proposition 8.2.5 and (8.31) mean $\theta = \hat{w}_{\mathrm{mas}}$. □

Proposition 8.2.5 also means that we can replace $\hat{w}_{\mathrm{mas}}$ in (PE) (presented in (8.3)) with the optimal value of (8.22). This results in the following equivalent reformulation of (PE):

$$\left.\begin{aligned} (\overline{\mathrm{PE}}) : \min_{\boldsymbol{u}, \boldsymbol{\varepsilon}_{\mathrm{e}}} \ & \int_{\Omega} (\hat{w}_{\mathrm{ref}}(\boldsymbol{\varepsilon}_{\mathrm{e}}) - \underline{\boldsymbol{p}} \cdot \boldsymbol{u}) \, \mathrm{d}\Omega - \int_{\Gamma_{\mathrm{N}}} \underline{\boldsymbol{t}} \cdot \boldsymbol{u} \mathrm{d}\Gamma \\ \mathrm{s.\,t.} \ & \boldsymbol{\varepsilon}_{\mathrm{e}} \preceq \frac{1}{2}(\nabla \boldsymbol{u}^{\mathrm{T}} + \nabla \boldsymbol{u}) \quad \text{in } \Omega, \\ & \boldsymbol{u} = \underline{\boldsymbol{u}} \quad \text{on } \Gamma_{\mathrm{D}}, \end{aligned}\right\} \tag{8.32}$$

where $\boldsymbol{u}$ and $\boldsymbol{\varepsilon}_{\mathrm{e}}$ are treated as the independent variables to be optimized. Note again that $\hat{w}_{\mathrm{ref}}$, defined by (8.12), is the strain energy function of the linear isotropic material and hence much more tractable than $\hat{w}_{\mathrm{mas}}$. The precise relation between (PE) in (8.3) and ($\overline{\mathrm{PE}}$) in (8.32) is established in the proposition below.

Proposition 8.2.7. *Suppose that* $\min(\mathrm{PE}) > -\infty$.[14] *Then* $(\overline{\mathrm{PE}})$ *is equivalent to* (PE) *in the following sense:*

(i) *Let* $(\bar{\boldsymbol{u}}, \bar{\boldsymbol{\varepsilon}})$ *be optimal for* (PE). *Define* $\bar{\boldsymbol{\varepsilon}}_{\mathrm{e}} \in \mathbb{S}^3$ *so that*

$$\hat{\boldsymbol{\sigma}}_{\mathrm{mas}}(\bar{\boldsymbol{\varepsilon}}) = (\lambda \operatorname{tr} \bar{\boldsymbol{\varepsilon}}_{\mathrm{e}})\boldsymbol{I} + 2\mu\bar{\boldsymbol{\varepsilon}}_{\mathrm{e}} \tag{8.33}$$

is satisfied with $\hat{\boldsymbol{\sigma}}_{\mathrm{mas}}$ *in* (8.30). *Then* $(\bar{\boldsymbol{u}}, \bar{\boldsymbol{\varepsilon}}_{\mathrm{e}})$ *is optimal for* $(\overline{\mathrm{PE}})$.

(ii) *Let* $(\bar{\boldsymbol{u}}, \bar{\boldsymbol{\varepsilon}}_{\mathrm{e}})$ *be optimal for* $(\overline{\mathrm{PE}})$. *Define* $\bar{\boldsymbol{\varepsilon}}$ *by* $\bar{\boldsymbol{\varepsilon}} = \hat{\boldsymbol{\varepsilon}}(\bar{\boldsymbol{u}})$ *with* $\hat{\boldsymbol{\varepsilon}}$ *in* (8.1). *Then* $(\bar{\boldsymbol{u}}, \bar{\boldsymbol{\varepsilon}})$ *is optimal for* (PE).

(iii) $\min(\mathrm{PE}) = \min(\overline{\mathrm{PE}})$.

Proof. We first restate (8.33) more explicitly. It follows from (8.19) and (8.20) that, for a given $\bar{\boldsymbol{\varepsilon}}$, $\bar{\boldsymbol{\varepsilon}}_{\mathrm{e}}$ satisfies (8.33) if and only if there exists $\bar{\boldsymbol{\varepsilon}}_{\mathrm{c}}$ and $\boldsymbol{\sigma}$ satisfying

$$\bar{\boldsymbol{\sigma}} = (\lambda \operatorname{tr} \bar{\boldsymbol{\varepsilon}}_{\mathrm{e}})\boldsymbol{I} + 2\mu\boldsymbol{\varepsilon}_{\mathrm{e}}, \tag{8.34a}$$

$$\bar{\boldsymbol{\varepsilon}} = \bar{\boldsymbol{\varepsilon}}_{\mathrm{e}} + \bar{\boldsymbol{\varepsilon}}_{\mathrm{c}}, \tag{8.34b}$$

$$\bar{\boldsymbol{\varepsilon}}_{\mathrm{c}} \succeq \boldsymbol{o}, \quad -\bar{\boldsymbol{\sigma}} \succeq \boldsymbol{o}, \quad \bar{\boldsymbol{\varepsilon}}_{\mathrm{c}} : (-\bar{\boldsymbol{\sigma}}) = 0. \tag{8.34c}$$

As stated before (see the claim above (8.30)), $(\bar{\boldsymbol{\varepsilon}}_{\mathrm{e}}, \bar{\boldsymbol{\varepsilon}}_{\mathrm{c}}, \bar{\boldsymbol{\sigma}})$ satisfying (8.34) exists uniquely. Furthermore, we reiterate that (8.34) coincides with the optimality condition for problem (8.22) as shown in Proposition 8.2.5.

It follows from Proposition 8.2.5 that we can rewrite (PE) as

$$\left.\begin{aligned} &\min_{\boldsymbol{u},\boldsymbol{\varepsilon}} \int_{\Omega} \left(\min_{\boldsymbol{\varepsilon}_{\mathrm{e}}} \{ \hat{w}_{\mathrm{ref}}(\boldsymbol{\varepsilon}_{\mathrm{e}}) \mid \boldsymbol{\varepsilon}_{\mathrm{e}} \preceq \boldsymbol{\varepsilon} \} - \underline{\boldsymbol{p}} \cdot \boldsymbol{u} \right) \mathrm{d}\Omega - \int_{\Gamma_{\mathrm{N}}} \underline{\boldsymbol{t}} \cdot \boldsymbol{u} \mathrm{d}\Gamma \\ &\mathrm{s.\,t.}\ \boldsymbol{\varepsilon} = \frac{1}{2}(\nabla \boldsymbol{u}^{\mathrm{T}} + \nabla \boldsymbol{u}) \quad \text{in } \Omega, \\ &\qquad \boldsymbol{u} = \underline{\boldsymbol{u}} \quad \text{on } \Gamma_{\mathrm{D}} \end{aligned}\right\} \tag{8.35}$$

without changing the optimal solution and the optimal value. Noting that $\boldsymbol{\varepsilon}_{\mathrm{e}}$ is subjected solely to the constraint $\boldsymbol{\varepsilon}_{\mathrm{e}} \preceq \boldsymbol{\varepsilon}$, problem (8.35) above can be further rewritten equivalently as

$$\left.\begin{aligned} &\min_{\boldsymbol{u},\boldsymbol{\varepsilon},\boldsymbol{\varepsilon}_{\mathrm{e}}} \int_{\Omega} (\hat{w}_{\mathrm{ref}}(\boldsymbol{\varepsilon}_{\mathrm{e}}) - \underline{\boldsymbol{p}} \cdot \boldsymbol{u}) \mathrm{d}\Omega - \int_{\Gamma_{\mathrm{N}}} \underline{\boldsymbol{t}} \cdot \boldsymbol{u} \mathrm{d}\Gamma \\ &\mathrm{s.\,t.} \quad \boldsymbol{\varepsilon}_{\mathrm{e}} \preceq \boldsymbol{\varepsilon} \quad \text{in } \Omega, \\ &\qquad\quad \boldsymbol{\varepsilon} = \frac{1}{2}(\nabla \boldsymbol{u}^{\mathrm{T}} + \nabla \boldsymbol{u}) \quad \text{in } \Omega, \\ &\qquad\quad \boldsymbol{u} = \underline{\boldsymbol{u}} \quad \text{on } \Gamma_{\mathrm{D}}, \end{aligned}\right\} \tag{8.36}$$

where $\boldsymbol{u}$, $\boldsymbol{\varepsilon}$, and $\boldsymbol{\varepsilon}_{\mathrm{e}}$ are considered independent variables. By eliminating $\boldsymbol{\varepsilon}$ from problem (8.36), we obtain $(\overline{\mathrm{PE}})$. This verifies assertion (iii).

[14] This value is not finite in general, especially for a loading condition dominated by tensile external forces. See Remark 8.3.7.

In assertion (i), the assumption (8.33) implies that $(\bar{\boldsymbol{u}}, \bar{\boldsymbol{\varepsilon}}, \bar{\boldsymbol{\varepsilon}}_{\mathrm{e}})$ satisfies (8.34), as mentioned before. This means that $(\bar{\boldsymbol{u}}, \bar{\boldsymbol{\varepsilon}}, \bar{\boldsymbol{\varepsilon}}_{\mathrm{e}})$ in (i) is feasible for problem (8.36) and hence for $(\overline{\mathrm{PE}})$. Moreover, it follows from Proposition 8.2.5 that $\hat{w}_{\mathrm{mas}}(\bar{\boldsymbol{\varepsilon}}) = \hat{w}_{\mathrm{ref}}(\bar{\boldsymbol{\varepsilon}}_{\mathrm{e}})$ holds. Consequently, assertion (iii) verifies that $(\bar{\boldsymbol{u}}, \bar{\boldsymbol{\varepsilon}}, \bar{\boldsymbol{\varepsilon}}_{\mathrm{e}})$ is optimal for $(\overline{\mathrm{PE}})$.

To see assertion (ii), recall that problems (8.35), (8.36), and $(\overline{\mathrm{PE}})$ are equivalent. Therefore, $(\bar{\boldsymbol{u}}, \bar{\boldsymbol{\varepsilon}})$ is optimal for (8.35), while $\bar{\boldsymbol{\varepsilon}}_{\mathrm{e}}$ is optimal for the inner problem involved in (8.35). From the optimality for the inner problem and Proposition 8.2.5 it follows that $\hat{w}_{\mathrm{mas}}(\bar{\boldsymbol{\varepsilon}}) = \hat{w}_{\mathrm{ref}}(\bar{\boldsymbol{\varepsilon}}_{\mathrm{e}})$ is satisfied. On the other hand, it is easy to verify that $(\bar{\boldsymbol{u}}, \bar{\boldsymbol{\varepsilon}})$ is feasible for (PE). Thus assertion (iii) verifies that $(\bar{\boldsymbol{u}}, \bar{\boldsymbol{\varepsilon}})$ is optimal for (PE). □

8.3 Principle of Complementary Energy for Masonry Structures

The principle of complementary energy for masonry structures is established by applying the Fenchel dual theory (section 2.2.5) to $(\overline{\mathrm{PE}})$ in (8.32).

8.3.1 Embedding to Fenchel form

We here show that $(\overline{\mathrm{PE}})$ can be embedded into the form of $(\mathrm{P_F})$ in (2.28),[15] which is the primal problem in the framework of Fenchel duality theory investigated in section 2.2.5. Before that, we first introduce the notion of some functional spaces, to which a solution of a boundary value problem in continuum mechanics belongs.

For the open subset Ω of $\mathbb{R}^3$, we denote by $L^2(\Omega)$ the space of (equivalence classes of) measurable functions v such that

$$\|v\|_{L^2(\Omega)} = \left(\int_\Omega |v(\boldsymbol{x})| \mathrm{d}\boldsymbol{x} \right)^{1/2} < +\infty,$$

where Ω is the interior of the domain occupied by the deformable body (at the undeformed configuration). Note that $L^2(\Omega)$ is a Hilbert space equipped with the scalar product

$$(v, w)_{L^2(\Omega)} = \int_\Omega v(\boldsymbol{x}) w(\boldsymbol{x}) \mathrm{d}\boldsymbol{x}.$$

In what follows, we suppose that the body force $\underline{\boldsymbol{p}}$ is given so that each component $\underline{p}_i$ belongs to $L^2(\Omega)$, and we write $\underline{\boldsymbol{p}} \in L^2(\Omega)^3$ with the notation

$$L^2(\Omega)^3 = \{\underline{\boldsymbol{p}} \mid \underline{p}_i \in L^2(\Omega) \ (i = 1, 2, 3)\}.$$

[15] See the orientation part of this chapter for convenience.

Furthermore, we denote by $L^2(\Omega)^{3\times 3}_{\mathbb{S}}$ the set of symmetric tensors, each component of which belongs to $L^2(\Omega)$; i.e.,

$$L^2(\Omega)^{3\times 3}_{\mathbb{S}} = \{\boldsymbol{\varepsilon} \mid \varepsilon_{ij} = \varepsilon_{ji} \in L^2(\Omega)\ (i,j = 1,2,3)\}.$$

Let m ($\neq 0$) be a positive integer, α_i ($i = 1,2,3$) be nonnegative integers, $\boldsymbol{\alpha} = (\alpha_1, \alpha_2, \alpha_3)$, and $|\boldsymbol{\alpha}| = \alpha_1 + \alpha_2 + \alpha_3$. The *Sobolev space* of order m on $L^2(\Omega)$, denoted $H^m(\Omega)$, is the space defined by

$$H^m(\Omega) = \{v \mid D^{\boldsymbol{\alpha}} v \in L^p(\Omega),\ |\boldsymbol{\alpha}| \leq m\},$$

where

$$D^{\boldsymbol{\alpha}} v = \frac{\partial^{|\boldsymbol{\alpha}|}}{\partial x_1^{\alpha_1} \partial x_2^{\alpha_2} \partial x_3^{\alpha_3}} v$$

is the partial derivative of v. Note that $H^m(\Omega)$ is a Hilbert space with the scalar product

$$(v, w)_{H^m(\Omega)} = \sum_{|\boldsymbol{\alpha}| \leq m} (D^{\boldsymbol{\alpha}} v, D^{\boldsymbol{\alpha}} w).$$

We denote by $H^{-m}(\Omega)$ the dual of $H^m(\Omega)$. Each component u_i of the displacement $\boldsymbol{u}$ is supposed to belong to $H^1(\Omega)$ in what follows. For simplicity, we write $\boldsymbol{u} \in H^1(\Omega)^3$, where

$$H^1(\Omega)^3 = \{\boldsymbol{u} \mid u_i \in H^1(\Omega)\ (i = 1,2,3)\}.$$

We prepare some notations for stating the boundary conditions rigorously with the use of the so-called trace theorem introduced below. For a positive real number s, which is not necessarily an integer, we can also define the Sobolev space $H^s(\Omega)$ by using the interpolation theory of Banach spaces.[16] Following Lions–Magenes [270, Eq. (1.7.10)], assume that $\Gamma = \operatorname{bd} \Omega$, which is a manifold of dimension 2, is once continuously differentiable, and that Ω lies locally on only one side of Γ. Then it is also possible to define the space $H^s(\Gamma)$ for a positive real number s [270, section 1.7.3]. Like the integer case, the dual of $H^s(\Gamma)$ is denoted by $H^{-s}(\Gamma)$. In what follows we use $H^{1/2}(\Gamma)$ and its dual $H^{-1/2}(\Gamma)$.

[16]See Lions–Magenes [270, section 1.2.1] for the interpolation theory of Banach spaces. For the definition of fractional Sobolev space based on the interpolation theory of Banach spaces, we refer to Lions–Magenes [270, section 1.9.1]; also Kikuchi–Oden [221, p. 15]. An alternative definition, in the special case when $\Omega = \mathbb{R}^n$, can be found in Duvaut–Lions [120, section 1.4.1], where the Fourier transformation is employed.

We further introduce the space $\mathrm{C}^k(\overline{\Omega})$ $(k = 0, 1, \ldots, \infty)$ for the closure $\overline{\Omega}$ of Ω. The space $\mathrm{C}^k(\overline{\Omega})$ is the set of functions f such that there exist an open set Ω_1 and an extension f_1 of f satisfying

$$\overline{\Omega} \subset \Omega_1, \quad f_1 \in \mathrm{C}^k(\Omega_1), \quad f_1|_{\Omega} = f;$$

see Oden–Demkowicz [341, p. 131].

With the preparation above, we can now introduce the *trace theorem*. Let γ be the operator defined by

$$\gamma(v) = v|_{\Gamma}, \quad \forall v \in \mathrm{C}^{\infty}(\overline{\Omega}).$$

Then the trace theorem asserts that γ can be extended uniquely to a continuous linear mapping, also denoted γ, from $H^1(\Omega)$ to $H^{1/2}(\Gamma)$.[17] Such a γ is called the *trace operator*. Moreover, the mapping $v \mapsto \gamma(v)$ is surjective from $H^1(\Omega)$ to $H^{1/2}(\Gamma)$. For $v \in H^1(\Omega)$, we call $\gamma(v)$ the *trace* of v on Γ. Note that the kernel of γ, denoted $H_0^1(\Omega)$, coincides with the closure in $H^1(\Omega)$ of the subspace of functions with compact support in Ω.[18]

The Dirichlet boundary condition, first introduced in section 8.2.1, is now formally stated by using the notion of trace of $\boldsymbol{u}$ on Γ. The corresponding trace operator, also denoted γ, is a mapping from $H^1(\Omega)^3$ to $H^{1/2}(\Gamma)^3$. Then the Dirichlet boundary condition is written as

$$\gamma(\boldsymbol{u}) = \underline{\boldsymbol{u}} \quad \text{on } \Gamma_{\mathrm{D}},$$

which makes sense provided that $\underline{\boldsymbol{u}} \in H^{1/2}(\Gamma)^3$. In what follows, we set

$$\underline{\boldsymbol{u}} = \boldsymbol{0} \quad \text{on } \Gamma_{\mathrm{D}} \tag{8.37}$$

to avoid technical difficulties. Moreover, we suppose for the Neumann boundary condition that $\underline{\boldsymbol{t}} \in L^2(\Gamma)^3$; for comprehensive treatments of boundary conditions, see Lions–Magenes [270, Chaps. 1 and 2] for general non-homogeneous boundary value problems, and Kikuchi–Oden [221, Chaps. 5, 6, and 8] for contact problems.

Now we are in a position to encode $(\overline{\mathrm{PE}})$ to the form of $(\mathrm{P_F})$ in (2.28). Define the primal variables and the associated spaces by

$$x = (\boldsymbol{u}, \boldsymbol{\varepsilon}_{\mathrm{e}}) \in \mathbb{V}, \quad z = \boldsymbol{Z} \in \mathbb{Y}, \tag{8.38a}$$

$$\mathbb{V} = H^1(\Omega)^3 \times L^2(\Omega)_{\mathbb{S}}^{3\times 3}, \tag{8.38b}$$

$$\mathbb{Y} = L^2(\Omega)_{\mathbb{S}}^{3\times 3}. \tag{8.38c}$$

[17] See Theorem 1.9.3 in Lions–Magenes [270]; also Theorem 5.3 in Kikuchi–Oden [221].

[18] See, e.g., Duvaut–Lions [120, p. 41].

Moreover, define $f : \mathbb{V} \to \mathbb{R} \cup \{+\infty\}$, $g : \mathbb{Y} \to \mathbb{R} \cup \{+\infty\}$, and $\Lambda : \mathbb{V} \to \mathbb{Y}$ by

$$f(x) = \begin{cases} \displaystyle\int_{\Omega} (\hat{w}_{\mathrm{ref}}(\boldsymbol{\varepsilon}_{\mathrm{e}}) - \underline{\boldsymbol{p}} \cdot \boldsymbol{u})\, \mathrm{d}\Omega - \int_{\Gamma_{\mathrm{N}}} \underline{\boldsymbol{t}} \cdot \gamma(\boldsymbol{u}) \mathrm{d}\Gamma & \text{if } \gamma(\boldsymbol{u}) = \boldsymbol{0} \text{ (on } \Gamma_{\mathrm{D}}), \\ +\infty & \text{otherwise,} \end{cases} \tag{8.39}$$

$$g(z) = \begin{cases} 0 & \text{if } \boldsymbol{Z} \preceq \boldsymbol{o} \text{ (in } \Omega), \\ +\infty & \text{otherwise,} \end{cases} \tag{8.40}$$

$$\Lambda x = \boldsymbol{\varepsilon}_{\mathrm{e}} - \frac{1}{2}(\nabla \boldsymbol{u}^{\mathrm{T}} + \nabla \boldsymbol{u}). \tag{8.41}$$

Note that the condition "$\boldsymbol{Z} \preceq \boldsymbol{o}$ (in Ω)" in (8.40) is interpreted to mean that $\boldsymbol{Z}$ is negative semidefinite almost everywhere on Ω. However, we omit such a qualifying phrase for simplicity.

It can be easily seen that f and g are proper convex functions, and that Λ is a continuous linear mapping. With this setting, ($\overline{\mathrm{PE}}$) in (8.32) is embedded into the form of ($\mathrm{P_F}$) in (2.28), which plays the role of the primal problem in the framework of the Fenchel duality. In what follows, we derive an explicit formulation of the dual problem ($\mathrm{P_F^*}$) in (2.31).

8.3.2 Dual problem

Let $\mathbb{V}^*$ and $\mathbb{Y}^*$ be the (topological) dual spaces of $\mathbb{V}$ and $\mathbb{Y}$, respectively. In accordance with (8.38), the variables and associated spaces for the Fenchel dual problem are given by[19]

$$x^* = (\boldsymbol{u}^*, \boldsymbol{\varepsilon}_{\mathrm{e}}^*), \quad z^* = \boldsymbol{Z}^*, \tag{8.42a}$$

$$\mathbb{V}^* = H^{-1}(\Omega)^3 \times L^2(\Omega)_{\mathbb{S}}^{3\times 3}, \tag{8.42b}$$

$$\mathbb{Y}^* = L^2(\Omega)_{\mathbb{S}}^{3\times 3}, \tag{8.42c}$$

where $x^* \in \mathbb{V}^*$ and $z^* \in \mathbb{Y}^*$.

For deriving an explicit formulation of ($\mathrm{P_F^*}$) in the situation considered, Proposition 8.3.2 and Proposition 8.3.4 will prepare explicit forms of $f^*(\Lambda^* z^*)$ and $g^*(z^*)$.[20] To this end, we introduce a generalized Green's formula for linear operators on Hilbert spaces, following Kikuchi–Oden [221, section 5.9]; see also Lions–Magenes [270, Chap. 2]. We note that, if Γ is sufficiently smooth, it is possible to show that a unit vector $\boldsymbol{n}$, outward and normal to

[19]Note that $\mathbb{V}^* \neq \mathbb{V}$, unlike in the case of finite-dimensional spaces in which $(\mathbb{R}^n)^* = \mathbb{R}^n$ holds. We see that $\mathbb{Y} = \mathbb{Y}^*$, because $(L^2(\Omega))^* = L^2(\Omega)$.

[20]The reader who is not interested in the details of derivation of f^*, g^*, and Λ^* can skip Proposition 8.3.2 and Proposition 8.3.4, which consist of quite straightforward application of the definitions of a conjugate function (see (2.13)) and an adjoint operator (see (2.30)).

boundary Γ, exists almost everywhere on Γ.[21] Moreover, we define the space $\mathcal{T}$ of symmetric tensors by[22]

$$\mathcal{T} = \{\boldsymbol{\tau} \in L^2(\Omega)^{3\times 3}_{\mathbb{S}} \mid \operatorname{div}\boldsymbol{\tau} \in L^2(\Omega)^3\}. \tag{8.43}$$

Now we state a variant of the trace theorem, with the (generalized) Green's formula provided as (8.44).

Theorem 8.3.1 (Kikuchi–Oden [221, Theorem 5.9]). *Suppose that Γ is sufficiently smooth.*[23] *Let $\boldsymbol{n}$ be the unit outward normal vector along $\Gamma = \operatorname{bd}\Omega$. Then there exists a uniquely determined linear continuous mapping π from $\mathcal{T}$ to $H^{-1/2}(\Gamma)^3$ such that*

$$\pi(\boldsymbol{\tau}) = \boldsymbol{\tau}|_{\Gamma}\boldsymbol{n}, \quad \text{if } \tau_{ij} \in \mathrm{C}^1(\overline{\Omega}) \ (i,j = 1,2,3)$$

and such that

$$\int_{\Omega} \boldsymbol{\tau} : (\nabla \boldsymbol{v})\mathrm{d}\Omega + \int_{\Omega} (\operatorname{div}\boldsymbol{\tau}) \cdot \boldsymbol{v}\mathrm{d}\Omega = \langle \pi(\boldsymbol{\tau}), \gamma(\boldsymbol{v})\rangle_{\Gamma}, \quad \forall \boldsymbol{\tau} \in \mathcal{T},\ \forall \boldsymbol{v} \in H^1(\Omega)^3, \tag{8.44}$$

where $\langle \cdot, \cdot \rangle$ denotes the duality pairing on $H^{-1/2}(\Gamma)^3 \times H^{1/2}(\Gamma)^3$ defined as

$$\langle \pi(\boldsymbol{\tau}), \gamma(\boldsymbol{v})\rangle_{\Gamma} = \int_{\Gamma} \pi(\boldsymbol{\tau}) \cdot \gamma(\boldsymbol{v})\mathrm{d}\Gamma.$$

Making use of Theorem 8.3.1, we can derive an explicit form of $f^*(\Lambda^* \boldsymbol{Z}^*)$ as follows.

Proposition 8.3.2. *For f in (8.39) and Λ in (8.41), we have that*

$$f^*(\Lambda^* \boldsymbol{Z}^*) = \begin{cases} \displaystyle\int_{\Omega} \hat{w}^*_{\mathrm{ref}}(\boldsymbol{Z}^*)\mathrm{d}\Omega & \text{if } -\operatorname{div}\boldsymbol{Z}^* = \underline{\boldsymbol{p}} \text{ (in } \Omega), \\ & \quad \boldsymbol{Z}^*\boldsymbol{n} = \underline{\boldsymbol{t}} \text{ (on } \Gamma_{\mathrm{N}}), \\ +\infty & \text{otherwise.} \end{cases} \tag{8.45}$$

Here, $\hat{w}^*_{\mathrm{ref}} : \mathbb{S}^3 \to \mathbb{R}$ is the conjugate function of $\hat{w}_{\mathrm{ref}}$ in (8.12), which is written explicitly as

$$\hat{w}^*_{\mathrm{ref}}(\boldsymbol{\varepsilon}^*_{\mathrm{e}}) = -\frac{\lambda}{4\mu(3\lambda + 2\mu)}(\operatorname{tr}\boldsymbol{\varepsilon}^*_{\mathrm{e}})^2 + \frac{1}{4\mu}\boldsymbol{\varepsilon}^*_{\mathrm{e}} : \boldsymbol{\varepsilon}^*_{\mathrm{e}}. \tag{8.46}$$

[21] See Theorem 5.4 in Kikuchi–Oden [221].

[22] See Kikuchi–Oden [221, Eq. (5.66)].

[23] More precisely, Ω is assumed to be a *Lipschitzian domain*; see [221, pp. 15–16] for the definition of Lipschitzian domain.

Proof. From (8.38a), (8.41), (8.42a) and the definition of the adjoint operator (see (2.30)), we have that

$$\begin{aligned}\langle \Lambda^* z^*, x\rangle &= \langle z^*, \Lambda x\rangle \\ &= \int_\Omega \boldsymbol{Z}^* : \left[\boldsymbol{\varepsilon}_{\mathrm{e}} - \frac{1}{2}(\nabla \boldsymbol{u}^{\mathrm{T}} + \nabla \boldsymbol{u})\right] \mathrm{d}\Omega \\ &= \int_\Omega (\boldsymbol{Z}^* : \boldsymbol{\varepsilon}_{\mathrm{e}} - \boldsymbol{Z}^* : \nabla \boldsymbol{u})\, \mathrm{d}\Omega. \end{aligned} \tag{8.47}$$

With the use of this relation, the definition of conjugate function (see (2.13)) to f in (8.39) yields

$$\begin{aligned} & f^*(\Lambda^* z^*) \\ &\quad = \sup\{\langle \Lambda^* z^*, x\rangle - f(x) \mid x \in \operatorname{dom} f\} \\ &\quad = \sup_{\boldsymbol{u}:\gamma(\boldsymbol{u})=\boldsymbol{0}\ (\text{on } \Gamma_{\mathrm{D}})} \left\{ -\langle \boldsymbol{Z}^*, \nabla \boldsymbol{u}\rangle + \int_\Omega \underline{\boldsymbol{p}} \cdot \boldsymbol{u} \mathrm{d}\Omega + \int_{\Gamma_{\mathrm{N}}} \underline{\boldsymbol{t}} \cdot \gamma(\boldsymbol{u}) \mathrm{d}\Gamma \right\} \\ &\qquad + \sup_{\boldsymbol{\varepsilon}_{\mathrm{e}}} \left\{ \langle \boldsymbol{Z}^*, \boldsymbol{\varepsilon}_{\mathrm{e}}\rangle - \int_\Omega \hat{w}_{\mathrm{ref}}(\boldsymbol{\varepsilon}_{\mathrm{e}}) \mathrm{d}\Omega \right\}. \end{aligned} \tag{8.48}$$

The first supremum in (8.48) is finite if and only if $\boldsymbol{Z}^* \in L^2(\Omega)^{3\times 3}_{\mathbb{S}}$ satisfies

$$\int_\Omega [-\boldsymbol{Z}^* : (\nabla \boldsymbol{v}) + \underline{\boldsymbol{p}} \cdot \boldsymbol{v}] \mathrm{d}\Omega + \int_{\Gamma_{\mathrm{N}}} \underline{\boldsymbol{t}} \cdot \gamma(\boldsymbol{v}) \mathrm{d}\Gamma = 0, \qquad \forall \boldsymbol{v} \in H^1(\Omega)^3 :\ \gamma(\boldsymbol{v}) = \boldsymbol{0}\ (\text{on } \Gamma_{\mathrm{D}}). \tag{8.49}$$

Condition (8.49) implies that

$$\operatorname{div} \boldsymbol{Z}^* + \underline{\boldsymbol{p}} = \boldsymbol{0} \quad \text{in } \Omega \tag{8.50}$$

in the sense of the distributions in Ω. Since the body force is assumed to satisfy $\underline{\boldsymbol{p}} \in L^2(\Omega)^3$ (see section 8.3.1), (8.50) implies $\boldsymbol{Z}^* \in \mathcal{T}$, where $\mathcal{T}$ is defined by (8.43). Therefore, from Theorem 8.3.1 we can define the trace $\pi(\boldsymbol{Z}^*)$ on Γ and have Green's formula

$$\int_\Omega \boldsymbol{Z}^* : (\nabla \boldsymbol{v}) \mathrm{d}\Omega + \int_\Omega (\operatorname{div} \boldsymbol{Z}^*) \cdot \boldsymbol{v} \mathrm{d}\Omega = \int_\Gamma \pi(\boldsymbol{Z}^*) \cdot \gamma(\boldsymbol{v}) \mathrm{d}\Omega. \tag{8.51}$$

Substituting (8.50) and (8.51) into (8.49) results in

$$\int_{\Gamma_{\mathrm{N}}} (\underline{\boldsymbol{t}} - \pi(\boldsymbol{Z}^*)) \cdot \gamma(\boldsymbol{v}) \mathrm{d}\Gamma - \int_{\Gamma_{\mathrm{D}}} \pi(\boldsymbol{Z}^*) \cdot \gamma(\boldsymbol{v}) \mathrm{d}\Gamma = 0, \qquad \forall \boldsymbol{v} \in H^1(\Omega)^3 :\ \gamma(\boldsymbol{v}) = \boldsymbol{0}\ (\text{on } \Gamma_{\mathrm{D}}),$$

which implies

$$\boldsymbol{Z}^* \boldsymbol{n} = \underline{\boldsymbol{t}} \quad \text{on } \Gamma_{\mathrm{N}}. \tag{8.52}$$

Conversely, (8.50) and (8.52) imply (8.49). Thus the first supremum in (8.48) is finite if and only if (8.50) and (8.52) are satisfied; in this case the supremum is equal to zero. It is immediate to reduce the second supremum in (8.48) to

$$\sup_{\boldsymbol{\varepsilon}_{\rm e}} \left\{ \langle \boldsymbol{Z}^*, \boldsymbol{\varepsilon}_{\rm e} \rangle - \int_\Omega \hat{w}_{\rm ref}(\boldsymbol{\varepsilon}_{\rm e}) {\rm d}\Omega \right\} = \int_\Omega \hat{w}^*_{\rm ref}(\boldsymbol{Z}^*) {\rm d}\Omega.$$

Thus we obtain (8.45).

By definition, $\hat{w}^*_{\rm ref}$ is written as

$$\hat{w}^*_{\rm ref}(\boldsymbol{Z}^*) = \sup_{\boldsymbol{\varepsilon}_{\rm e}} \{ \boldsymbol{Z}^* : \boldsymbol{\varepsilon}_{\rm e} - \hat{w}_{\rm ref}(\boldsymbol{\varepsilon}_{\rm e}) \}, \tag{8.53}$$

where $\hat{w}$ is given by (8.12). Since $\hat{w}$ is a convex function, the supremum in (8.53) is attained when $\boldsymbol{\varepsilon}_{\rm e}$ satisfies the stationarity condition of the objective function, which is written as

$$\boldsymbol{Z}^* - (\lambda \operatorname{tr} \boldsymbol{\varepsilon}_{\rm e}) \boldsymbol{I} - 2\mu \boldsymbol{\varepsilon}_{\rm e} = \boldsymbol{o}.$$

Since the condition above is nothing but the constitutive law in the linear isotropic elasticity, its inversion is known as

$$\boldsymbol{\varepsilon}_{\rm e} = -\frac{\lambda}{2\mu(3\lambda + 2\mu)} (\operatorname{tr} \boldsymbol{Z}^*) \boldsymbol{I} + \frac{1}{2\mu} \boldsymbol{Z}^*.$$

Substitution of this equation into the objective function of (8.53) eliminates $\boldsymbol{\varepsilon}_{\rm e}$, and as a consequence (8.46) is obtained. □

Remark 8.3.3. The latter half of the proof of Proposition 8.3.2 is nothing but a conventional procedure of the Legendre transformation of the strain energy function $\hat{w}_{\rm ref}$, and $\hat{w}^*_{\rm ref}$ amounts to the function known as the *complementary strain energy function*. ■

Proposition 8.3.4. *The conjugate function* $g^* : \mathbb{Y}^* \to \mathbb{R} \cup \{+\infty\}$ *of* g *defined by* (8.40) *is written as*

$$g^*(\boldsymbol{Z}^*) = \begin{cases} 0 & \text{if } \boldsymbol{Z}^* \succeq \boldsymbol{o} \text{ (in } \Omega), \\ +\infty & \text{otherwise.} \end{cases} \tag{8.54}$$

Proof. This result is an immediate consequence of applying Fact 1.3.17 (ii) to

$$\begin{aligned} g^*(\boldsymbol{Z}^*) &= \sup_{\boldsymbol{Z}} \{ \langle \boldsymbol{Z}^*, \boldsymbol{Z} \rangle \mid \boldsymbol{Z} \in \operatorname{dom} g \} \\ &= -\inf_{\boldsymbol{Z}} \{ \langle \boldsymbol{Z}^*, -\boldsymbol{Z} \rangle \mid -\boldsymbol{Z} \succeq \boldsymbol{o} \}. \end{aligned}$$ □

Approaching closer the Fenchel dual problem in the form of (2.31), the result of Proposition 8.3.4 is rewritten as

$$-g^*(-\boldsymbol{Z}^*) = \begin{cases} 0 & \text{if } -\boldsymbol{Z}^* \succeq \boldsymbol{o} \text{ (in } \Omega), \\ -\infty & \text{otherwise.} \end{cases}$$

By rewriting the variables as[24]

$$\boldsymbol{\sigma} = \boldsymbol{Z}^*, \tag{8.55}$$

$(\mathrm{P}_{\mathrm{F}}^*)$ in (2.31) is explicitly written as

$$\left.\begin{aligned} (\overline{\mathrm{CE}}) : \max_{\boldsymbol{\sigma}} \ & -\int_{\Omega} \hat{w}_{\mathrm{ref}}^*(\boldsymbol{\sigma}) \mathrm{d}\Omega \\ \text{s.t.} \ & -\operatorname{div} \boldsymbol{\sigma} = \underline{\boldsymbol{p}} \quad \text{in } \Omega, \\ & -\boldsymbol{\sigma} \succeq \boldsymbol{o} \quad \text{in } \Omega, \\ & \boldsymbol{\sigma n} = \underline{\boldsymbol{t}} \quad \text{on } \Gamma_{\mathrm{N}}, \end{aligned}\right\} \tag{8.56}$$

where $\hat{w}_{\mathrm{ref}}^*$ is defined by (8.46), which is the complementary strain energy function in the linear isotropic elasticity.

Thus the Fenchel dual of $(\overline{\mathrm{PE}})$ in (8.32) is obtained as $(\overline{\mathrm{CE}})$ above. In the following section we apply the Fenchel duality theory to the primal-dual pair of $(\overline{\mathrm{PE}})$ and $(\overline{\mathrm{CE}})$.

Remark 8.3.5. We have developed the Fenchel duality theory in Chapter 2 in the finite-dimensional setting. In contrast, the framework of the continuum mechanics, with which the investigation of this chapter is proceeding, requires the notion of optimization problems in the infinite-dimensional spaces. The Fenchel duality in the infinite-dimensional setting requires more technically careful treatment of functional spaces to which the data belong and in which the solution is sought. However, we do not deal with this issue in detail. The principal interest of our current work is to discuss a modern treatment of the nonsmoothness property of the constitutive law, and subsequently the optimization problems are to be solved by applying the conventional finite-dimensional discretization. For comprehensive treatments of the duality theory in infinite-dimensional setting, see, e.g., Allaire [10], Ekeland–Temam [121] and Lions–Magenes [270]. ■

8.3.3 Duality and optimality

We thus obtain the primal-dual pair of the conic optimization problems, say, $(\overline{\mathrm{PE}})$ in (8.32) and $(\overline{\mathrm{CE}})$ in (8.56), in the framework of the Fenchel duality. Our next concern is to apply the duality and optimality established for $(\mathrm{P_F})$ and $(\mathrm{P}_{\mathrm{F}}^*)$ (see Proposition 2.2.9 and Proposition 2.2.5) to $(\overline{\mathrm{PE}})$ and $(\overline{\mathrm{CE}})$ that are under consideration.[25] The first result is the strong duality between $(\overline{\mathrm{PE}})$ and $(\overline{\mathrm{CE}})$, which is obtained by applying Proposition 2.2.9.

Proposition 8.3.6. *Suppose that there exists $\boldsymbol{\sigma} \in L^2(\Omega)_{\mathbb{S}}^{3\times 3}$ satisfying*

$$\begin{aligned} -\operatorname{div} \boldsymbol{\sigma} = \underline{\boldsymbol{p}}, \ \boldsymbol{\sigma} \prec \boldsymbol{o} \quad & \text{in } \Omega, \\ \boldsymbol{\sigma n} = \underline{\boldsymbol{t}} \quad & \text{on } \Gamma_{\mathrm{D}}. \end{aligned}$$

[24] It shall be verified in section 8.3.3 that $\boldsymbol{Z}^*$ truly corresponds to the stress tensor.

[25] We still suppose (8.37) in continuation of the investigation of section 8.3.1.

Assume that the given boundary condition is smooth enough. Then, each of $(\overline{\mathrm{PE}})$ *and* $(\overline{\mathrm{CE}})$ *has an optimal solution, and* $\min(\overline{\mathrm{PE}}) = \max(\overline{\mathrm{CE}})$.

Proof. It suffices to show that, under the assumptions postulated here, $(\overline{\mathrm{PE}})$ and $(\overline{\mathrm{CE}})$ satisfy the assumptions required in Proposition 2.2.12, because $(\overline{\mathrm{PE}})$ and $(\overline{\mathrm{CE}})$ correspond to $(\mathrm{P_F})$ and $(\mathrm{P_F^*})$, respectively.

From the definitions of f and g (in (8.39) and (8.40)), it is straightforward to see that

$$\begin{aligned}\mathrm{ri}(\mathrm{dom}\, f) &= \{(\boldsymbol{u}, \boldsymbol{\varepsilon}_{\mathrm{e}}) \mid \boldsymbol{u} = \mathbf{0} \text{ (on } \Gamma_{\mathrm{D}})\},\\ \mathrm{ri}(\mathrm{dom}\, g) &= \{\boldsymbol{Z} \mid \boldsymbol{Z} \prec \boldsymbol{o} \text{ (in } \Omega)\}.\end{aligned}$$

For any given $\boldsymbol{u}$ satisfying $\boldsymbol{u} = \mathbf{0}$ (on Γ_{D}), we can choose an $\boldsymbol{\varepsilon}_{\mathrm{e}}$ satisfying

$$\boldsymbol{\varepsilon}_{\mathrm{e}} \prec \frac{1}{2}(\nabla \boldsymbol{u}^{\mathrm{T}} + \nabla \boldsymbol{u}); \tag{8.57}$$

for example, we may choose $\boldsymbol{\varepsilon}_{\mathrm{e}} = -\alpha \boldsymbol{I}$ with sufficiently large α. Since the action of Λ is defined by (8.41), condition (8.57) can be interpreted as $\Lambda(\boldsymbol{u}, \boldsymbol{\varepsilon}_{\mathrm{e}}) \in \mathrm{ri}(\mathrm{dom}\, g)$, and thus assumption (i) required in Proposition 2.2.12 is always satisfied.

We next examine assumption (ii) of Proposition 2.2.12; i.e.,

$$\exists \boldsymbol{Z}^* \in \mathrm{ri}(\mathrm{dom}\, g^*) : \quad -\Lambda^* \boldsymbol{Z}^* \in \mathrm{ri}(\mathrm{dom}\, f^*),$$

or, equivalently,

$$\exists \boldsymbol{Z}^* : \quad -\boldsymbol{Z}^* \in \mathrm{ri}(\mathrm{dom}\, g^*), \ \Lambda^* \boldsymbol{Z}^* \in \mathrm{ri}(\mathrm{dom}\, f^*). \tag{8.58}$$

It follows from (8.45) that we obtain

$$\begin{aligned}&\Lambda^* \boldsymbol{Z}^* \in \mathrm{ri}(\mathrm{dom}\, f^*)\\ &\quad\Leftrightarrow \quad -\mathrm{div}\, \boldsymbol{Z}^* = \underline{\boldsymbol{p}} \text{ (in } \Omega), \ \boldsymbol{Z}^* \boldsymbol{n} = \underline{\boldsymbol{t}} \text{ (on } \Gamma_{\mathrm{N}}),\end{aligned}$$

while (8.54) yields

$$\mathrm{ri}(\mathrm{dom}\, g^*) = \{\boldsymbol{Z}^* \mid \boldsymbol{Z}^* \succ \boldsymbol{o} \text{ (in } \Omega)\}.$$

Therefore, by rewriting the variable $\boldsymbol{Z}^*$ as $\boldsymbol{\sigma}$ according to (8.55), condition (8.58) is equivalently rewritten as

$$\exists \boldsymbol{\sigma} : \quad -\boldsymbol{\sigma} \succ \boldsymbol{o}, \ -\mathrm{div}\, \boldsymbol{\sigma} = \underline{\boldsymbol{p}} \text{ (in } \Omega), \ \boldsymbol{\sigma n} = \underline{\boldsymbol{t}} \text{ (on } \Gamma_{\mathrm{N}}),$$

which is guaranteed by the assumption considered in this proposition. Thus assumption (ii) required in Proposition 2.2.12 is satisfied. □

Remark 8.3.7. Recall that we assume the boundedness of the objective function of $(\overline{\mathrm{PE}})$ for establishing Proposition 8.2.7. This assumption corresponds to the constraint qualification for $(\overline{\mathrm{CE}})$ in Proposition 8.3.6. Indeed, for some loading conditions, $(\overline{\mathrm{CE}})$ has no feasible solution, while the objective function of $(\overline{\mathrm{PE}})$ becomes unbounded. Illustrative examples include a rectangular (or cuboid) body subjected to the uniaxial tension load. ■

The optimality conditions for $(\overline{\mathrm{PE}})$ and $(\overline{\mathrm{CE}})$ are given below.

Proposition 8.3.8. *Under the hypotheses in Proposition 8.3.6,* $(\bar{\boldsymbol{u}}, \bar{\boldsymbol{\varepsilon}}_{\mathrm{e}})$ *and* $\bar{\boldsymbol{\sigma}}$ *are optimal for* $(\overline{\mathrm{PE}})$ *and* $(\overline{\mathrm{CE}})$, *respectively, if and only if they satisfy (in the weak sense*[26]*)*

$$\bar{\boldsymbol{\sigma}} = \lambda(\operatorname{tr}\bar{\boldsymbol{\varepsilon}}_{\mathrm{e}})\boldsymbol{I} + 2\mu\bar{\boldsymbol{\varepsilon}}_{\mathrm{e}} \quad \text{in } \Omega, \tag{8.59a}$$

$$-\operatorname{div}\bar{\boldsymbol{\sigma}} = \underline{\boldsymbol{p}} \quad \text{in } \Omega, \tag{8.59b}$$

$$\bar{\boldsymbol{u}} = \boldsymbol{0} \quad \text{on } \Gamma_{\mathrm{D}}; \qquad \bar{\boldsymbol{\sigma}}\boldsymbol{n} = \underline{\boldsymbol{t}} \quad \text{on } \Gamma_{\mathrm{N}}, \tag{8.59c}$$

$$\bar{\boldsymbol{\varepsilon}}_{\mathrm{e}} \preceq \hat{\boldsymbol{\varepsilon}}(\bar{\boldsymbol{u}}), \ \bar{\boldsymbol{\sigma}} \preceq \boldsymbol{o} \quad \text{in } \Omega, \tag{8.59d}$$

$$(\bar{\boldsymbol{\varepsilon}}_{\mathrm{e}} - \hat{\boldsymbol{\varepsilon}}(\bar{\boldsymbol{u}})) : \bar{\boldsymbol{\sigma}} = 0 \quad \text{in } \Omega, \tag{8.59e}$$

where $\hat{\boldsymbol{\varepsilon}}$ *is defined by* (8.1).

Proof. The hypotheses postulated here guarantee the assumption required in Proposition 2.2.9. For deriving the optimality condition (8.59), we here employ the formulation in Proposition 2.2.9 (b).

For considering the first equation in Proposition 2.2.9 (b), we gather the results of $f(x)$, $f^*(\Lambda^* z^*)$, and $\langle \Lambda^* z^*, x\rangle$ from section 8.3.1 and section 8.3.2. By definition (8.39) of f, we see that if $\boldsymbol{u} = \boldsymbol{0}$ (on Γ_{D}) in (8.59c) is satisfied, then f is finite (and the converse is also true) and

$$f(x) = \int_{\Omega} (\hat{w}_{\mathrm{ref}}(\boldsymbol{\varepsilon}_{\mathrm{e}}) - \underline{\boldsymbol{p}} \cdot \boldsymbol{u}) \,\mathrm{d}\Omega - \int_{\Gamma_{\mathrm{N}}} \underline{\boldsymbol{t}} \cdot \gamma(\boldsymbol{u}) \mathrm{d}\Gamma.$$

Recall that f^* is given in Proposition 8.3.2. Rewrite the variable $\boldsymbol{Z}^*$ as $\boldsymbol{\sigma}$[27] to see that $f^*(\Lambda^* z^*)$ is finite if and only if (8.59b) and the condition on Γ_{N} in (8.59c) are satisfied. Furthermore, when they are satisfied, $f^*(\Lambda^* z^*)$ takes the value

$$f^*(\Lambda^* z^*) = \int_{\Omega} \hat{w}^*_{\mathrm{ref}}(\boldsymbol{\sigma}) \mathrm{d}\Omega.$$

From (8.47) and (8.51), the pairing $\langle \Lambda^* z^*, x\rangle$ is written with the current variables as

$$\langle \Lambda^* z^*, x\rangle = \int_{\Omega} (\operatorname{div}\boldsymbol{\sigma}) \cdot \boldsymbol{u} \mathrm{d}\Omega - \int_{\Gamma} (\boldsymbol{\sigma}\boldsymbol{n}) \cdot \gamma(\boldsymbol{u}) \mathrm{d}\Gamma + \int_{\Omega} \boldsymbol{\sigma} : \boldsymbol{\varepsilon}_{\mathrm{e}} \mathrm{d}\Omega.$$

Thus, $f(x) + f^*(\Lambda^* z^*)$ is finite if and only if (8.59b) and (8.59c) are satisfied. In this case, the equation $f(x) + f^*(\Lambda^* z^*) = \langle \Lambda^* z^*, x\rangle$ is reduced to

$$\int_{\Omega} (\hat{w}_{\mathrm{ref}}(\boldsymbol{\varepsilon}_{\mathrm{e}}) + \hat{w}^*_{\mathrm{ref}}(\boldsymbol{\sigma})) \,\mathrm{d}\Omega = \int_{\Omega} \boldsymbol{\sigma} : \boldsymbol{\varepsilon}_{\mathrm{e}} \mathrm{d}\Omega.$$

This condition is equivalent to $\boldsymbol{\sigma} = \dfrac{\partial \hat{w}_{\mathrm{ref}}}{\partial \boldsymbol{\varepsilon}_{\mathrm{e}}}(\boldsymbol{\varepsilon}_{\mathrm{e}})$ (see Proposition 2.1.12) and thence to (8.59a).

[26] We do not state precisely in which classes we seek the solutions $(\boldsymbol{u}, \boldsymbol{\varepsilon}_{\mathrm{e}})$ and $\boldsymbol{\sigma}$ to (8.59). A rigorous treatment of such a issue is beyond the aim of this book.
[27] See (8.55).

Second, for considering the second equation in Proposition 2.2.9 (b), we recall that $g(\Lambda x)$ is defined by (8.40) and (8.41), while $g^*(z^*)$ is given by Proposition 8.3.4. Thus, if (and only if) (8.59d) is satisfied, then both $g(\Lambda x)$ and $g^*(-z^*)$ are finite, and $g(\Lambda x) = g^*(-z^*) = 0$. Finally the expression $\langle -z^*, \Lambda x\rangle$ with the current variables is obtained from (8.41) and $z^* = \boldsymbol{Z}^* = \boldsymbol{\sigma}$ as

$$\langle -z^*, \Lambda x\rangle = -\int_{\Omega} \boldsymbol{\sigma} : \left[\boldsymbol{\varepsilon}_{\mathrm{e}} - \frac{1}{2}(\nabla \boldsymbol{u}^{\mathrm{T}} + \nabla \boldsymbol{u})\right] \mathrm{d}\Omega,$$

which is finite. Consequently $g(\Lambda x) + g^*(-z^*)$ is finite if and only if (8.59d) is satisfied. Moreover, in such a case the equation $g(\Lambda x) + g^*(-z^*) = \langle -z^*, \Lambda x\rangle$ is reduced to

$$\int_{\Omega} \boldsymbol{\sigma} : \left[\boldsymbol{\varepsilon}_{\mathrm{e}} - \frac{1}{2}(\nabla \boldsymbol{u}^{\mathrm{T}} + \nabla \boldsymbol{u})\right] \mathrm{d}\Omega = 0.$$

This condition is equivalent to (8.59e) in the weak sense. □

Since (8.59a) corresponds to the constitutive law of the no-tension body, say, (8.13), the variable $\boldsymbol{\sigma}$ can be interpreted as the stress tensor. Then condition (8.59b) can be read as the force-balance equation. In accordance with this observation, we formally state the principle of complementary energy for the equilibrium of masonry structures in the following section.

8.3.4 Principle of complementary energy

As a major consequence of the analysis that has evolved through the past sections, we here establish the minimum principle of the complementary energy for masonry structures consisting of no-tension material.

Conventionally the principle of the complementary energy is stated in the form of a minimization problem, although $(\overline{\mathrm{CE}})$ in (8.56) is formulated as the maximization problem in accordance with the fashion of the duality theory of optimization. By changing the sign of the objective function to transform maximization to minimization, $(\overline{\mathrm{CE}})$ is rewritten as

$$\left.\begin{aligned}
(\mathrm{CE}) : \min_{\boldsymbol{\sigma}} \ & \int_{\Omega} \hat{w}^*_{\mathrm{ref}}(\boldsymbol{\sigma}) \mathrm{d}\Omega \\
\text{s.t.} \ & -\operatorname{div} \boldsymbol{\sigma} = \underline{\boldsymbol{p}} \quad \text{in } \Omega, \\
& \boldsymbol{\sigma} \preceq \boldsymbol{o} \quad \text{in } \Omega, \\
& \boldsymbol{\sigma n} = \underline{\boldsymbol{t}} \quad \text{on } \Gamma_{\mathrm{N}}.
\end{aligned}\right\} \qquad (8.60)$$

Here, $\hat{w}^*_{\mathrm{ref}}$ is the complementary strain energy function in the linear isotropic elasticity, as given in (8.46). The main result stated below guarantees that the stress tensor at the equilibrium state can be obtained as the optimal solution of (CE) in (8.60). The minimum principle of the total potential energy, say, (PE) in (8.3),[28] is simultaneously verified, although it was recognized as a

[28] In continuation of section 8.3.1,we still suppose that $\underline{\boldsymbol{u}} = \boldsymbol{0}$ (on Γ_{D}).

hypothesis in section 8.2.1 to launch the analysis at the beginning of this chapter.

Proposition 8.3.9. *Under the hypotheses in Proposition 8.3.6,* $(\bar{\boldsymbol{u}}, \bar{\boldsymbol{\varepsilon}})$ *and* $\bar{\boldsymbol{\sigma}}$ *are the optimal solutions of* (PE) *and* (CE), *respectively, if and only if they correspond to the variables*[29] *at the equilibrium state of the masonry structure. Moreover,* $\min(\mathrm{PE}) = -\min(\mathrm{CE})$.

Proof. Since the only difference between (CE) and $(\overline{\mathrm{CE}})$ is the sign of the objective function, the necessary and sufficient condition for the optimality of (CE) is given by (8.59) in Proposition 8.3.8. From this observation and Proposition 8.2.7 (iii) it follows that Proposition 8.3.6 implies $\min(\mathrm{PE}) = -\min(\mathrm{CE})$.

It follows from (8.20) and (8.30) that conditions (8.59a), (8.59d), and (8.59e) are equivalent to $\bar{\boldsymbol{\sigma}} = \hat{\boldsymbol{\sigma}}_{\mathrm{mas}}(\hat{\boldsymbol{\varepsilon}}(\bar{\boldsymbol{u}}))$, which is the constitutive law of no-tension material. The optimality condition of (PE) is obtained by substituting the relation in Proposition 8.2.7 (ii) to the optimality condition of $(\overline{\mathrm{PE}})$ given by (8.59). Thus we can conclude that $(\bar{\boldsymbol{u}}, \bar{\boldsymbol{\varepsilon}})$ and $\bar{\boldsymbol{\sigma}}$ are optimal for (PE) and (CE), respectively, if and only if they satisfy

$$\bar{\boldsymbol{\sigma}} = \hat{\boldsymbol{\sigma}}_{\mathrm{mas}}(\hat{\boldsymbol{\varepsilon}}(\bar{\boldsymbol{u}})) \quad \text{in } \Omega, \tag{8.61a}$$

$$\bar{\boldsymbol{\varepsilon}} = \frac{1}{2}(\nabla \bar{\boldsymbol{u}}^{\mathrm{T}} + \nabla \bar{\boldsymbol{u}}) \quad \text{in } \Omega, \tag{8.61b}$$

$$-\operatorname{div} \bar{\boldsymbol{\sigma}} = \underline{\boldsymbol{p}} \quad \text{in } \Omega, \tag{8.61c}$$

$$\bar{\boldsymbol{u}} = \boldsymbol{0} \quad \text{on } \Gamma_{\mathrm{D}}; \qquad \bar{\boldsymbol{\sigma}}\boldsymbol{n} = \underline{\boldsymbol{t}} \quad \text{on } \Gamma_{\mathrm{N}}. \tag{8.61d}$$

Here, (8.61) clearly coincides with the boundary value problem for the equilibrium of a masonry structure. This guarantees that $\bar{\boldsymbol{u}}$ and $\bar{\boldsymbol{\varepsilon}}$ correspond to the displacement and strain at the equilibrium state, while $\bar{\boldsymbol{\sigma}}$ corresponds to the stress at the same equilibrium state. □

Remark 8.3.10. Since $\hat{w}^*_{\mathrm{ref}}$ defined by (8.46) is strictly convex, so is the objective function of (CE) in (8.60). Therefore, the optimal solution of (CE), if it exists, is unique. In contrast, the optimal solution of (PE) is not necessarily unique, because $\hat{w}_{\mathrm{mas}}$ is not strictly convex.[30] Indeed, the uniqueness of the displacement field $\boldsymbol{u}$ in the equilibrium depends upon the applied load. The simplest example is the square wall illustrated in Figure 8.4 with $\alpha = 0$ (see section 8.4.2.1): The wall is subjected to the uniaxial compression, and thence the equilibrated stress is uniquely determined as $(\sigma_x, \sigma_y, \tau_{xy}) = (0, q_y, 0)$. For the reference elastic body, the corresponding strain is $(\varepsilon_x, \varepsilon_y, \gamma_{xy}) = (-(\nu/E)q_y, q_y/E, 0)$, where E and ν are Young's modulus and Poisson's ratio, respectively. For the no-tension body, any amount of crack can open in the x-direction, because no stress takes place in the x-direction. In other words, any $(\varepsilon_x, \varepsilon_y, \gamma_{xy})$ in $\{(\varepsilon_x, q_y/E, 0) \mid \varepsilon_x \geq -(\nu/E)q_y\}$ causes the same stress state $(\sigma_x, \sigma_y, \tau_{xy}) = (0, q_y, 0)$ and thence is the equilibrium solution. ■

[29] As we saw in section 8.2.1 and section 8.3.3, $\boldsymbol{u}$ and $\boldsymbol{\varepsilon}$ denote the displacement vector and the linear strain tensor, while $\boldsymbol{\sigma}$ is the stress tensor.

[30] It is easy to confirm this claim: Given two different $\boldsymbol{\varepsilon}_1$, $\boldsymbol{\varepsilon}_2 \succ \boldsymbol{o}$, which mean cracks open in all the principal directions, we have that $\hat{w}_{\mathrm{mas}}(\boldsymbol{\varepsilon}_1) = \hat{w}_{\mathrm{mas}}(\boldsymbol{\varepsilon}_2) = \hat{w}_{\mathrm{mas}}((\boldsymbol{\varepsilon}_1 + \boldsymbol{\varepsilon}_2)/2) = 0$ because of $(\boldsymbol{\varepsilon}_1 + \boldsymbol{\varepsilon}_2)/2 \succ \boldsymbol{o}$.

8.4 Numerical Aspects

The equilibrium configuration is primarily identified as an optimizer of (PE) in (8.3), which is the original form of the minimization problem of the total potential energy. In contrast, now we know that (PE) is equivalently reformulated as $(\overline{\text{PE}})$, which is in the form of the conic optimization problem. As we have already seen, $(\overline{\text{PE}})$ can completely enjoy the Fenchel duality theory. Besides this theoretical benefit, $(\overline{\text{PE}})$ has an advantage over (PE) from the viewpoint of numerical computation as follows.

In a conventional methodology of the computational mechanics, we aim to solve the system of equations stemming from the finite element discretization of the weak form of (PE). The crucial difficulty of this approach arises from the complication of the constitutive law of the no-tension model, which appears as $\hat{w}_{\text{mas}}$ in (PE). Due to the complication of $\hat{w}_{\text{mas}}$, the numerical solution for the obtained system of equations requires some sort of iteration procedure. One of popular strategies for dealing with such an iteration procedure is a return-mapping method [141, 297, 397], which was originally developed for elastoplastic problems [413].

In contrast, $(\overline{\text{PE}})$ is in the form of conic optimization. Hence, from the viewpoint of modern optimization, it is natural to solve this convex optimization problem directly. The numerical solution in this direction does *not* involve the iteration procedure arising from the nonsmoothness in the constitutive law of the no-tension material. Indeed, the strain energy function $\hat{w}_{\text{ref}}$ involved in $(\overline{\text{PE}})$ is that of the linear isotropic material. The nonsmoothness property is transfered to the constraints of $(\overline{\text{PE}})$, which can be treated efficiently within the modern computational optimization methods introduced in Chapter 6.

In section 8.4.1 we approximate the variational problem $(\overline{\text{PE}})$ by applying a spatial discretization within the framework of the conventional finite element method. It is shown that this discretization yields an SDP problem as a natural discretization of $(\overline{\text{PE}})$. Numerical results are shown in section 8.4.2, where the obtained SDP problem is solved using the primal-dual interior-point method.

8.4.1 Spatial discretization

For handling the problem numerically, $(\overline{\text{PE}})$ in (8.32) should be discretized. In what follows, we approximate the fields of the variables, $\boldsymbol{u}$ and $\boldsymbol{\varepsilon}_{\text{e}}$, following the conventional finite element methodology to reduce $(\overline{\text{PE}})$ to a finite-dimensional SDP problem.

Suppose that the domain Ω of the continuous body is subdivided into the n^{E} finite elements, each of which occupies the domain Ω^e, i.e., $\Omega = \bigcup_{e=1}^{n^{\text{E}}} \Omega^e$. In the finite element procedure, the vector field of displacements $\boldsymbol{u}$ is approximated by using interpolation functions. As an interpolation scheme, the con-

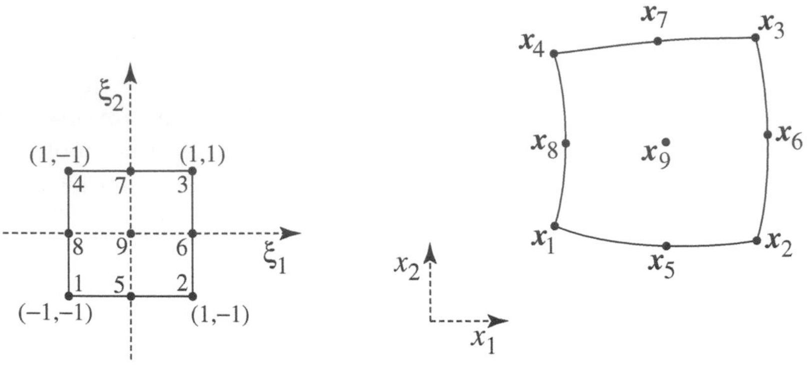

(a) Reference element Ω° and the natural coordinate system (ξ_1, ξ_2).

(b) Actual geometry Ω^{e} and the global coordinate system (x_1, x_2).

FIGURE 8.3: The nine-noded isoparametric quadrilateral element.

cept of *isoparametric element* is widely used, in which the geometry and the displacement of an element are approximated by using the same interpolation functions. We briefly introduce the standard finite element procedure based on the isoparametric concept; see, e.g., Wriggers [457] and Zienkiewicz–Taylor [472], for fundamentals of the finite element method.

For example, a planar isoparametric quadrilateral element is shown in Figure 8.3. Within the isotropic concept, we first consider the normalized reference element Ω°, defined with respect to the *natural coordinate* system. For example, the normalized reference element for the nine-noded isoparametric element is defined by

$$\Omega^{\circ} = \{\boldsymbol{\xi} \in \mathbb{R}^2 \mid -1 \le \xi_i \le 1 \ (i = 1, 2)\}$$

as shown in Figure 8.3(a), where ξ_1 and ξ_2 are the axes of the natural coordinate system. The continuous body is discretized into the elements as already stated. Figure 8.3(b) shows the actual geometry[31] of the element Ω^{e}, where (x_1, x_2) is the *global coordinate system*.

Let n^{S} denote the number of nodes for an element, where n^{S} depends on the approximation order as well as the choice of interpolation functions. If we use the Lagrange polynomials of order 2, then a quadrilateral element has $n^{\mathrm{S}} = 9$ nodes. As shown in Figure 8.3(b), we denote by $\boldsymbol{x}_I$ the location vector of the node I, which is specified with respect to the global coordinate system.

[31] Which corresponds to the undeformed configuration.

Then we interpolate the geometry of the element Ω^e as

$$\boldsymbol{x} = \sum_{I=1}^{n^{\rm S}} N_I(\boldsymbol{\xi})\boldsymbol{x}_I \quad (\text{for } \boldsymbol{x} \in \Omega^e,\ \boldsymbol{\xi} \in \Omega^\circ), \tag{8.62}$$

where N_I $(I = 1, \dots, n^{\rm S})$ are the interpolation functions (often called *shape functions*).[32] The vector field of displacements $\boldsymbol{u}(\boldsymbol{x})$ in Ω^e is then approximated by using the same interpolation functions as

$$\boldsymbol{u}(\boldsymbol{x}) \simeq \boldsymbol{u}^e(\boldsymbol{x}) := \sum_{I=1}^{n^{\rm S}} N_I(\boldsymbol{\xi})\boldsymbol{u}_I \quad (\text{for } \boldsymbol{x} \in \Omega^e,\ \boldsymbol{\xi} \in \Omega^\circ), \tag{8.63}$$

where $\boldsymbol{u}_I$ is the unknown nodal displacement of the node I. Thus the unknown vector field $\boldsymbol{u}(\boldsymbol{x})$ $(\boldsymbol{x} \in \Omega^e)$ is approximately represented in terms of the finitely many unknowns $\boldsymbol{u}_1, \dots, \boldsymbol{u}_{n^{\rm S}}$. In accordance with (8.1) and (8.63), the approximation of the strain tensor $\boldsymbol{\varepsilon}(\boldsymbol{x})$, denoted $\boldsymbol{\varepsilon}^e(\boldsymbol{x})$, for $\boldsymbol{x} \in \Omega^e$ is obtained as

$$\begin{aligned}\boldsymbol{\varepsilon}^e(\boldsymbol{x}) &= \frac{1}{2}(\nabla \boldsymbol{u}^e(\boldsymbol{x})^{\rm T} + \nabla \boldsymbol{u}^e(\boldsymbol{x})) \\ &= \frac{1}{2}\sum_{I=1}^{n^{\rm S}} \left[\left(\frac{\partial N_I(\boldsymbol{\xi})}{\partial \boldsymbol{x}} \boldsymbol{u}_I \right)^{\rm T} + \left(\frac{\partial N_I(\boldsymbol{\xi})}{\partial \boldsymbol{x}} \boldsymbol{u}_I \right) \right].\end{aligned} \tag{8.64}$$

Notice here that the relation between $\boldsymbol{x}$ and $\boldsymbol{\xi}$ is given by (8.62).

Let $\tilde{\boldsymbol{u}} \in \mathbb{R}^d$ denote the vector of the nodal displacements of the discretized structure, where d is the number of degrees of freedom[33]. We denote by $\tilde{\Gamma}_{\rm D}$ the set of indices of degrees of freedom corresponding to the boundary $\Gamma_{\rm D}$, on which the displacements are prescribed. Accordingly, the constraints on the prescribed displacements are written as

$$\tilde{u}_j = 0, \quad \forall j \in \tilde{\Gamma}_{\rm D}. \tag{8.65}$$

Let $\tilde{\Gamma}_{\rm N}$ denote the complement of $\tilde{\Gamma}_{\rm D}$ with respect to the set of all the indices of degrees of freedom; i.e., $\tilde{\Gamma}_{\rm N}$ is the set of degrees of freedom where external loads are applied. We denote by $(\underline{p}_j | j \in \tilde{\Gamma}_{\rm N})$ the vector of the equivalent external nodal loads.

We next consider the integration of the strain energy, i.e.,

$$\int_{\Omega} \hat{w}_{\rm ref}(\boldsymbol{\varepsilon}) {\rm d}\Omega = \sum_{e=1}^{n^{\rm E}} \int_{\Omega^e} \hat{w}_{\rm ref}(\boldsymbol{\varepsilon}) {\rm d}\Omega^e, \tag{8.66}$$

[32] The so-called Lagrange, serendipity, and hierarchic polynomials are widely used; see, e.g., Oñate [346] and Wriggers [457].

[33] From now on we indicate by $\tilde{(\cdot)}$ the approximation (or, the discretization) of the corresponding variable within the finite element methodology.

which appears in the objective function of ($\overline{\mathrm{PE}}$) in (8.32). The integration in (8.66) is performed for each e in the parameter space of the reference element Ω°. Turning toward this end, the integral is transformed from the initial configuration to the reference configuration as

$$\int_{\Omega^e} \hat{w}_{\mathrm{ref}}(\boldsymbol{\varepsilon}) \mathrm{d}\Omega^e = \int_{\Omega^{\circ}} \hat{w}_{\mathrm{ref}}(\boldsymbol{\xi}) \det \boldsymbol{J}^e(\boldsymbol{\xi}) \mathrm{d}\Omega^{\circ}, \tag{8.67}$$

where the Jacobian matrix $\boldsymbol{J}^e$ is defined by $\mathrm{d}\boldsymbol{x} = \boldsymbol{J}^e \mathrm{d}\boldsymbol{\xi}$ in Ω^e. Mostly the integration over Ω° is performed numerically, which yields for the integral in (8.67) the approximation

$$\int_{\Omega^{\circ}} \hat{w}_{\mathrm{ref}}(\boldsymbol{\varepsilon}(\boldsymbol{\xi})) \det \boldsymbol{J}^e(\boldsymbol{\xi}) \mathrm{d}\Omega^{\circ} \simeq \sum_{q=1}^{n^{\mathrm{G}}} \hat{w}_{\mathrm{ref}}(\boldsymbol{\xi}_q) \det \boldsymbol{J}^e(\boldsymbol{\xi}_q) \rho_q. \tag{8.68}$$

As usual, we apply the Gauss quadrature with n^{G} evaluation points, thence for each $q = 1, \dots, n^{\mathrm{G}}$ we denote by $\boldsymbol{\xi}_q$ the coordinate of the Gauss evaluation point, and ρ_q is the associated weighting factor. By using the conventional *Voigt notation*, the strain and stress are put in vector forms as

$$\mathbf{vec}(\boldsymbol{\varepsilon}) = \begin{bmatrix} \varepsilon_{11} \\ \varepsilon_{22} \\ \varepsilon_{33} \\ 2\varepsilon_{12} \\ 2\varepsilon_{23} \\ 2\varepsilon_{13} \end{bmatrix}, \quad \mathbf{vec}(\boldsymbol{\sigma}) = \begin{bmatrix} \sigma_{11} \\ \sigma_{22} \\ \sigma_{33} \\ \sigma_{12} \\ \sigma_{23} \\ \sigma_{13} \end{bmatrix}. \tag{8.69}$$

Then the strain energy in (8.68) is written as

$$\hat{w}_{\mathrm{ref}}(\boldsymbol{\varepsilon}(\boldsymbol{\xi}_q)) = \frac{1}{2} \mathbf{vec}(\boldsymbol{\varepsilon}(\boldsymbol{\xi}_q))^{\mathrm{T}} C \, \mathbf{vec}(\boldsymbol{\varepsilon}(\boldsymbol{\xi}_q)), \tag{8.70}$$

where $C \in \mathbb{S}^6$ is the *elasticity tensor* in a matrix form. By the matrix $\tilde{\varepsilon}^{qe} \in \mathbb{S}^3$ we represent the value of the strain tensor at the point $\boldsymbol{x}(\boldsymbol{\xi}_q)$ in Ω^e.[34] From (8.64), $\tilde{\varepsilon}^{qe}$ is written in terms of the nodal displacements as

$$\tilde{\varepsilon}^{qe} = \frac{1}{2} \sum_{I=1}^{n^{\mathrm{S}}} \left[\left(\frac{\partial N_I}{\partial \boldsymbol{x}}(\boldsymbol{\xi}_q) \boldsymbol{u}_I \right)^{\mathrm{T}} + \left(\frac{\partial N_I}{\partial \boldsymbol{x}}(\boldsymbol{\xi}_q) \boldsymbol{u}_I \right) \right],$$

which is further rewritten compactly as

$$\mathbf{vec}(\tilde{\varepsilon}^{qe}) = B^{qe} \tilde{\boldsymbol{u}}, \tag{8.71}$$

[34]The reader should not confuse the superscripts q and e of $\tilde{\varepsilon}^{qe}$ with the suffixes for representing components of a matrix; the index of a component is written as a subscript. So, for each $q = 1, \dots, n^{\mathrm{G}}$ and for each $e = 1, \dots, n^{\mathrm{E}}$, we have a 3×3 matrix $\tilde{\varepsilon}^{qe}$.

where $B^{qe} \in \mathbb{R}^{6\times d}$ is a constant matrix. By using this matrix $\tilde{\varepsilon}^{qe}$, and from (8.70), each term in the summation of the right-hand side of (8.68) is reduced to

$$\hat{w}_{\mathrm{ref}}(\boldsymbol{\varepsilon}(\boldsymbol{\xi}_q)) \det \boldsymbol{J}^e(\boldsymbol{\xi}_q)\rho_q = \frac{\tilde{\rho}^{qe}}{2} \mathbf{vec}(\tilde{\varepsilon}^{qe})^{\mathrm{T}} C \, \mathbf{vec}(\tilde{\varepsilon}^{qe}), \tag{8.72}$$

where we write

$$\tilde{\rho}^{qe} = \rho_q \det \boldsymbol{J}^e(\boldsymbol{\xi}_q)$$

for simplicity. Consequently, from (8.67), (8.68), and (8.72), the strain energy in (8.66) is approximated by

$$\int_{\Omega} \hat{w}_{\mathrm{ref}}(\boldsymbol{\varepsilon}) \mathrm{d}\Omega \simeq \sum_{e=1}^{n^{\mathrm{E}}} \sum_{q=1}^{n^{\mathrm{G}}} \frac{\tilde{\rho}^{qe}}{2} \mathbf{vec}(\tilde{\varepsilon}^{qe})^{\mathrm{T}} C \, \mathbf{vec}(\tilde{\varepsilon}^{qe}), \tag{8.73}$$

where $\tilde{\varepsilon}^{qe}$ should satisfy the compatibility relation in (8.71).

Our next concern is how to deal with the positive-semidefinite constraints of $(\overline{\mathrm{PE}})$ in (8.32). We represent the value $\boldsymbol{\varepsilon}_{\mathrm{e}}$ at the point $\boldsymbol{x}(\boldsymbol{\xi}_q)$ in Ω^e by a matrix $\tilde{\varepsilon}_{\mathrm{e}}^{qe} \in \mathbb{S}^3$. The constraint $\boldsymbol{\varepsilon}_{\mathrm{e}} \preceq \frac{1}{2}(\nabla \boldsymbol{u}^{\mathrm{T}} + \nabla \boldsymbol{u})$ is thus imposed at each Gauss evaluation point as

$$\tilde{\varepsilon}_{\mathrm{e}}^{qe} \succeq \boldsymbol{\varepsilon}^e(\boldsymbol{x}(\boldsymbol{\xi}_q)), \quad \forall q \ (\text{in } \Omega^e), \tag{8.74}$$

where $\boldsymbol{\varepsilon}^e$ is the approximated strain tensor introduced in (8.64). By using (8.71), the constraints in (8.74) are reduced to

$$\tilde{\varepsilon}_{\mathrm{e}}^{qe} \succeq \mathrm{Mat}(B^{qe}\tilde{\boldsymbol{u}}), \quad \forall q; \ \forall e, \tag{8.75}$$

where $\mathrm{Mat}(\cdot)$ is the inverse operation of $\mathbf{vec}(\cdot)$ in (8.69); i.e., applying $\mathrm{Mat}(\cdot)$ to the strain vector in the Voigt notation yields the strain tensor in a matrix form.

Consequently, the results in (8.65), (8.73), and (8.75) yield for $(\overline{\mathrm{PE}})$ in (8.32) the discretization

$$\left.\begin{array}{lll} (\overline{\mathrm{PE}}^{\mathrm{FE}}): & \displaystyle\min_{\tilde{u}_j, \tilde{\varepsilon}_{\mathrm{e}}^{qe}} & \displaystyle\sum_{e=1}^{n^{\mathrm{E}}} \sum_{q=1}^{n^{\mathrm{G}}} \frac{\tilde{\rho}^{qe}}{2} \mathbf{vec}(\tilde{\varepsilon}_{\mathrm{e}}^{qe})^{\mathrm{T}} C \, \mathbf{vec}(\tilde{\varepsilon}_{\mathrm{e}}^{qe}) - \sum_{j \in \tilde{\Gamma}_{\mathrm{N}}} \tilde{\underline{p}}_j \tilde{u}_j \\ & \mathrm{s.t.} & -\tilde{\varepsilon}_{\mathrm{e}}^{qe} + \mathrm{Mat}(B^{qe}\tilde{\boldsymbol{u}}) \succeq O, \quad \forall q; \ \forall e, \\ & & \tilde{u}_j = 0, \quad \forall j \in \tilde{\Gamma}_{\mathrm{D}}. \end{array}\right\} \tag{8.76}$$

Here, the vector $\tilde{\boldsymbol{u}} \in \mathbb{R}^d$ and the symmetric matrices $\tilde{\varepsilon}_{\mathrm{e}}^{qe} \in \mathbb{S}^3$ $(\forall q; \ \forall e)$ are considered as the independent variables.

Remark 8.4.1. It should be claimed that $(\overline{\mathrm{PE}}^{\mathrm{FE}})$ in (8.76) is a convex optimization problem as follows. Since C is the matrix form of the elasticity tensor, it is certainly positive definite, and thence the objective function is a convex quadratic function in terms of $\tilde{\varepsilon}_{\mathrm{e}}^{qe}$ and $\tilde{u}_j$. With regard to the constraints, we easily see that $\mathrm{Mat}(B^{qe}\tilde{\boldsymbol{u}})$ is a (matrix-valued) linear function in terms of $\tilde{\boldsymbol{u}}$, because B^{qe} is a constant matrix and $\mathrm{Mat}(\cdot)$ is a linear mapping from $\mathbb{R}^6$ onto $\mathbb{S}^3$. Hence, the condition $-\tilde{\varepsilon}_{\mathrm{e}}^{qe} + \mathrm{Mat}(B^{qe}\tilde{\boldsymbol{u}}) \succeq O$ is a linear matrix inequality, which is a convex constraint (see section 1.5). Thus ($\overline{\mathrm{PE}}^{\mathrm{FE}}$) is the minimization problem of a convex quadratic function over a linear matrix inequality and can be solved efficiently by using the primal-dual interior-point method. In Remark 8.4.2, we shall further reformulate ($\overline{\mathrm{PE}}^{\mathrm{FE}}$) as a standard form of the conic optimization. ■

Remark 8.4.2. We show an explicit reduction of ($\overline{\mathrm{PE}}^{\mathrm{FE}}$) to a standard form of conic optimization (see section 1.6 for the definition of the standard form of conic optimization). In particular, ($\overline{\mathrm{PE}}^{\mathrm{FE}}$) is embedded into the dual standard form, say, problem (1.27).

The objective function of ($\overline{\mathrm{PE}}^{\mathrm{FE}}$) should be transformed to a linear function. Concerning the quadratic term, we aim at minimizing

$$\frac{1}{2}\,\mathbf{vec}(\tilde{\varepsilon}_{\mathrm{e}}^{qe})^{\mathrm{T}} C\, \mathbf{vec}(\tilde{\varepsilon}_{\mathrm{e}}^{qe}).$$

This minimization is equivalently rewritten as minimizing an upper bound for the quadratic term, i.e., the minimization of an auxiliary variable $\tilde{w}^{qe}$ under the constraint

$$\tilde{w}^{qe} \geq \frac{1}{2}\,\mathbf{vec}(\tilde{\varepsilon}_{\mathrm{e}}^{qe})^{\mathrm{T}} C\, \mathbf{vec}(\tilde{\varepsilon}_{\mathrm{e}}^{qe}). \tag{8.77}$$

Furthermore, since $C \in \mathbb{S}^6$ is positive definite, there exists a matrix $G \in \mathbb{R}^{6\times 6}$ satisfying $C = G^{\mathrm{T}}G$ (see Fact 1.3.7). For example, we can choose G as the square root of C (see Definition 1.3.4). Then the convex quadratic inequality (8.77) can be expressed as a second-order cone constraint as (see Fact 1.5.2 (b))

$$(\tilde{w}^{qe}/2) + 1 \geq \left\| \begin{bmatrix} (\tilde{w}^{qe}/2) - 1 \\ G\,\mathbf{vec}(\tilde{\varepsilon}_{\mathrm{e}}^{qe}) \end{bmatrix} \right\|. \tag{8.78}$$

This inequality means that the eight-dimensional vector

$$\begin{bmatrix} (\tilde{w}^{qe}/2) + 1 \\ (\tilde{w}^{qe}/2) - 1 \\ G\,\mathbf{vec}(\tilde{\varepsilon}_{\mathrm{e}}^{qe}) \end{bmatrix}$$

is included in the second-order cone (see Definition 1.4.1).

As a consequence, ($\overline{\mathrm{PE}}^{\mathrm{FE}}$) in (8.76) can be rewritten as

$$\left.\begin{aligned} (\overline{\mathrm{PE}}^{\mathrm{FE}}_{\mathrm{cone}}):\ & \min_{\tilde{u}_j, \tilde{w}^{qe}, \tilde{\varepsilon}_{\mathrm{e}}^{qe}} && \sum_{e=1}^{n^{\mathrm{E}}}\sum_{q=1}^{n^{\mathrm{G}}} \tilde{\rho}^{qe}\tilde{w}^{qe} - \sum_{j\in\tilde{\Gamma}_{\mathrm{N}}} \tilde{p}\tilde{u}_j \\ & \mathrm{s.\,t.} && \tilde{u}_j \in \mathbb{R}_+,\ -\tilde{u}_j \in \mathbb{R}_+, \quad \forall j \in \tilde{\Gamma}_{\mathrm{D}}, \\ & && \begin{bmatrix} (\tilde{w}^{qe}/2) + 1 \\ (\tilde{w}^{qe}/2) - 1 \\ G\,\mathbf{vec}(\tilde{\varepsilon}_{\mathrm{e}}^{qe}) \end{bmatrix} \in \mathbb{L}^8, \quad \forall q;\ \forall e, \\ & && -\tilde{\varepsilon}_{\mathrm{e}}^{qe} + \mathrm{Mat}(B^{qe}\tilde{\boldsymbol{u}}) \in \mathbb{S}^3_+, \quad \forall q;\ \forall e. \end{aligned}\right\} \tag{8.79}$$

This problem is in the form of the standard conic optimization (1.27), where the corresponding convex cone is given by $\mathcal{C} = \mathcal{C}^* = \mathbb{R}_+^{2|\tilde{\Gamma}_{\mathrm{D}}|} \times (\mathbb{L}^8)^{n^{\mathrm{E}}n^{\mathrm{G}}} \times (\mathbb{S}^3_+)^{n^{\mathrm{E}}n^{\mathrm{G}}}$. ■

Remark 8.4.3. The problem ($\overline{\mathrm{PE}}^{\mathrm{FE}}_{\mathrm{cone}}$) in (8.79) is an optimization with linear inequalities, second-order cone inequalities, and linear matrix inequalities. Since a second-order cone constraint can be expressed as a positive-semidefinite constraint (see Fact 1.5.1), ($\overline{\mathrm{PE}}^{\mathrm{FE}}_{\mathrm{cone}}$) can further be reduced to an SDP problem; showing an explicit reformulation is omitted. In this sense, each of ($\overline{\mathrm{PE}}^{\mathrm{FE}}_{\mathrm{cone}}$) and ($\overline{\mathrm{PE}}^{\mathrm{FE}}$) is solved as an SDP problem.

For general SDP problems, many polynomial time algorithms have been proposed; the primal-dual interior-point method is an outstanding one. Thus ($\overline{\mathrm{PE}}^{\mathrm{FE}}_{\mathrm{cone}}$) can be solved within polynomial time in terms of n^{E} and d, i.e., the number of finite elements and the number of degrees of freedom of displacements.[35] Although the constitutive law of no-tension material possesses nonsmoothness properties, solution process of ($\overline{\mathrm{PE}}^{\mathrm{FE}}_{\mathrm{cone}}$) based on the primal-dual interior-point method does not involve any iterative procedure for determining the *correct* crack states. Therefore, the computational effort required to solve ($\overline{\mathrm{PE}}^{\mathrm{FE}}_{\mathrm{cone}}$) depends not on the problem setting but only on the size of the problem, and the convergence to the solution is theoretically guaranteed independently of the existence/absence of cracks. This is regarded as the great advantage, from the viewpoint of numerical solution, of the conic optimization formulation presented throughout the book. ■

8.4.2 Examples

The equilibrium configurations of masonry structures are computed based on the convex optimization formulations presented. All the simulations are performed solving ($\overline{\mathrm{PE}}^{\mathrm{FE}}_{\mathrm{cone}}$) in (8.79) by using the primal-dual interior-point method. Among several software packages that implement the interior-point method (e.g., SDPA [459], SDPT3 [435], CSDP [57]), we use SeDuMi Ver. 1.05 [424] on MATLAB® Ver. 7.9 [305] to solve ($\overline{\mathrm{PE}}^{\mathrm{FE}}_{\mathrm{cone}}$).

In all the examples the masonry structures are supposed to be subjected to plane-stress conditions and are discretized by linear triangular elements. Young's modulus and Poisson's ratio are set equal to 1 GPa and 0.2, respectively. The thickness of the structures is 10 mm.

8.4.2.1 Square wall under uniformly distributed loads

The first example is a square masonry wall in the literature [141, 291], which is subjected to a vertical dead load and a horizontal live load. The boundary conditions are shown in Figure 8.4, where $L = 1.0\,\mathrm{m}$. A uniformly distributed vertical load of $q_y = 1.8 \times 10^3\,\mathrm{kN/m^2}$ is applied to the top edge. The left edge is subjected to a horizontal load of $q_x = \alpha \bar{q}_x$, where α is a loading parameter, and $\bar{q}_x = 1.0 \times 10^3\,\mathrm{kN/m^2}$. The wall is discretized by $2 \times (10 \times 10) = 200$ triangular elements.

Figure 8.5 depicts the variation of the horizontal displacement, denoted δ, at the top-right corner with respect to the loading parameter α in the range of

[35] The number of integration point, n^{G}, can normally be considered as a constant.

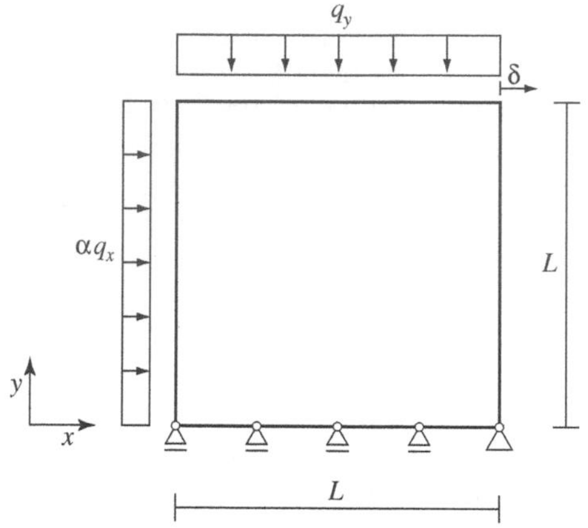

FIGURE 8.4: Geometry and boundary conditions for the example of a square masonry wall.

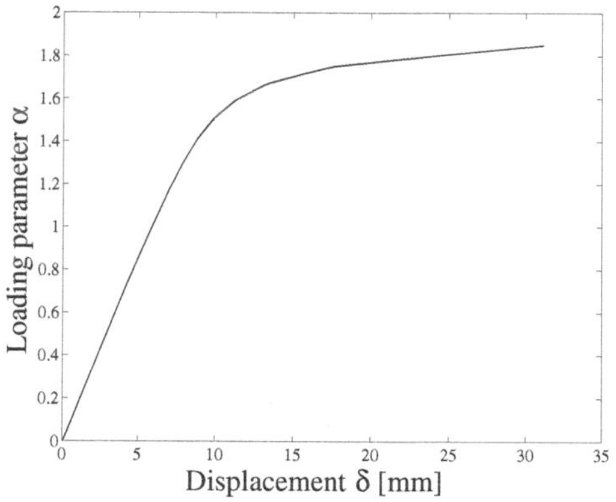

FIGURE 8.5: Variation of the horizontal displacement δ of the square wall with respect to the loading parameter α.

$[0, 1.85]$. In Figure 8.6 we illustrate the solution obtained for $\alpha = 1.85$, which corresponds to the equilibrium state at the incipient collapse; Figures 8.6(a) and 8.6(b) show the contours of the maximal and minimal principal stresses; Figure 8.6(c) shows the contours of the maximal principal value of the crack strain ε_c, together with the deformed configuration of the wall, in which we can observed that cracks open along the bottom edge. Note that the minimal principal value of ε_c is equal to zero everywhere, which means that, at each point where crack occurs, crack opens only in the direction of the maximum principal stress.

Remark 8.4.4. Figure 8.5 may strike the reader as being a result obtained by per-

(a) First principal stress.

(b) Second principal stress.

(c) First principal crack strain and deformation.

FIGURE 8.6: Computational results of the square wall example at $\alpha = 1.85$. (a) maximal principal stress; (b) minimal principal stress; (c) maximal principal value of ε_{c} and the deformed configuration.

forming the incremental analysis, but that is not the case. Since the inelastic strain is assumed to be non-dissipative, the equilibrium state is path independent. Moreover, the primal-dual interior-point method does not require the initial solution, which is near the equilibrium state. Thus, one SDP problem is solved for each loading condition independently. ■

Remark 8.4.5. In [141, 291], besides the loads considered in Figure 8.4, a relatively small distributed compressive loads are supposed to be applied to the both vertical edges to make $\alpha = 0$. As explored in Remark 8.3.10, there exist infinitely many displacement fields in the equilibrium with $\alpha = 0$. Such indeterminacy of the displacements causes the difficulty in numerical computation, especially for a method based on the tangent stiffness. The additional loads on the vertical edges were considered to make the tangent stiffness matrix at $\alpha = 0$ positive definite. In

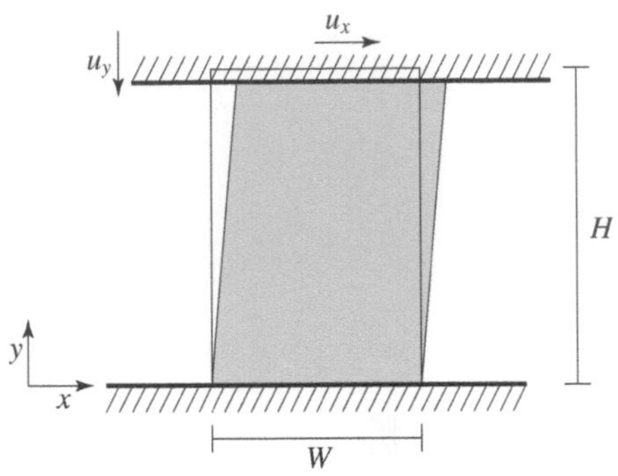

FIGURE 8.7: Geometry and boundary conditions of the shear panel example.

contrast, the presented method based on the SDP formulation is not affected by the indeterminacy of the displacements; by solving ($\overline{\mathrm{PE}}^{\mathrm{FE}}_{\mathrm{cone}}$) using the primal-dual interior-point method we obtain one of the solutions without any difficulty. ■

8.4.2.2 Shear panel

We next consider the shear panel represented in Figure 8.7. Similar examples can found in the literature, e.g., [49, 297]. The rectangular panel, with the dimensions of $W = 1.0\,\mathrm{m}$ and $H = 2.0\,\mathrm{m}$, is uniformly discretized by $2 \times (25 \times 250) = 12{,}500$ elements.

Figure 8.8 depicts obtained equilibrium solutions for $u_y = 2.0\,\mathrm{mm}$ and $u_x = 0.0$, 2.0, and 10.0 mm, where we show the contours of the two principal stresses as well as the maximal principal value of the crack strain ε_{c}. It is found that the minimal principal value of ε_{c} vanishes almost everywhere. It is observed from Figure 8.8 that the complementarity condition between the minimal principal stress and the maximal crack strain is satisfied at each point. The crack strain shown in Figure 8.8(c) suggests the localization of the damage zone to a narrow band.

8.4.2.3 Clamped beam in bending

A masonry beam is simulated by the clamped rectangular body illustrated in Figure 8.10, where $W = 5.0\,\mathrm{m}$, $H = 1.0\,\mathrm{m}$. The body is discretized by $2 \times (100 \times 20) = 4000$ elements. The body is first prestressed in the negative direction of the x-axis by a displacement u_x. Then this displacement is held fixed for the subsequent application of the uniformly distributed vertical load of αq_0 at to the top edge, where $q_0 = 1.0 \times 10^4\,\mathrm{kN/m^2}$.

We investigate the variation of the vertical displacement, denoted δ, at the middle point of the top edge. In Figure 8.10 we show the variations of δ with respect to the loading parameter α for three cases, i.e., $u_x = 1.0\,\mathrm{mm}$,

First principal stress Second principal stress First principal crack strain

(a) Solution for $(u_x, u_y) = (0, 2.0)$ mm.

(b) Solution for $(u_x, u_y) = (2.0, 2.0)$ mm.

(c) Solution for $(u_x, u_y) = (10.0, 2.0)$ mm.

FIGURE 8.8: Computational results of the shear panel example.

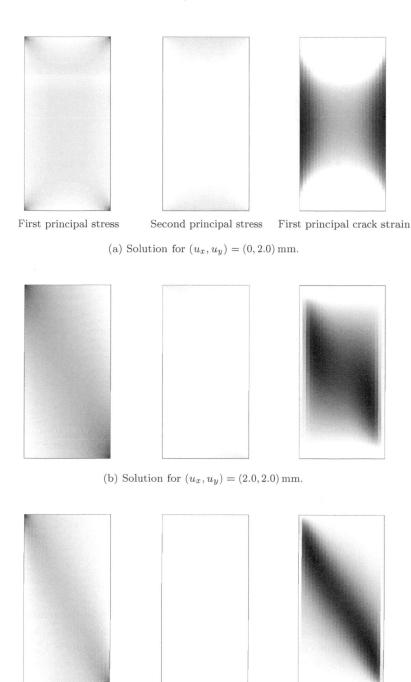

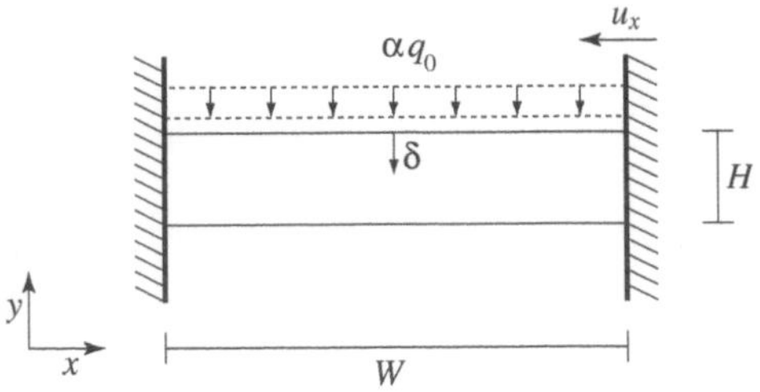

FIGURE 8.9: Geometry and boundary conditions of the clamped beam example.

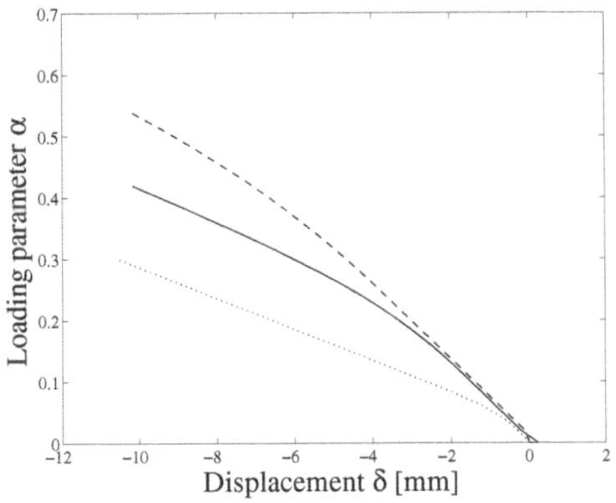

FIGURE 8.10: Variation of the vertical displacement δ of the clamped beam with respect to the loading parameter α; '······': $u_x = 1.0\,\mathrm{mm}$, '——': $u_x = 5.0\,\mathrm{mm}$, '- - -': $u_x = 10.0\,\mathrm{mm}$.

5.0 mm, and 10.0 mm. It is observed that the stiffness of the masonry body is increased by introducing a larger prestress. For the case of $u_x = 5.0\,\mathrm{mm}$ the development of cracks is shown in Figure 8.11 and Figure 8.12. Note that, at $u_x = 1.0\,\mathrm{mm}$ in Figure 8.11, the minimal principal value of ε_{c} vanishes everywhere. It can be observed in Figure 8.12 that cracks open in both of two principal directions along the bottom edge. Moreover, an arch-like system appears, as seen in the contours of the maximal principal stress to transmit the vertical external load to the supports.

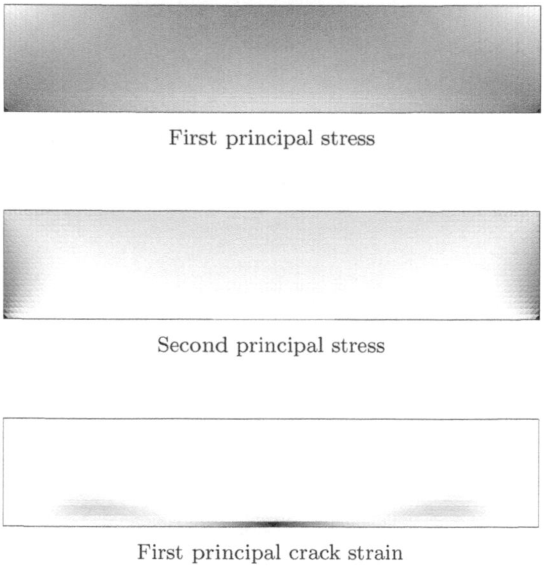

FIGURE 8.11: Computational results of the clamped beam example. Solution for $(u_x, \alpha) = (5.0\,\text{mm}, 0.1)$.

8.5 Notes

1. Although the concept of no-tension materials dates back to the 18th century (see Di Pasquale [118]), the study of masonry structures using the no-tension model was initiated in the 1960s by Heyman [176, 177], who performed the limit analysis of masonry structures. This pioneering work was followed by in-depth discussions on the constitutive law, early ones of which were mostly done in Italy; see, e.g., Panzeca–Polizzotto [358] and the references therein. Some formulations of the constitutive law by splitting the strain into elastic and inelastic parts as well as necessary conditions for the equilibrium under several particular loading conditions were investigated by Del Piero [106] and Di Pasquale [118]; Giaquinta–Giusti [145] investigated the plane-stress case in particular.

For a continuum consisting of no-tension material, the existence and convergence of the solution to the equilibrium problem have recently been revisited by Lucchesi–Šilhavý–Zani [280] and Padovani [350]. The applicability of the fundamental theorems of the limit analysis was studied by Braides–Chiadò Piat [63] and Del Piero [107].

As refinements of the crude no-tension model, the rate-dependent form

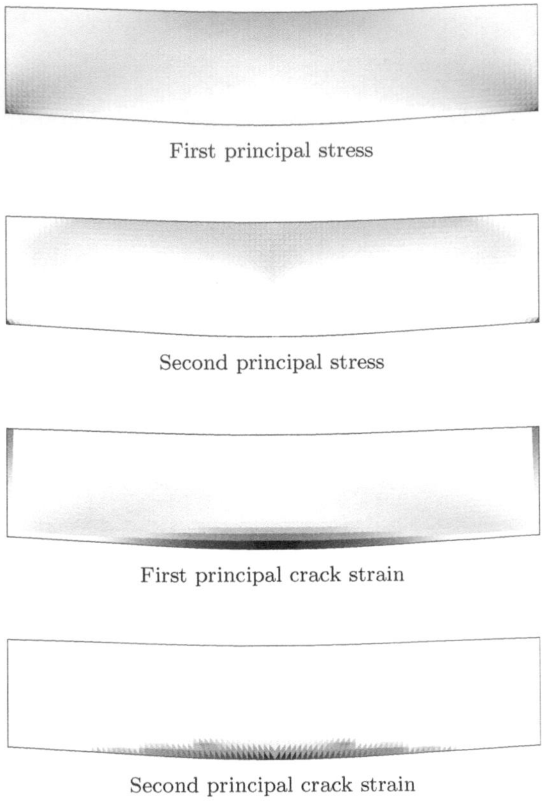

FIGURE 8.12: Computational results of the clamped beam example (continued). Solution for $(u_x, \alpha) = (5.0\,\mathrm{mm}, 1.0)$.

was introduced by Angelillo [16], and the thermal dilatation was considered by Padovani [349]. Extensions to orthotropic damage models, in the plane-stress case, are in Berto–Saetta–Scotta–Vitaliani [49] and Lourenço–Rots–Blaauwendraad [273]. Homogenization techniques for the no-tension model and more sophisticated constitutive models have been discussed by many authors. See Braides–Chiadò Piat [63], Cecchi–Di Marco [72], Kawa–Pietruszczak–Shieh-Beygi [219], Luciano–Sacco [281], Marfia–Sacco [296], and Sacco [397]. An interesting extension of the no-tension solid to the model of granular materials can be found in Nguyen–Duhamel–Nedjar [337].

2. Concerning numerical methods for analysis of masonry structures, a Newton–Raphson method based on the tangent stiffness was proposed by Lucchesi–Padovani–Pagni [276] for the no-tension model. Subsequently, this method was extended to constitutive laws considering bounded tensile strength by Lucchesi–Padovani–Pasquinelli [277] and Lucchesi–Padovani–Zani [279]. Alfano–

Rosati–Valoroso [7] proposed to enhance the Newton–Raphson method for the no-tension model by incorporating the line search and the tangent-secant approach. Numerical analyses of no-tension solids subjected to thermal loads and thermodynamics were performed by Padovani–Pasquinelli–Zani [351] and Lucchesi–Padovani–Pasquinelli [278], respectively.

3. The complementarity condition between the stress tensor and crack strain tensor can also be captured as the normality rule in the plasticity theory, in which the plastic strain rate is postulated to be normal to the boundary of admissible set of the stress tensor. For plasticity the admissible set of stress is defined by the yield function, while for the no-tension solid the stress should belong to the set of positive-semidefinite matrices. Noting this similarity, numerical algorithms for elastoplasticity have been extended to no-tension solids. One of the earliest contributions is by Maier–Nappi [291], who approximated the admissible set of stress, i.e., the positive-semidefinite cone, by a polyhedral cone. Thus the constraints on the stress tensor are approximated by some linear inequalities, and accordingly the minimization problem of the complementary energy results in the quadratic programming (QP) problem.[36]

By virtue of the analogy of the plasticity, the return-mapping method was also extended to the numerical analysis of no-tension solids. See Fuschi–Giambanco–Rizzo [141], Marfia–Sacco [297], and Sacco [397].

4. For the no-tension model, the convexity of the strain energy $\hat{w}_{\rm mas}$, established in Proposition 8.2.6, is not a new result: A proof for the plane-stress case was given by Giaquinta–Giusti [145] showing that the Hessian of $\hat{w}_{\rm mas}$ is positive semidefinite, and a more general case is given in Anzellotti [19]. In section 8.2.3 we presented a simple proof using Proposition 8.2.5.

5. The minimization problem of the complementary energy, say, (CE) in section 8.3.4, can also be found in the literature [100, 145]; for example, Cuomo–Ventura [100] derived the same formulation based on the Hellinger–Reissner principle. We presented an alternative derivation using the modern optimization theory.

The discussion made in Remark 8.3.10 concerning the absence of uniqueness for the kinematic variables in the equilibrium state, despite the uniqueness of the stress, can be found in [19, 100, 145]. Although the nonuniqueness in the displacement and strain fields may cause the difficulty in the numerical computation for existing methods, the presented method can successfully find a solution, as explained in Remark 8.4.5.

Since the boundary of the feasible set of (CE) is nonsmooth, some numerical methods have been proposed by regularizing this nonsmoothness property: Maier–Nappi [291] performed a linear approximation of the positive-

[36] More precisely, by QP we mean that a minimization problem of a convex quadratic function under some linear inequality constraints; see (1.31) for the definition.

semidefinite cone, and Cuomo–Ventura [100] presented a regularization based on the augmented Lagrangian method. From the viewpoint of the conic optimization, however, (CE) (and also ($\overline{\mathrm{PE}}$)) can be solved directly with the primal-dual interior-point method without any difficulty. To the author's knowledge, no convex optimization reformulation of the minimization problem of the total potential energy, say, ($\overline{\mathrm{PE}}$) in section 8.2.3, is known; our exposition is based on [205].

Chapter 9

Planar Membranes

Orientation In this chapter we deal with static analyses of membrane structures consisting of the structural elements, which possess negligibly small compressive and flexural stiffnesses. The investigation here proceeds with the same format as Chapter 8. We first show that the minimization problem of the potential energy for membranes can be equivalently rewritten as a convex optimization problem subjected to conic constraints. The obtained problem is then embedded into the form

$$(\mathrm{P_F}): \quad \inf_x \{f(x) + g(\Lambda x) \mid x \in \mathbb{V}\}, \qquad \text{(see (2.28))}$$

where $f : \mathbb{V} \to \mathbb{R} \cup \{+\infty\}$ and $g : \mathbb{Y} \to \mathbb{R} \cup \{+\infty\}$ are closed proper convex functions, and $\Lambda : \mathbb{V} \to \mathbb{Y}$ is a continuous linear mapping. We then derive the explicit form of the Fenchel dual problem

$$(\mathrm{P_F^*}): \quad \sup_{z^*} \{-f^*(\Lambda^* z^*) - g^*(-z^*) \mid z^* \in \mathbb{Y}^*\}, \qquad \text{(see (2.31))}$$

where $\mathbb{V}^*$ and $\mathbb{Y}^*$ are dual to $\mathbb{V}$ and $\mathbb{Y}$, respectively. We shall show that $(\mathrm{P_F^*})$ corresponds to the minimization problem of the complementary energy for membranes by applying Proposition 2.2.9 (the strong duality) and Proposition 2.2.5 (the optimality condition). The distinguished result of this chapter is that the primal-dual energy principles for membranes considering the geometrical nonlinearity can be embedded to the convex optimization problems over the positive-semidefinite cones. For the notation concerning tensor algebra, see section 8.1.1.

9.1 Introduction

Thin membranes have very small and, in general, negligible flexural stiffness and can be idealized as no-compression materials. Compressive stress are handled via changes in membrane geometry called *wrinkling*. Analysis of wrinkling is important predicting the structural response of membranes. Applications of thin structures that can be modeled as membranes vary widely (see, e.g., a survey by Jenkins [194]), including the structures in the fields of civil engineering and architecture (e.g., prestressed membrane roof of a dome structure [15, 232, 398, 450]), aeronautics (e.g., light aircraft [27]), mechanical engineering (e.g., diaphragm valves or cloth draping [82]), and biomechanics (e.g., skin [102, 302] or blood vessels [185]).

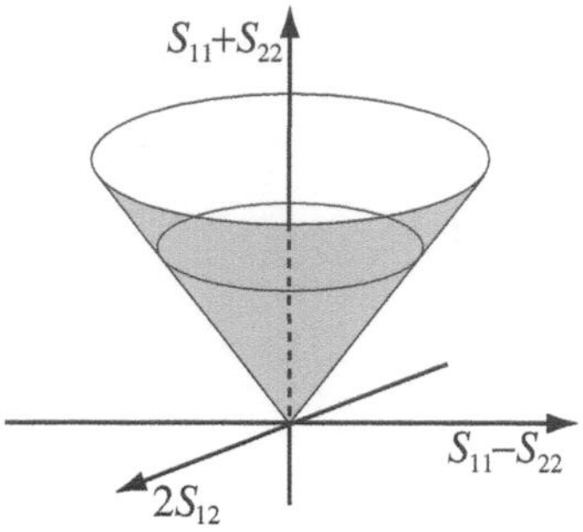

FIGURE 9.1: The set of positive-semidefinite symmetric stress tensor $\boldsymbol{S} = (S_{ij}) \in \mathbb{S}^2$.

Since the compressive stiffness is assumed to be zero while the tensile stiffness has a positive value, membranes apparently possess nonsmoothness property in the constitutive law (stress–strain relation). Accordingly, the response of membranes is distinguished as either *taut*, *wrinkling*, or *slack*. Furthermore, the nonsmoothness property also appears at the boundary of the admissible set of the stress tensor (and also that of the wrinkle strain tensor as seen below), which can be recognized as follows.

The absence of the resistance to compressive stresses means that every principal stress cannot be negative. Let $\boldsymbol{S} = (S_{ij}) \in \mathbb{S}^2$ denote the stress tensor, which is the 2×2 symmetric tensor representing the stress in a membrane. We denote by $\lambda_i(\boldsymbol{S})$ $(i = 1, 2)$ the principal stresses, i.e., the eigenvalues of $\boldsymbol{S}$. Then $\lambda_i(\boldsymbol{S})$'s are nonnegative if and only if $\boldsymbol{S} \succeq \boldsymbol{o}$. On the other hand, this condition is also equivalent to

$$\lambda_1(\boldsymbol{S}) + \lambda_2(\boldsymbol{S}) \geq 0, \quad \lambda_1(\boldsymbol{S})\lambda_2(\boldsymbol{S}) \geq 0,$$

which in turn yields for the elements of $\boldsymbol{S}$ the condition

$$S_{11} \geq 0, \quad S_{22} \geq 0, \quad S_{11}S_{22} - S_{12}^2 \geq 0, \tag{9.1}$$

because $\det \boldsymbol{S} = \lambda_1(\boldsymbol{S})\lambda_2(\boldsymbol{S})$. It follows from Fact 1.5.3 that (9.1) is equivalently rewritten as

$$S_{11} + S_{22} \geq \left\| \begin{bmatrix} S_{11} - S_{22} \\ 2S_{12} \end{bmatrix} \right\|,$$

which is a second-order cone constraint. Thus the admissible set of $\boldsymbol{S}$, say, $\{\boldsymbol{S} \in \mathbb{S}^2 \mid \boldsymbol{S} \succeq \boldsymbol{o}\}$, is as illustrated in Figure 9.1, and its boundary possesses nonsmoothness; more illustrations of admissible sets can be found in Epstein–Forcinito [124, Figures 1–5].

This chapter is concerned with the static analysis of planar membranes. After a brief discussion concerning the infinitesimal theory of planar membranes in section 9.2, we investigate in section 9.3 through section 9.5 the geometrically nonlinear theory. In section 9.3 we see that the minimum principle of the potential energy can be reduced to a form of conic optimization, and the duality of energy principles is the subject of section 9.4. A solution method for the equilibrium analysis is presented in section 9.5 with some numerical examples.

We follow the notation introduced in section 8.1.1 for tensor algebra. Particularly, since we restrict ourselves to membranes in a two-dimensional plane, we can ignore the distinction between the covariant and contravariant components of a tensor; hence, for example, we write $\boldsymbol{\alpha} \in \mathbb{R}^{2\times 2}$ if $\boldsymbol{\alpha}$ is a tensor of order two. For two tensors $\boldsymbol{\alpha}$, $\boldsymbol{\beta} \in \mathbb{R}^{2\times 2}$, we represent their scalar product by $\boldsymbol{\alpha} : \boldsymbol{\beta} = \boldsymbol{\alpha} \bullet \boldsymbol{\beta} = \sum_{i=1}^{2}\sum_{j=1}^{2} \alpha_{ij}\beta_{ij}$. Similarly, the scalar product of two vectors $\boldsymbol{p}$, $\boldsymbol{q} \in \mathbb{R}^2$ is denoted by $\boldsymbol{p} \cdot \boldsymbol{q} = \boldsymbol{p}^{\mathrm{T}}\boldsymbol{q} = \sum_{i=1}^{2} p_i q_i$.

9.2 Analysis in Small Deformation

This section briefly discusses the variational principles for membranes in the infinitesimal theory. The principles of the potential and complementary energies are investigated in a way similar to section 8.2 and section 8.3, respectively. The result in this section is to be compared with the study of the geometrically nonlinear case developed in section 9.3.

The reader who is not particularly interested in the membrane theory without considering the geometrical nonlinearity can pass over this section.

9.2.1 Principle of potential energy in small deformation

Consider an elastic body placed in two-dimensional space, consisting of an isotropic membrane material and occupying the domain $\overline{\Omega} := \mathrm{cl}\,\Omega$, where $\Omega \subset \mathbb{R}^2$ is a specified bounded and connected open set. As in Chapter 8, we denote by $\boldsymbol{u}(\boldsymbol{x}) \in \mathbb{R}^2$ the displacement vector at the point $\boldsymbol{x} \in \overline{\Omega}$. The compatibility relation between $\boldsymbol{u}$ and the linear strain tensor $\boldsymbol{\varepsilon} \in \mathbb{S}^2$ is given by (8.1), i.e.,

$$\boldsymbol{\varepsilon} = \hat{\boldsymbol{\varepsilon}}(\boldsymbol{u}) := \frac{1}{2}(\nabla \boldsymbol{u}^{\mathrm{T}} + \nabla \boldsymbol{u}). \tag{9.2}$$

We denote by $\boldsymbol{\sigma} \in \mathbb{S}^2$ the stress tensor corresponding to $\boldsymbol{\varepsilon}$.

We next describe the constitutive relation of the membrane. Roughly speaking, we assume that the membrane behaves like a conventional thin elastic body under tensile normal stresses and suddenly loosens when we try to apply a compressive normal stress. The membrane is said to be in the *taut state* when it is subjected to tensile normal stresses and behaves totally as an elastic body. Otherwise, the membrane is either in the *wrinkled state* or in the *slack state*: in the wrinkled state the membrane is subjected to the uniaxial tensile stress, and no stress is sustained in the slack state. More precisely, we adopt the following physical hypotheses, which are often postulated in the analysis of membranes:

(i) The flexural stiffness is negligible.

(ii) Any compressive stress cannot be sustained.

(iii) The strain $\boldsymbol{\varepsilon}$ can be additively decomposed to the two parts: the elastic and inelastic strains.

(iv) The inelastic strain corresponds to the reversible slack, which can take place perpendicular to the plane of nonpositive principal stress.

(v) The tensile stress is caused elastically solely by the elastic stress.

These hypotheses are restated formally as follows. Hypothesis (i) means that the internal force of the membrane is in the plane-stress state. It is physically observed that a membrane forms out-of-plane waves when it slacks, but the bending moment caused by such a out-of-plane deformation is neglected. Hypothesis (ii) is rewritten in terms of the eigenvalues $\lambda_i(\boldsymbol{\sigma})$ of $\boldsymbol{\sigma}$ as[1]

$$\lambda_i(\boldsymbol{\sigma}) \geq 0, \quad i = 1, 2, \tag{9.3}$$

where we set $\lambda_1(\boldsymbol{\sigma}) \geq \lambda_2(\boldsymbol{\sigma})$. According to hypothesis (iii), we decompose the total strain $\boldsymbol{\varepsilon}$ as[2]

$$\boldsymbol{\varepsilon} = \boldsymbol{\varepsilon}_{\mathrm{e}} + \boldsymbol{\varepsilon}_{\mathrm{w}}. \tag{9.4}$$

Here, $\boldsymbol{\varepsilon}_{\mathrm{e}}$ is the elastic strain, and $\boldsymbol{\varepsilon}_{\mathrm{w}}$ is the inelastic strain, also called the *wrinkling strain* because of its physical meaning given by hypothesis (iv). Hypothesis (iv) states some properties simultaneously. The wrinkling behavior is assumed to be reversible, which means that the constitutive law does not change even if the membrane undergoes repetition of releasing and reintroducing the stresses. The wrinkling strain $\boldsymbol{\varepsilon}_{\mathrm{w}}$ is regarded as the measure of "shortening" of the membrane due to slack,[3] and hence we have that

$$\omega_i(\boldsymbol{\varepsilon}_{\mathrm{w}}) \leq 0, \quad i = 1, 2, \tag{9.5}$$

[1] Because of this property, the constitutive law considered here is regarded as a counterpart of the *no-tension material* studied in section 8.2.2, although only a few authors, e.g., Contri–Schrefler [94], use the term *no-compression material* to mean the membrane material.

[2] In section 9.2.2 we shall see that this decomposition is determined uniquely.

[3] This rough sketch of the meaning of $\boldsymbol{\varepsilon}_{\mathrm{w}}$ should be interpreted in the sense of *smeared* way in the context of continuum mechanics.

where $\omega_i(\boldsymbol{\varepsilon}_{\rm w})$ $(\omega_1(\boldsymbol{\varepsilon}_{\rm w}) \geq \omega_2(\boldsymbol{\varepsilon}_{\rm w}))$ are the eigenvalues of $\boldsymbol{\varepsilon}_{\rm w}$. To see the relation between $\boldsymbol{\varepsilon}_{\rm w}$ and $\boldsymbol{\sigma}$, we consider the following three stress states:

$$\omega_1(\boldsymbol{\varepsilon}_{\rm w}) > 0, \quad \omega_2(\boldsymbol{\varepsilon}_{\rm w}) > 0, \tag{9.6a}$$

$$\omega_1(\boldsymbol{\varepsilon}_{\rm w}) > 0, \quad \omega_2(\boldsymbol{\varepsilon}_{\rm w}) = 0, \tag{9.6b}$$

$$\omega_1(\boldsymbol{\varepsilon}_{\rm w}) = 0, \quad \omega_2(\boldsymbol{\varepsilon}_{\rm w}) = 0 \tag{9.6c}$$

in accordance with (9.3). Then hypothesis (iv) requires that $\boldsymbol{\varepsilon}_{\rm w}$ should satisfy[4]

$$\begin{cases} \omega_1(\boldsymbol{\varepsilon}_{\rm w}) = 0, \omega_2(\boldsymbol{\varepsilon}_{\rm w}) = 0 & \text{if (9.6a)} \quad \text{[taut]}, \\ \omega_1(\boldsymbol{\varepsilon}_{\rm w}) = 0, \omega_2(\boldsymbol{\varepsilon}_{\rm w}) \leq 0 & \text{if (9.6b)} \quad \text{[wrinkled]}, \\ \omega_1(\boldsymbol{\varepsilon}_{\rm w}) \leq 0, \omega_2(\boldsymbol{\varepsilon}_{\rm w}) \leq 0 & \text{if (9.6c)} \quad \text{[slack]}. \end{cases} \tag{9.7}$$

Let $\boldsymbol{\phi}_i \in \mathbb{R}^2$ $(i = 1, 2)$ denote the orthonormal eigenvectors of $\boldsymbol{\sigma}$ corresponding to $\lambda_i(\boldsymbol{\sigma})$. In the case of (9.6b), hypothesis (iv) means that a slack can take place in the direction of $\boldsymbol{\phi}_2$. Therefore, in the second case of (9.7), $\omega_2(\boldsymbol{\varepsilon}_{\rm w}) < 0$ can occur in the direction of $\boldsymbol{\phi}_2$, which means that $\boldsymbol{\phi}_2$ is also an eigenvector of $\boldsymbol{\varepsilon}_{\rm w}$. In the case of (9.6a) we have $\boldsymbol{\varepsilon}_{\rm w} = \boldsymbol{o}$, while in the case of (9.6c) we have $\boldsymbol{\sigma} = \boldsymbol{o}$. Thus, in all cases, we have that[5]

$$\boldsymbol{\sigma} \text{ and } \boldsymbol{\varepsilon}_{\rm w} \text{ are coaxial.} \tag{9.8}$$

Moreover, from (9.6) and (9.7), we see that

$$\lambda_i(\boldsymbol{\sigma})\omega_i(\boldsymbol{\varepsilon}_{\rm w}) = 0, \quad i = 1, 2. \tag{9.9}$$

Finally, hypothesis (v), together with the reversibility assumption in (iv), means that there exists a strain energy function, denoted $\hat{w}_{\rm mem} : \mathbb{S}^2 \to \mathbb{R}$, with which the constitutive reflation of the membrane is described as

$$\boldsymbol{\sigma} \in \partial \hat{w}_{\rm mem}(\boldsymbol{\varepsilon}). \tag{9.10}$$

Moreover, this relation is equivalently rewritten as

$$\boldsymbol{\sigma} \in \partial \hat{w}_{\rm ref}(\boldsymbol{\varepsilon}_{\rm e}) \tag{9.11}$$

with some $\hat{w}_{\rm ref} : \mathbb{S}^2 \to \mathbb{R}$, because $\boldsymbol{\sigma}$ depends only on the elastic strain $\boldsymbol{\varepsilon}_{\rm e}$. Thus the constitutive law of a membrane is completely described by (9.3)–(9.11).

[4]Physically a wrinkle occurs when $\omega_1(\boldsymbol{\varepsilon}_{\rm w}) = 0$ and $\omega_2(\boldsymbol{\varepsilon}_{\rm w}) < 0$. Therefore, the second case of (9.7) should be interpreted as "possibly wrinkled" more precisely. Similarly, the pure slack state corresponds to $\omega_1(\boldsymbol{\varepsilon}_{\rm w}) < 0$ and $\omega_2(\boldsymbol{\varepsilon}_{\rm w}) < 0$, and hence the third case of (9.7) should be read as "possibly slack" more precisely. The case in which $\boldsymbol{\sigma} = \boldsymbol{o}$ and $\boldsymbol{\varepsilon}_{\rm w} = \boldsymbol{o}$ is regarded as a "neutral" state.

[5]We mean by (9.8) that $\boldsymbol{\sigma}$ and $\boldsymbol{\varepsilon}_{\rm w}$ share the same system of eigenvectors.

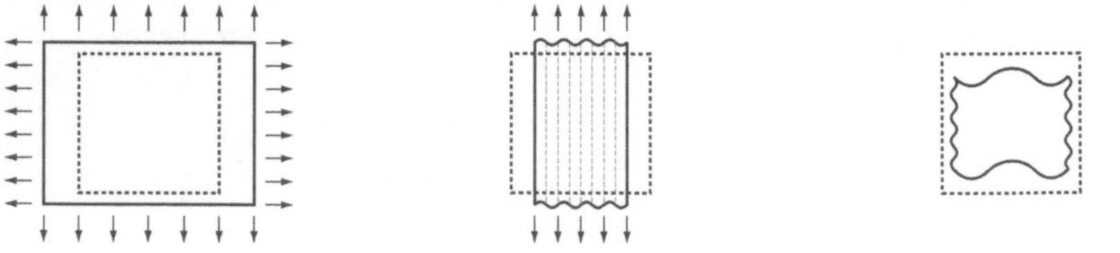

(a) Taut state. (b) Wrinkled state. (c) Slack state.

FIGURE 9.2: Three states of isotropic membranes. '......': undeformed configuration; '——': deformed configuration.

Associated with $\hat{w}_{\rm ref}$ in (9.11), we introduce a fictitious elastic body, whose response is given by

$$\boldsymbol{\sigma} \in \partial \hat{w}_{\rm ref}(\boldsymbol{\varepsilon}). \tag{9.12}$$

This fictitious body is called a *reference elastic body*,[6] like the case of no-tension material investigated in section 8.2.2. For a given strain $\boldsymbol{\varepsilon}$, the response $\boldsymbol{\sigma}$ of the reference elastic body of a membrane is obtained by artificially eliminating slacks. In other words, the strain energy (density) of the reference body is evaluated for a structure consisting of the same material of the membrane but thick enough so that no slack takes place.

In what follows, we assume linear isotropy of the membrane. It is natural that the corresponding reference elastic body is the conventional linear isotropic material, the strain energy function of which is given by

$$\hat{w}_{\rm ref}(\boldsymbol{\varepsilon}) = \frac{1}{2}\lambda(\operatorname{tr}\boldsymbol{\varepsilon})^2 + \mu\boldsymbol{\varepsilon} : \boldsymbol{\varepsilon}, \tag{9.13}$$

where λ and μ are the Lamé moduli. The response function $\hat{\boldsymbol{\sigma}}_{\rm ref}$ is written explicitly as

$$\hat{\boldsymbol{\sigma}}_{\rm ref}(\boldsymbol{\varepsilon}) := \frac{\partial \hat{w}_{\rm ref}}{\partial \boldsymbol{\varepsilon}}(\boldsymbol{\varepsilon}) = (\lambda \operatorname{tr}\boldsymbol{\varepsilon})\boldsymbol{I} + 2\mu\boldsymbol{\varepsilon}. \tag{9.14}$$

Consequently the constitutive law (9.11) of the membrane is written as

$$\boldsymbol{\sigma} = \hat{\boldsymbol{\sigma}}_{\rm ref}(\boldsymbol{\varepsilon}_{\rm e}) := (\lambda \operatorname{tr}\boldsymbol{\varepsilon}_{\rm e})\boldsymbol{I} + 2\mu\boldsymbol{\varepsilon}_{\rm e}. \tag{9.15}$$

The isotropy implies that, as seen in (9.15), $\boldsymbol{\sigma}$ and $\boldsymbol{\varepsilon}_{\rm e}$ are coaxial. From this observation, together with (9.4) and (9.8), we see that all of $\boldsymbol{\sigma}$, $\boldsymbol{\varepsilon}$, $\boldsymbol{\varepsilon}_{\rm e}$, and $\boldsymbol{\varepsilon}_{\rm c}$ are coaxial. Therefore, in an isotropic membrane, if a wrinkle takes

[6] The reference elastic body for a membrane is also called the *fictitious elastic body* or the *original material* by some authors; see, e.g., Jarasjarunngkiat–Wüchner–Bletzinger [191].

place, if any, in the direction of the principal strain.[7] The three states of isotropic membranes, i.e., the taut, wrinkled, and slack states, are illustrated in Figure 9.2.

We thus see that the constitutive relation of a linear isotropic membrane is characterized by (9.3)–(9.9) and (9.15). It should be clear that this relation is abstractly written as (9.10) with $\hat{w}_{\rm mem}$. The minimization problem of the total potential energy is then formulated as

$$\left.\begin{aligned} (\mathrm{PE}) : \min_{\boldsymbol{u},\boldsymbol{\varepsilon}} & \int_{\Omega} (\hat{w}_{\rm mem}(\boldsymbol{\varepsilon}) - \underline{\boldsymbol{p}} \cdot \boldsymbol{u})\, \mathrm{d}\Omega - \int_{\Gamma_{\rm N}} \underline{\boldsymbol{t}} \cdot \boldsymbol{u} \mathrm{d}\Gamma \\ \mathrm{s.\,t.}\ & \boldsymbol{\varepsilon} = \frac{1}{2}(\nabla \boldsymbol{u}^{\rm T} + \nabla \boldsymbol{u}) \quad \text{in } \Omega, \\ & \boldsymbol{u} = \underline{\boldsymbol{u}} \quad \text{on } \Gamma_{\rm D}, \end{aligned}\right\} \tag{9.16}$$

where $\Gamma_{\rm N} \cup \Gamma_D$ is a partition of the boundary $\Gamma := \mathrm{bd}\,\Omega$ of the membrane, $\underline{\boldsymbol{t}} : \Gamma_{\rm N} \to \mathbb{R}^2$ is the specified traction per unit area in the reference configuration, $\underline{\boldsymbol{u}} : \Gamma_{\rm D} \to \mathbb{R}^2$ is prescribed displacement, and $\underline{\boldsymbol{p}} : \Omega \to \mathbb{R}^2$ is the body force per unit volume in the reference configuration.

In a manner similar to section 8.3, the optimality and duality of (PE) in (9.16) can be studied within the framework of Fenchel duality theory. The first step toward our study consists of reformulating (PE) into a conic optimization problem, which is the subject of the next section.

9.2.2 Conic optimization formulation

In the preceding section the constitutive relation of a membrane is given by (9.10), which is concretely written as (9.3)–(9.9) and (9.15). It is easy to see that this constitutive relation is equivalently rewritten as[8]

$$\boldsymbol{\sigma} = (\lambda \operatorname{tr} \boldsymbol{\varepsilon}_{\rm e})\boldsymbol{I} + 2\mu \boldsymbol{\varepsilon}_{\rm e}, \tag{9.17a}$$

$$\boldsymbol{\varepsilon} = \boldsymbol{\varepsilon}_{\rm e} + \boldsymbol{\varepsilon}_{\rm w}, \tag{9.17b}$$

$$-\boldsymbol{\varepsilon}_{\rm w} \succeq \boldsymbol{o}, \quad \boldsymbol{\sigma} \succeq \boldsymbol{o}, \quad (-\boldsymbol{\varepsilon}_{\rm w}) : \boldsymbol{\sigma} = 0. \tag{9.17c}$$

Furthermore, by using $\hat{w}_{\rm ref}$ in (9.13), $\hat{w}_{\rm mem}$ is rewritten as follows.

Proposition 9.2.1. *Suppose that $\boldsymbol{\varepsilon} \in \mathbb{S}^2$ is given, and consider*

$$\left.\begin{aligned} \min_{\boldsymbol{\varepsilon}_{\rm e}}\ & \hat{w}_{\rm ref}(\boldsymbol{\varepsilon}_{\rm e}) \\ \mathrm{s.\,t.}\ & \boldsymbol{\varepsilon}_{\rm e} \succeq \boldsymbol{\varepsilon}, \end{aligned}\right\} \tag{9.18}$$

where $\boldsymbol{\varepsilon}_{\rm e} \in \mathbb{S}^2$ is the variable. Then $\boldsymbol{\varepsilon}_{\rm e}$ is optimal for (9.18) if and only if there exist $\boldsymbol{\sigma}$ and $\boldsymbol{\varepsilon}_{\rm w}$ satisfying (9.17). Moreover, the optimal value of (9.18) is equal to $\hat{w}_{\rm mem}(\boldsymbol{\varepsilon})$.

---

[7] Which is not true for an anisotropic membrane in general.

[8] This assertion can be shown in a manner similar to Proposition 8.2.3.

Proof. Analogous to the proof of Proposition 8.2.5. □

Remark 9.2.2. From Proposition 9.2.1 one can easily show that $\hat{w}_{\rm mem}$ is a convex function. The convexity of $\hat{w}_{\rm mem}$ verifies that the equilibrium state corresponds to the global minimum solution of (PE) in (9.16). ■

Since problem (9.18) considered in Proposition 9.2.1 is a minimization of a strictly convex function under a convex constraint, its optimal solution $\boldsymbol{\varepsilon}_{\rm e}$ exists uniquely. This means that, for a given $\boldsymbol{\varepsilon}$, the decomposition in (9.17b) is determined uniquely, and thence $\boldsymbol{\sigma}$ is also determined uniquely from (9.17a). For this reason there exists a function $\hat{\boldsymbol{\sigma}}_{\rm mem} : \mathbb{S}^2 \to \mathbb{S}^2$ such that

$$\boldsymbol{\sigma} = \hat{\boldsymbol{\sigma}}_{\rm mem}(\boldsymbol{\varepsilon}) \quad \Leftrightarrow \quad (9.17). \tag{9.19}$$

Proposition 9.2.1 gives an alternative (and tractable) expression of $\hat{w}_{\rm mem}$ in (PE). As a consequence, (PE) is equivalently rewritten as

$$\left.\begin{aligned} (\overline{\rm PE}) : \min_{\boldsymbol{u},\boldsymbol{\varepsilon}_{\rm e}} \ & \int_{\Omega} (\hat{w}_{\rm ref}(\boldsymbol{\varepsilon}_{\rm e}) - \underline{\boldsymbol{p}} \cdot \boldsymbol{u}) \,{\rm d}\Omega - \int_{\Gamma_{\rm N}} \underline{\boldsymbol{t}} \cdot \boldsymbol{u} {\rm d}\Gamma \\ {\rm s.\,t.} \ & \boldsymbol{\varepsilon}_{\rm e} \succeq \frac{1}{2}(\nabla \boldsymbol{u}^{\rm T} + \nabla \boldsymbol{u}) \quad \text{in } \Omega, \\ & \boldsymbol{u} = \underline{\boldsymbol{u}} \quad \text{on } \Gamma_{\rm D}. \end{aligned}\right\} \tag{9.20}$$

More precisely, the relation between (PE) and $(\overline{\rm PE})$ is stated in the proposition below, which can be shown in a manner similar to Proposition 8.2.7.

Proposition 9.2.3. *Suppose that* min (PE) $> -\infty$.[9] *Then* $(\overline{\rm PE})$ *is equivalent to* (PE) *in the following sense:*

(i) *Let* $(\bar{\boldsymbol{u}}, \bar{\boldsymbol{\varepsilon}})$ *be optimal for* (PE). *Define* $\bar{\boldsymbol{\varepsilon}}_{\rm e} \in \mathbb{S}^2$ *so that*

$$\hat{\boldsymbol{\sigma}}_{\rm mem}(\bar{\boldsymbol{\varepsilon}}) = \hat{\boldsymbol{\sigma}}_{\rm ref}(\bar{\boldsymbol{\varepsilon}}_{\rm e})$$

is satisfied. Then $(\bar{\boldsymbol{u}}, \bar{\boldsymbol{\varepsilon}}_{\rm e})$ *optimal for* $(\overline{\rm PE})$.

(ii) *Let* $(\bar{\boldsymbol{u}}, \bar{\boldsymbol{\varepsilon}}_{\rm e})$ *be optimal for* $(\overline{\rm PE})$, *and define* $\bar{\boldsymbol{\varepsilon}}$ *by* $\bar{\boldsymbol{\varepsilon}} = \hat{\boldsymbol{\varepsilon}}(\bar{\boldsymbol{u}})$ *with* $\hat{\boldsymbol{\varepsilon}}$ *in* (9.2). *Then* $(\bar{\boldsymbol{u}}, \bar{\boldsymbol{\varepsilon}})$ *is optimal for* (PE).

(iii) min (PE) = min $(\overline{\rm PE})$.

Proposition 9.2.3 asserts that $(\overline{\rm PE})$ in (9.20) is a conic optimization reformulation of (PE) in (9.16). The next step of our study is to embed $(\overline{\rm PE})$ into the form of $({\rm P_F})$ in (2.28)[10] so that we can employ the Fenchel duality theory.

[9]This value is not finite in general. Illustrative examples include a square membrane structure subjected to the uniaxial compression load.

[10]See the orientation part of this chapter for convenience.

9.2.3 Principle of complementary energy in small deformation

The Fenchel duality theory is applied to $(\overline{\mathrm{PE}})$ to study the duality and optimality of the energy principles for linear membranes.

9.2.3.1 Fenchel dual problem

We derive the Fenchel dual problem of $(\overline{\mathrm{PE}})$ in (9.20). To begin, we show that $(\overline{\mathrm{PE}})$ can be embedded into the form of $(\mathrm{P_F})$ in (2.28).

The situation under consideration is almost identical to what is studied in section 8.3.1 for masonry structures; i.e., the primal variables x and the variable spaces, $\mathbb{V}$ and $\mathbb{Y}$, are defined by (8.38), and f and Λ are given by (8.39) and (8.41), respectively. The only difference is the definition of g, which was defined by (8.40) but in the present situation should read

$$g(\boldsymbol{Z}) = \begin{cases} 0 & \text{if } \boldsymbol{Z} \succeq \boldsymbol{o} \text{ (in } \Omega), \\ +\infty & \text{otherwise.} \end{cases} \tag{9.21}$$

Then $(\overline{\mathrm{PE}})$ in (9.20) is written in the form of $(\mathrm{P_F})$ in (2.28).

Turning toward obtaining an an explicit formulation of the dual problem $(\mathrm{P_F^*})$ in (2.31) in the present situation, we first observe that the dual spaces $\mathbb{V}^*$ and $\mathbb{Y}^*$ are given by (8.42) and that $f^*(\Lambda^* \boldsymbol{Z}^*)$ is given by (8.45) in Proposition 8.3.2. For g in (9.21), in a manner similar to Proposition 8.3.4, its conjugate function g^* is obtained as

$$g^*(-\boldsymbol{Z}^*) = \begin{cases} 0 & \text{if } \boldsymbol{Z}^* \preceq \boldsymbol{o} \text{ (in } \Omega), \\ +\infty & \text{otherwise.} \end{cases} \tag{9.22}$$

In accordance with (8.55), the variables are rewritten as $\boldsymbol{\sigma} = \boldsymbol{Z}^*$, and then $(\mathrm{P_F^*})$ in (2.31) is explicitly written as

$$\left.\begin{aligned} (\overline{\mathrm{CE}}) : \max_{\boldsymbol{\sigma}} \ & -\int_{\Omega} \hat{w}_{\mathrm{ref}}^*(\boldsymbol{\sigma}) \mathrm{d}\Omega + \int_{\Gamma_{\mathrm{D}}} (\boldsymbol{\sigma}\boldsymbol{n}) \cdot \underline{\boldsymbol{u}} \mathrm{d}\Gamma \\ \text{s.t.} \ & -\operatorname{div} \boldsymbol{\sigma} = \underline{\boldsymbol{p}} \quad \text{in } \Omega, \\ & \boldsymbol{\sigma} \succeq \boldsymbol{o} \quad \text{in } \Omega, \\ & \boldsymbol{\sigma}\boldsymbol{n} = \underline{\boldsymbol{t}} \quad \text{on } \Gamma_{\mathrm{N}}, \end{aligned}\right\} \tag{9.23}$$

where the conjugate function $\hat{w}_{\mathrm{ref}}^*$ of $\hat{w}_{\mathrm{ref}}$ is explicitly given by (8.46), which is the complementary strain energy function of the linear isotropic body.

9.2.3.2 Principle of complementary energy

In a manner similar to Proposition 8.3.6, the relation between $(\overline{\mathrm{PE}})$ and $(\overline{\mathrm{CE}})$ under consideration is obtained from the strong duality theorem (Proposition 2.2.9) as follows.

Proposition 9.2.4. *Suppose that there exists* $\boldsymbol{\sigma} \in L^2(\Omega)^{3\times 3}_{\mathbb{S}}$ *satisfying*

$$\begin{aligned} -\operatorname{div} \boldsymbol{\sigma} = \underline{\boldsymbol{p}}, \ \boldsymbol{\sigma} \succ \boldsymbol{o} \quad & \text{in } \Omega, \\ \boldsymbol{\sigma}\boldsymbol{n} = \underline{\boldsymbol{t}} \qquad\qquad & \text{on } \Gamma_{\mathrm{D}}. \end{aligned}$$

Assume that the given boundary condition is smooth enough. Then, each of $(\overline{\mathrm{PE}})$ *and* $(\overline{\mathrm{CE}})$ *has an optimal solution, and* $\min(\overline{\mathrm{PE}}) = \max(\overline{\mathrm{CE}})$ *holds.*

Remark 9.2.5. Recall that, in Proposition 9.2.3, we made an assumption that the objective function of $(\overline{\mathrm{PE}})$ is bounded. This assumption corresponds to the constraint qualification for $(\overline{\mathrm{CE}})$ under consideration in Proposition 9.2.4. Indeed, the feasibility of $(\overline{\mathrm{CE}})$ depends on the loading condition, and $(\overline{\mathrm{PE}})$ becomes unbounded if $(\overline{\mathrm{CE}})$ is infeasible. The simplest infeasible example of $(\overline{\mathrm{CE}})$ is a rectangular membrane subjected to the uniaxial compression. Even in such a case, however, the minimization problem of the complementary energy becomes feasible if the finite deformation is taken account, which shall be discussed in section 9.4. ■

In parallel with Proposition 8.3.8 for masonry structures, the optimality conditions of $(\overline{\mathrm{PE}})$ and $(\overline{\mathrm{CE}})$ for linear membranes are obtained by applying the general result established in Proposition 2.2.9 to the situation under consideration as follows.

Proposition 9.2.6. *Under the hypotheses in Proposition 9.2.4,* $(\bar{\boldsymbol{u}}, \bar{\boldsymbol{\varepsilon}})$ *and* $\bar{\boldsymbol{\sigma}}$ *are optimal for* $(\overline{\mathrm{PE}})$ *and* $(\overline{\mathrm{CE}})$*, respectively, if and only if they satisfy (in the weak sense)*

$$\begin{aligned} & \bar{\boldsymbol{\sigma}} = \lambda(\operatorname{tr} \bar{\boldsymbol{\varepsilon}}_{\mathrm{e}})\boldsymbol{I} + 2\mu\bar{\boldsymbol{\varepsilon}}_{\mathrm{e}} \quad \text{in } \Omega, \\ & -\operatorname{div} \bar{\boldsymbol{\sigma}} = \underline{\boldsymbol{p}} \quad \text{in } \Omega, \\ & \bar{\boldsymbol{u}} = \underline{\boldsymbol{u}} \quad \text{on } \Gamma_{\mathrm{D}}; \qquad \bar{\boldsymbol{\sigma}}\boldsymbol{n} = \underline{\boldsymbol{t}} \quad \text{on } \Gamma_{\mathrm{N}}, \\ & \bar{\boldsymbol{\varepsilon}}_{\mathrm{e}} \succeq \hat{\boldsymbol{\varepsilon}}(\bar{\boldsymbol{u}}), \ \bar{\boldsymbol{\sigma}} \succeq \boldsymbol{o}, \ (\bar{\boldsymbol{\varepsilon}}_{\mathrm{e}} - \hat{\boldsymbol{\varepsilon}}(\bar{\boldsymbol{u}})) : \bar{\boldsymbol{\sigma}} = 0 \quad \text{in } \Omega, \end{aligned}$$

where $\hat{\boldsymbol{\varepsilon}}$ *is given by* (9.2).

In a conventional way, the principle of the complementary energy is stated in the form of a minimization problem, by changing the sign of the objective function of $(\overline{\mathrm{CE}})$ in (9.23) as

$$\left.\begin{aligned} (\mathrm{CE}) : \min_{\boldsymbol{\sigma}} \ & \int_{\Omega} \hat{w}^*_{\mathrm{ref}}(\boldsymbol{\sigma}) \mathrm{d}\Omega - \int_{\Gamma_{\mathrm{D}}} (\boldsymbol{\sigma}\boldsymbol{n}) \cdot \underline{\boldsymbol{u}} \mathrm{d}\Gamma \\ \text{s.t.} \ & -\operatorname{div} \boldsymbol{\sigma} = \underline{\boldsymbol{p}} \quad \text{in } \Omega, \\ & \boldsymbol{\sigma} \succeq \boldsymbol{o} \quad \text{in } \Omega, \\ & \boldsymbol{\sigma}\boldsymbol{n} = \underline{\boldsymbol{t}} \quad \text{on } \Gamma_{\mathrm{N}}. \end{aligned}\right\} \tag{9.24}$$

From Proposition 9.2.6 and the equivalence of (CE) and $(\overline{\mathrm{CE}})$, we obtain the following main result, which guarantees that the stress tensor at the equilibrium state can be obtained as the optimal solution of (CE) in (8.60).[11]

[11]The proof is analogous to Proposition 8.3.9 and hence is omitted.

Proposition 9.2.7. *Under the hypotheses in Proposition 9.2.4, the optimal solutions of* (PE) *and* (CE)*, denoted* $(\bar{\boldsymbol{u}}, \bar{\varepsilon})$ *and* $\bar{\boldsymbol{\sigma}}$*, are in the equilibrium. Moreover,* min (PE) = − min (CE).

Remark 9.2.8. It is easy to see that the objective function of (CE) in (9.24) is strictly convex, which implies that if there is an optimal solution of (CE) then it is unique. In contrast, $\hat{w}_{\rm mem}$ is not strictly convex, and the optimal solution of (PE) is not necessarily unique. Indeed, the uniqueness of the displacement field $\boldsymbol{u}$ in the equilibrium depends upon the applied load, as discussed in Remark 8.3.10 for masonry structures. ■

Remark 9.2.9. In a manner similar to masonry structures, we can show that $(\overline{\rm PE})$ in (9.20) is approximated as a finite-dimensional SDP problem by applying a finite-element discretization performed in section 8.4.1. The Only difference between the membrane and the masonry structure is the sign of the positive-semidefinite constraint; i.e., we now have the constraint

$$\tilde{\varepsilon}_{\rm e}^{qe} - {\rm Mat}(B^{qe}\tilde{\boldsymbol{u}}) \succeq O$$

instead of $\tilde{\varepsilon}_{\rm e}^{qe} - {\rm Mat}(B^{qe}\tilde{\boldsymbol{u}}) \preceq O$ for masonry structures (see $(\overline{\rm PE}_{\rm cone}^{\rm FE})$ in (8.79)). Thus a discretization of $(\overline{\rm PE})$ in (9.20) for membranes results in thc conic optimization problem as

$$\left.\begin{array}{lll}(\overline{\rm PE}_{\rm cone}^{\rm FE}): & \displaystyle\min_{\tilde{u}_j, \tilde{w}^{qe}, \tilde{\varepsilon}_{\rm e}^{qe}} & \displaystyle\sum_{e=1}^{n^{\rm E}} \sum_{q=1}^{n^{\rm G}} \tilde{\rho}^{qe} \tilde{w}^{qe} - \sum_{j \in \tilde{\Gamma}_{\rm N}} \underline{\tilde{p}} \tilde{u}_j \\ & {\rm s.\,t.} & \tilde{u}_j = \underline{\tilde{u}}_j, \quad \forall j \in \tilde{\Gamma}_{\rm D}, \\ & & (\tilde{w}^{qe}/2) + 1 \geq \left\| \begin{bmatrix} (\tilde{w}^{qe}/2) - 1 \\ G\, \mathbf{vec}(\tilde{\varepsilon}_{\rm e}^{qe}) \end{bmatrix} \right\|, \quad \forall q;\ \forall e, \\ & & \tilde{\varepsilon}_{\rm e}^{qe} - {\rm Mat}(B^{qe}\tilde{\boldsymbol{u}}) \succeq O, \quad \forall q;\ \forall e, \end{array}\right\} \tag{9.25}$$

where G is a constant matrix obtained from the elasticity tensor; see (8.78). The problem above is an SDP problem, which involves the linear inequality constraints, second-order cone constraints, and positive-semidefinite constraints. ■

Remark 9.2.10. For planar membranes undergoing small deformations, $(\overline{\rm PE}_{\rm cone}^{\rm FE})$ in (9.25) can be further reduced to an SOCP problem. Define $\tilde{Y}^{qe} \in \mathbb{S}^2$ by

$$\tilde{Y}^{qe} = \tilde{\varepsilon}_{\rm e}^{qe} - {\rm Mat}(B^{qe}\tilde{\boldsymbol{u}})$$

for simplicity. Then, for each (q, e), the positive-semidefinite constraint of $(\overline{\rm PE}_{\rm cone}^{\rm FE})$ is written as $\tilde{Y}^{qe} \succeq O$, which is equivalent to

$$\det \tilde{Y}^{qe} \geq 0, \quad {\rm tr}\, \tilde{Y}^{qe} \geq 0, \tag{9.26}$$

because $\tilde{Y}^{qe}$ is a 2×2 symmetric matrix. In terms of the components $\tilde{Y}_{ij}^{qe}$ of $\tilde{Y}^{qe}$, condition (9.26) can be written as

$$\tilde{Y}_{11}^{qe} \tilde{Y}_{22}^{qe} \geq (\tilde{Y}_{12}^{qe})^2, \quad \tilde{Y}_{11}^{qe} \geq 0, \quad \tilde{Y}_{22}^{qe} \geq 0.$$

Moreover, by using Fact 1.5.3, this condition is equivalently reduced to the following second-order cone constraint as

$$\tilde{Y}_{11}^{qe} + \tilde{Y}_{22}^{qe} \geq \left\| \begin{bmatrix} \tilde{Y}_{11}^{qe} - \tilde{Y}_{22}^{qe} \\ \tilde{Y}_{12}^{qe}/2 \end{bmatrix} \right\|. \tag{9.27}$$

Thus the positive-semidefinite constraints of $(\overline{\mathrm{PE}}^{\mathrm{FE}}_{\mathrm{cone}})$ in (9.25) are rewritten as the second-order cone constraints (9.27), and thence $(\overline{\mathrm{PE}}^{\mathrm{FE}}_{\mathrm{cone}})$ is consequently reduced to an SOCP problem.[12] ■

9.3 Principle of Potential Energy for Membranes

From this point we suppose that the membrane undergoes the finite deformation within the two-dimensional space, while the strain is assumed to be small. The notation used here for the large deformation theory basically follows a conventional one found in [90, 184, 457].

Let $\Omega \subset \mathbb{R}^2$ be a bounded, open, and connected set with a sufficiently smooth boundary $\Gamma := \mathrm{bd}\,\Omega$.[13] We consider an elastic body (membrane) occupying $\overline{\Omega} := \mathrm{cl}\,\Omega$, where the set $\overline{\Omega}$ is called the reference configuration. A deformation of the elastic body is expressed by a vector field $\varphi : \overline{\Omega} \to \mathbb{R}^2$, as illustrated in Figure 9.3. The deformed configuration is defined by $\overline{\Omega}^{\varphi} := \varphi(\overline{\Omega})$. We denote by

$$\boldsymbol{u}(\boldsymbol{x}) = \boldsymbol{\varphi}(\boldsymbol{x}) - \boldsymbol{x} \tag{9.28}$$

the displacement vector at each point $\boldsymbol{x} \in \overline{\Omega}$.

9.3.1 Constitutive law

Let $\boldsymbol{F}(\boldsymbol{x}) \in \mathbb{R}^{2\times 2}$ denote the *deformation gradient* defined by

$$\boldsymbol{F}(\boldsymbol{x}) := \nabla\boldsymbol{\varphi}(\boldsymbol{x}).$$

We say that the membrane is *elastic* if at each point $\boldsymbol{x}^{\varphi} := \boldsymbol{\varphi}(\boldsymbol{x}) \in \Omega^{\varphi}$ of the deformed configuration the *Cauchy stress* tensor $\boldsymbol{T}^{\varphi}(\boldsymbol{x}^{\varphi})$ is completely determined only by $\boldsymbol{F}(\boldsymbol{x})$ at the corresponding point $\boldsymbol{x} \in \Omega$. Hence, for an elastic membrane, there exists a function $\hat{\boldsymbol{T}}^{\varphi}_{\mathrm{mem}} : \Omega \times \mathbb{R}^{2\times 2} \to \mathbb{S}^2$ that expresses $\boldsymbol{T}^{\varphi}(\boldsymbol{x}^{\varphi})$ through a constitutive relation of the form

$$\boldsymbol{T}^{\varphi}(\boldsymbol{x}^{\varphi}) = \hat{\boldsymbol{T}}^{\varphi}_{\mathrm{mem}}(\boldsymbol{x}, \boldsymbol{F}(\boldsymbol{x})). \tag{9.29}$$

---

[12] In section 8.4.1 we investigated a masonry structure in the three-dimensional space. Therefore $(\overline{\mathrm{PE}}^{\mathrm{FE}}_{\mathrm{cone}})$ in (8.79) involves the positive-semidefinite constraints of 3×3 symmetric matrices. A positive-semidefinite constraint of the order 3 cannot be reduced to the second-order cone constraint, which means that $(\overline{\mathrm{PE}}^{\mathrm{FE}}_{\mathrm{cone}})$ for a three-dimensional masonry structure is not an SOCP but an SDP problem.

[13] More precisely, Ω is assumed to be a Lipschitzian domain; see [221, pp. 15–16] for its definition.

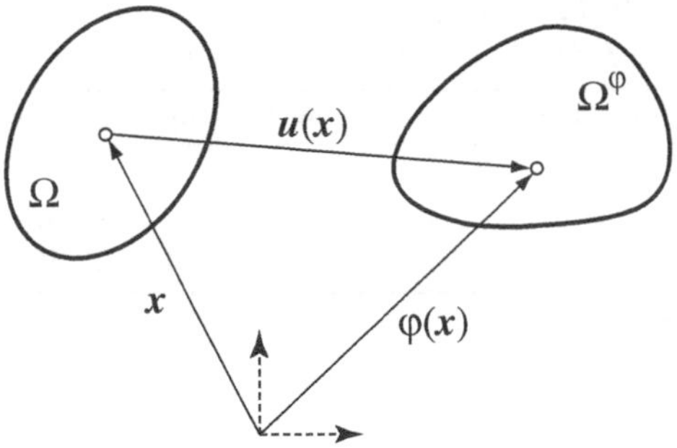

FIGURE 9.3: Deformation of a continuum media.

To describe the constitutive relation of a membrane, we adopt the hypotheses introduced in section 9.2.1 for the infinitesimal theory. However, within the framework of the geometrically nonlinear theory, we have to take care of the fact that the wrinkling phenomena are observed at the deformed configuration, which should be clearly distinguished from the reference configuration. Therefore, we begin by considering the wrinkling phenomena at the deformed configuration. Thus the physical hypotheses listed below should be considered for the Cauchy stress $\boldsymbol{T}^{\varphi}(\boldsymbol{x}^{\varphi})$ and its work-conjugate strain, i.e., the Almansi strain. The *Almansi strain* tensor, denoted $\boldsymbol{A}(\boldsymbol{x}^{\varphi}) \in \mathbb{S}^2$, at the point $\boldsymbol{x}^{\varphi}$ is defined by

$$\boldsymbol{A}(\boldsymbol{x}^{\varphi}) = \frac{1}{2}(\boldsymbol{I} - \boldsymbol{F}(\boldsymbol{x})^{-\mathrm{T}}\boldsymbol{F}(\boldsymbol{x})^{-1}). \tag{9.30}$$

Now we state the hypotheses, which are basically same as the ones considered in the infinitesimal theory:

(i) The flexural stiffness is negligible.

(ii) Any compressive stress cannot be sustained.

(iii) The strain can be additively decomposed to the two parts: the elastic and inelastic strains.

(iv) The inelastic strain corresponds to the reversible slack, which can take place perpendicular to the plane of nonpositive principal stress *at the deformed configuration.*

(v) The tensile stress is caused elastically solely by the elastic stress.

Note that in hypothesis (iv) a slack (or wrinkle) is to be in the principal direction of $\boldsymbol{T}^{\varphi}$.

The same observation as that in section 9.2.1 can be developed, where $\boldsymbol{\sigma}$ and $\boldsymbol{\varepsilon}$ are replaced with $\boldsymbol{T}^{\varphi}$ and $\boldsymbol{A}$, respectively. From hypothesis (ii), we

TABLE 9.1: Wrinkle criterion based on the principal stresses $\lambda_i(\boldsymbol{T}^\varphi)$ $(i = 1, 2)$, where $\lambda_1(\boldsymbol{T}^\varphi) \geq \lambda_2(\boldsymbol{T}^\varphi)$.

principal stresses		wrinkling	membrane state
$\lambda_1(\boldsymbol{T}^\varphi) > 0$	$\lambda_2(\boldsymbol{T}^\varphi) > 0$	no	taut (Figure 9.2(a))
$\lambda_1(\boldsymbol{T}^\varphi) > 0$	$\lambda_2(\boldsymbol{T}^\varphi) = 0$	one axial	wrinkled (Figure 9.2(b))
$\lambda_1(\boldsymbol{T}^\varphi) = 0$	$\lambda_2(\boldsymbol{T}^\varphi) = 0$	two axial	slack (Figure 9.2(c))

obtain

$$\lambda_i(\boldsymbol{T}^\varphi) \geq 0, \quad i = 1, 2 \tag{9.31}$$

for the eigenvalues of $\boldsymbol{T}^\varphi$, where $\lambda_1(\boldsymbol{T}^\varphi) \geq \lambda_2(\boldsymbol{T}^\varphi)$. Furthermore, from hypothesis (iv), the winkling criterion can be given in terms of the principal Cauchy stresses as shown in Table 9.1 (see, e.g., [191, 380]).

Remark 9.3.1. Instead of the wrinkling criterion in Table 9.1, the slackness can also be defined in terms of the largest principal value of the Almansi strain tensor $\boldsymbol{A}$. Let $\omega_i(\boldsymbol{A})$ $(\omega_1(\boldsymbol{A}) \geq \omega_2(\boldsymbol{A}))$ denote the eigenvalues of $\boldsymbol{A}$. Then a membrane is slack if and only if $\omega_1(\boldsymbol{A}) \leq 0$. Conversely, $\omega_1(\boldsymbol{A}) > 0$ if the membrane is either wrinkled or taut. However, the tautness cannot be determined correctly based on the principal strains; indeed, $\omega_2(\boldsymbol{A}) > 0$ does not necessarily hold in the taut state because of the effect of Poisson's ratio [191, p. 776]. ■

According to hypothesis (iii), we decompose $\boldsymbol{A}(\boldsymbol{x}^\varphi)$ as

$$\boldsymbol{A}(\boldsymbol{x}^\varphi) = \boldsymbol{A}_{\mathrm{e}}(\boldsymbol{x}^\varphi) + \boldsymbol{A}_{\mathrm{w}}(\boldsymbol{x}^\varphi), \tag{9.32}$$

where $\boldsymbol{A}_{\mathrm{e}}$ and $\boldsymbol{A}_{\mathrm{w}}$ are the elastic strain and inelastic (i.e., wrinkling) strain, respectively. Since the wrinkling strain $\boldsymbol{A}_{\mathrm{w}}$ is regarded as the measure of "shortening" of the membrane due to slack, it should satisfy

$$\omega_i(\boldsymbol{A}_{\mathrm{w}}) \leq 0, \quad i = 1, 2, \tag{9.33}$$

where $\omega_1(\boldsymbol{A}_{\mathrm{w}}) \geq \omega_2(\boldsymbol{A}_{\mathrm{w}})$. Furthermore, from hypothesis (iv) we can show that

$$\boldsymbol{T}^\varphi \text{ and } \boldsymbol{A}_{\mathrm{w}} \text{ are coaxial}, \tag{9.34}$$

$$\lambda_i(\boldsymbol{T}^\varphi)\omega_i(\boldsymbol{A}_{\mathrm{w}}) = 0, \quad i = 1, 2; \tag{9.35}$$

see the discussions in section 9.2.1 for more detail. It follows from Fact 1.3.20 that (9.31), (9.33), (9.34), and (9.35) are equivalently rewritten as

$$-\boldsymbol{A}_{\mathrm{w}}(\boldsymbol{x}^\varphi) \succeq \boldsymbol{o}, \quad \boldsymbol{T}^\varphi(\boldsymbol{x}^\varphi) \succeq \boldsymbol{o}, \quad -\boldsymbol{A}_{\mathrm{w}}(\boldsymbol{x}^\varphi) : \boldsymbol{T}^\varphi(\boldsymbol{x}^\varphi) = 0, \tag{9.36}$$

which is a consequence of hypotheses (i)–(iv).

For a wrinkled state, the decomposition of $\boldsymbol{A}$ in (9.32), together with (9.36), is illustrated in Figure 9.4. It is seen from (9.34) and (9.35) that the principal

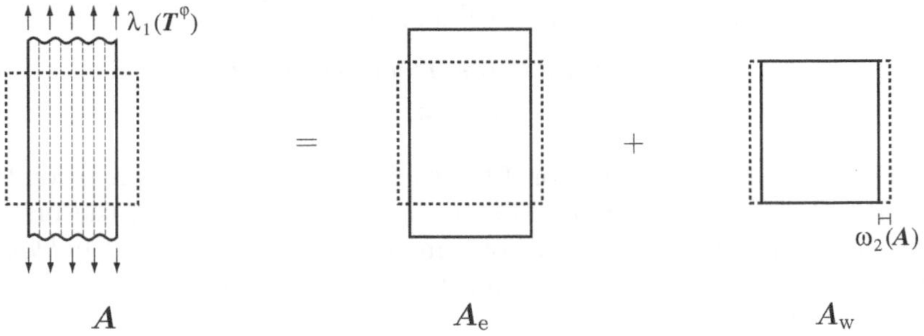

FIGURE 9.4: A schematic exposition of the decomposition (9.32) of the Almansi strain $\boldsymbol{A}$ to the elastic part $\boldsymbol{A}_{\mathrm{e}}$ and the wrinkling part $\boldsymbol{A}_{\mathrm{w}}$ in the case of an isotropic membrane in a wrinkled state.

direction of the Cauchy stress corresponding to $\lambda_1(\boldsymbol{T}^\varphi) > 0$ coincides with the direction along the wrinkle (which is the principal direction of the Almansi strain corresponding to $\omega_1(\boldsymbol{A}_{\mathrm{w}}) = 0$), while the other principal direction of the stress (which corresponds to $\lambda_2(\boldsymbol{T}^\varphi) = 0$) coincides with the direction transverse to the wrinkle (which is the principal direction corresponding to $\omega_1(\boldsymbol{A}_{\mathrm{w}}) < 0$). From hypothesis (v) (and as seen below), $\boldsymbol{A}_{\mathrm{e}}$ is the strain corresponding to $\boldsymbol{T}^\varphi$ when the wrinkle is artificially removed. Note that $\boldsymbol{A}_{\mathrm{e}}$ is not positive semidefinite in general, because of Poisson's effect.

Remark 9.3.2. Consider a wrinkled state. As shown in Figure 9.36, the direction along the wrinkle should coincide with the principal direction of the tensile principal Cauchy stress, and the direction transverse to the wrinkle should coincide with the principal direction of the vanishing principal Cauchy stress. This condition, expressed in terms of the principal values and principal directions of $\boldsymbol{T}^\varphi$ and $\boldsymbol{A}_{\mathrm{w}}$, has been used widely to describe the wrinkled state (see, e.g., [274, pp. 1021–1024]). In contrast, condition (9.36) gives a unified expression of all the three membrane states in Table 9.1. Moreover, it is also advantageous that the coincidence condition of the principal directions of stress and strain does not appear explicitly in (9.36) but is included as a consequence of the complementarity condition over positive-semidefinite cone. ■

Naturally it is simpler to consider the constitutive relation in the reference configuration. This motivates us to rewrite condition (9.36) in terms of the second Piola–Kirchhoff stress tensor, denoted $\boldsymbol{S}$, and the Green–Lagrange strain, denoted $\boldsymbol{E}$.

The relation between the *second Piola–Kirchhoff stress* tensor $\boldsymbol{S}(\boldsymbol{x})$ and the Cauchy stress tensor $\boldsymbol{T}^\varphi(\boldsymbol{x}^\varphi)$ is given by

$$\boldsymbol{S}(\boldsymbol{x}) = J(\boldsymbol{x})\boldsymbol{F}(\boldsymbol{x})^{-1}\boldsymbol{T}^\varphi(\boldsymbol{x}^\varphi)\boldsymbol{F}(\boldsymbol{x})^{-\mathrm{T}}, \tag{9.37}$$

where $J(\boldsymbol{x}) = \det \boldsymbol{F}(\boldsymbol{x})$ is the Jacobian determinant. The *first Piola–Kirchhoff stress* tensor $\boldsymbol{\Pi}(\boldsymbol{x})$ is related to $\boldsymbol{T}^{\varphi}$ and $\boldsymbol{S}$ as

$$\boldsymbol{\Pi}(\boldsymbol{x}) = J(\boldsymbol{x})\boldsymbol{T}^{\varphi}(\boldsymbol{x}^{\varphi})\boldsymbol{F}(\boldsymbol{x})^{-\mathrm{T}} \tag{9.38}$$

$$= \boldsymbol{F}(\boldsymbol{x})\boldsymbol{S}(\boldsymbol{x}). \tag{9.39}$$

Since the Cauchy stress tensor $\boldsymbol{T}^{\varphi}(\boldsymbol{x}^{\varphi})$ is symmetric, so is the second Piola–Kirchhoff stress tensor $\boldsymbol{S}(\boldsymbol{x})$ as seen in (9.37). In contrast, the first Piola–Kirchhoff stress tensor $\boldsymbol{\Pi}(\boldsymbol{x})$ is not symmetric in general. The *Kirchhoff stress tensor*, denoted $\boldsymbol{\tau}^{\varphi}(\boldsymbol{x}^{\varphi})$, is a symmetric tensor defined by

$$\boldsymbol{\tau}^{\varphi}(\boldsymbol{x}^{\varphi}) = J(\boldsymbol{x})\boldsymbol{T}^{\varphi}(\boldsymbol{x}^{\varphi}). \tag{9.40}$$

The *Green–Lagrange strain* tensor, denoted $\boldsymbol{E}(\boldsymbol{x})$, is a symmetric tensor written in terms of the Almansi strain tensor $\boldsymbol{A}$ as (see (9.30))

$$\begin{aligned} \boldsymbol{E}(\boldsymbol{x}) &= \frac{1}{2}(\nabla \boldsymbol{u}^{\mathrm{T}} + \nabla \boldsymbol{u} + \nabla \boldsymbol{u}^{\mathrm{T}}\nabla \boldsymbol{u}) \\ &= \frac{1}{2}(\boldsymbol{F}(\boldsymbol{x})^{\mathrm{T}}\boldsymbol{F}(\boldsymbol{x}) - \boldsymbol{I}) \\ &= \boldsymbol{F}(\boldsymbol{x})^{\mathrm{T}}\boldsymbol{A}(\boldsymbol{x}^{\varphi})\boldsymbol{F}(\boldsymbol{x}). \end{aligned} \tag{9.41}$$

In accordance with the additive decomposition (9.32) of $\boldsymbol{A}$ into $\boldsymbol{A}_{\mathrm{e}}$ and $\boldsymbol{A}_{\mathrm{w}}$, the elastic and wrinkling parts of the Green–Lagrange strain tensor are defined by

$$\boldsymbol{E}_{\mathrm{e}}(\boldsymbol{x}) = \boldsymbol{F}(\boldsymbol{x})^{\mathrm{T}}\boldsymbol{A}_{\mathrm{e}}(\boldsymbol{x}^{\varphi})\boldsymbol{F}(\boldsymbol{x}), \tag{9.42}$$

$$\boldsymbol{E}_{\mathrm{w}}(\boldsymbol{x}) = \boldsymbol{F}(\boldsymbol{x})^{\mathrm{T}}\boldsymbol{A}_{\mathrm{w}}(\boldsymbol{x}^{\varphi})\boldsymbol{F}(\boldsymbol{x}), \tag{9.43}$$

which lead to the additive decomposition of $\boldsymbol{E}$ in the form of

$$\boldsymbol{E}(\boldsymbol{x}) = \boldsymbol{E}_{\mathrm{e}}(\boldsymbol{x}) + \boldsymbol{E}_{\mathrm{w}}(\boldsymbol{x}). \tag{9.44}$$

We first show that the complementarity condition over positive-semidefinite cone in (9.36) is inherited by a pair of $\boldsymbol{E}_{\mathrm{w}}$ and $\boldsymbol{S}$ defined by (9.37) and (9.43).

Proposition 9.3.3. *$\boldsymbol{A}_{\mathrm{w}}$ and $\boldsymbol{T}^{\varphi}$ satisfy* (9.36) *if and only if $\boldsymbol{E}_{\mathrm{w}}$ and $\boldsymbol{S}$ satisfy*

$$-\boldsymbol{E}_{\mathrm{w}} \succeq \boldsymbol{o}, \quad \boldsymbol{S} \succeq \boldsymbol{o}, \quad -\boldsymbol{E}_{\mathrm{w}} : \boldsymbol{S} = 0. \tag{9.45}$$

Proof. Since $\boldsymbol{F}$ is nonsingular, it is immediate from definition (9.43) of $\boldsymbol{E}_{\mathrm{w}}$ that $-\boldsymbol{E}_{\mathrm{w}} \succeq \boldsymbol{o}$ if and only if $-\boldsymbol{A}_{\mathrm{w}} \succeq \boldsymbol{o}$. Similarly, it follows from (9.37) that $\boldsymbol{S} \succeq \boldsymbol{o}$ if and only if $\boldsymbol{T}^{\varphi} \succeq \boldsymbol{o}$.

Since $\boldsymbol{A}_{\mathrm{w}}$ is a symmetric tensor, the definition (1.3.12) of inner product of matrices yields

$$\begin{aligned} \boldsymbol{E}_{\mathrm{w}} \bullet \boldsymbol{S} &= (\boldsymbol{F}^{\mathrm{T}}\boldsymbol{A}_{\mathrm{w}}\boldsymbol{F}) \bullet (J\boldsymbol{F}^{-1}\boldsymbol{T}^{\varphi}\boldsymbol{F}^{-\mathrm{T}}) \\ &= J\,\mathrm{tr}(\boldsymbol{F}^{\mathrm{T}}\boldsymbol{A}_{\mathrm{w}}\boldsymbol{F}\boldsymbol{F}^{-1}\boldsymbol{T}^{\varphi}\boldsymbol{F}^{-\mathrm{T}}) \\ &= J\,\mathrm{tr}((\boldsymbol{F}^{\mathrm{T}}\boldsymbol{A}_{\mathrm{w}}\boldsymbol{T}^{\varphi})\boldsymbol{F}^{-\mathrm{T}}) \\ &= J\,\mathrm{tr}(\boldsymbol{F}^{-\mathrm{T}}(\boldsymbol{F}^{\mathrm{T}}\boldsymbol{A}_{\mathrm{w}}\boldsymbol{T}^{\varphi})) \\ &= J\boldsymbol{A}_{\mathrm{w}} \bullet \boldsymbol{T}^{\varphi}. \end{aligned}$$

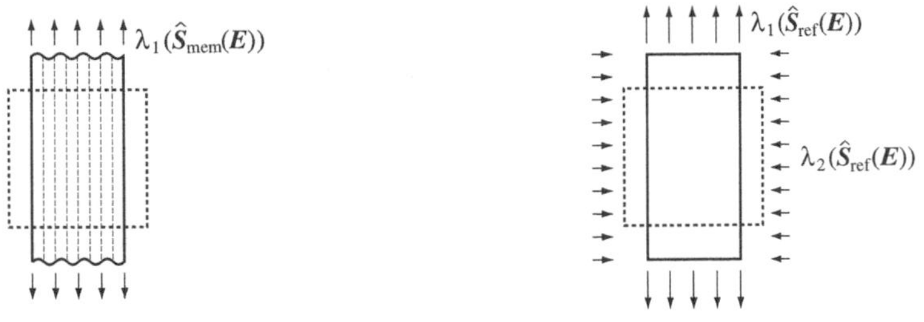

(a) $\hat{\boldsymbol{S}}_{\text{mem}}(\boldsymbol{E})$ (b) $\hat{\boldsymbol{S}}_{\text{ref}}(\boldsymbol{E})$

FIGURE 9.5: The response of an isotropic membrane in (9.46) and the response of the corresponding reference elastic body in (9.48) for a given strain $\boldsymbol{E}$.

Hence, $\boldsymbol{E}_{\text{w}} : \boldsymbol{S} = 0$ if and only if $\boldsymbol{A}_{\text{w}} : \boldsymbol{T}^{\varphi} = 0$, which concludes the proof. □

By introducing the conventional axiom of material frame indifference and the assumptions that the membrane material is homogeneous and that the reference configuration is a natural state (i.e., $\hat{\boldsymbol{T}}^{\varphi}_{\text{mem}}(\boldsymbol{x}, \boldsymbol{I}) = \boldsymbol{o}$ for all $\boldsymbol{x} \in \Omega$), we can show that there exists a mapping $\hat{\boldsymbol{S}}_{\text{mem}} : \mathbb{S}^2 \rightarrow \mathbb{S}^2$ such that (9.29) is alternatively written as (see, e.g., [90, Chap. 3])

$$\boldsymbol{S}(\boldsymbol{x}) = \hat{\boldsymbol{S}}_{\text{mem}}(\boldsymbol{E}(\boldsymbol{x})), \tag{9.46}$$

where the relation between $\boldsymbol{S}$ and $\boldsymbol{T}^{\varphi}$ is given by (9.37). It follows from hypothesis (v) that $\boldsymbol{S}$ depends only on $\boldsymbol{E}_{\text{e}}$, which means that there exists a mapping $\hat{\boldsymbol{S}}_{\text{ref}} : \mathbb{S}^2 \rightarrow \mathbb{S}^2$ satisfying

$$\hat{\boldsymbol{S}}_{\text{ref}}(\boldsymbol{E}_{\text{e}}) = \hat{\boldsymbol{S}}_{\text{mem}}(\boldsymbol{E}) \tag{9.47}$$

for any $\boldsymbol{E}$. Note that such a mapping is not unique, but we choose the simplest one that does not possess the nonsmoothness property due to the no-compression property of the membrane; a concrete example is given right below.

In what follows we assume that $\hat{\boldsymbol{S}}_{\text{ref}}$ is strictly monotone.[14] The reference elastic body, introduced in section 9.2.1 for the membrane in the infinitesimal theory, is a fictitious body, the constitutive relation of which is given by

$$\boldsymbol{S} = \hat{\boldsymbol{S}}_{\text{ref}}(\boldsymbol{E}). \tag{9.48}$$

[14]See section 2.1.2 for the definition of strict monotonicity of functions.

For simplicity, we assume that $\hat{\boldsymbol{S}}_{\rm ref}$ is a linear function; i.e., it can be written as[15]

$$\hat{\boldsymbol{S}}_{\rm ref}(\boldsymbol{E}) = \boldsymbol{C} : \boldsymbol{E} \tag{9.49}$$

with a constant fourth-order symmetric tensor $\boldsymbol{C}$, called the elasticity tensor. In other words, the reference elastic body is assumed to be linearly elastic; i.e., the membrane is assumed to behaves as a linear elastic body if we eliminate wrinkles artificially. For a given $\boldsymbol{E}$, Figure 9.5 depicts the stress states of an isotropic membrane $\hat{\boldsymbol{S}}_{\rm mem}(\boldsymbol{E})$ and that of the corresponding reference elastic body $\hat{\boldsymbol{S}}_{\rm ref}(\boldsymbol{E})$. Note that the reference elastic body can possibly undergo compressive stresses as a result of forced elimination of wrinkles.

Now we are in a position to describe the constitutive relation of the membrane in terms of $\boldsymbol{S}$ and $\boldsymbol{E}$. Recall that conditions (9.32) and (9.36) are equivalently rewritten as (9.44) and (9.45), respectively. Therefore, in conjunction with together with (9.46) and (9.47), the constitutive relation in terms of $\boldsymbol{S}$ and $\boldsymbol{E}$ can be written as

$$\boldsymbol{S} = \hat{\boldsymbol{S}}_{\rm ref}(\boldsymbol{E}_{\rm e}), \tag{9.50a}$$
$$\boldsymbol{E} = \boldsymbol{E}_{\rm e} + \boldsymbol{E}_{\rm w}, \tag{9.50b}$$
$$-\boldsymbol{E}_{\rm w} \succeq \boldsymbol{o}, \quad \boldsymbol{S} \succeq \boldsymbol{o}, \quad -\boldsymbol{E}_{\rm w} : \boldsymbol{S} = 0. \tag{9.50c}$$

The stored energy functions, associated with the stress response functions $\hat{\boldsymbol{S}}_{\rm mem}$ and $\hat{\boldsymbol{S}}_{\rm ref}$, are defined by

$$\hat{w}_{\rm mem}(\boldsymbol{E}) = \int \hat{\boldsymbol{S}}_{\rm mem}(\boldsymbol{E}) : {\rm d}\boldsymbol{E}, \tag{9.51}$$
$$\hat{w}_{\rm ref}(\boldsymbol{E}) = \int \hat{\boldsymbol{S}}_{\rm ref}(\boldsymbol{E}) : {\rm d}\boldsymbol{E} = \frac{1}{2}\boldsymbol{E} : \boldsymbol{C} : \boldsymbol{E}, \tag{9.52}$$

where $\boldsymbol{E} : \boldsymbol{C} : \boldsymbol{E} = \hat{\boldsymbol{S}}_{\rm ref}(\boldsymbol{E}) : \boldsymbol{E}$ with (9.49). It is obvious from the mechanical reason that $\hat{w}_{\rm mem}(\boldsymbol{o}) = 0$ and $\hat{w}_{\rm ref}(\boldsymbol{o}) = 0$. Since $\hat{\boldsymbol{S}}_{\rm ref}$ is supposed to be strictly monotone, $\hat{w}_{\rm ref}$ is strictly convex (see Fact 2.1.2). The constitutive law is then written in terms of $\hat{w}_{\rm mem}$ as[16]

$$\boldsymbol{S} \in \partial \hat{w}_{\rm mem}(\boldsymbol{E}) \quad \Leftrightarrow \quad \exists \boldsymbol{E}_{\rm e}, \boldsymbol{E}_{\rm w} \in \mathbb{S}^2 : \ (9.50)\ . \tag{9.53}$$

[15]The relation (9.49) is alternatively written in a componentwise manner as

$$S_{ij} = \sum_{p=1}^{2} \sum_{q=1}^{2} C_{ijpq} E_{pq}$$

with $\boldsymbol{S} = \hat{\boldsymbol{S}}_{\rm ref}(\boldsymbol{E})$ and $\boldsymbol{S} = (S_{ij})$.

[16]We shall see that $\boldsymbol{S}$ in (9.53) is determined uniquely for a given $\boldsymbol{E}$; see the exposition given right below Proposition 9.3.4.

Example 9.1
In the geometrically nonlinear theory, the material behavior is often described by an isotropic linear relation between $\boldsymbol{S}$ and $\boldsymbol{E}$ as

$$\boldsymbol{S} = (\lambda \operatorname{tr} \boldsymbol{E})\boldsymbol{I} + 2\mu \boldsymbol{E}. \tag{9.54}$$

The material obeying the constitutive (9.54) is called the *St. Venant material*,[17] where λ and μ in (9.54) correspond to the Lamé constants of Hooke's law in the infinitesimal theory.

Suppose that the membrane is isotropic and that a linear relation between $\boldsymbol{S}$ and $\boldsymbol{E}$ is observed when the membrane is subjected to tensile stresses. Naturally the reference elastic body is assumed to obey the St. Venant material law, i.e.,

$$\hat{\boldsymbol{S}}_{\mathrm{ref}}(\boldsymbol{E}) = (\lambda \operatorname{tr} \boldsymbol{E})\boldsymbol{I} + 2\mu \boldsymbol{E},$$

with which the response of the membrane is described by a relation between $\boldsymbol{S}$ and $\boldsymbol{E}$ given through (9.50). The stored energy function for the reference elastic body is then obtained according to (9.52) as

$$\hat{w}_{\mathrm{ref}}(\boldsymbol{E}) = \frac{1}{2}\boldsymbol{E} : \boldsymbol{C} : \boldsymbol{E} = \frac{1}{2}\lambda(\operatorname{tr} \boldsymbol{E})^2 + \mu \boldsymbol{E} : \boldsymbol{E}.$$

Thus the reference elastic body does not inherit the no-compression property at all. Particularly, for the St. Venant material, $\hat{\boldsymbol{S}}_{\mathrm{ref}}$ is a linear function, while $\hat{\boldsymbol{S}}_{\mathrm{mem}}$ is certainly a nonsmooth (and hence is nonlinear) function. $\square$

The stored energy function $\hat{w}_{\mathrm{mem}}$ in (9.53) is not represented explicitly, which is a major source of difficulty in the analysis of membranes. We next establish a tractable representation of $\hat{w}_{\mathrm{mem}}$ alternative to (9.53), which plays a key role in reformulating the minimization problem of the total potential energy into a tractable form.

Proposition 9.3.4. *Suppose that $\boldsymbol{E} \in \mathbb{S}^2$ is given. Consider the optimization problem*

$$\left.\begin{array}{ll} \displaystyle\min_{\boldsymbol{E}_{\mathrm{e}} \in \mathbb{S}^2} & \hat{w}_{\mathrm{ref}}(\boldsymbol{E}_{\mathrm{e}}) \\ \mathrm{s.\,t.} & \boldsymbol{E}_{\mathrm{e}} \succeq \boldsymbol{E} \end{array}\right\} \tag{9.55}$$

in the variable $\boldsymbol{E}_{\mathrm{e}} \in \mathbb{S}^2$. Then $\bar{\boldsymbol{E}}_{\mathrm{e}}$ is optimal if and only if there exists $\boldsymbol{S}$ and $\boldsymbol{E}_{\mathrm{w}}$ satisfying (9.50). *Moreover, the optimal value is equal to $\hat{w}_{\mathrm{mem}}(\boldsymbol{E})$, where $\hat{w}_{\mathrm{mem}}$ is defined by* (9.53).

[17] The St. Venant material [457, p. 40] is also called the St. Venant–Kirchhoff material. See, e.g., Ciarlet [90, p. 130] and Wriggers [457, p. 40].

Proof. Problem (9.55) consists of the convex objective function and the convex feasible set with a nonempty interior. Hence, we apply Proposition 2.2.9 (b) to problem (9.55) to show that its optimality condition coincides with (9.50).

The remainder of this proof is very similar to the proof of Proposition 8.2.5. Specifically, we can encode problem (9.55) into the form of ($\mathrm{P_F}$) in (2.28), i.e., $\inf_{x}\{f(x) + g(\Lambda x) \mid x \in \mathbb{V}\}$, by defining $f : \mathbb{V} \to \mathbb{R}$, $g : \mathbb{Y} \to \mathbb{R} \cup \{+\infty\}$, and $\Lambda : \mathbb{V} \to \mathbb{Y}$ by

$$f(\boldsymbol{E}_{\mathrm{e}}) = \hat{w}_{\mathrm{ref}}(\boldsymbol{E}_{\mathrm{e}}) = \frac{1}{2}\boldsymbol{E}_{\mathrm{e}} : \boldsymbol{C} : \boldsymbol{E}_{\mathrm{e}},$$
$$g(\boldsymbol{Z}) = \begin{cases} 0 & \text{if } \boldsymbol{Z} - \boldsymbol{E} \succeq \boldsymbol{o}, \\ +\infty & \text{otherwise,} \end{cases}$$
$$\Lambda \boldsymbol{E}_{\mathrm{e}} = \boldsymbol{E}_{\mathrm{e}}$$

with

$$x = \boldsymbol{E}_{\mathrm{e}} \in \mathbb{V}, \quad z = \boldsymbol{Z} \in \mathbb{Y}, \quad \mathbb{V} = \mathbb{Y} = \mathbb{S}^2.$$

Write the dual variable and space, in the setting of Proposition 2.2.9 (b), as

$$z^* = \boldsymbol{Z}^* \in \mathbb{Y}^*, \quad \mathbb{Y}^* = \mathbb{S}^2$$

to see that the first equation in Proposition 2.2.9 (b) can be rewritten as

$$\bar{\boldsymbol{Z}}^* = \frac{\partial f}{\partial \boldsymbol{E}_{\mathrm{e}}}(\bar{\boldsymbol{E}}_{\mathrm{e}}) = \boldsymbol{C} : \bar{\boldsymbol{E}}_{\mathrm{e}}. \tag{9.56}$$

The second equation in Proposition 2.2.9 (b) is equivalent to

$$\bar{\boldsymbol{E}}_{\mathrm{e}} - \boldsymbol{E} \succeq \boldsymbol{o}, \quad \bar{\boldsymbol{Z}}^* \succeq \boldsymbol{o}, \quad (\bar{\boldsymbol{E}}_{\mathrm{e}} - \boldsymbol{E}) : \bar{\boldsymbol{Z}}^* = 0, \tag{9.57}$$

because the conjugate function of g is obtained by using Fact 1.3.17 (ii) as

$$g^*(\boldsymbol{Z}^*) = \sup_{\boldsymbol{Z} \in \mathbb{S}^3} \{\boldsymbol{Z}^* : \boldsymbol{Z} \mid \boldsymbol{Z} - \boldsymbol{E} \succeq \boldsymbol{o}\} = \begin{cases} \boldsymbol{E} : \boldsymbol{Z}^* & \text{if } -\boldsymbol{Z}^* \succeq \boldsymbol{o}, \\ +\infty & \text{otherwise.} \end{cases} \tag{9.58}$$

Consequently, $\bar{\boldsymbol{E}}_{\mathrm{e}}$ is optimal for (9.55) if and only if there exists $\bar{\boldsymbol{Z}}^*$ satisfying (9.56) and (9.57). These optimality conditions are rewritten as (9.50) by putting

$$-\bar{\boldsymbol{E}}_{\mathrm{w}} := \bar{\boldsymbol{E}}_{\mathrm{e}} - \boldsymbol{E}, \quad \bar{\boldsymbol{S}} := \bar{\boldsymbol{Z}}^*.$$

It is then immediate from (9.53) to see that the optimal value of (9.55) coincides with $\hat{w}_{\mathrm{mem}}(\boldsymbol{E})$. □

The latter assertion of Proposition 9.3.4 is explicitly written as

$$\hat{w}_{\mathrm{mem}}(\boldsymbol{E}) = \min_{\boldsymbol{E}_{\mathrm{e}} \in \mathbb{S}^2} \{\hat{w}_{\mathrm{ref}}(\boldsymbol{E}_{\mathrm{e}}) \mid \boldsymbol{E}_{\mathrm{e}} \succeq \boldsymbol{E}\}. \tag{9.59}$$

By using this expression, one can easily show that $\hat{w}_{\rm mem}$ is convex in a manner similar to Proposition 8.2.6. Moreover, the optimal solution of the problem in (9.59) is unique, since $\hat{w}_{\rm ref}$ is strictly convex. Therefore, $\bar{\boldsymbol{E}}_{\rm e}$ in Proposition 9.3.4 is determined uniquely, and thence the decomposition of $\boldsymbol{E}$ in (9.50b) is also determined uniquely. Consequently, it is verified that $\boldsymbol{S}$, $\boldsymbol{E}_{\rm e}$, and $\boldsymbol{E}_{\rm w}$ in (9.53) are determined uniquely. For this reason, there exists a function $\hat{\boldsymbol{S}}_{\rm mem}$ such that

$$\boldsymbol{S} = \hat{\boldsymbol{S}}_{\rm mem}(\boldsymbol{E}) \quad \Leftrightarrow \quad (9.50). \tag{9.60}$$

9.3.2 Principle of potential energy

Let the vector field $\underline{\boldsymbol{p}} : \Omega \to \mathbb{R}^2$ denote (the density of) the applied body force per unit volume in the reference configuration. Consider a partition of Γ into two disjoint subsets as $\Gamma = \Gamma_{\rm N} \cup \Gamma_{\rm D}$. We denote by $\underline{\boldsymbol{t}} : \Gamma_{\rm N} \to \mathbb{R}^2$ the specified (density of) traction (or the surface force) per unit area in the reference configuration. The deformation on $\Gamma_{\rm D}$ is prescribed as $\boldsymbol{\varphi} = \underline{\boldsymbol{\varphi}}$. Then the minimization problem of the total potential energy is formulated in terms of the displacement field $\boldsymbol{u}$ as

$$\left.\begin{aligned}
&\min_{\boldsymbol{u},\boldsymbol{E}} \int_{\Omega} \hat{w}_{\rm mem}(\boldsymbol{E}){\rm d}\Omega - \int_{\Omega} \underline{\boldsymbol{p}} \cdot \boldsymbol{u}{\rm d}\Omega - \int_{\Gamma_{\rm N}} \underline{\boldsymbol{t}} \cdot \boldsymbol{u}{\rm d}\Gamma \\
&\ {\rm s.\,t.}\ \boldsymbol{E} = \frac{1}{2}(\nabla\boldsymbol{u}^{\rm T} + \nabla\boldsymbol{u} + \nabla\boldsymbol{u}^{\rm T}\nabla\boldsymbol{u}) \quad \text{in } \Omega, \\
&\qquad u_j = \underline{u}_j \quad \text{on } \Gamma_{\rm D},
\end{aligned}\right\}$$

where the Green–Lagrange strain $\boldsymbol{E}$ is defined by (9.41). Alternatively, from (9.28) the formulation in terms of the deformation $\boldsymbol{\varphi}$ results in

$$\left.\begin{aligned}
({\rm PE}) : &\min_{\boldsymbol{\varphi},\boldsymbol{E}} \int_{\Omega} \hat{w}_{\rm mem}(\boldsymbol{E}){\rm d}\Omega \\
&\qquad - \int_{\Omega} \underline{\boldsymbol{p}} \cdot (\boldsymbol{\varphi} - \boldsymbol{x}){\rm d}\Omega - \int_{\Gamma_{\rm N}} \underline{\boldsymbol{t}} \cdot (\boldsymbol{\varphi} - \boldsymbol{x}){\rm d}\Gamma \\
&{\rm s.\,t.}\ \boldsymbol{E} = \frac{1}{2}(\nabla\boldsymbol{\varphi}^{\rm T}\nabla\boldsymbol{\varphi} - \boldsymbol{I}) \quad \text{in } \Omega, \\
&\qquad \boldsymbol{\varphi} = \underline{\boldsymbol{\varphi}} \quad \text{on } \Gamma_{\rm D}.
\end{aligned}\right\} \tag{9.61}$$

Thus the equilibrium configuration of the membrane corresponds to a stationary point of (PE) in (9.61).

By virtue of Proposition 9.3.4, the problem above can be reformulated as the following one without changing the optimal solution:

$$\left.\begin{aligned}
(\overline{\rm PE}) : &\min_{\boldsymbol{\varphi},\boldsymbol{E}_{\rm e}} \int_{\Omega} \hat{w}_{\rm ref}(\boldsymbol{E}_{\rm e}){\rm d}\Omega - \int_{\Omega} \underline{\boldsymbol{p}} \cdot (\boldsymbol{\varphi} - \boldsymbol{x}){\rm d}\Omega - \int_{\Gamma_{\rm N}} \underline{\boldsymbol{t}} \cdot (\boldsymbol{\varphi} - \boldsymbol{x}){\rm d}\Gamma \\
&{\rm s.\,t.}\ \boldsymbol{E}_{\rm e} \succeq \frac{1}{2}(\nabla\boldsymbol{\varphi}^{\rm T}\nabla\boldsymbol{\varphi} - \boldsymbol{I}) \quad \text{in } \Omega, \\
&\qquad \boldsymbol{\varphi} = \underline{\boldsymbol{\varphi}} \quad \text{on } \Gamma_{\rm D}.
\end{aligned}\right\} \tag{9.62}$$

More precisely, the optimal solutions of (PE) in (9.61) and $(\overline{\text{PE}})$ in (9.62) are related to each other as stated in Proposition 9.3.5 below.

Proposition 9.3.5. $(\overline{\text{PE}})$ *is equivalent to* (PE) *in the following sense:*

(i) *Let* $(\bar{\varphi}, \bar{\boldsymbol{E}}_{\rm e})$ *be optimal for* $(\overline{\text{PE}})$, *and define* $\bar{\boldsymbol{E}}$ *by* $\bar{\boldsymbol{E}} = \frac{1}{2}(\nabla\bar{\varphi}^{\rm T}\nabla\bar{\varphi} - \boldsymbol{I})$. *Then* $(\bar{\varphi}, \bar{\boldsymbol{E}})$ *is optimal for* (PE).

(ii) *Let* $(\bar{\varphi}, \bar{\boldsymbol{E}})$ *be optimal for* (PE). *Define* $\bar{\boldsymbol{E}}_{\rm e} \in \mathbb{S}^2$ *so that*

$$\hat{\boldsymbol{S}}_{\rm mem}(\bar{\boldsymbol{E}}) = \hat{\boldsymbol{S}}_{\rm ref}(\bar{\boldsymbol{E}}_{\rm e}) \tag{9.63}$$

is satisfied with $\hat{\boldsymbol{S}}_{\rm ref}$ *and* $\hat{\boldsymbol{S}}_{\rm mem}$ *in* (9.48) *and* (9.60), *respectively. Then* $(\bar{\varphi}, \bar{\boldsymbol{E}}_{\rm e})$ *is optimal for* $(\overline{\text{PE}})$.

(iii) (PE) *and* $(\overline{\text{PE}})$ *share the same optimal value.*

Proof. This result essentially relies on the fact that substitution of the argument in Proposition 9.3.4 into (PE) yields $(\overline{\text{PE}})$. A proof is given in a manner analogous to Proposition 8.2.7 and is omitted here. □

Remark 9.3.6. We can actually show slightly more than Proposition 9.3.5 (ii): If $(\bar{\varphi}, \bar{\boldsymbol{E}})$ is a stationary point for problem (9.61), then $(\bar{\varphi}, \bar{\boldsymbol{E}}_{\rm e})$ satisfying

$$\exists \bar{\boldsymbol{S}} \in \mathbb{S}^2 : \quad \bar{\boldsymbol{E}}_{\rm e} - \bar{\boldsymbol{E}} \succeq \boldsymbol{o}, \ \bar{\boldsymbol{S}} \succeq \boldsymbol{o}, \ (\bar{\boldsymbol{E}}_{\rm e} - \bar{\boldsymbol{E}}) : \bar{\boldsymbol{s}} = 0$$

is an optimal solution of problem (9.62). We postpone this issue to section 9.4.4, in which we shall show that a necessary and sufficient condition for the optimality of $(\overline{\text{PE}})$ coincides precisely with the boundary value problem (9.115) for the equilibrium of the membrane. Hence any stationary point of (PE), i.e., a solution of the boundary value problem, is optimal for $(\overline{\text{PE}})$. As an important (and rather surprising) consequence, the potential energy attains the minimum in (PE) for any equilibrium configuration, which is not usually the case in finite deformation theory. See Proposition 9.4.12 in section 9.4.4 for more account. ■

9.4 Principle of Complementary Energy for Membranes

The importance of $(\overline{\text{PE}})$ in (9.62) stems from the fact that this problem is written in the form of conic optimization problem, and hence it enjoys the theoretical and numerical benefits of the convex optimization problems. In this section the principle of complementary energy for membranes is established by applying the results of the Fenchel dual theory (in section 2.2.5) to $(\overline{\text{PE}})$.

9.4.1 Embedding to Fenchel form

The subsequent sections are devoted to investigating ($\overline{\text{PE}}$) in (9.62) by using the framework of the Fenchel duality introduced in section 2.2.5. To begin, we show that ($\overline{\text{PE}}$) can be embedded into the form of (P_F) in (2.28),[18] which plays a role of the primal problem in the framework of the Fenchel duality theory.

A key issue is the treatment of the positive-semidefinite constraint in ($\overline{\text{PE}}$). From the lemma on the Schur complement (see Fact 1.3.10), the positive-semidefinite constraint can be reduced to

$$\boldsymbol{E}_\text{e} \succeq \frac{1}{2}(\nabla\boldsymbol{\varphi}^\text{T}\nabla\boldsymbol{\varphi} - \boldsymbol{I}) \quad \Leftrightarrow \quad \begin{bmatrix} \boldsymbol{E}_\text{e} + (1/2)\boldsymbol{I} & -(1/2)\nabla\boldsymbol{\varphi}^\text{T} \\ -(1/2)\nabla\boldsymbol{\varphi} & (1/2)\boldsymbol{I} \end{bmatrix} \succeq \boldsymbol{o}.$$

Hence, we can equivalently rewrite ($\overline{\text{PE}}$) as

$$\left.\begin{aligned} &\min_{\boldsymbol{\varphi},\boldsymbol{E}_\text{e}} \int_\Omega \hat{w}_\text{ref}(\boldsymbol{E}_\text{e})\text{d}\Omega - \int_\Omega \underline{\boldsymbol{p}}\cdot(\boldsymbol{\varphi}-\boldsymbol{x})\text{d}\Omega - \int_{\Gamma_\text{N}} \underline{\boldsymbol{t}}\cdot(\boldsymbol{\varphi}-\boldsymbol{x})\text{d}\Gamma \\ &\text{s.t.} \ \begin{bmatrix} \boldsymbol{E}_\text{e} + (1/2)\boldsymbol{I} & -(1/2)\nabla\boldsymbol{\varphi}^\text{T} \\ -(1/2)\nabla\boldsymbol{\varphi} & (1/2)\boldsymbol{I} \end{bmatrix} \succeq \boldsymbol{o} \quad \text{in } \Omega, \\ &\qquad \boldsymbol{\varphi} = \underline{\boldsymbol{\varphi}} \quad \text{on } \Gamma_\text{D}. \end{aligned}\right\} \tag{9.64}$$

In what follows, we suppose that the Dirichlet boundary condition is given as

$$\underline{\boldsymbol{\varphi}} = \boldsymbol{x} \quad \text{on } \Gamma_\text{D} \tag{9.65}$$

to avoid technical difficulties. Moreover, we suppose for the Neumann boundary condition that $\underline{\boldsymbol{t}} \in L^2(\Gamma)^3$; see Lions–Magenes [270, Chaps. 1 and 2] for comprehensive treatments of non-homogeneous boundary value problems. Each component $\underline{p}_i$ of $\underline{\boldsymbol{p}}$ is supposed to belong to $L^2(\Omega)$, and we write $\underline{\boldsymbol{p}} \in L^2(\Omega)^2$ with the notation

$$L^2(\Omega)^2 = \{\underline{\boldsymbol{p}} \mid \underline{p}_i \in L^2(\Omega) \ (i=1,2)\}.$$

Concerning the variables of problem (9.64), we set $\boldsymbol{\varphi} \in H^1(\Omega)^2$ and $\boldsymbol{E}_\text{e} \in L^2(\Omega)^{2\times 2}_\mathbb{S}$ with

$$\begin{aligned} H^1(\Omega)^2 &= \{\boldsymbol{u} \mid u_i \in H^1(\Omega) \ (i=1,2)\}, \\ L^2(\Omega)^{2\times 2}_\mathbb{S} &= \{\boldsymbol{E}_\text{e} \mid E_{\text{e}ij} = E_{\text{e}ji} \in L^2(\Omega) \ (i,j=1,2)\}; \end{aligned}$$

see section 8.3.1 for more details of the definitions of these functional spaces.

In the setting above, we define the primal variables and spaces by

$$x = (\boldsymbol{\varphi}, \boldsymbol{E}_\text{e}) \in \mathbb{V}, \tag{9.66a}$$

$$\mathbb{V} = H^1(\Omega)^2 \times L^2(\Omega)^{2\times 2}_\mathbb{S}, \tag{9.66b}$$

$$\mathbb{Y} = L^2(\Omega)^{4\times 4}_\mathbb{S} \tag{9.66c}$$

[18]See the orientation part of this chapter for convenience.

in accordance with the notation in the Fenchel duality in section 2.2.5. Subsequently, define $f : \mathbb{V} \to \mathbb{R} \cup \{+\infty\}$, $g : \mathbb{Y} \to \mathbb{R} \cup \{+\infty\}$, and $\Lambda : \mathbb{V} \to \mathbb{Y}$ by[19]

$$f(x) = \begin{cases} \displaystyle\int_{\Omega} \hat{w}_{\mathrm{ref}}(\boldsymbol{E}_{\mathrm{e}}) \mathrm{d}\Omega - \int_{\Omega} \underline{\boldsymbol{p}} \cdot (\boldsymbol{\varphi} - \boldsymbol{x}) \mathrm{d}\Omega & \\ \displaystyle\quad - \int_{\Gamma_{\mathrm{N}}} \underline{\boldsymbol{t}} \cdot (\gamma(\boldsymbol{\varphi}) - \boldsymbol{x}) \mathrm{d}\Gamma & \text{if } \gamma(\boldsymbol{\varphi}) = \boldsymbol{x} \text{ (on } \Gamma_{\mathrm{D}}), \\ +\infty & \text{otherwise,} \end{cases} \tag{9.67}$$

$$g(z) = \begin{cases} 0 & \text{if } \begin{bmatrix} \boldsymbol{Z}_{11} + (1/2)\boldsymbol{I} & \boldsymbol{Z}_{21}^{\mathrm{T}} \\ \boldsymbol{Z}_{21} & \boldsymbol{Z}_{22} + (1/2)\boldsymbol{I} \end{bmatrix} \succeq \boldsymbol{o} \text{ (in } \Omega), \\ +\infty & \text{otherwise,} \end{cases} \tag{9.68}$$

$$\Lambda x = \begin{bmatrix} \boldsymbol{E}_{\mathrm{e}} & -(1/2)\nabla\boldsymbol{\varphi}^{\mathrm{T}} \\ -(1/2)\nabla\boldsymbol{\varphi} & \boldsymbol{o} \end{bmatrix}, \tag{9.69}$$

where γ in (9.67) is the trace operator from $H^1(\Omega)^2$ to $H^{1/2}(\Gamma)^2$; see section 8.3.1. Notice here that the element of $\mathbb{Y}$ is denoted by

$$\mathbb{Y} \ni z = \boldsymbol{Z} = \begin{bmatrix} \boldsymbol{Z}_{11} & \boldsymbol{Z}_{21}^{\mathrm{T}} \\ \boldsymbol{Z}_{21} & \boldsymbol{Z}_{22} \end{bmatrix}. \tag{9.70}$$

We thus encode problem (9.64) (and hence ($\overline{\mathrm{PE}}$) also) into the form of problem (2.28). In the next section we describe the dual problem (2.31) in the present situation.

9.4.2 Dual problem

It follows from (9.66) that the variables and associated spaces for the Fenchel dual problem are given by

$$x^* = (\boldsymbol{\varphi}^*, \boldsymbol{E}_{\mathrm{e}}^*), \quad z^* = \boldsymbol{Z}^* = \begin{bmatrix} \boldsymbol{Z}_{11}^* & \boldsymbol{Z}_{21}^{*\mathrm{T}} \\ \boldsymbol{Z}_{21}^* & \boldsymbol{Z}_{22}^* \end{bmatrix}, \tag{9.71a}$$

$$\mathbb{V}^* = H^{-1}(\Omega)^2 \times L^2(\Omega)_{\mathbb{S}}^{2\times 2}, \tag{9.71b}$$

$$\mathbb{Y}^* = L^2(\Omega)_{\mathbb{S}}^{4\times 4} = \mathbb{Y}. \tag{9.71c}$$

We next derive the explicit formulation of problem ($\mathrm{P}_{\mathrm{F}}^*$) in (2.31),[20] which is dual to (P_{F}) in (2.28). Turning toward this end, we calculate the explicit forms of $f^*(\Lambda^* z^*)$ and $g^*(z^*)$ in the following two propositions.[21]

[19] The reader henceforth should not confuse x with $\boldsymbol{x}$; $x \in \mathbb{V}$ is the variable of the Fenchel primal problem (P_{F}), while $\boldsymbol{x} \in \mathbb{R}^2$ is the vector pointing to a location within the body of membrane $\overline{\Omega}$. Therefore, $\boldsymbol{x}$ is not treated as a variable of the optimization problem.

[20] See the orientation part of this chapter for convenience.

[21] The reader who is not especially interested in the calculations of the conjugate functions and adjoint operators can pass over Proposition 9.4.1 and Proposition 9.4.2.

Proposition 9.4.1. *For f in (9.67) and Λ in (9.69), we have that*

$$f^*(\Lambda^* \boldsymbol{Z}^*) = \begin{cases} \displaystyle\int_\Omega \hat{w}^*_{\text{ref}}(\boldsymbol{Z}^*_{11})\mathrm{d}\Omega - \int_{\Gamma_\mathrm{D}} (\boldsymbol{Z}^*_{21}\boldsymbol{n}) \cdot \boldsymbol{x}\mathrm{d}\Gamma & \\ \qquad \displaystyle - \int_\Omega \underline{\boldsymbol{p}} \cdot \boldsymbol{x}\mathrm{d}\Omega - \int_{\Gamma_\mathrm{N}} \underline{\boldsymbol{t}} \cdot \boldsymbol{x}\mathrm{d}\Gamma & \text{if } -\operatorname{div} \boldsymbol{Z}^*_{21} = \underline{\boldsymbol{p}} \text{ (in } \Omega), \\ & \boldsymbol{Z}^*_{21}\boldsymbol{n} = \underline{\boldsymbol{t}} \text{ (on } \Gamma_\mathrm{N}), \\ +\infty & \text{otherwise,} \end{cases}$$

where $\hat{w}^*_{\text{ref}} : \mathbb{S}^2 \to \mathbb{R}$ is the conjugate function of $\hat{w}_{\text{ref}}$.

Proof. From (9.66a), (9.69), and (9.71a), we have that

$$\begin{aligned} \langle \Lambda^* z^*, x \rangle &= \langle z^*, \Lambda \rangle \\ &= \int_\Omega \begin{bmatrix} \boldsymbol{Z}^*_{11} & \boldsymbol{Z}^{*\mathrm{T}}_{21} \\ \boldsymbol{Z}^*_{21} & \boldsymbol{Z}^*_{22} \end{bmatrix} \bullet \begin{bmatrix} \boldsymbol{E}_\mathrm{e} & -(1/2)\nabla\boldsymbol{\varphi}^\mathrm{T} \\ -(1/2)\nabla\boldsymbol{\varphi} & \boldsymbol{o} \end{bmatrix} \mathrm{d}\Omega \\ &= \int_\Omega (\boldsymbol{Z}^*_{11} \bullet \boldsymbol{E}_\mathrm{e} - \boldsymbol{Z}^*_{21} \bullet \nabla\boldsymbol{\varphi})\, \mathrm{d}\Omega. \end{aligned} \tag{9.72}$$

Direct application of the definition of conjugate function (see (2.13)) to f in (9.67) yields

$$\begin{aligned} f^*(\Lambda^* z^*) &= \sup\{\langle \Lambda^* z^*, x\rangle - f(x) \mid x \in \operatorname{dom} f\} \\ &= \sup_{\substack{\boldsymbol{\varphi}:\gamma(\boldsymbol{\varphi})=\boldsymbol{x} \\ (\text{on } \Gamma_\mathrm{D})}} \left\{ -\langle \boldsymbol{Z}^*_{21}, \nabla\boldsymbol{\varphi}\rangle + \int_\Omega \underline{\boldsymbol{p}} \cdot (\boldsymbol{\varphi} - \boldsymbol{x})\mathrm{d}\Omega + \int_{\Gamma_\mathrm{N}} \underline{\boldsymbol{t}} \cdot (\gamma(\boldsymbol{\varphi}) - \boldsymbol{x})\mathrm{d}\Gamma \right\} \\ &\quad + \sup_{\boldsymbol{E}_\mathrm{e}} \left\{ \langle \boldsymbol{Z}^*_{11}, \boldsymbol{E}_\mathrm{e}\rangle - \int_\Omega \hat{w}_{\text{ref}}(\boldsymbol{E}_\mathrm{e})\mathrm{d}\Omega \right\} \end{aligned} \tag{9.73}$$

$$\begin{aligned} &= \sup_{\substack{\boldsymbol{u}:\gamma(\boldsymbol{u})=\boldsymbol{0} \\ (\text{on } \Gamma_\mathrm{D})}} \left\{ -\langle \boldsymbol{Z}^*_{21}, \nabla\boldsymbol{u}\rangle + \int_\Omega \underline{\boldsymbol{p}} \cdot \boldsymbol{u}\mathrm{d}\Omega + \int_{\Gamma_\mathrm{N}} \underline{\boldsymbol{t}} \cdot \gamma(\boldsymbol{u})\mathrm{d}\Gamma \right\} - \langle \boldsymbol{Z}^*_{21}, \boldsymbol{I}\rangle \\ &\quad + \sup_{\boldsymbol{E}_\mathrm{e}} \left\{ \langle \boldsymbol{Z}^*_{11}, \boldsymbol{E}_\mathrm{e}\rangle - \int_\Omega \hat{w}_{\text{ref}}(\boldsymbol{E}_\mathrm{e})\mathrm{d}\Omega \right\}, \end{aligned} \tag{9.74}$$

where relation (9.28) was used. The first supremum in (9.74) is finite if and only if $\boldsymbol{Z}^*_{21} \in L^2(\Omega)^{2\times 2}_{\mathbb{S}}$ satisfies

$$\int_\Omega \left[-\boldsymbol{Z}^*_{21} : (\nabla\boldsymbol{v}) + \underline{\boldsymbol{p}} \cdot \boldsymbol{v} \right] \mathrm{d}\Omega + \int_{\Gamma_\mathrm{N}} \underline{\boldsymbol{t}} \cdot \gamma(\boldsymbol{v})\mathrm{d}\Gamma = 0, \qquad \forall \boldsymbol{v} \in H^1(\Omega)^2 : \gamma(\boldsymbol{v}) = \boldsymbol{0} \text{ (on } \Gamma_\mathrm{D}). \tag{9.75}$$

Condition (9.75) implies that

$$\operatorname{div} \boldsymbol{Z}^*_{21} + \underline{\boldsymbol{p}} = \boldsymbol{0} \quad \text{in } \Omega \tag{9.76}$$

in the sense of the distributions in Ω. Since we assume $\underline{\boldsymbol{p}} \in L^2(\Omega)^2$, we see that (9.76) implies $\boldsymbol{Z}^*_{21} \in \mathcal{T}$ with $\mathcal{T}$ in (8.43). Therefore, from

Theorem 8.3.1 we can define the continuous linear mapping $\pi : \mathcal{T} \to H^{-1/2}(\Gamma)^2$, and obtain Green's formula

$$\int_\Omega \boldsymbol{Z}_{21}^* : (\nabla \boldsymbol{v}) \mathrm{d}\Omega + \int_\Omega (\mathrm{div}\, \boldsymbol{Z}_{21}^*) \cdot \boldsymbol{v} \mathrm{d}\Omega = \int_\Gamma \pi(\boldsymbol{Z}_{21}^*) \cdot \gamma(\boldsymbol{v}) \mathrm{d}\Omega. \quad (9.77)$$

Substituting (9.76) and (9.77) into (9.75) results in

$$\int_{\Gamma_\mathrm{N}} (\underline{\boldsymbol{t}} - \pi(\boldsymbol{Z}_{21}^*)) \cdot \gamma(\boldsymbol{v}) \mathrm{d}\Gamma - \int_{\Gamma_\mathrm{D}} \pi(\boldsymbol{Z}_{21}^*) \cdot \gamma(\boldsymbol{v}) \mathrm{d}\Gamma = 0,$$
$$\forall \boldsymbol{v} \in H^1(\Omega)^2 : \gamma(\boldsymbol{v}) = \boldsymbol{0} \text{ (on } \Gamma_\mathrm{D}),$$

which implies

$$\boldsymbol{Z}_{21}^* \boldsymbol{n} = \underline{\boldsymbol{t}} \quad \text{on } \Gamma_\mathrm{N}. \quad (9.78)$$

Conversely, (9.76) and (9.78) imply (9.75). Thus the first supremum in (9.74) is finite if and only if (9.76) and (9.78) are satisfied. If that is the case, by using Green's formula

$$\int_\Omega \boldsymbol{Z}_{21}^* : (\nabla \boldsymbol{\varphi}) \mathrm{d}\Omega + \int_\Omega (\mathrm{div}\, \boldsymbol{Z}_{21}^*) \cdot \boldsymbol{\varphi} \mathrm{d}\Omega = \int_\Gamma \pi(\boldsymbol{Z}_{21}^*) \cdot \gamma(\boldsymbol{\varphi}) \mathrm{d}\Omega \quad (9.79)$$

and by substituting (9.76) and (9.78) we see that the first supremum in (9.73) is reduced to

$$\sup_{\substack{\boldsymbol{\varphi}:\gamma(\boldsymbol{\varphi})=\boldsymbol{x} \\ (\text{on } \Gamma_\mathrm{D})}} \left\{ \int_{\Gamma_\mathrm{D}} -\pi(\boldsymbol{Z}_{21}^*) \cdot \gamma(\boldsymbol{\varphi}) \mathrm{d}\Gamma \right\} - \int_\Omega \underline{\boldsymbol{p}} \cdot \boldsymbol{x} \mathrm{d}\Omega - \int_{\Gamma_\mathrm{N}} \underline{\boldsymbol{t}} \cdot \boldsymbol{x} \mathrm{d}\Gamma$$
$$= - \int_{\Gamma_\mathrm{D}} \pi(\boldsymbol{Z}_{21}^*) \cdot \boldsymbol{x} \mathrm{d}\Gamma - \int_\Omega \underline{\boldsymbol{p}} \cdot \boldsymbol{x} \mathrm{d}\Omega - \int_{\Gamma_\mathrm{N}} \underline{\boldsymbol{t}} \cdot \boldsymbol{x} \mathrm{d}\Gamma. \quad (9.80)$$

On the other hand, it is immediately evident that the second supremum in (9.73) is reduced to

$$\sup_{\boldsymbol{E}_\mathrm{e}} \left\{ \langle \boldsymbol{Z}_{11}^*, \boldsymbol{E}_\mathrm{e} \rangle - \int_\Omega \hat{w}_\mathrm{ref}(\boldsymbol{E}_\mathrm{e}) \mathrm{d}\Omega \right\} = \int_\Omega \hat{w}_\mathrm{ref}^*(\boldsymbol{Z}_{11}^*) \mathrm{d}\Omega. \quad (9.81)$$

The assertion of this proposition is then obtained from (9.80) and (9.81). □

Proposition 9.4.2. *The conjugate function $g^* : \mathbb{Y}^* \to \mathbb{R} \cup \{+\infty\}$ of g in (9.68) is written as*

$$g^*(\boldsymbol{Z}^*) = \begin{cases} -\displaystyle\int_\Omega \frac{1}{2} \mathrm{tr}(\boldsymbol{Z}_{11}^* + \boldsymbol{Z}_{22}^*) \mathrm{d}\Omega & \text{if } -\boldsymbol{Z}^* \succeq \boldsymbol{o} \text{ (in } \Omega), \\ +\infty & \text{otherwise.} \end{cases}$$

Proof. For simplicity, we write

$$\hat{\boldsymbol{Z}}(\boldsymbol{Z}) = \begin{bmatrix} \boldsymbol{Z}_{11} + (1/2)\boldsymbol{I} & \boldsymbol{Z}_{21}^\mathrm{T} \\ \boldsymbol{Z}_{21} & \boldsymbol{Z}_{22} + (1/2)\boldsymbol{I} \end{bmatrix}. \quad (9.82)$$

By applying the definition of the conjugate function to g in (9.68), we obtain

$$\begin{aligned}
g^*(z^*) &= \sup\{\langle z^*, z\rangle \mid z \in \operatorname{dom} g\} \\
&= \sup_{\boldsymbol{Z}} \left\{ \int_\Omega \boldsymbol{Z}^* : \boldsymbol{Z} \mathrm{d}\Omega \mid \hat{\boldsymbol{Z}}(\boldsymbol{Z}) \succeq \boldsymbol{o} \ (\text{in } \Omega) \right\} \\
&= \sup_{\boldsymbol{Z}} \left\{ \int_\Omega \boldsymbol{Z}^* : \hat{\boldsymbol{Z}}(\boldsymbol{Z}) \mathrm{d}\Omega \mid \hat{\boldsymbol{Z}}(\boldsymbol{Z}) \succeq \boldsymbol{o} \ (\text{in } \Omega) \right\} \\
&\quad - \int_\Omega (\boldsymbol{Z}_{11}^* + \boldsymbol{Z}_{22}^*) : (1/2)\boldsymbol{I} \mathrm{d}\Omega \\
&= -\int_\Omega \inf_{\boldsymbol{Z}} \{(-\boldsymbol{Z}^*) : \hat{\boldsymbol{Z}}(\boldsymbol{Z}) \mid \hat{\boldsymbol{Z}}(\boldsymbol{Z}) \succeq \boldsymbol{o}\} \mathrm{d}\Omega - \int_\Omega \frac{1}{2} \operatorname{tr}(\boldsymbol{Z}_{11}^* + \boldsymbol{Z}_{22}^*) \mathrm{d}\Omega.
\end{aligned}$$

Then the conclusion follows from Fact 1.3.17 (ii). □

We thus obtain the explicit forms of $f^*(\Lambda^* z^*)$ and $g^*(z^*)$ in Proposition 9.4.1 and Proposition 9.4.2, respectively. Consequently, the Fenchel dual problem (P_F^*) in (2.31), i.e., $\sup_{z^*}\{-f^*(\Lambda^* z^*) - g^*(-z^*) \mid z^* \in \mathbb{Y}^*\}$, can be written explicitly as

$$\left.\begin{aligned}
\max_{\boldsymbol{Z}^*} \ & -\int_\Omega \hat{w}_{\mathrm{ref}}^*(\boldsymbol{Z}_{11}^*) \mathrm{d}\Omega - \int_\Omega \frac{1}{2} \operatorname{tr}(\boldsymbol{Z}_{11}^* + \boldsymbol{Z}_{22}^*) \mathrm{d}\Omega \\
& + \int_{\Gamma_\mathrm{D}} (\boldsymbol{Z}_{21}^* \boldsymbol{n}) \cdot \boldsymbol{x} \mathrm{d}\Gamma + \int_\Omega \underline{\boldsymbol{p}} \cdot \boldsymbol{x} \mathrm{d}\Omega + \int_{\Gamma_\mathrm{N}} \underline{\boldsymbol{t}} \cdot \boldsymbol{x} \mathrm{d}\Gamma \\
\text{s.t.} \ & -\operatorname{div} \boldsymbol{Z}_{21}^* = \underline{\boldsymbol{p}} \quad \text{in } \Omega, \\
& \begin{bmatrix} \boldsymbol{Z}_{11}^* & \boldsymbol{Z}_{21}^{*\mathrm{T}} \\ \boldsymbol{Z}_{21}^* & \boldsymbol{Z}_{22}^* \end{bmatrix} \succeq \boldsymbol{o} \quad \text{in } \Omega, \\
& \boldsymbol{Z}_{21}^* \boldsymbol{n} = \underline{\boldsymbol{t}} \quad \text{on } \Gamma_\mathrm{N}.
\end{aligned}\right\} \tag{9.83}$$

By rewriting the variables as[22]

$$\boldsymbol{S} = \boldsymbol{Z}_{11}^*, \quad \boldsymbol{\Pi} = \boldsymbol{Z}_{21}^*, \quad \boldsymbol{\tau}^\varphi = \boldsymbol{Z}_{22}^*, \tag{9.84}$$

the problem above is rewritten as

$$\left.\begin{aligned}
(\overline{\mathrm{CE}}): \max_{\boldsymbol{S}, \boldsymbol{\Pi}, \boldsymbol{\tau}^\varphi} \ & -\int_\Omega \hat{w}_{\mathrm{ref}}^*(\boldsymbol{S}) \mathrm{d}\Omega - \int_\Omega \frac{1}{2} \operatorname{tr}(\boldsymbol{S} + \boldsymbol{\tau}^\varphi) \mathrm{d}\Omega \\
& + \int_{\Gamma_\mathrm{D}} (\boldsymbol{\Pi} \boldsymbol{n}) \cdot \boldsymbol{x} \mathrm{d}\Gamma + \int_\Omega \underline{\boldsymbol{p}} \cdot \boldsymbol{x} \mathrm{d}\Omega + \int_{\Gamma_\mathrm{N}} \underline{\boldsymbol{t}} \cdot \boldsymbol{x} \mathrm{d}\Gamma \\
\text{s.t.} \quad & -\operatorname{div} \boldsymbol{\Pi} = \underline{\boldsymbol{p}} \quad \text{in } \Omega, \\
& \begin{bmatrix} \boldsymbol{S} & \boldsymbol{\Pi}^\mathrm{T} \\ \boldsymbol{\Pi} & \boldsymbol{\tau}^\varphi \end{bmatrix} \succeq \boldsymbol{o} \quad \text{in } \Omega, \\
& \boldsymbol{\Pi} \boldsymbol{n} = \underline{\boldsymbol{t}} \quad \text{on } \Gamma_\mathrm{N}.
\end{aligned}\right\} \tag{9.85}$$

[22] It shall be verified in section 9.4.3 that $\boldsymbol{S}$, $\boldsymbol{\Pi}$, and $\boldsymbol{\tau}^\varphi$ truly correspond to the first Piola–Kirchhoff stress tensor, the second Piola–Kirchhoff stress tensor, and the Kirchhoff stress tensor, respectively.

Remark 9.4.3. In the objective function of $(\overline{\text{CE}})$, the last two terms

$$\int_{\Omega} \underline{\boldsymbol{p}} \cdot \boldsymbol{x} \mathrm{d}\Omega + \int_{\Gamma_{\mathrm{N}}} \underline{\boldsymbol{t}} \cdot \boldsymbol{x} \mathrm{d}\Gamma$$

are constant. Note that $\boldsymbol{x}$ represents the point of the reference configuration. Although the optimal solution does not change even if we omit these constant terms, we prefer to leave them in the objective function. Indeed, with these terms, as shown in Proposition 9.4.4, the optimal value of $(\overline{\text{CE}})$ coincides with $(\overline{\text{PE}})$. ■

9.4.3 Duality and optimality

We now have the primal-dual pair of conic optimization problems: $(\overline{\text{PE}})$ in (9.62) and $(\overline{\text{CE}})$ in (9.85). This pair is settled in parallel with the Fenchel primal and dual problems, i.e., (P_{F}) in (2.28) and (P_{F}^*) in (2.31). Our next concern is to study the optimality conditions for $(\overline{\text{PE}})$ and $(\overline{\text{CE}})$ by applying the general results of the Fenchel duality.[23] The goal of this analysis is to show that, at the optimal solution of $(\overline{\text{CE}})$, the variables $\boldsymbol{\Pi}$, $\boldsymbol{S}$, and $\boldsymbol{\tau}^{\varphi}$ correspond to the first Piola–Kirchhoff stress, the second Piola–Kirchhoff stress, and the Kirchhoff stress, respectively, at the equilibrium state of the membrane. To begin, we establish the strong duality between $(\overline{\text{PE}})$ and $(\overline{\text{CE}})$ by applying Proposition 2.2.9.

Proposition 9.4.4. *Suppose that the boundary condition given on $\Gamma_{\mathrm{D}} \cup \Gamma_{\mathrm{N}}$ is smooth enough. Then, each of $(\overline{\text{PE}})$ and $(\overline{\text{CE}})$ has an optimal solution, and* $\min(\overline{\text{PE}}) = \max(\overline{\text{CE}})$.

Proof. It suffices to show that $(\overline{\text{PE}})$ and $(\overline{\text{CE}})$ satisfy the assumptions required in Proposition 2.2.12, because $(\overline{\text{PE}})$ and $(\overline{\text{CE}})$ correspond to (P_{F}) and (P_{F}^*), respectively.

From the definitions of f and g (in (9.67) and (9.68)), it is straightforward to see that

$$\begin{aligned} \operatorname{ri}(\operatorname{dom} f) &= \{(\boldsymbol{\varphi}, \boldsymbol{E}_{\mathrm{e}}) \mid \boldsymbol{\varphi} = \boldsymbol{x} \ (\text{on } \Gamma_{\mathrm{D}})\}, \\ \operatorname{ri}(\operatorname{dom} g) &= \left\{ \boldsymbol{Z} \mid \begin{bmatrix} \boldsymbol{Z}_{11} + (1/2)\boldsymbol{I} & \boldsymbol{Z}_{21}^{\mathrm{T}} \\ \boldsymbol{Z}_{21} & \boldsymbol{Z}_{22} + (1/2)\boldsymbol{I} \end{bmatrix} \succ \boldsymbol{o} \ (\text{in } \Omega) \right\}. \end{aligned}$$

Recall that the action of Λ is defined by (9.69). Assumption (i) required in Proposition 2.2.12, i.e., the condition

$$\exists x \in \operatorname{ri}(\operatorname{dom} f): \quad \Lambda x \in (\operatorname{ri} \operatorname{dom} g), \tag{9.86}$$

[23] As stated Remark 8.3.5 with regard to analysis of the duality in the principles for masonry structures, the discussion on the duality in the infinite-dimensional setting requires more careful treatment of functional spaces than the finite-dimensional case introduced in Chapter 2. However, as in the case of masonry structures, we shall not deal with this issue in detail but will suppose that the boundary conditions are given to be sufficiently smooth. Especially, we suppose (9.65) for the Dirichlet boundary condition.

is equivalent to the condition that there exists $(\boldsymbol{\varphi}, \boldsymbol{E}_{\rm e})$ satisfying $\boldsymbol{\varphi} = \boldsymbol{x}$ (on $\Gamma_{\rm D}$) and

$$\begin{bmatrix} \boldsymbol{E}_{\rm e} + (1/2)\boldsymbol{I} & -(1/2)\nabla\boldsymbol{\varphi}^{\rm T} \\ -(1/2)\nabla\boldsymbol{\varphi} & (1/2)\boldsymbol{I} \end{bmatrix} \succ \boldsymbol{o} \quad \text{in } \Omega. \tag{9.87}$$

However, it is easy to see that, for any $\boldsymbol{\varphi}$ satisfying $\boldsymbol{\varphi} = \boldsymbol{x}$ (on $\Gamma_{\rm D}$), we can always choose an $\boldsymbol{E}_{\rm e}$ satisfying (9.87). For example, $\boldsymbol{E}_{\rm e} = \alpha\boldsymbol{I}$ with sufficiently large α satisfies (9.87), because (9.87) is equivalent to $\boldsymbol{E}_{\rm e} \succ (1/2)(\nabla\boldsymbol{\varphi}^{\rm T}\nabla\boldsymbol{\varphi} - \boldsymbol{I})$ (in Ω); see Fact 1.3.10. Thus assumption (i) required in Proposition 2.2.12 is always satisfied.

We next consider assumption (ii) required in Proposition 2.2.12, which is equivalently rewritten as

$$\exists \boldsymbol{Z}^* : \quad -\boldsymbol{Z}^* \in \operatorname{ri}(\operatorname{dom} g^*), \ \Lambda^*\boldsymbol{Z}^* \in \operatorname{ri}(\operatorname{dom} f^*). \tag{9.88}$$

From the result of Proposition 9.4.1, we have that

$$\begin{aligned} &\Lambda^*\boldsymbol{Z}^* \in \operatorname{ri}(\operatorname{dom} f^*) \\ &\quad\Leftrightarrow \quad -\operatorname{div} \boldsymbol{Z}^*_{21} = \underline{\boldsymbol{p}} \text{ (in } \Omega), \ \boldsymbol{Z}^*_{21}\boldsymbol{n} = \underline{\boldsymbol{t}} \text{ (on } \Gamma_{\rm N}). \end{aligned} \tag{9.89}$$

The result of Proposition 9.4.2 means

$$\operatorname{ri}(\operatorname{dom} g^*) = \{\boldsymbol{Z}^* \mid -\boldsymbol{Z}^* \succ \boldsymbol{o} \text{ (in } \Omega)\}. \tag{9.90}$$

By rewriting the variables in (9.89) and (9.90) according to (9.84), we see that (9.88) is satisfied if and only if there exists a 3-tuple of $\boldsymbol{S}$, $\boldsymbol{\Pi}$, and $\boldsymbol{\tau}^{\varphi}$ satisfying

$$\begin{bmatrix} \boldsymbol{S} & \boldsymbol{\Pi}^{\rm T} \\ \boldsymbol{\Pi} & \boldsymbol{\tau}^{\varphi} \end{bmatrix} \succ \boldsymbol{o} \text{ (in } \Omega), \tag{9.91}$$

$$-\operatorname{div} \boldsymbol{\Pi} = \underline{\boldsymbol{p}} \text{ (in } \Omega), \quad \boldsymbol{\Pi}\boldsymbol{n} = \underline{\boldsymbol{t}} \text{ (on } \Gamma_{\rm N}). \tag{9.92}$$

It is easy to see that, for any $\boldsymbol{\Pi}$ satisfying (9.92), we can always choose $\boldsymbol{S}$ and $\boldsymbol{\tau}^{\varphi}$ so that (9.91) is satisfied (e.g., $\boldsymbol{S} = \boldsymbol{\tau}^{\varphi} = \alpha\boldsymbol{I}$ with sufficiently large α). □

Remark 9.4.5. In Proposition 9.4.4 and the subsequent results we assume only that the given boundary conditions are smooth enough. This is not the case in the small deformation theory. As discussed in Remark 9.2.5 (in section 9.2.3.2), the existence of solution in the small deformation theory depends on the loading condition. In contrast, Proposition 9.4.4 asserts that, even for the case in which the small deformation theory lacks a solution, the equilibrium solution can exist by considering the large deformation (as far as the boundary conditions are smooth enough). ■

We next derive the optimality conditions for ($\overline{\rm PE}$) and ($\overline{\rm CE}$) by applying Proposition 2.2.9.

Proposition 9.4.6. *Under the hypotheses in Proposition 9.4.4, the following statements hold:*

(i) $(\bar{\varphi}, \bar{\boldsymbol{E}}_{\rm e})$ *and* $(\bar{\boldsymbol{S}}, \bar{\boldsymbol{\Pi}}, \bar{\tau}^{\varphi})$ *are optimal for* $(\overline{\rm PE})$ *and* $(\overline{\rm CE})$*, respectively, if and only if they satisfy (in the weak sense)*

$$\bar{\boldsymbol{S}} = \hat{\boldsymbol{S}}_{\rm ref}(\bar{\boldsymbol{E}}_{\rm e}) \quad \text{in } \Omega, \tag{9.93a}$$

$$-\operatorname{div} \bar{\boldsymbol{\Pi}} = \underline{\boldsymbol{p}} \quad \text{in } \Omega, \tag{9.93b}$$

$$\bar{\varphi} = \underline{\boldsymbol{x}} \quad \text{on } \Gamma_{\rm D}; \qquad \bar{\boldsymbol{\Pi}}\boldsymbol{n} = \underline{\boldsymbol{t}} \quad \text{on } \Gamma_{\rm N}, \tag{9.93c}$$

$$\frac{1}{2}\begin{bmatrix} 2\bar{\boldsymbol{E}}_{\rm e} + \boldsymbol{I} & -\nabla\bar{\varphi}^{\rm T} \\ -\nabla\bar{\varphi} & \boldsymbol{I} \end{bmatrix} \succeq \boldsymbol{o}, \quad \begin{bmatrix} \bar{\boldsymbol{S}} & \bar{\boldsymbol{\Pi}}^{\rm T} \\ \bar{\boldsymbol{\Pi}} & \bar{\tau}^{\varphi} \end{bmatrix} \succeq \boldsymbol{o} \quad \text{in } \Omega, \tag{9.93d}$$

$$\frac{1}{2}\begin{bmatrix} 2\bar{\boldsymbol{E}}_{\rm e} + \boldsymbol{I} & -\nabla\bar{\varphi}^{\rm T} \\ -\nabla\bar{\varphi} & \boldsymbol{I} \end{bmatrix} \bullet \begin{bmatrix} \bar{\boldsymbol{S}} & \bar{\boldsymbol{\Pi}}^{\rm T} \\ \bar{\boldsymbol{\Pi}} & \bar{\tau}^{\varphi} \end{bmatrix} = 0 \quad \text{in } \Omega. \tag{9.93e}$$

(ii) *Optimal solutions* $(\bar{\varphi}, \bar{\boldsymbol{E}}_{\rm e})$ *and* $(\bar{\boldsymbol{S}}, \bar{\boldsymbol{\Pi}}, \bar{\tau}^{\varphi})$ *satisfy*

$$\bar{\boldsymbol{E}}_{\rm e} - \hat{\boldsymbol{E}}(\bar{\varphi}) \succeq \boldsymbol{o}, \quad \bar{\boldsymbol{S}} \succeq \boldsymbol{o}, \quad (\bar{\boldsymbol{E}}_{\rm e} - \hat{\boldsymbol{E}}(\bar{\varphi})) \bullet \bar{\boldsymbol{S}} = 0 \tag{9.94}$$

at each point in Ω*, where* $\hat{\boldsymbol{E}}(\bar{\varphi}) = \frac{1}{2}(\nabla\bar{\varphi}^{\rm T}\nabla\bar{\varphi} - \boldsymbol{I})$.

Proof. Assertion (ii) follows from condition (9.93a) in assertion (i) and Proposition 9.3.5 (ii) with $\hat{\boldsymbol{S}}_{\rm ref}$ in (9.60), because (9.94) is reduced to (9.50) by putting $\boldsymbol{E}_{\rm w} = \bar{\boldsymbol{E}}_{\rm e} - \hat{\boldsymbol{E}}(\bar{\varphi})$. It thus suffices to show assertion (i). For this purpose we make use of the formulation in Proposition 2.2.9 (b).

To see the first equation of Proposition 2.2.9 (b), we use the results on $f(x)$, $f^*(\Lambda^* z^*)$, and $\langle \lambda^* z^*, x\rangle$ studied in section 9.4.1 and section 9.4.2. By the definition in (9.67), if the boundary condition on $\Gamma_{\rm D}$ in (9.93c) is satisfied, then f is finite (and the converse is also true) and

$$f(x) = \int_{\Omega} \hat{w}_{\rm ref}(\boldsymbol{E}_{\rm e}) {\rm d}\Omega - \int_{\Omega} \underline{\boldsymbol{p}} \cdot (\boldsymbol{\varphi} - \boldsymbol{x}) {\rm d}\Omega - \int_{\Gamma_{\rm N}} \underline{\boldsymbol{t}} \cdot (\gamma(\boldsymbol{\varphi}) - \boldsymbol{x}) {\rm d}\Gamma.$$

Noting that $f^*(\Lambda^* \boldsymbol{Z}^*)$ is given in Proposition 9.4.1 and rewriting the variable according to (9.84), one can see that $f^*(\Lambda^* \boldsymbol{Z}^*)$ is finite if and only if (9.93b) and the condition on $\Gamma_{\rm N}$ in (9.93c) are satisfied. Moreover, if these conditions are satisfied, then $f^*(\Lambda^* \boldsymbol{Z}^*)$ takes the value

$$f^*(\Lambda^* z^*) = \int_{\Omega} \hat{w}^*_{\rm ref}(\boldsymbol{S}) {\rm d}\Omega - \int_{\Gamma_{\rm D}} \pi(\boldsymbol{\Pi}) \cdot \boldsymbol{x} {\rm d}\Gamma - \int_{\Omega} \underline{\boldsymbol{p}} \cdot \boldsymbol{x} {\rm d}\Omega - \int_{\Gamma_{\rm N}} \underline{\boldsymbol{t}} \cdot \boldsymbol{x} {\rm d}\Gamma.$$

It follows from (9.72) and (9.79) that the pairing $\langle \Lambda^* z^*, x\rangle$ is written with the current notation of variables as

$$\langle \Lambda^* z^*, x\rangle = \int_{\Omega} (\operatorname{div} \boldsymbol{\Pi}) \cdot \boldsymbol{\varphi} {\rm d}\Omega - \int_{\Gamma} \pi(\boldsymbol{\Pi}) \cdot \gamma(\boldsymbol{\varphi}) {\rm d}\Gamma + \int_{\Omega} \boldsymbol{S} \bullet \boldsymbol{E}_{\rm e} {\rm d}\Omega.$$

Hence, $f(x) + f^*(\Lambda^* z^*)$ is finite if and only if (9.93b) and (9.93c) are satisfied. Moreover, if that is the case, the equation $f(x) + f^*(\Lambda^* z^*) = \langle x, \Lambda^* z^*\rangle$ is reduced to

$$\int_{\Omega} \hat{w}_{\rm ref}(\boldsymbol{E}_{\rm e}) {\rm d}\Omega + \int_{\Omega} \hat{w}^*_{\rm ref}(\boldsymbol{S}) {\rm d}\Omega = \int_{\Omega} \boldsymbol{S} \bullet \boldsymbol{E}_{\rm e} {\rm d}\Omega, \tag{9.95}$$

which is equivalent to (9.93a).[24]

Second, for consideration of the second equation in Proposition 2.2.9 (b), we recall that $g(\Lambda x)$ is defined by (9.68) and (9.69), while $g(z^*)$ is given by Proposition 9.4.2. Thus, if (and only if) (9.93d) is satisfied, then both $g(\Lambda x)$ and $g^*(-z^*)$ are finite, $g(\Lambda x) + g^*(-z^*) = \frac{1}{2}\int_\Omega \mathrm{tr}(\boldsymbol{S} + \boldsymbol{\tau}^\varphi)\mathrm{d}\Omega$. Finally the expression of $\langle -z^*, \Lambda x\rangle$ with the current variables is obtained from (9.69) and (9.70) as

$$\langle -z^*, \Lambda x\rangle = -\int_\Omega \begin{bmatrix} \boldsymbol{E}_\mathrm{e} & -(1/2)\nabla\boldsymbol{\varphi}^\mathrm{T} \\ -(1/2)\nabla\boldsymbol{\varphi} & \boldsymbol{o} \end{bmatrix} \bullet \begin{bmatrix} \boldsymbol{S} & \boldsymbol{\Pi}^\mathrm{T} \\ \boldsymbol{\Pi} & \boldsymbol{\tau}^\varphi \end{bmatrix} \mathrm{d}\Omega.$$

Consequently, $g(\Lambda x)+g^*(-z^*)$ is finite if and only if (9.93d) is satisfied, and in such a case the equation $g(\Lambda x)+g^*(-z^*) = \langle -z^*, \Lambda x\rangle$ is reduced to

$$\frac{1}{2}\int_\Omega \mathrm{tr}(\boldsymbol{S} + \boldsymbol{\tau}^\varphi)\mathrm{d}\Omega + \frac{1}{2}\int_\Omega \begin{bmatrix} 2\boldsymbol{E}_\mathrm{e} & -\nabla\boldsymbol{\varphi}^\mathrm{T} \\ -\nabla\boldsymbol{\varphi} & \boldsymbol{o} \end{bmatrix} \bullet \begin{bmatrix} \boldsymbol{S} & \boldsymbol{\Pi}^\mathrm{T} \\ \boldsymbol{\Pi} & \boldsymbol{\tau}^\varphi \end{bmatrix} \mathrm{d}\Omega = 0,$$

which is equivalent to (9.93e) in the weak sense. □

Note that, in the optimality condition (9.93), conditions (9.93d) and (9.93e) form the complementarity condition over 4×4 positive-semidefinite cone.[25] In the following, we shall see that this complementarity condition has a crucial meaning from a mechanical point of view: from this complementarity condition it is guaranteed that the first and the second Piola–Kirchhoff stress tensors satisfy the compatibility condition with the deformation of the body; see Proposition 9.4.10. For exploring this issue, we further investigate the relation between the dual variables at the optimal solution of ($\overline{\mathrm{CE}}$).

Proposition 9.4.7. *If $(\bar{\boldsymbol{S}}, \bar{\boldsymbol{\Pi}}, \bar{\boldsymbol{\tau}}^\varphi)$ is optimal for ($\overline{\mathrm{CE}}$), then it satisfies*

$$\bar{\boldsymbol{\tau}}^\varphi = \bar{\boldsymbol{\Pi}}\bar{\boldsymbol{S}}^+\bar{\boldsymbol{\Pi}}^\mathrm{T}, \tag{9.96}$$

$$\bar{\boldsymbol{\Pi}}(\boldsymbol{I} - \bar{\boldsymbol{S}}^+\bar{\boldsymbol{S}}) = \boldsymbol{o}, \tag{9.97}$$

where $\bar{\boldsymbol{S}}^+$ means the pseudo-inverse of $\bar{\boldsymbol{S}}$.

Proof. It follows from Fact 1.3.11 that the positive-semidefinite constraint of ($\overline{\mathrm{CE}}$) in (9.85) is equivalently rewritten as

$$\boldsymbol{S} \succeq \boldsymbol{o}, \quad (\boldsymbol{I} - \boldsymbol{S}\boldsymbol{S}^+)\boldsymbol{\Pi}^\mathrm{T} = \boldsymbol{o}, \tag{9.98}$$

$$\boldsymbol{\tau}^\varphi \succeq \boldsymbol{\Pi}\boldsymbol{S}^+\boldsymbol{\Pi}^\mathrm{T}. \tag{9.99}$$

Since $\boldsymbol{S}$ is a symmetric tensor, the last equality in (9.98) is equivalent to (9.97).

If we replace the positive-semidefinite constraint in ($\overline{\mathrm{CE}}$) with (9.98) and (9.99), only (9.99) is the constraint on $\boldsymbol{\tau}^\varphi$, and in the objective

[24]The assertion of this equivalence follows from Proposition 2.1.12.

[25]In contrast, it is noted that the constitutive law of a membrane is formulated by using (9.50c), which is the complementarity condition over 2×2 positive-semidefinite cone.

function $\boldsymbol{\tau}^{\varphi}$ appears only in the term $-\int_{\Omega}(1/2)\operatorname{tr}\boldsymbol{\tau}^{\varphi}\mathrm{d}\Omega$. Therefore, in terms of $\boldsymbol{\tau}^{\varphi}$, $(\overline{\mathrm{CE}})$ is regarded as the minimization of $\operatorname{tr}\boldsymbol{\tau}^{\varphi}$ over the constraint (9.99).

Motivated by this observation, consider the problem

$$\min_{\boldsymbol{\tau}^{\varphi}\in\mathbb{S}^2}\{\operatorname{tr}\boldsymbol{\tau}^{\varphi} \mid \boldsymbol{\tau}^{\varphi}\succeq \boldsymbol{\Pi}\boldsymbol{S}^{+}\boldsymbol{\Pi}^{\mathrm{T}}\}, \tag{9.100}$$

where only $\boldsymbol{\tau}^{\varphi}$ is treated as the variable. This problem is the minimization of a convex function over a convex feasible set (with a nonempty interior). Hence, we apply Proposition 2.2.9 (b) to derive its optimality condition.

To this end, we encode problem (9.100) into the form of $(\mathrm{P_F})$ in (2.28), i.e., $\inf_{x}\{f(x)+g(\Lambda x) \mid x\in\mathbb{V}\}$, by defining $f:\mathbb{V}\to\mathbb{R}$, $g:\mathbb{Y}\to\mathbb{R}\cup\{+\infty\}$, and $\Lambda:\mathbb{V}\to\mathbb{Y}$ by

$$\begin{aligned}
f(\boldsymbol{\tau}^{\varphi}) &= \boldsymbol{I}\bullet\boldsymbol{\tau}^{\varphi},\\
g(\boldsymbol{Z}) &= \begin{cases} 0 & \text{if } \boldsymbol{Z}-\boldsymbol{\Pi}\boldsymbol{S}^{+}\boldsymbol{\Pi}^{\mathrm{T}}\succeq\boldsymbol{o},\\ +\infty & \text{otherwise,}\end{cases}\\
\Lambda\boldsymbol{\tau}^{\varphi} &= \boldsymbol{\tau}^{\varphi}
\end{aligned}$$

with

$$x=\boldsymbol{\tau}^{\varphi}\in\mathbb{V},\quad z=\boldsymbol{Z}\in\mathbb{Y},\quad \mathbb{V}=\mathbb{Y}=\mathbb{S}^2.$$

In a manner similar to (9.58) in Proposition 9.3.4, we obtain

$$g^*(\boldsymbol{Z}^*) = \begin{cases} (\boldsymbol{\Pi}\boldsymbol{S}^{+}\boldsymbol{\Pi}^{\mathrm{T}})\bullet\boldsymbol{Z}^* & \text{if } -\boldsymbol{Z}^*\succeq\boldsymbol{o},\\ +\infty & \text{otherwise.}\end{cases}$$

Consequently, the optimality condition, say, Proposition 2.2.9 (b), is rewritten for problem (9.100) as

$$\begin{aligned}
&\bar{\boldsymbol{Z}}^* = \boldsymbol{I},\\
&\bar{\boldsymbol{\tau}}^{\varphi}-\boldsymbol{\Pi}\boldsymbol{S}^{+}\boldsymbol{\Pi}^{\mathrm{T}}\succeq\boldsymbol{o},\quad \bar{\boldsymbol{Z}}^*\succeq\boldsymbol{o},\quad (\bar{\boldsymbol{\tau}}^{\varphi}-\boldsymbol{\Pi}\boldsymbol{S}^{+}\boldsymbol{\Pi}^{\mathrm{T}})\bullet\bar{\boldsymbol{Z}}^*=0.
\end{aligned}$$

Thus we see that the trace of the positive-semidefinite matrix $\bar{\boldsymbol{\tau}}^{\varphi}-\boldsymbol{\Pi}\boldsymbol{S}^{+}\boldsymbol{\Pi}^{\mathrm{T}}$ is equal to zero at the optimal solution, which yields (9.96). □

The following result is obtained as an immediate consequence of Proposition 9.4.7.

Corollary 9.4.8. *Suppose that $(\bar{\boldsymbol{S}},\bar{\boldsymbol{\Pi}},\bar{\boldsymbol{\tau}}^{\varphi})$ is optimal for $(\overline{\mathrm{CE}})$. Then the following statements hold:*

(a) *If $\bar{\boldsymbol{S}}\succ\boldsymbol{o}$, then $\bar{\boldsymbol{\tau}}^{\varphi}=\bar{\boldsymbol{\Pi}}\bar{\boldsymbol{S}}^{-1}\bar{\boldsymbol{\Pi}}^{\mathrm{T}}$.*

(b) *If $\bar{\boldsymbol{S}}=\boldsymbol{o}$, then $\bar{\boldsymbol{\tau}}^{\varphi}=\boldsymbol{o}$ and $\bar{\boldsymbol{\Pi}}=\boldsymbol{o}$.*

(c) *Otherwise, there exist* $s_1 \in \mathbb{R}$, $\boldsymbol{\phi}_1 \in \mathbb{R}^2$, *and* $\boldsymbol{\psi} \in \mathbb{R}^2$ *satisfying*

$$\bar{\boldsymbol{S}} = s_1 \boldsymbol{\phi}_1 \boldsymbol{\phi}_1^{\mathrm{T}} \quad (s_1 > 0, \ \|\boldsymbol{\phi}_1\| = 1), \tag{9.101a}$$

$$\bar{\boldsymbol{\Pi}} = \boldsymbol{\psi} \boldsymbol{\phi}_1^{\mathrm{T}}, \tag{9.101b}$$

$$\bar{\boldsymbol{\tau}}^{\varphi} = \frac{1}{s_1} \boldsymbol{\psi} \boldsymbol{\psi}^{\mathrm{T}}. \tag{9.101c}$$

Proof. Assertion (a) immediately follows from (9.96) of Proposition 9.4.7. To see assertion (b), from $\bar{\boldsymbol{S}}^{+} = \boldsymbol{o}$ and Proposition 9.4.7 we obtain $\bar{\boldsymbol{\tau}}^{\varphi} = \boldsymbol{o}$. Therefore, the positive-semidefinite constraint of ($\overline{\mathrm{CE}}$) is reduced to $\begin{bmatrix} \boldsymbol{o} & \boldsymbol{\Pi}^{\mathrm{T}} \\ \boldsymbol{\Pi} & \boldsymbol{o} \end{bmatrix} \succeq \boldsymbol{o}$, which implies $\boldsymbol{\Pi} = \boldsymbol{o}$.

Assertion (c) is obtained as follows. Since $\bar{\boldsymbol{S}} \not\succ \boldsymbol{o}$ and $\bar{\boldsymbol{S}} \neq \boldsymbol{o}$, it satisfies (9.101a) and

$$\bar{\boldsymbol{S}} \boldsymbol{\phi}_2 = \mathbf{0},$$

where $\boldsymbol{\phi}_1$ and $\boldsymbol{\phi}_2$ are eigenvectors of $\bar{\boldsymbol{S}}$ satisfying $\|\boldsymbol{\phi}_1\| = \|\boldsymbol{\phi}_2\| = 1$ and $\boldsymbol{\phi}_1^{\mathrm{T}} \boldsymbol{\phi}_2 = 0$. For the positive-semidefinite constraint of ($\overline{\mathrm{CE}}$), consider the following quadratic form:

$$\begin{bmatrix} \boldsymbol{\phi}_2 \\ \boldsymbol{\eta} \end{bmatrix}^{\mathrm{T}} \begin{bmatrix} \bar{\boldsymbol{S}} & \boldsymbol{\Pi}^{\mathrm{T}} \\ \boldsymbol{\Pi} & \boldsymbol{\tau}^{\varphi} \end{bmatrix} \begin{bmatrix} \boldsymbol{\phi}_2 \\ \boldsymbol{\eta} \end{bmatrix} = 2\boldsymbol{\eta}^{\mathrm{T}} \boldsymbol{\Pi} \boldsymbol{\phi}_2 + \boldsymbol{\eta}^{\mathrm{T}} \boldsymbol{\tau}^{\varphi} \boldsymbol{\eta}, \tag{9.102}$$

where $\boldsymbol{\eta} \in \mathbb{R}^2$. Since $\boldsymbol{\tau}^{\varphi} \succeq \boldsymbol{o}$, (9.102) is nonnegative for any $\boldsymbol{\eta}$ if and only if $\boldsymbol{\Pi} \boldsymbol{\phi}_2 = \mathbf{0}$ and thence the optimal $\bar{\boldsymbol{\Pi}}$ should be written in the form of (9.101b). By using

$$\bar{\boldsymbol{S}}^{+} = \frac{1}{s_1} \boldsymbol{\phi}_1 \boldsymbol{\phi}_1^{\mathrm{T}}$$

and (9.96) of Proposition 9.4.7, simple calculations yield

$$\boldsymbol{\tau}^{\varphi} = \bar{\boldsymbol{\Pi}} \bar{\boldsymbol{S}}^{+} \bar{\boldsymbol{\Pi}}^{\mathrm{T}} = (\boldsymbol{\psi} \boldsymbol{\phi}_1^{\mathrm{T}})[(1/s_1) \boldsymbol{\phi}_1 \boldsymbol{\phi}_1^{\mathrm{T}}](\boldsymbol{\phi} \boldsymbol{\psi}_1^{\mathrm{T}}) = (1/s_1) \boldsymbol{\psi} \boldsymbol{\psi}^{\mathrm{T}},$$

which concludes the proof. □

Remark 9.4.9. From a mechanical point of view, the three cases studied in Corollary 9.4.8 correspond to the classification of the wrinkle states of a membrane (see Table 9.1 in section 9.3.1). Case (a) of Corollary 9.4.8 corresponds to the taut state, Case (b) is the slack state, and case (c) is the wrinkling state. ■

Corollary 9.4.8, together with the optimality conditions in Proposition 9.4.6, suggests the physical interpretation of the dual variables $\boldsymbol{\Pi}$, $\boldsymbol{S}$, and $\boldsymbol{\tau}^{\varphi}$. Since $\boldsymbol{S}$ at the optimal solution is given by (9.93a), i.e., the constitutive law of the membrane, $\boldsymbol{S}$ is regarded as the second Piola–Kirchhoff stress tensor. The force-balance equation in (9.93a) suggests that $\boldsymbol{\Pi}$ corresponds to the first Piola–Kirchhoff stress tensor. To see what $\boldsymbol{\tau}^{\varphi}$ means, consider case (a) of

Corollary 9.4.8. If we postulate that (9.39) is satisfied, then we obtain

$$\begin{aligned} \tau^{\varphi} &= \bar{\boldsymbol{\Pi}}\bar{\boldsymbol{S}}^{-1}\bar{\boldsymbol{\Pi}}^{\mathrm{T}} \\ &= \bar{\boldsymbol{\Pi}}\bar{\boldsymbol{S}}^{-1}(\bar{\boldsymbol{F}}\bar{\boldsymbol{S}}) \quad \text{[from (9.39)]} \\ &= \bar{\boldsymbol{\Pi}}\bar{\boldsymbol{F}} \\ &= J\boldsymbol{T}^{\varphi}. \quad \text{[from (9.38)]} \end{aligned} \tag{9.103}$$

It then follows from this relation and (9.38) and (9.40) that τ^{φ} and $\boldsymbol{T}^{\varphi}$ correspond to the Kirchhoff stress tensor and the Cauchy stress tensor, respectively. Thus, if (9.39) is shown to hold, then all the variables of $(\overline{\mathrm{CE}})$ are interpreted as the static variables. Thus the key issue is to show that $\boldsymbol{\Pi}$ and $\boldsymbol{S}$ are related by (9.39) to each other through $\boldsymbol{F} = \nabla\varphi$; note that (9.39) is not included as a constraint in $(\overline{\mathrm{CE}})$ or $(\overline{\mathrm{PE}})$, and hence arbitrarily chosen feasible solutions of $(\overline{\mathrm{PE}})$ and $(\overline{\mathrm{CE}})$ do *not* necessarily satisfy (9.39). Rather surprisingly, the complementarity condition (9.93e), together with the positive-semidefinite constraints (9.93d), guarantees that the optimal φ, $\boldsymbol{S}$, and $\boldsymbol{\Pi}$ should satisfy (9.39), which shall be shown in Proposition 9.4.10.

Making use of Proposition 9.4.7 and Corollary 9.4.8, we now provide the interpretation of the complementarity condition in (9.93e) over the positive-semidefinite cone. This complementarity condition gives the relation between the primal variable φ and the dual variables $\boldsymbol{\Pi}$ and $\boldsymbol{S}$.

Proposition 9.4.10. *Suppose that $(\bar{\varphi}, \bar{\boldsymbol{E}}_{\mathrm{e}})$ and $(\bar{\boldsymbol{S}}, \bar{\boldsymbol{\Pi}}, \bar{\tau}^{\varphi})$ are optimal for $(\overline{\mathrm{PE}})$ and $(\overline{\mathrm{CE}})$, respectively. Then $\bar{\boldsymbol{\Pi}} = (\nabla\bar{\varphi})\bar{\boldsymbol{S}}$ holds.*

Proof. We write $\bar{\boldsymbol{F}} = \nabla\bar{\varphi}$ for simplicity. Define δ by

$$\delta = (2\bar{\boldsymbol{E}}_{\mathrm{e}} + \boldsymbol{I}) \bullet \bar{\boldsymbol{S}} - 2\bar{\boldsymbol{F}} \bullet \bar{\boldsymbol{\Pi}} + \boldsymbol{I} \bullet \bar{\tau}^{\varphi} \tag{9.104}$$

so that the complementarity condition (9.93e) is written simply as

$$\delta = 0. \tag{9.105}$$

By using the notation in Proposition 9.4.6 (ii), we obtain

$$2\hat{\boldsymbol{E}}(\bar{\varphi}) + \boldsymbol{I} = \bar{\boldsymbol{F}}^{\mathrm{T}}\bar{\boldsymbol{F}},$$

from which, along with Proposition 9.4.6 (ii), it follows that the first term of (9.104) is reduced to

$$\begin{aligned} (2\bar{\boldsymbol{E}}_{\mathrm{e}} + \boldsymbol{I}) \bullet \bar{\boldsymbol{S}} &= [2(\bar{\boldsymbol{E}}_{\mathrm{e}} - \hat{\boldsymbol{E}}(\bar{\varphi})) + (2\hat{\boldsymbol{E}}(\bar{\varphi}) + \boldsymbol{I})] \bullet \bar{\boldsymbol{S}} \\ &= 2(\bar{\boldsymbol{E}}_{\mathrm{e}} - \hat{\boldsymbol{E}}(\bar{\varphi})) \bullet \bar{\boldsymbol{S}} + (\bar{\boldsymbol{F}}^{\mathrm{T}}\bar{\boldsymbol{F}}) \bullet \bar{\boldsymbol{S}} \\ &= (\bar{\boldsymbol{F}}^{\mathrm{T}}\bar{\boldsymbol{F}}) \bullet \bar{\boldsymbol{S}}. \end{aligned}$$

Substituting this relation into (9.104) yields

$$\delta = \bar{\boldsymbol{S}} \bullet (\bar{\boldsymbol{F}}^{\mathrm{T}}\bar{\boldsymbol{F}}) - 2\bar{\boldsymbol{\Pi}} \bullet \bar{\boldsymbol{F}} + \boldsymbol{I} \bullet \bar{\tau}^{\varphi}. \tag{9.106}$$

The proof proceeds by considering the following three cases.

(Case 1): Suppose that $\bar{\boldsymbol{S}} \succ \boldsymbol{o}$. We further restate δ in (9.106) as follows. For the first term of (9.106), simple calculations and Fact 1.3.13 yield

$$\bar{\boldsymbol{S}} \bullet (\bar{\boldsymbol{F}}^{\mathrm{T}} \bar{\boldsymbol{F}}) = (\bar{\boldsymbol{S}} \bar{\boldsymbol{F}}^{\mathrm{T}})^{\mathrm{T}} \bullet \bar{\boldsymbol{F}} = (\bar{\boldsymbol{F}} \bar{\boldsymbol{S}}) \bullet \bar{\boldsymbol{F}}. \tag{9.107}$$

By using Fact 1.3.13 and the condition (9.97) of Proposition 9.4.7, the second term of (9.106) is reduced to

$$\begin{aligned}\bar{\boldsymbol{\Pi}} \bullet \bar{\boldsymbol{F}} &= \bar{\boldsymbol{F}}^{\mathrm{T}} \bullet \bar{\boldsymbol{\Pi}}^{\mathrm{T}} = \bar{\boldsymbol{F}}^{\mathrm{T}} \bullet (\bar{\boldsymbol{S}} \bar{\boldsymbol{S}}^{-1} \bar{\boldsymbol{\Pi}}^{\mathrm{T}}) \\ &= (\bar{\boldsymbol{S}} \bar{\boldsymbol{F}}^{\mathrm{T}}) \bullet (\bar{\boldsymbol{S}}^{-1} \bar{\boldsymbol{\Pi}}^{\mathrm{T}}) = (\bar{\boldsymbol{F}} \bar{\boldsymbol{S}}) \bullet (\bar{\boldsymbol{\Pi}} \bar{\boldsymbol{S}}^{-1}),\end{aligned} \tag{9.108}$$

because $\bar{\boldsymbol{S}}^{+} = \bar{\boldsymbol{S}}^{-1}$ for $\bar{\boldsymbol{S}} \succ \boldsymbol{o}$. Similarly, from calculations using Fact 1.3.13, the last term of (9.106) is reduced to

$$\begin{aligned}\boldsymbol{I} \bullet \bar{\tau}^{\varphi} &= \boldsymbol{I} \bullet (\bar{\boldsymbol{\Pi}} \bar{\boldsymbol{S}}^{-1} \bar{\boldsymbol{\Pi}}^{\mathrm{T}}) \\ &= (\bar{\boldsymbol{\Pi}}^{\mathrm{T}} \boldsymbol{I}) \bullet (\bar{\boldsymbol{S}}^{-1} \bar{\boldsymbol{\Pi}}^{\mathrm{T}}) = \bar{\boldsymbol{\Pi}} \bullet (\bar{\boldsymbol{\Pi}} \bar{\boldsymbol{S}}^{-1}).\end{aligned} \tag{9.109}$$

Thus, from (9.107)–(9.109), (9.106) is reduced to

$$\begin{aligned}\delta &= (\bar{\boldsymbol{F}} \bar{\boldsymbol{S}}) \bullet \bar{\boldsymbol{F}} - \bar{\boldsymbol{\Pi}} \bullet \bar{\boldsymbol{F}} - (\bar{\boldsymbol{F}} \bar{\boldsymbol{S}}) \bullet (\bar{\boldsymbol{\Pi}} \bar{\boldsymbol{S}}^{-1}) + \bar{\boldsymbol{\Pi}} \bullet (\bar{\boldsymbol{\Pi}} \bar{\boldsymbol{S}}^{-1}) \\ &- (\bar{\boldsymbol{\Pi}} - \bar{\boldsymbol{F}} \bar{\boldsymbol{S}}) \bullet (\bar{\boldsymbol{\Pi}} \bar{\boldsymbol{S}}^{-1} - \bar{\boldsymbol{F}}) \\ &= (\bar{\boldsymbol{\Pi}} - \bar{\boldsymbol{F}} \bar{\boldsymbol{S}}) \bullet [(\bar{\boldsymbol{\Pi}} - \bar{\boldsymbol{F}} \bar{\boldsymbol{S}}) \bar{\boldsymbol{S}}^{-1}] \\ &= [(\bar{\boldsymbol{\Pi}} - \bar{\boldsymbol{F}} \bar{\boldsymbol{S}})^{\mathrm{T}} (\bar{\boldsymbol{\Pi}} - \bar{\boldsymbol{F}} \bar{\boldsymbol{S}})] \bullet \bar{\boldsymbol{S}}^{-1},\end{aligned} \tag{9.110}$$

where the last equality follows from Fact 1.3.13. Since $(\bar{\boldsymbol{\Pi}} - \bar{\boldsymbol{F}} \bar{\boldsymbol{S}})^{\mathrm{T}} (\bar{\boldsymbol{\Pi}} - \bar{\boldsymbol{F}} \bar{\boldsymbol{S}}) \succeq \boldsymbol{o}$ and $\bar{\boldsymbol{S}}^{-1} \succ \boldsymbol{o}$, we see in (9.110) that the condition $\delta = 0$ is regarded as the complementarity condition over the positive-semidefinite cone. Therefore, from Proposition 1.3.20, $\bar{\boldsymbol{S}}^{-1} \succ \boldsymbol{o}$ implies that (9.105) is satisfied if and only if $\bar{\boldsymbol{F}}$, $\bar{\boldsymbol{S}}$, and $\bar{\boldsymbol{\Pi}}$ satisfy

$$\bar{\boldsymbol{\Pi}} - \bar{\boldsymbol{F}} \bar{\boldsymbol{S}} = \boldsymbol{o}.$$

(Case 2): Suppose that $\bar{\boldsymbol{S}} = \boldsymbol{o}$. We then obtain Corollary 9.4.8 (b), and nothing is to be proved.

(Case 3): Suppose that $\bar{\boldsymbol{S}} \not\succ \boldsymbol{o}$ and $\bar{\boldsymbol{S}} \neq \boldsymbol{o}$, which corresponds to the case investigated in Corollary 9.4.8 (c). Substituting (9.101) into (9.106) yields

$$\begin{aligned}\delta &= s_1(\boldsymbol{\phi}_1 \boldsymbol{\phi}_1^{\mathrm{T}}) \bullet (\bar{\boldsymbol{F}}^{\mathrm{T}} \bar{\boldsymbol{F}}) - 2(\boldsymbol{\psi} \boldsymbol{\phi}_1^{\mathrm{T}}) \bullet \bar{\boldsymbol{F}} + \boldsymbol{I} \bullet \left(\frac{1}{s_1} \boldsymbol{\psi} \boldsymbol{\psi}^{\mathrm{T}} \right) \\ &= s_1 (\bar{\boldsymbol{F}} \boldsymbol{\phi}_1)^{\mathrm{T}} (\bar{\boldsymbol{F}} \boldsymbol{\phi}_1) - 2 \boldsymbol{\psi}^{\mathrm{T}} \bar{\boldsymbol{F}} \boldsymbol{\phi}_1 + \frac{1}{s_1} \boldsymbol{\psi}^{\mathrm{T}} \boldsymbol{\psi} \\ &= \frac{1}{s_1} (s_1 \bar{\boldsymbol{F}} \boldsymbol{\phi}_1 - \boldsymbol{\psi})^{\mathrm{T}} (s_1 \bar{\boldsymbol{F}} \boldsymbol{\phi}_1 - \boldsymbol{\psi}).\end{aligned}$$

Therefore, (9.105) is satisfied if and only if $\bar{\boldsymbol{F}}$ satisfies

$$\boldsymbol{\psi} = s_1 \bar{\boldsymbol{F}} \boldsymbol{\phi}_1. \tag{9.111}$$

By substituting this into (9.101b) and then using (9.101a), we obtain

$$\bar{\boldsymbol{\Pi}} = \bar{\boldsymbol{\psi}} \boldsymbol{\phi}_1^{\mathrm{T}} = s_1 \bar{\boldsymbol{F}} \boldsymbol{\phi}_1 \boldsymbol{\phi}_1^{\mathrm{T}} = \bar{\boldsymbol{F}} (s_1 \boldsymbol{\phi}_1 \boldsymbol{\phi}_1^{\mathrm{T}}) = \bar{\boldsymbol{F}} \bar{\boldsymbol{S}},$$

which concludes the proof. □

From the proposition above and (9.103), we can conclude that the dual variables correspond to static variables as follows:

- $\boldsymbol{S}$: the second Piola–Kirchhoff stress tensor
- $\boldsymbol{\Pi}$: the first Piola–Kirchhoff stress tensor
- $\boldsymbol{\tau}^{\varphi}$: the Kirchhoff stress tensor

See also the discussion around (9.103).

Remark 9.4.11. In the proof of Proposition 9.4.10 (Case 3), we can see that the vector $\boldsymbol{\psi}$ in Corollary 9.4.8 (c) has been determined as (9.111). Accordingly, condition (9.101) in Corollary 9.4.8 (c) is reduced to

$$\bar{\boldsymbol{S}} = s_1 \boldsymbol{\phi}_1 \boldsymbol{\phi}_1^{\mathrm{T}} \quad (s_1 > 0,\ \|\boldsymbol{\phi}_1\| = 1), \tag{9.112a}$$

$$\bar{\boldsymbol{\Pi}} = s_1 (\nabla \bar{\boldsymbol{\varphi}}) \boldsymbol{\phi}_1 \boldsymbol{\phi}_1^{\mathrm{T}}, \tag{9.112b}$$

$$\bar{\boldsymbol{\tau}}^{\varphi} = s_1 [(\nabla \bar{\boldsymbol{\varphi}}) \boldsymbol{\phi}_1][(\nabla \bar{\boldsymbol{\varphi}}) \boldsymbol{\phi}_1]^{\mathrm{T}}. \tag{9.112c}$$

By using (9.103), the corresponding Cauchy stress tensor is given by

$$\bar{\boldsymbol{T}}^{\varphi} = \frac{s_1}{J} [(\nabla \bar{\boldsymbol{\varphi}}) \boldsymbol{\phi}_1][(\nabla \bar{\boldsymbol{\varphi}}) \boldsymbol{\phi}_1]^{\mathrm{T}}. \tag{9.113}$$

Thus $(\nabla \bar{\boldsymbol{\varphi}}) \boldsymbol{\phi}_1$ is the eigenvector corresponding to the eigenvalue $s_1 \|(\nabla \bar{\boldsymbol{\varphi}}) \boldsymbol{\phi}_1\|^2 / J$ of $\bar{\boldsymbol{T}}^{\varphi}$, and the other eigenvalue of $\bar{\boldsymbol{T}}^{\varphi}$ is equal to zero. This means that $(\nabla \bar{\boldsymbol{\varphi}}) \boldsymbol{\phi}_1$ corresponds to the direction along the wrinkle (at the deformed configuration). Note that $(\nabla \bar{\boldsymbol{\varphi}}) \boldsymbol{\phi}_2$ is not the direction transverse to the wrinkle in general; the transverse direction should be orthogonal to $(\nabla \bar{\boldsymbol{\varphi}}) \boldsymbol{\phi}_1$. Consequently, $\boldsymbol{\phi}_1$ corresponds to the direction along the wrinkle at the reference configuration, while $\boldsymbol{\phi}_2$ is not the transverse direction. Thus at the reference configuration the directions along and transverse to the wrinkle are not orthogonal to each other. Mention of this fact can be found in, e.g., [274, section 2.4]. ■

9.4.4 Principle of complementary energy

As a consequence of the quite long analysis through in section 9.3.1 through section 9.4.3, we establish in this section the minimum principle of the complementary energy for membrane structures. The duality between $(\overline{\mathrm{PE}})$ in (9.62) and $(\overline{\mathrm{CE}})$ in (9.85) certainly plays a key role.[26]

Since the principle of the complementary energy is conventionally stated in the form of a minimization problem, we exchange the sign of the objective

[26]Recall that we suppose (9.65) for avoiding technical difficulties arising from nonhomogeneous boundary condition in a boundary value problem; see Lions–Magenes [270, Chaps. 1 and 2] for comprehensive treatments of nonhomogeneous boundary value problems.

function of $(\overline{\text{CE}})$. Moreover, by using Proposition 9.4.7 we replace the positive-semidefinite constraint in $(\overline{\text{CE}})$ with more direct constraints, which results in

$$\left.\begin{aligned}
(\text{CE}): \min_{\boldsymbol{S},\boldsymbol{\Pi},\boldsymbol{\tau}^{\varphi}} \quad & \int_{\Omega} \hat{w}^{*}_{\text{ref}}(\boldsymbol{S})\mathrm{d}\Omega + \int_{\Omega} \frac{1}{2}\operatorname{tr}(\boldsymbol{S}+\boldsymbol{\tau}^{\varphi})\mathrm{d}\Omega \\
& \quad - \int_{\Gamma_{\text{D}}} (\boldsymbol{\Pi}\boldsymbol{n})\cdot\boldsymbol{x}\mathrm{d}\Gamma - \int_{\Omega} \underline{\boldsymbol{p}}\cdot\boldsymbol{x}\mathrm{d}\Omega - \int_{\Gamma_{\text{N}}} \underline{\boldsymbol{t}}\cdot\boldsymbol{x}\mathrm{d}\Gamma \\
\text{s.t.} \quad & -\operatorname{div}\boldsymbol{\Pi} = \underline{\boldsymbol{p}} \quad \text{in } \Omega, \\
& \boldsymbol{\tau}^{\varphi} = \boldsymbol{\Pi}\boldsymbol{S}^{+}\boldsymbol{\Pi}^{\text{T}}, \ \boldsymbol{\Pi}(\boldsymbol{I}-\boldsymbol{S}^{+}\boldsymbol{S}) = \boldsymbol{o} \quad \text{in } \Omega, \\
& \boldsymbol{S} \succeq \boldsymbol{o} \quad \text{in } \Omega, \\
& \boldsymbol{\Pi}\boldsymbol{n} = \underline{\boldsymbol{t}} \quad \text{on } \Gamma_{\text{N}}.
\end{aligned}\right\} \tag{9.114}$$

The main result stated in the following guarantees that the stress tensors at the equilibrium state can be obtained as the optimal solution of (CE) in (9.114). The *minimum* principle of the total potential energy, say, (PE) in (9.61), is simultaneously confirmed formally particularly for membranes, which is rather surprising because in general, e.g., for plates consisting of the St. Venant material, only a *stationarity* condition of the potential energy can be reached in the large deformation theory.

Proposition 9.4.12. *Under the hypotheses in Proposition 9.4.4,* $(\bar{\boldsymbol{\varphi}}, \bar{\boldsymbol{E}})$ *and* $(\bar{\boldsymbol{S}}, \bar{\boldsymbol{\Pi}}, \bar{\boldsymbol{\tau}}^{\varphi})$ *are optimal solutions of* (PE) *and* (CE), *respectively, if and only if they correspond to the variables*[27] *at the equilibrium state of the membrane. Moreover,* $\min(\text{PE}) = -\min(\text{CE})$.

Proof. The proof is based on the optimality conditions for $(\overline{\text{PE}})$ and $(\overline{\text{CE}})$ given by (9.93) in Proposition 9.4.6. Note that the optimality condition for (PE) is obtained by substituting the relation in Proposition 9.3.5 (i) to (9.93). For (CE), the lemma on the Schur complement (see Fact 1.3.10) implies that any feasible solution of (CE) is feasible for $(\overline{\text{CE}})$. Conversely, it follows from Proposition 9.4.7 that the optimal solution of $(\overline{\text{CE}})$ is feasible for (CE). Consequently, the optimality condition for (CE) is given by (9.93), and hence $\min(\text{CE}) = -\max(\overline{\text{CE}})$. From this relation and Proposition 9.3.5 (iii) it follows that Proposition 9.4.4 yields $\min(\text{PE}) = -\min(\text{CE})$.

Moreover, it follows from (9.60) that conditions (9.93a) and (9.94) of Proposition 9.4.6 are equivalent to $\bar{\boldsymbol{S}} = \hat{\boldsymbol{S}}_{\text{mem}}(\bar{\boldsymbol{E}})$. From Proposition 9.4.10 we can see that $\boldsymbol{\Pi} = (\nabla\boldsymbol{\varphi})\boldsymbol{S}$ is also satisfied by the optimal solutions of (PE) and (CE). We thus conclude that $(\bar{\boldsymbol{\varphi}}, \bar{\boldsymbol{E}})$ and $(\bar{\boldsymbol{S}}, \bar{\boldsymbol{\Pi}}, \bar{\boldsymbol{\tau}}^{\varphi})$ are optimal for (PE) and (CE), respectively, if and only if

[27] As shown in section 9.3.2 and section 9.4.3, $\boldsymbol{\varphi}$ and $\boldsymbol{E}$ denote the deformation and the Green–Lagrange strain tensor, while $\boldsymbol{S}$, $\boldsymbol{\Pi}$, and $\boldsymbol{\tau}^{\varphi}$ are the second Piola–Kirchhoff stress, the first Piola–Kirchhoff stress, and the Kirchhoff stress tensors, respectively.

they satisfy

$$\bar{\boldsymbol{S}} = \hat{\boldsymbol{S}}_{\rm mem}(\bar{\boldsymbol{E}}) \quad \text{in } \Omega, \tag{9.115a}$$
$$\bar{\boldsymbol{E}} = \frac{1}{2}(\nabla\bar{\boldsymbol{\varphi}}^{\rm T}\nabla\bar{\boldsymbol{\varphi}} - \boldsymbol{I}) \quad \text{in } \Omega, \tag{9.115b}$$
$$\bar{\boldsymbol{\Pi}} = (\nabla\bar{\boldsymbol{\varphi}})\bar{\boldsymbol{S}} \quad \text{in } \Omega, \tag{9.115c}$$
$$\bar{\boldsymbol{\tau}}^{\varphi} = \bar{\boldsymbol{\Pi}}\bar{\boldsymbol{S}}^{+}\bar{\boldsymbol{\Pi}}^{\rm T} \quad \text{in } \Omega, \tag{9.115d}$$
$$-\operatorname{div}\bar{\boldsymbol{\Pi}} = \underline{\boldsymbol{p}} \quad \text{in } \Omega, \tag{9.115e}$$
$$\bar{\boldsymbol{\varphi}} = \boldsymbol{x} \quad \text{on } \Gamma_{\rm D}; \qquad \bar{\boldsymbol{\Pi}}\boldsymbol{n} = \underline{\boldsymbol{t}} \quad \text{on } \Gamma_{\rm N}. \tag{9.115f}$$

Note that (9.115a) implies $\bar{\boldsymbol{S}} \succeq \boldsymbol{o}$, and hence $\bar{\boldsymbol{\tau}}^{\varphi}$ defined by (9.115d) satisfies $\bar{\boldsymbol{\tau}}^{\varphi} - \bar{\boldsymbol{\Pi}}\bar{\boldsymbol{S}}^{+}\bar{\boldsymbol{\Pi}}^{\rm T} \succeq \boldsymbol{o}$. Moreover, $\boldsymbol{\Pi}(\boldsymbol{I} - \boldsymbol{S}^{+}\boldsymbol{S}) = \boldsymbol{o}$ in (CE) is guaranteed to be satisfied from (9.115c) and hence is omitted from (9.115).

Now we can clearly see that (9.115) forms the boundary value problem for the equilibrium of a membrane. Condition (9.115a) corresponds to the constitutive law of the membrane, (9.115b) is the compatibility condition of the kinematic variables, (9.115c) and (9.115d) give the compatibility conditions between the static variables, (9.115e) is the force-balance equation at the deformed configuration, and the boundary condition is given by (9.115f). Thus solving (PE) and (CE) each is equivalent to solving (9.115). Therefore, $(\bar{\boldsymbol{\varphi}}, \bar{\boldsymbol{E}})$ and $(\bar{\boldsymbol{S}}, \bar{\boldsymbol{\Pi}}, \bar{\boldsymbol{\tau}}^{\varphi})$ correspond to the kinematic and static variables at the equilibrium state. □

Like (PE),[28] (CE) is a nonconvex optimization problem. In other words, $(\overline{\text{CE}})$ is regarded as a convex reformulation of (CE). Thus, through the analysis performed in section 9.4.1 through section 9.4.3, we have enjoyed the rewards stemming from the convexity of $(\overline{\text{PE}})$ and $(\overline{\text{CE}})$.

It is emphasized that (CE) does not include any kinematic variables (i.e., deformations or strains). In general geometrically nonlinear problems, the variational principle cannot be formulated only in terms of static variables (i.e., stresses). This is because the force-balance equation depends on the deformation. In other words, the first and the second Piola–Kirchhoff stress tensors should be compatible with the deformation, and the two stresses cannot take arbitrary values independently; the relation of these stresses is given through kinematic variables such as the deformation gradient. In contrast, in (CE) for membranes, those two stresses are regarded as independent variables in the optimization problem, and it is guaranteed that the compatibility with the deformation, say, (9.115c), is satisfied automatically at the optimal solution.

Remark 9.4.13. The last two terms, $-\int_{\Omega}\underline{\boldsymbol{p}}\cdot\boldsymbol{x}\,{\rm d}\Omega - \int_{\Gamma_{\rm N}}\underline{\boldsymbol{t}}\cdot\boldsymbol{x}\,{\rm d}\Gamma$, in the objective function of (CE) are constant and depend on the reference configuration $\boldsymbol{x}$. Although we can ignore those terms from the viewpoint of numerical optimization, the presence of these terms means physically that the (value of) complementary energy in the

[28] In (PE), the compatibility condition $\boldsymbol{E} = \frac{1}{2}(\nabla\boldsymbol{\varphi}^{\rm T}\nabla\boldsymbol{\varphi} - \boldsymbol{I})$ is a nonconvex constraint.

large deformation theory depends on the reference configuration, which is not the case in small deformation theory (see (9.24) in section 9.2.3.2). ■

9.5 Numerical Aspects

In the past sections we have enjoyed theoretical advantages of $(\overline{\mathrm{PE}})$ and $(\overline{\mathrm{CE}})$, over (PE) and (CE), stemming from their conic optimization formulations. This section demonstrates further rewards of $(\overline{\mathrm{PE}})$ and $(\overline{\mathrm{CE}})$ from the viewpoint of the numerical solution.

It is noted that the geometrical nonlinearity considered does not appear in the problem formulation explicitly in $(\overline{\mathrm{PE}})$ and $(\overline{\mathrm{CE}})$[29] but is *hidden* in the positive-semidefinite constraints. At the optimal solution the complementarity condition over positive-semidefinite cone is satisfied, and this eventually verifies the compatibility considering the geometrical nonlinearity, as studied in Proposition 9.4.10. Most numerical methods for conic optimization aim at solving the optimality condition (9.93) in Proposition 9.4.6, and the primary nonlinear property to be dealt with is involved in the complementarity condition (9.93e). The search direction at each step of the iterative procedure is defined by the Newton equation obtained by linearizing (9.93e). Thus the iterative procedure of our approach differs from that in the conventional nonlinear equilibrium analysis; in the latter approach the total deformation process is subdivided into sequential loading steps and the source of nonlinearity of the governing equations solved at each loading step is the compatibility relation.

In section 9.5.1 we apply the conventional finite-element discretization to the variational problem $(\overline{\mathrm{PE}})$ to obtain an SDP problem with finitely many variables and constraints. Thus the equilibrium configuration undergoing the large deformation can be computed by solving solely one conic optimization problem. Numerical results are presented in section 9.5.2, where the obtained SDP problem is solved using the primal-dual interior-point method.

9.5.1 Spatial discretization

For numerical solution $(\overline{\mathrm{PE}})$ in (9.62) is discretized by applying the conventional finite element methodology. This discretization can be performed in a manner similar to the case of masonry structures discussed in section 8.4.1, and hence we here describe only essential points briefly.

[29]In contrast, (PE) in (9.61) involves the equality constraint representing the compatibility between the Green–Lagrange strain $\boldsymbol{E}$ and the deformation $\boldsymbol{\varphi}$, while (CE) in (9.114) involves the nonlinear equality constraints between the first Piola–Kirchhoff stress $\boldsymbol{\Pi}$, the second Piola–Kirchhoff stress $\boldsymbol{S}$, and the Kirchhoff stress $\boldsymbol{\tau}^{\boldsymbol{\varphi}}$.

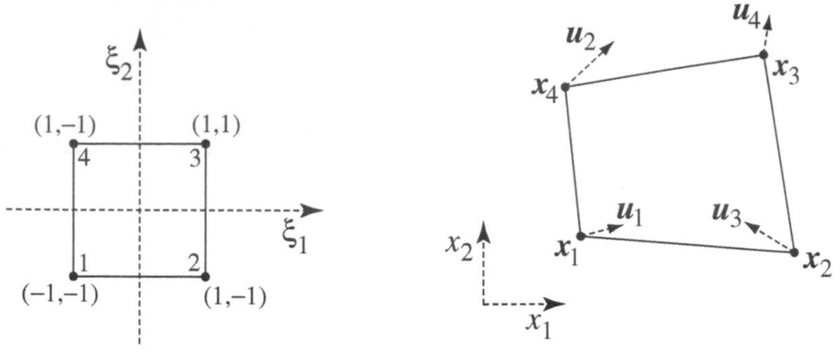

(a) Reference element Ω° and the natural coordinate system (ξ_1, ξ_2).

(b) Actual geometry Ω^{e} and the global coordinate system (x_1, x_2).

FIGURE 9.6: The four-noded isoparametric quadrilateral element.

The domain Ω of the membrane is subdivided as $\Omega = \bigcup_{e=1}^{n^{\mathrm{E}}} \Omega^{e}$, where n^{E} is the number of finite elements. We approximate the vector field $\boldsymbol{u}$ in Ω^{e} by using interpolation functions. Based on the concept of isoparametric elements, as done in section 8.4.1, we consider a normalized reference element Ω° defined with respect to the natural coordinate system (ξ_1, ξ_2); see Figure 9.6(a) for an example of the normalized reference element of the four-noded planar isoparametric quadrilateral element. Then the actual geometry of the element Ω^{e} is interpolated as

$$\boldsymbol{x} = \sum_{I=1}^{n^{\mathrm{S}}} N_I(\boldsymbol{\xi}) \boldsymbol{x}_I \quad (\text{for } \boldsymbol{x} \in \Omega^{e},\ \boldsymbol{\xi} \in \Omega^{\circ}),$$

where N_I is the interpolation function called the *shape function*, and $\boldsymbol{x}_I$ are the location vector of the node I with respect to the global coordinate system (x_1, x_2), as illustrated in Figure 9.6(b) with $n^{\mathrm{S}} = 4$. The vector of displacement field $\boldsymbol{u}(\boldsymbol{x})$ for $\boldsymbol{x} \in \Omega^{e}$ is approximated by using the same interpolation functions as

$$\boldsymbol{u}(\boldsymbol{x}) \simeq \boldsymbol{u}^{e}(\boldsymbol{x}) := \sum_{I=1}^{n^{\mathrm{S}}} N_I(\boldsymbol{\xi}) \boldsymbol{u}_I \quad (\text{for } \boldsymbol{x} \in \Omega^{e},\ \boldsymbol{\xi} \in \Omega^{\circ}), \tag{9.116}$$

where $\boldsymbol{u}_I \in \mathbb{R}^2$ $(I = 1, \dots, n^{\mathrm{S}})$ are the nodal displacement vectors of Ω^{e}, as shown in Figure 9.6(b). The deformation gradient $\boldsymbol{F}(\boldsymbol{x}) = \nabla \boldsymbol{\varphi}(\boldsymbol{x})$ at $\boldsymbol{x} \in \Omega^{e}$

is then approximated by $\boldsymbol{F}^{\mathrm{e}}(\boldsymbol{x})$ as

$$\boldsymbol{F}(\boldsymbol{x}) \simeq \boldsymbol{F}^{\mathrm{e}}(\boldsymbol{x}) = \nabla(\boldsymbol{u}^{\mathrm{e}}(\boldsymbol{x}) + \boldsymbol{x}) = \sum_{I=1}^{n^{\mathrm{S}}} \frac{\partial N_I(\boldsymbol{\xi})}{\partial \boldsymbol{x}} \boldsymbol{u}_I + \boldsymbol{I}, \tag{9.117}$$

where (9.28) was used.

As done in section 8.4.1, let $\tilde{\boldsymbol{u}} \in \mathbb{R}^d$ denote the vector of the nodal displacements of the discretized structure, where d is the number of degrees of freedom. We denote by $\tilde{\Gamma}_{\mathrm{D}}$ the set of indices of degrees of freedom corresponding to the boundary Γ_{D}. Then the constraints on the prescribed displacements are written as

$$\tilde{u}_j = \underline{\tilde{u}}_j, \quad \forall j \in \tilde{\Gamma}_{\mathrm{D}}.$$

Let $\tilde{\Gamma}_{\mathrm{N}}$ denote the set of indices of degrees of freedom corresponding to $\Gamma_{\mathrm{N}} \cup \Omega$; i.e., $\tilde{\Gamma}_{\mathrm{N}}$ is the set of degrees of freedom where external loads are prescribed. We denote by $(\underline{p}_j | j \in \tilde{\Gamma}_{\mathrm{N}})$ the vector of the equivalent external nodal loads.

As usual, we employ the Gauss quadrature for evaluating the total potential energy. By the matrix $\tilde{F}^{qe} \in \mathbb{R}^{2\times 2}$ we represent the value of the deformation gradient at the point $\boldsymbol{x}(\boldsymbol{\xi}_q)$ in Ω^e, where $\boldsymbol{\xi}_q$ denotes the coordinate of the Gauss evaluation point with respect to the natural coordinate system. From (9.117), $\tilde{F}^{qe}$ is written in terms of the nodal displacements as

$$\tilde{F}^{qe} = \sum_{I=1}^{n^{\mathrm{S}}} \frac{\partial N_I}{\partial \boldsymbol{x}}(\boldsymbol{\xi}_q)\boldsymbol{u}_I + I.$$

This equation can be rewritten compactly in the form of

$$\tilde{F}^{qe} = \mathrm{Mat}(B^{qe}\tilde{\boldsymbol{u}}) + I, \tag{9.118}$$

where $B^{qe} \in \mathbb{R}^{4\times d}$ is a constant matrix. By $\mathrm{Mat}(\cdot)$ we mean the operator that transforms a four-dimensional vector to a 2×2 matrix.[30] Likewise, by the matrix $\tilde{E}^{qe} \in \mathbb{S}^2$ we represent the value of the Green–Lagrange strain tensor at $\boldsymbol{x}(\boldsymbol{\xi}_q)$ in Ω^e. Then, in a manner similar to (8.73), the strain energy stored in the membrane is approximated by

$$\int_{\Omega} \hat{w}_{\mathrm{ref}}(\boldsymbol{E}_{\mathrm{e}})\mathrm{d}\Omega \simeq \sum_{e=1}^{n^{\mathrm{E}}} \sum_{q=1}^{n^{\mathrm{G}}} \frac{\tilde{\rho}^{qe}}{2} \,\mathbf{vec}(\tilde{E}_{\mathrm{e}}^{qe})^{\mathrm{T}} C \,\mathbf{vec}(\tilde{E}_{\mathrm{e}}^{qe}), \tag{9.119}$$

[30]Since $\tilde{F}^{qe}$ is not a symmetric matrix, the definition of $\mathrm{Mat}(\cdot)$ used here differs from that in section 8.4.1. In (8.75) $\mathrm{Mat}(\cdot)$ is considered to transform a vector to a symmetric matrix in a way consistent with the Voigt notation in (8.69), while in (9.118) the four-dimensional vector $B^{qe}\tilde{\boldsymbol{u}}$ is transformed to a 2×2 non-symmetric matrix.

where $C \in \mathbb{S}^3$ is the elasticity tensor in the matrix form, and $\mathbf{vec}(\tilde{E}_{\mathrm{e}}^{qe}) \in \mathbb{R}^3$ is designated in the Voigt notation similar to (8.69), i.e.,

$$\mathbf{vec}(\tilde{E}_{\mathrm{e}}^{qe}) = \begin{bmatrix} \tilde{E}_{\mathrm{e}11}^{qe} \\ \tilde{E}_{\mathrm{e}22}^{qe} \\ 2\tilde{E}_{\mathrm{e}12}^{qe} \end{bmatrix}.$$

The positive-semidefinite constraint of $(\overline{\mathrm{PE}})$ in (9.62) is required to be satisfied at each Gauss evaluation point as

$$\tilde{E}_{\mathrm{e}}^{qe} \succeq \frac{1}{2}[(\tilde{F}^{qe})^{\mathrm{T}}(\tilde{F}^{qe}) - I], \quad \forall q;\ \forall e,$$

which is equivalently rewritten as[31]

$$\begin{bmatrix} 2\tilde{E}_{\mathrm{e}}^{qe} + I & -(\tilde{F}^{qe})^{\mathrm{T}} \\ -\tilde{F}^{qe} & I \end{bmatrix} \succeq O, \quad \forall q;\ \forall e. \tag{9.120}$$

Consequently, (9.118), (9.119), and (9.120) yield for $(\overline{\mathrm{PE}})$ the discretization

$$\left.\begin{aligned} (\overline{\mathrm{PE}}^{\mathrm{FE}}):\ \min_{\tilde{u}_j, \tilde{E}_{\mathrm{e}}^{qe}, \tilde{F}^{qe}} \quad & \sum_{e=1}^{n^{\mathrm{E}}} \sum_{q=1}^{n^{\mathrm{G}}} \frac{\tilde{\rho}^{qe}}{2} \mathbf{vec}(\tilde{E}_{\mathrm{e}}^{qe})^{\mathrm{T}} C\, \mathbf{vec}(\tilde{E}_{\mathrm{e}}^{qe}) - \sum_{j \in \tilde{\Gamma}_{\mathrm{N}}} \underline{\tilde{p}}_j \tilde{u}_j \\ \mathrm{s.t.} \quad & \begin{bmatrix} 2\tilde{E}_{\mathrm{e}}^{qe} + I & -(\tilde{F}^{qe})^{\mathrm{T}} \\ -\tilde{F}^{qe} & I \end{bmatrix} \succeq O, \quad \forall q;\ \forall e, \\ & \tilde{F}^{qe} = \mathrm{Mat}(B^{qe} \tilde{\boldsymbol{u}}) + I, \quad \forall q;\ \forall e, \\ & \tilde{u}_j = \underline{\tilde{u}}_j, \quad \forall j \in \tilde{\Gamma}_{\mathrm{D}}, \end{aligned}\right\} \tag{9.121}$$

where the vector $\tilde{\boldsymbol{u}} \in \mathbb{R}^d$, the symmetric matrices $\tilde{E}_{\mathrm{e}}^{qe} \in \mathbb{S}^2$ $(\forall q;\ \forall e)$, and the 2×2 matrices $\tilde{F}_{\mathrm{e}}^{qe} \in \mathbb{R}^{2\times 2}$ $(\forall q;\ \forall e)$ are the independent variables.[32] We can see that $(\overline{\mathrm{PE}}^{\mathrm{FE}})$ in (9.121) is a convex optimization problem as follows. Since C is positive semidefinite, the objective function is a convex quadratic function in $\tilde{E}_{\mathrm{e}}^{qe}$ and $\tilde{u}_j$. The constraints consist of linear matrix inequalities and linear equality constraints, and hence the feasible set is convex.

Remark 9.5.1. We can further reduce $(\overline{\mathrm{PE}}^{\mathrm{FE}})$ to a standard form of conic optimization[33] by introducing some extra variables in a manner similar to Remark 8.4.2. Indeed, minimizing $\frac{\tilde{\rho}^{qe}}{2} \mathbf{vec}(\tilde{E}_{\mathrm{e}}^{qe})^{\mathrm{T}} C\, \mathbf{vec}(\tilde{E}_{\mathrm{e}}^{qe})$ in $(\overline{\mathrm{PE}}^{\mathrm{FE}})$ is equivalent to minimizing $\tilde{w}^{qe}$ over the constraint

$$\tilde{w}^{qe} \geq \frac{\tilde{\rho}^{qe}}{2} \mathbf{vec}(\tilde{E}_{\mathrm{e}}^{qe})^{\mathrm{T}} C\, \mathbf{vec}(\tilde{E}_{\mathrm{e}}^{qe}),$$

[31] See problem (9.64) and expositions there.

[32] Simple substitution eliminates $\tilde{F}^{qe}$ from problem (9.121), but we leave this variable for simplifying presentation of the problem formulation.

[33] See problems (1.26) and (1.27) in section 1.6 for the definition of the standard form of conic optimization.

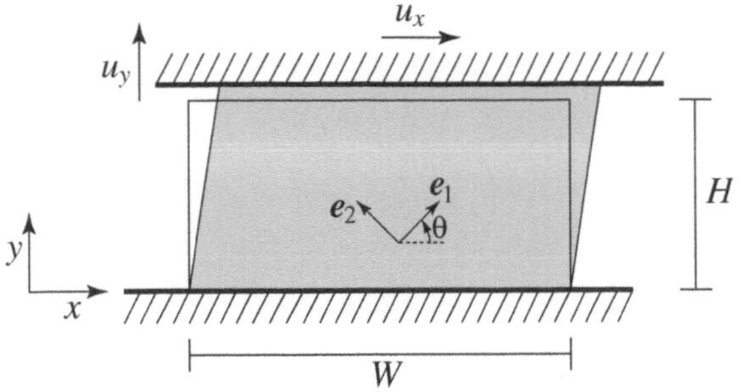

FIGURE 9.7: Geometry and boundary conditions for the shear test example.

and this inequality can be rewritten as a second-order cone constraint as demonstrated in (8.78). As a consequence, $(\overline{\mathrm{PE}}^{\mathrm{FE}})$ is reformulated as a minimization of a linear function over some second-order cone constraints and positive-semidefinite constraints.

A second-order cone constraint can be expressed as a positive-semidefinite constraint (see Fact 1.5.1); $(\overline{\mathrm{PE}}^{\mathrm{FE}})$ can be further reduced to an SDP problem. As discussed in Remark 8.4.3 for masonry structures, an SDP problem can be solved by using the primal-dual interior-point method efficiently. The solution process of this approach does *not* involve any procedure for determining the consistent wrinkle states. The convergence to the equilibrium configuration is theoretically guaranteed, and the computational effort required to solve $(\overline{\mathrm{PE}}^{\mathrm{FE}}_{\mathrm{cone}})$ depends not on the problem setting such as loading conditions but only on the size of the problem. This is regarded as a crucial advantage of the conic optimization approach from a viewpoint of the numerical solution. ■

9.5.2 Examples

The equilibrium configurations of planar membranes are computed based on the conic optimization formulations presented. All the simulations are performed by solving $(\overline{\mathrm{PE}}^{\mathrm{FE}})$ in (9.121) by using SeDuMi Ver. 1.05 [424], which is an implementation of the primal-dual interior-point method. In all the examples the membranes are supposed to be subjected to plane-stress conditions and are discretized by linear triangular elements.

9.5.2.1 Shear test of rectangular membrane

This section presents a shear test of a two-dimensional rectangular membrane as a benchmark example [191, 380, 455]. Figure 9.7 defines the geometry of the membrane, where $W = 200.0\,\mathrm{mm}$ and $H = 100.0\,\mathrm{mm}$. The top and bottom edges are fixed. The membrane is first prestressed in the x-direction by a displacement of $u_y = 1.0\,\mathrm{mm}$, and then this displacement is held fixed

(a) First principal stress.

(b) Second principal stress.

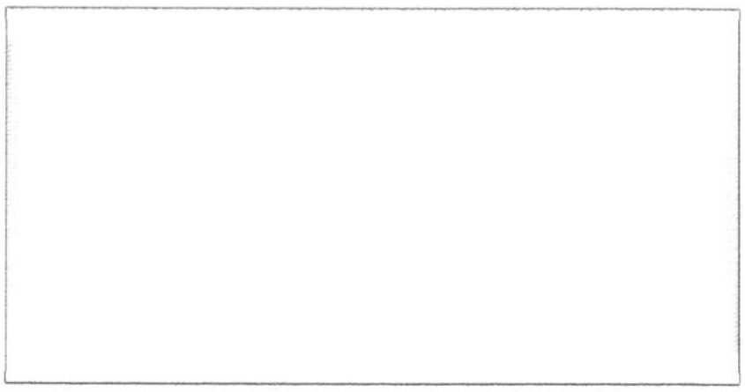

(c) First principal wrinkling strain.

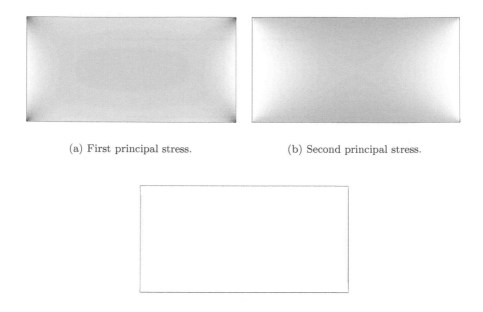

FIGURE 9.8: Computational results of the shear test example of the isotropic membrane: the solution for $(u_x, u_y) = (0.0, 1.0)\,\mathrm{mm}$.

for the subsequent shear loading up to $u_x = 5.0\,\mathrm{mm}$ in the y-direction. The membrane is uniformly discretized by $2 \times (100 \times 80) = 16{,}000$ linear triangular elements.

9.5.2.1.1 Isotropic material As an isotropic material, similar to the examples studied in [191, 455], we choose that the thickness is $t = 25\,\mu\mathrm{m}$, Young's modulus is $E = 3500\,\mathrm{N/mm}^2$, and Poisson's ratio is $\rho = 0.31$.

The solutions obtained for $u_x = 0.0\,\mathrm{mm}$, $u_x = 1.0\,\mathrm{mm}$, and $u_x = 5.0\,\mathrm{mm}$ are shown in Figure 9.8, Figure 9.9, and Figure 9.10, respectively, in which the distributions of the two principal stresses and the maximal principal value of the wrinkling strain are depicted. Note that for the stress and the strain we here show the Cauchy stress tensor and the Almansi strain tensor on the deformed equilibrium configurations. In all the cases, the minimal principal value of the wrinkling strain vanishes almost everywhere. In Figure 9.8(c) wrinkles are found on both the vertical edges. The complementarity condition between the stress and the wrinkling strain is confirmed, for example, between Figure 9.8(b) and Figure 9.8(c). The development of wrinkles is observed in the figure series Figure 9.8(c) to Figure 9.10(c).

As a sort of imperfection we next suppose that the shape of the top edge is

(a) First principal stress.

(b) Second principal stress.

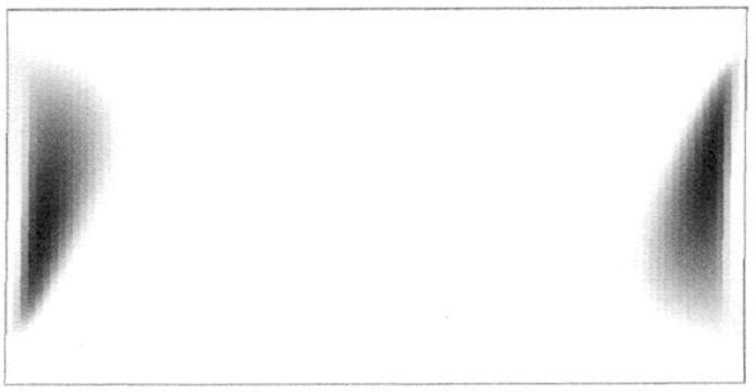

(c) First principal wrinkling strain.

FIGURE 9.9: Computational results of the shear test example of the isotropic membrane: the solution for $(u_x, u_y) = (1.0, 1.0)$ mm.

supposed to be statistically distributed, where the prescribed displacement u_y is assumed to obey the normal distribution with mean 1.0 mm and the standard deviation 5.0×10^{-2} mm. The horizontal displacement u_x is assumed to be prescribed precisely. Note that the imperfection of thickness is considered in [380] instead of the prescribed displacements.

Figure 9.11 shows the development of the wrinkles, in terms of the maximal principal value of the wrinkling strain. In $u_x = 0$ there are two wrinkles mostly due to Poisson's effect. Several wrinkling zones are observed in Figure 9.11(b), which grow visibly as u_x increases. Finally the wrinkles almost coalesce as shown in Figure 9.11(d).

9.5.2.1.2 Orthotropic material The shear test is now performed for an orthotropic extension of the St. Venant material [191, 274]. The constitutive law between the second Piola–Kirchhoff stress tensor and the Green–Lagrange strain tensor is given by

$$\begin{bmatrix} S_{11} \\ S_{22} \\ S_{12} \end{bmatrix} = \begin{bmatrix} 1/E_1^{\mathrm{Y}} & -\nu_{12}/E_2^{\mathrm{Y}} & 0 \\ -\nu_{21}/E_1^{\mathrm{Y}} & 1/E_2^{\mathrm{Y}} & 0 \\ 0 & 0 & 1/2G \end{bmatrix}^{-1} \begin{bmatrix} E_{11} \\ E_{22} \\ 2E_{12} \end{bmatrix}, \tag{9.122}$$

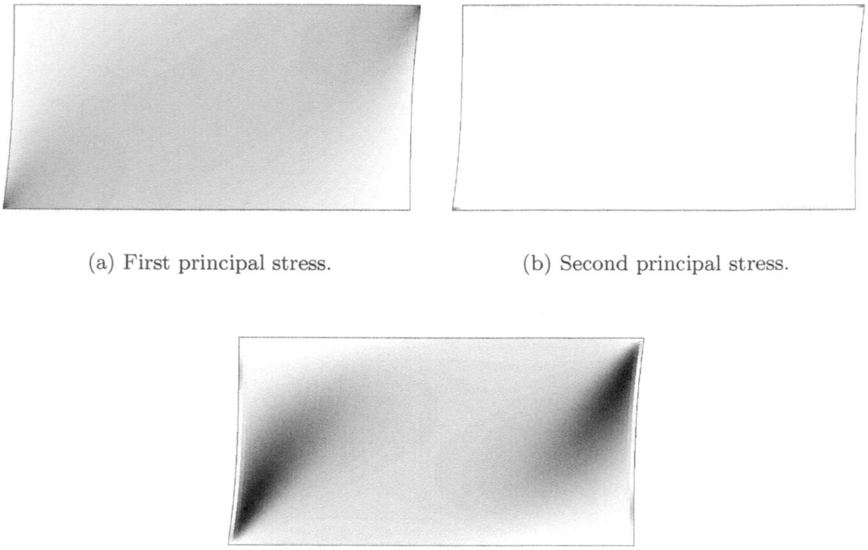

(a) First principal stress.

(b) Second principal stress.

(c) First principal wrinkling strain.

FIGURE 9.10: Computational results of the shear test example of the isotropic membrane: the solution for $(u_x, u_y) = (5.0, 1.0)\,\mathrm{mm}$.

where the material constants are $E_1^{\mathrm{Y}} = 1.0 \times 10^5\,\mathrm{Pa}$, $E_2^{\mathrm{Y}} = 10.0 \times 10^5\,\mathrm{Pa}$, $\nu_{12} = 0.3$, $\nu_{12}E_1^{\mathrm{Y}} = \nu_{21}E_2^{\mathrm{Y}}$, and $G = 0.4 \times 10^5\,\mathrm{Pa}$. The first fiber direction, corresponding to E_1, is supposed to coincide with $\boldsymbol{e}_1$, which is given by an angle of θ with respect to a horizontal line, as shown in Figure 9.7. The geometry and the loading conditions are identical to the isotropic case in Figure 9.11, which incorporates the imperfection of the vertical displacement; i.e., u_y is supposed to obey the normal distribution with mean $1.0\,\mathrm{mm}$ and the standard deviation $5.0 \times 10^{-2}\,\mathrm{mm}$. Note that the horizontal displacement u_x is supposed to be prescribed precisely.

In Figure 9.12 and Figure 9.13, we show the maximal principal stress (in terms of the Cauchy stress) and the maximal principal wrinkling strain (in terms of the Almansi strain) for each case of $\theta = 0$, $\pi/4$, $\pi/2$, and $3\pi/4$. The wrinkling behaviors in the cases of $\theta = 0$ and $\theta = 3\pi/4$ are roughly similar to the isotropic case (see Figure 9.11(c)). In contrast, the wrinkling patterns observed in the cases of $\theta = \pi/4$ and $\theta = \pi/2$ are quite different from the isotropic case.

Remark 9.5.2. The figure series Figure 9.8–Figure 9.10 may be taken by some readers as a result of incremental analyses consisting of sequential loading steps, each of which is small enough. In fact, although such a sequential procedure is usually

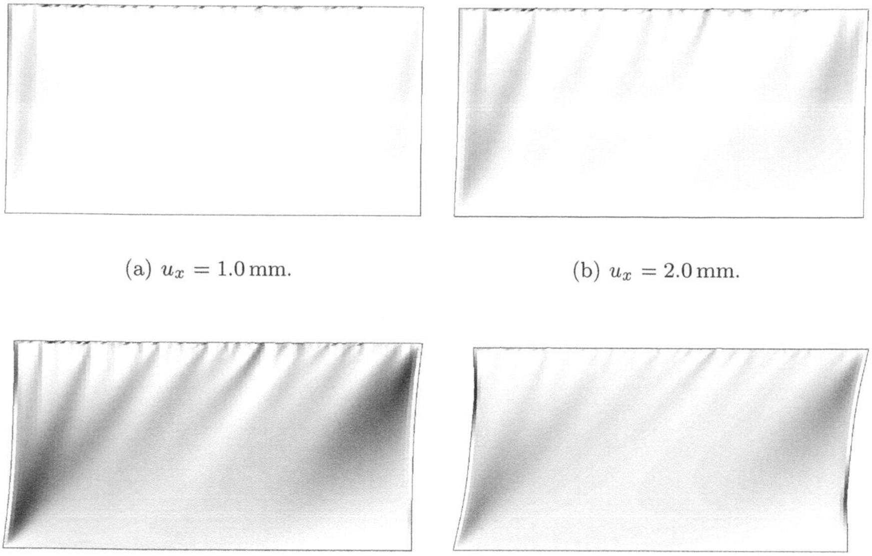

(a) $u_x = 1.0\,\text{mm}$.

(b) $u_x = 2.0\,\text{mm}$.

(c) $u_x = 5.0\,\text{mm}$.

(d) $u_x = 10.0\,\text{mm}$.

FIGURE 9.11: Development of wrinkles for the shear test example of the isotropic membrane with randomly generated vertical displacements.

carried out to perform the large deformation analysis, the methodology presented here does *not* require the sequential incremental analysis with respect to the loading steps. The equilibrium configuration at the specified loading condition can be obtained by solving a single convex optimization problem. A numerical solution of the optimization problem certainly involves the iterative procedure, but the iterations performed by the optimization algorithm (typically the interior-point method) do not correspond to the loading steps at all. Thus our solution process is quite different from the conventional one in the large deformation analysis.

All the numerical examples of membranes are solved using SeDuMi Ver. 1.05 [424], which is an implementation of the primal-dual interior-point method. We always use the initial solution automatically supplied by the solver, which is irrelevant to the reference configuration of Figure 9.7. Then SeDuMi requires 41, 46, and 48 iterations to find the solutions in Figure 9.8, Figure 9.9, and Figure 9.10, respectively. In the examples with the imperfection, the solutions in Figure 9.11(a)–Figure 9.11(d) are obtained after 49, 47, 48, and 48 iterations, respectively. Thus the computational cost required is independent of the magnitudes of the prescribed displacements, which is the distinguished aspect of the methodology presented here. See also Remark 9.5.4. ■

Remark 9.5.3. The computational cost of our method is also independent of the anisotropy of the membrane, although for the conventional method it is known that the absence of the isotropic nature sometimes increases the computational cost

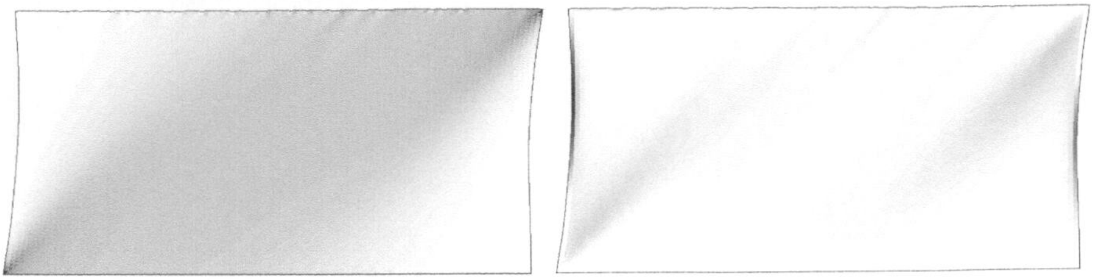

First principal stress　　　First principal wrinkling strain

(a) Solution for $\theta = \pi/4$.

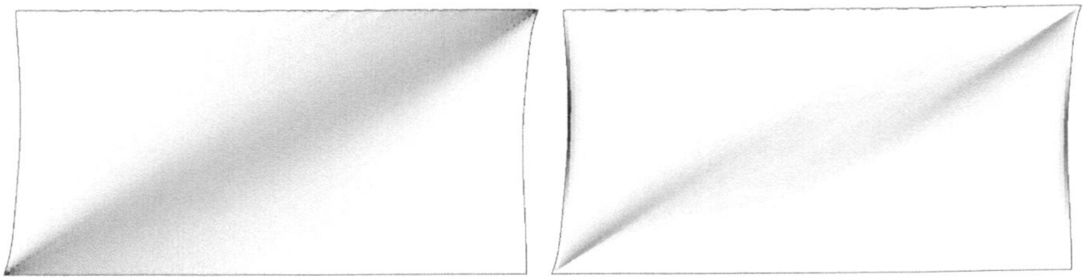

(b) Solution for $\theta = \pi/2$.

FIGURE 9.12: Computational results of the shear test example of the orthotropic membrane: the solutions for $u_x = 5.0\,\mathrm{mm}$ and $(\mathbf{E}(u_y), \mathbf{var}(u_y)^{1/2}) = (1.0, 5.0 \times 10^{-2})\,\mathrm{mm}$.

drastically. Concerning the orthotropic examples, the solutions in Figure 9.12(a)–Figure 9.13(b) are obtained after 43, 45, 42, and 41 iterations, respectively, using SeDuMi [424]. Thus the significant difference of the computational time is observed compared with the isotropic cases in Remark 9.5.2, which is one keen advantage to our approach. ■

9.5.2.2 Annulus membrane under in-plane torsion

As a second example we consider an annulus membrane illustrated in Figure 9.14, which is examined to demonstrate performance of various solution methods [191, 254, 274, 391]. An experimental study (in the physical sense) of the annulus membrane subjected to a similar situation can be found in [312].

The membrane is attached to a rigid disk at the inner edge and a fixed guard ring at the outer edge. The inner rigid disk is rotated counter-clockwise with an angle of 10°. The plane-stress condition is assumed. The outer and inner radii are $r_{\mathrm{out}} = 12.5\,\mathrm{m}$ and $r_{\mathrm{in}} = 5.0\,\mathrm{m}$. The thickness of the membrane is $25\,\mu\mathrm{m}$. The membrane is discretized by 240 linear triangular elements.

We simulate both isotropic and orthotropic materials. In the case of an

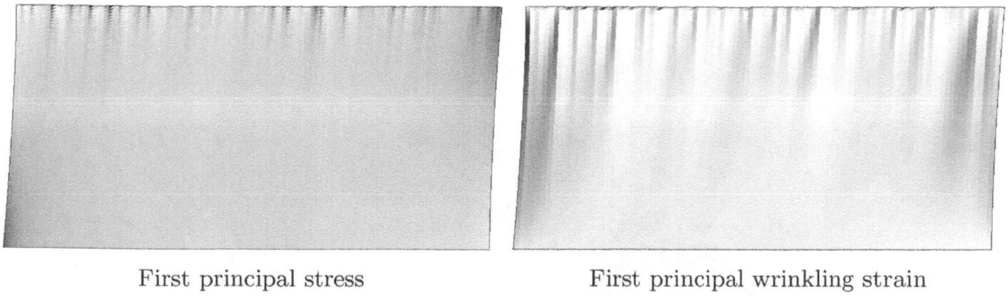

(a) Solution for $\theta = 3\pi/4$.

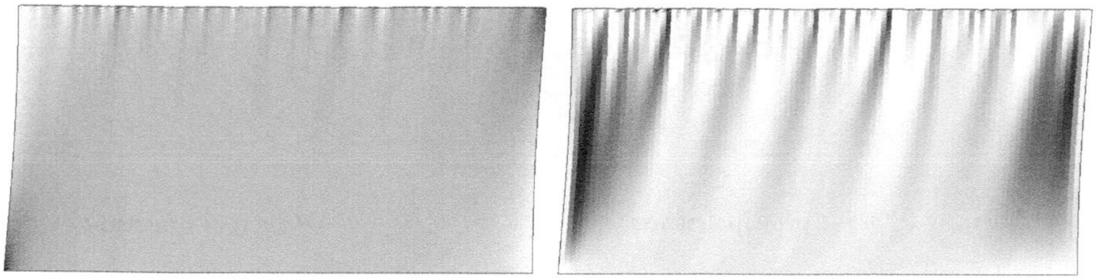

(b) Solution for $\theta = 0$.

FIGURE 9.13: Computational results of the shear test example of the orthotropic membrane: the solutions for $u_x = 5.0\,\text{mm}$ and $(\mathbf{E}(u_y), \mathbf{var}(u_y)^{1/2}) = (1.0, 5.0 \times 10^{-2})\,\text{mm}$ (continued).

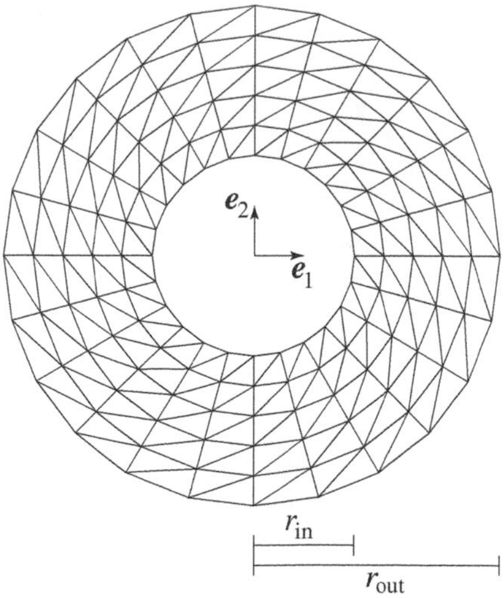

FIGURE 9.14: Geometry of an annulus membrane.

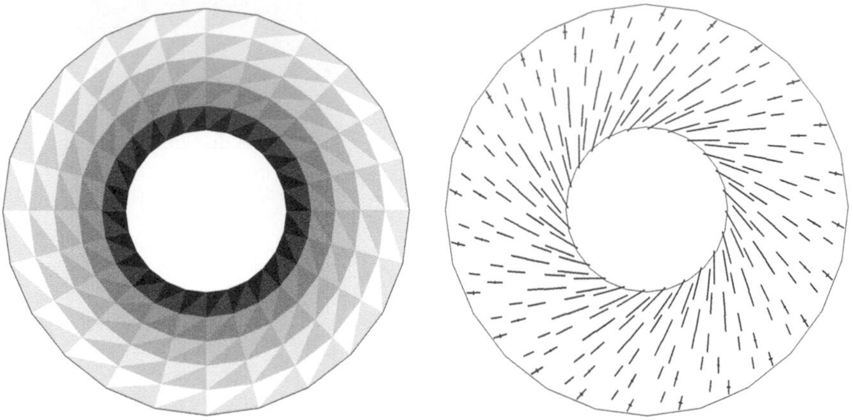

First principal stress

Principal directions

(a) Isotropic material.

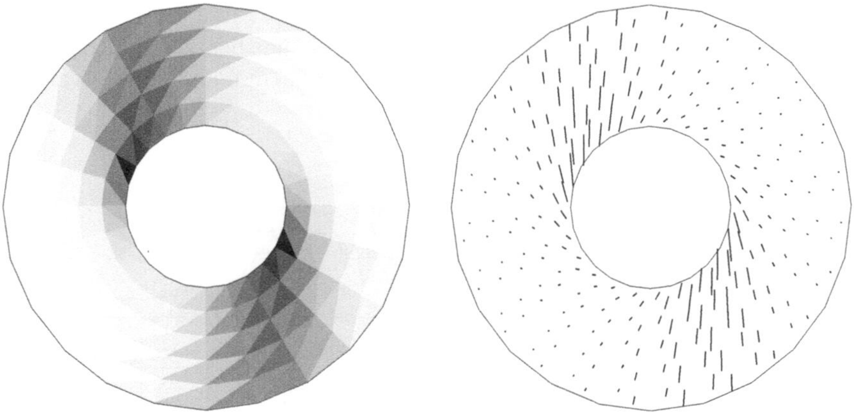

(b) Orthotropic material.

FIGURE 9.15: Maximal principal stress and two principal directions of the annulus membrane.

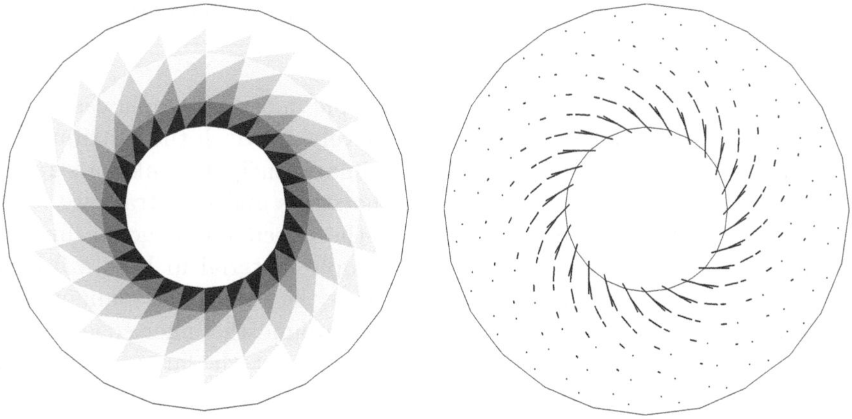

First principal stress Principal directions

(a) Isotropic material.

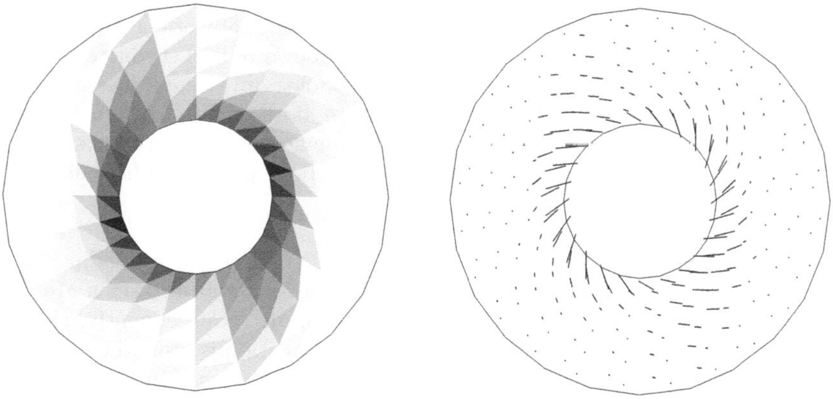

(b) Orthotropic material.

FIGURE 9.16: Maximal principal wrinkling strain and two principal directions of the annulus membrane.

isotropic material we assume the St. Venant material with Young's modulus $E = 1.0 \times 10^5$ Pa and Poisson's ratio $\nu = 0.4$. For an orthotropic case, an orthotropic extension of St. Venant material, defined by (9.122), is supposed, where $E_1^{\rm Y} = 1.0 \times 10^5$ Pa, $E_2^{\rm Y} = 10.0 \times 10^5$ Pa, $\nu_{12} = 0.3$, $\nu_{12} E_1^{\rm Y} = \nu_{21} E_2^{\rm Y}$, and $G = 0.35 \times 10^5$ Pa. The warp and fill directions of textile are assumed to coincide with $\boldsymbol{e}_1$ and $\boldsymbol{e}_2$, respectively, in the reference configuration (Figure 9.14).

The solutions obtained for the isotropic and orthotropic cases are shown in Figure 9.15 and Figure 9.16, respectively, in which the distributions of the maximal principal stress and the maximal principal wrinkling strain are depicted for each case. Note that for the stress and the strain we here depict the second Piola–Kirchhoff stress and the Green–Lagrange strain on the reference configurations. For the orthotropic material in Figure 9.15(a), it is observed that the stress distributes with the rotational symmetry, and the principal stress attains the largest value around the inner edge. By contrast, for the orthotropic case shown in Figure 9.15(b), the large values of the principal stresses concentrate along the direction of the vertical direction, i.e., the stiffer fiber direction.

Remark 9.5.4. The equilibrium analysis with the geometrical nonlinearity is usually realized by performing the incremental analyses sequentially, in which the final boundary condition is achieved after a finite number of loading steps. At each loading step the system of nonlinear equations is solved to find the incremental solution. In contrast, it is emphasized that the methodology presented does *not* require such incremental analyses. The equilibrium state at the final loading condition can be obtained by solving one SDP problem. A numerical solution of the SDP certainly requires the iterative procedure, but the iterations performed here are irrelevant to the loading steps.

For example, in [191, section 4.2] the final solution corresponding to the angle of 10° is achieved after 10 loading steps, where several iterations are required at each loading step to solve the nonlinear equations. Consequently, the number of total iterations needed to obtain the final deformation is around 50 for an isotropic case and is sometimes much more for an orthotropic case. However, in the present methodology based on the convex optimization, the final solution corresponding to the angle of 10° can be obtained by solving solely one conic optimization problem. The corresponding problem is solved by using the primal-dual interior-point method (SeDuMi Ver. 1.05 [424]) with 18 iterations for the isotropic case and 21 iterations for the orthotropic case. It is also noted that the number of iterations required for the orthotropic case is not much larger than that for the isotropic case, while for the conventional method the absence of isotropic property usually causes slow convergence; see, e.g., the discussion in [380, section 3.4] and the numerical example in [191, section 4.2] indicating around 25% increase of the number of iterations compared with the isotropic case. See Remark 9.5.3 for a comparison of the isotropic and orthotropic cases for the shear test example. ■

9.6 Notes

1. Probably the earliest study of the mechanics of wrinkling membranes dates back to Wagner [449] published in 1929, in which the concept of *tension field* was introduced to simplify the post-buckling behavior of very thin shear panels used in aircraft construction. Particularly, in [449] the flexural stiffness and the (post-buckling) planar compressive stiffness of the isotropic pannel are assumed to be negligible, and the problem then is simplified to finding the magnitude and direction of uniaxial stress distribution. The tension field theory was subsequently developed for analytical investigation of wrinkling membranes (see Steigmann [416] for a survey of the early development of theory). Although its applications rely upon case-by-case analyses, the tension field theory has been discussed by many authors for various cases, including anisotropic materials [294] and large deformations [102]; see Danielson–Natarajan [102], Mansfield [293, 294], Steigmann [416], Steigmann–Pipkin [419], and Zak [463] for details. However, these analytical approaches based on the tension field theory are not applicable to membranes with arbitrary shapes and arbitrary loading conditions.

2. Mainstream contemporary numerical methods, in conjunction with finite element methods, may be divided into two categories: One is based on a modification of the kinematic relations, and the other aims at modifying the constitutive relation. A less popular, but no less important, approach compared with these two methods is based on a modification of the strain energy function, similar to the methodology developed in this chapter.

3. The first attempts at the approach relying on a modification of the kinematic relations are Roddeman [391] and Roddeman–Drukker–Oomens–Janssen [392, 393], in which a method based on a modification scheme of the deformation gradient was proposed. For the isotropic Mooney–Rivlin material a modification scheme of the principal stretches was proposed by Wu [458]. Originally considered by Roddeman–Drukker–Oomens–Janssen [392] in 1987, the modified deformation gradient has been employed by many authors cited below and defined for the deformation gradient $\boldsymbol{F}$ as

$$\boldsymbol{F}_{\rm e} = (\boldsymbol{I} - \beta \boldsymbol{d}_{\rm w} \boldsymbol{d}_{\rm w}^{\rm T})\boldsymbol{F} \quad (\beta \leq 0),$$

where $\boldsymbol{F}_{\rm e}$ corresponds to the *elastic part* of the deformation gradient in accordance with the terminology used in this chapter, $|\beta|$ represents the amount of the wrinkle, and $\boldsymbol{d}_{\rm w}$ is the unit vector along which the principal Cauchy stress vanishes, i.e., $\boldsymbol{d}_{\rm w}$ is in the direction transverse to the winkle. An extension of this approach to the curved membranes is in Muttin [329]. A method for detecting the wrinkling direction was proposed by Kang–Im [200, 201], in which the wrinkling direction is computed by solving the nonlinear equations in terms of the modified Green–Lagrange strain. In this method, the interval $[0, 180^\circ)$ is first divided into finitely many intervals, and then a root-finding algorithm for solving a nonlinear equation is performed over the candidate small interval.The system of nonlinear equations for finding the wrinkling direction involving the second Piola–Kirchhoff stress is in Lu–Accorsi–Leonard [274] and is in the form of

$$\boldsymbol{d}_{\mathrm{w}}^{\mathrm{T}} \hat{\boldsymbol{S}}_{\mathrm{ref}}(\boldsymbol{F}_{\mathrm{e}}) \boldsymbol{d}_{\mathrm{w}} = 0, \quad \boldsymbol{d}_{\mathrm{t}}^{\mathrm{T}} \hat{\boldsymbol{S}}_{\mathrm{ref}}(\boldsymbol{F}_{\mathrm{e}}) \boldsymbol{d}_{\mathrm{w}} = 0,$$

where $\boldsymbol{d}_{\mathrm{t}}$ is a unit vector orthogonal to $\boldsymbol{d}_{\mathrm{w}}$. Use of the Newton method to the similar formulation is in Schoop–Taenzer–Hornig [403]. This formulation has been applied to find the analytical solutions of a few specific examples of membrane [187] and numerical analysis of elastic-plastic membranes [188]. An iterative procedure for finding the wrinkling strain $\boldsymbol{E}_{\mathrm{w}}$ was proposed by Raible–Tegeler–Löhnert–Wriggers [380] and is based on the decomposition (9.44) of the Green–Lagrange strain.

4. The second major methodology is based on a modification of the constitutive relation, in which the elasticity tensor is modified so that the compressive stresses are eliminated. In Contri–Schrefler [94] a trial-and-error approach was performed by using a simple scheme for eliminating the principal compressive stress based on Mohr's circle. The concept of *effective elasticity tensor*, which is a modification of the constitutive law to eliminate the compressive stresses, has been discussed by many authors, e.g., Ding–Yang [117], Fujikake–Kojima–Fukushima [138], and Miller–Hedgepeth–Weingarten–Das–Kahyai [311], in which the Young modulus and the Poisson ratio of the membrane are considered variables depending on the stress state. Liu–Jenkins–Schur [271] introduced a penalty parameter to prevent numerical instability due to the vanishing stiffness in the direction orthogonal to the wrinkle. Further refinements on this modified material model with the penalty are in Jarasjarunngkiat–Wüchner–Bletzinger [191], Rossi–Lazzari–Vitaliani–Oñate [395], and Valdés–Miquel–Oñate [442]. The reader may refer to Miyazaki [313] for a survey of these modified material models.

As studied in section 9.3.1, the wrinkling strain $\boldsymbol{E}_{\mathrm{w}}$ and the stress $\boldsymbol{S}$ share the same principal directions. For the isotropic case, particularly, these directions also coincide with the principal directions of the strain $\boldsymbol{E}$, and hence in the direction along the wrinkle the tensile strain energy attains the maximum value. Oriented by this fact, Lee–Youn [254] considered the maximization problem of the tensile strain energy to determine the wrinkle direction in the course of a modified material model approach.

It follows from (9.44) and Proposition 9.3.4 that the strain energy function $\hat{w}_{\mathrm{mem}}$ of the membrane is alternatively written as

$$\begin{aligned} \hat{w}_{\mathrm{mem}}(\boldsymbol{E}) &= \min_{\boldsymbol{E}_{\mathrm{e}} \in \mathbb{S}^2} \{\hat{w}_{\mathrm{ref}}(\boldsymbol{E}_{\mathrm{e}}) \mid \boldsymbol{E}_{\mathrm{e}} \succeq \boldsymbol{E}\} \\ &= \min_{\boldsymbol{E}_{\mathrm{w}} \in \mathbb{S}^2} \{\hat{w}_{\mathrm{ref}}(\boldsymbol{E} - \boldsymbol{E}_{\mathrm{w}}) \mid -\boldsymbol{E}_{\mathrm{w}} \succeq \boldsymbol{o}\} \qquad (9.123) \end{aligned}$$

with $\hat{w}_{\text{ref}}$ defined by (9.52). If we know that the membrane is in the wrinkle state at the point under consideration, then $\boldsymbol{E}_{\text{w}}$ is known to be the rank-1 matrix, written in the form of $\boldsymbol{E}_{\text{w}} = \beta' \boldsymbol{\phi}\boldsymbol{\phi}^{\text{T}}$, because one of the eigenvalues of $\boldsymbol{E}_{\text{w}}$ vanishes. Moreover, if the direction $\boldsymbol{\phi}$ is known in advance, then (9.123) is reduced to

$$\begin{aligned}\hat{w}_{\text{mem}}(\boldsymbol{E}) &= \min_{\beta' \in \mathbb{R}} \{\hat{w}_{\text{ref}}(\boldsymbol{E} - \beta' \boldsymbol{\phi}\boldsymbol{\phi}^{\text{T}}) \mid \beta' \leq 0\} \\ &= \min_{\beta' \in \mathbb{R}} \{\hat{w}_{\text{ref}}(\boldsymbol{E} - \beta' \boldsymbol{\phi}\boldsymbol{\phi}^{\text{T}})\}, \qquad (9.124)\end{aligned}$$

since the strain energy does not increase by introducing the nonzero amount of wrinkle $\beta' < 0$, as far as $\boldsymbol{E}$ correctly corresponds to the wrinkle state. For the isotropic case Akita–Nakashino–Natori–Park [5] used this relation, where (9.124) was looked upon as a projection of $\boldsymbol{E}$ onto the direction $\boldsymbol{\phi}\boldsymbol{\phi}^{\text{T}}$ with respect to the norm induced by $\hat{w}_{\text{ref}}$ and a modified constitutive law was presented based on the projection. An orthotropic extension of this approach is in Jarasjarunngkiat–Wüchner–Bletzinger [192]. Thus several approaches partially used the framework presented in this chapter, but its entire advantage can be enjoyed by the conic optimization approach.

5. The third approach is based on the so-called *relaxed strain energy function*, coined by Pipkin [369], in which the out-of-plane deformation of wrinkles is treated in terms of the nonlinearity of constitutive law by modifying the strain energy function. All the assumptions postulated in the tension field theory are satisfied simply by replacing the conventional strain energy function with the relaxed strain energy function. Following the explicit formulation of the relaxed energy function for the isotropic neo-Hookean material given by Pipkin [369] in terms of the principal stretches, the theoretical developments of the relaxed energy function are in Pipkin [370, 371, 372], and Steigmann–Pipkin [418], with a particular interest in the neo-Hookean material [163, 418]. Although the relaxed energy function is defined as a quasi-convexification of the strain energy function [369, 372], its explicit formulation is not known in general. As particular cases, the formulations for the linear elastic material are in Barsotti–Ligarò–Royer-Carfagni [34] and Pipkin [370]; the Varga material are in Steigmann [417]; the St. Venant material are in Gil–Bonet [146]; and the Fung model of the anisotropic tissue are in Massabò–Gambarotta [302].

This approach has several interesting features. The problem can be solved using the in-plane membrane field variables without referring to out-of-plane variables; the theory uses a variational approach and therefore overcomes the difficulties of other theories in the constitutive law involving the change of slack/wrinkle/taut states. Thus the relaxed-energy approach shares a similar idea with the approach optimal value of problem (9.55) considered in Proposition 9.3.4; i.e., if we write

$$\bar{w}(\boldsymbol{E}) = \min_{\boldsymbol{E}_{\text{e}} \in \mathbb{S}^2} \{\hat{w}_{\text{ref}}(\boldsymbol{E}_{\text{e}}) \mid \boldsymbol{E}_{\text{e}} \succeq \boldsymbol{E}\},$$

then $\bar{w}$ serves as a relaxed strain energy function. The remarkable difference of the present analysis from the relaxed-energy method is that we "transfer" the inequality from the strain-energy level to the kinematic-relation level, as seen in (9.62). By substituting $\bar{w}$ into the problem (9.61) we obtain a bi-level optimization problem that is attacked in the relaxed-energy methodology, but we prefer to treat the single-level optimization problem in (9.62) within the framework of the conic optimization methodology.

Jenkins–Leonard [195] applied the theory of relaxed energy density to the viscoelastic analysis of membranes. For isotropic membranes consisting of the neo-Hookean material, the associated relaxed strain energy is known to have a closed-form representation (Pipkin [369, section 9]). Based on this representation of the relaxed strain energy, Haseganu–Steigmann [163] performed a numerical analysis of isotropic membranes in conjunction with the dynamic relaxation method (in which the equilibrium state to be found is regarded as a long-time limit of a damped dynamic response of the structure). An extension of the concept of relaxed energy from the viewpoint of saturated elasticity is in Epstein [123], with which a numerical algorithm based on the dynamic relaxation method was presented for anisotropic membranes by Epstein–Forcinito [124]. Chen–Sun–Wu–Yuen [82] accounted for weak resistance of the membrane against wrinkling due to the small bending rigidity by adding a strain energy due to the weak compressive stiffness, called the "wrinkling energy," to the original relaxed strain energy in Pipkin [372]. Then the minimization problem of the potential energy is solved by using the conjugate gradient method. Mosler [325] proposed a combination method of bi-level minimization of potential energy based on the relaxed strain energy [372]. By adapting the notation in this chapter, the decomposition of the right Cauchy–Green tensor $\boldsymbol{C}$ $(= \boldsymbol{F}^{\rm T}\boldsymbol{F})$, in the fashion of (9.44), is written as

$$\boldsymbol{C} = \boldsymbol{C}_{\rm e} + \boldsymbol{C}_{\rm w}.$$

The negative semidefiniteness of the wrinkle part $\boldsymbol{C}_{\rm w}$ yields the expression

$$\boldsymbol{C}_{\rm w} = -{\beta_1}^2 \boldsymbol{\phi}_1 \boldsymbol{\phi}_1^{\rm T} - {\beta_2}^2 \boldsymbol{\phi}_2 \boldsymbol{\phi}_2^{\rm T},$$

where $\boldsymbol{\phi}_1$, $\boldsymbol{\phi}_2$ are orthonormal basis vectors. At each evaluation point of the quadrature, we define $\bar{\beta}_i$ and $\bar{\boldsymbol{\phi}}_i$ $(i = 1, 2)$ by solving the following lower-level optimization problem:

$$(\bar{\beta}_i, \bar{\boldsymbol{\phi}}_i) = \arg \min_{\substack{\|\boldsymbol{\phi}_i\|=1 \\ \boldsymbol{\phi}_1^{\rm T}\boldsymbol{\phi}_2 = 0}} \hat{w}_{\rm ref}(\hat{\boldsymbol{E}}(\boldsymbol{C} - \boldsymbol{C}_{\rm w})),$$

where $\hat{\boldsymbol{E}}(\boldsymbol{C}) = (1/2)(\boldsymbol{C} - \boldsymbol{I})$. Then the relaxed strain energy is approximated by

$$\hat{w}_{\rm relax}(\boldsymbol{C}) = \hat{w}_{\rm ref}(\hat{\boldsymbol{E}}(\boldsymbol{C} - \bar{\boldsymbol{C}}_{\rm w})),$$

where $\bar{C}_{\mathrm{w}} = -\sum_{i=1}^{2} \bar{\beta}_i^2 \bar{\phi}_i \bar{\phi}_i^{\mathrm{T}}$. The upper-level optimization problem, which attempts to minimize the total potential energy defined by using $\hat{w}_{\mathrm{relax}}$, is solved by Mosler [325]. An extension of this approach to inelastic membranes is due to Mosler–Cirak [326]. In contrast, this chapter presented, by virtue of the framework of conic optimization, a single-level optimization problem for finding the equilibrium state.

6. An alternative approach, closely related to the one explored in this chapter, is a formulation based on the nonlinear complementarity problem in Jeong–Kwak [197]. In the complementarity condition between $\boldsymbol{E}_{\mathrm{w}}$ and $\boldsymbol{S}$, given in Proposition 9.3.3, it follows from Fact 1.3.20 that $\boldsymbol{E}_{\mathrm{w}}$ and $\boldsymbol{S}$ share the same system of eigenvectors, and their principal values satisfy the complementarity condition, i.e.,

$$-\lambda_i(\boldsymbol{E}_{\mathrm{w}}) \geq 0, \quad \lambda_i(\boldsymbol{S}) \geq 0, \quad \lambda_i(\boldsymbol{E}_{\mathrm{w}})\lambda_i(\boldsymbol{S}) = 0 \quad (i = 1, 2).$$

Therefore, by formulating the equilibrium equations with respect to the principal stress axes, the governing equations for the equilibrium of a membrane are reduced to the nonlinear complementarity problem. This formulation is in Jeong–Kwak [197], who solved the nonlinear complementarity problem numerically based on a sequential approximation as a linear complementarity problem (LCP).

7. As seen in **2**–**6**, various numerical methods have been proposed for wrinkling analysis of membranes. Nevertheless, many authors restrict themselves to the isotropic case, and the applicability of the methods to anisotropic cases is still unclear with a few exceptions, e.g., [124, 191, 200, 274, 380, 442]. The approach presented in this chapter does not depend on the material anisotropy, in both theoretical and numerical aspects.

8. Concerning the infinitesimal theory in section 9.2, the result presented in Proposition 9.2.1 is not new; the same statement is found in Pipkin [371, section 5]. The principle of complementary energy, say, (CE) in (9.24), is also essentially the same as that in Pipkin [371]. Remark 9.2.2 gives an alternative proof to the result originally in Pipkin [371].

In the geometrically nonlinear framework, the concept of the complementary energy principle was investigated by Pipkin [372], where the strain energy function w is regarded as a function of the right Cauchy–Green deformation tensor. An explicit form of the complementary energy principle, however, was not derived. In section 9.4, as an original result, we derived an explicit form of the complementary energy principle (see (9.114)) involving only the static variables and showed that the *minimum* of the complementary energy is attained the equilibrium solution. See Remark 5.1.10 for a short survey on the complementary energy principles for "usual" structures, such as elastic bodies or trusses, that do not possess unilateral properties; also, section 5.4 includes information of more references concerning the principles of complementary energy in the geometrically nonlinear theory.

The numerical solution presented in section 9.5 is the first method with the guaranteed convergence for finding the equilibrium state of a membrane in the geometrically nonlinear framework. Generally speaking, introducing wrinkling behavior often causes numerical instabilities (see, e.g., [271, section 4]) and/or slow convergence (see, e.g., [274, 395]). A conclusion drawn from the analysis of the present work, however, asserts that introducing wrinkling brings convexity to the energy principles. Under large deformation, the principle of the potential energy for elastic bodies is not convex due to the geometrical nonlinearity, and there exist more than one stationary point in general; the principle for membranes is convex and the equilibrium solution is obtained by finding the minimum point. This sheds new light on the nonsmooth mechanics via the convex optimization methodology introduced in the present work.

Chapter 10

Frictional Contact Problems

Orientation In this chapter we consider deformations of a linear elastic body whose boundaries, at least a part of boundaries, may be possibly subjected to conditions of *contact* and *friction*. The notion of *contact* is a combination of *contiguity* and *touch*, the two notions designated to be dual. The existence of friction differentiates stick and sliding motions. Thus the contact problems involving friction inevitably includes nonsmoothness properties and is a suitable material for studying nonsmooth mechanics, as we shall see throughout this chapter.

The principal goal of this chapter is the investigation of the quasistatic problem in terms of the complementarity condition subjected to conic constraints, in which the kinematic and static variables are subjected to the inclusions sharing a same form.

10.1 Friction Law

Consider an elastic structure discretized into finite elements and a rigid body with a sufficiently flat surface, as illustrated in Figure 10.1. Throughout this chapter, we assume that the displacement of the elastic body is small and that the rigid body is fixed in the physical three-dimensional space.

Suppose that the set of indices of nodes is partitioned into the three disjoint sets, $\mathcal{P}_{\mathrm{D}}$, $\mathcal{P}_{\mathrm{N}}$, and $\mathcal{P}_{\mathrm{c}}$. The displacement is prescribed for a node belonging to $\mathcal{P}_{\mathrm{D}}$, while the force applied to a node belonging to $\mathcal{P}_{\mathrm{N}}$ is prescribed. We thus consider the standard boundary conditions, i.e., the Dirichlet (or *kinematic*) and Neumann (or *static*) boundary conditions, for the nodes belonging to $\mathcal{P}_{\mathrm{D}}$ and $\mathcal{P}_{\mathrm{N}}$, respectively. Besides these standard boundary conditions, we consider a condition called a *contact boundary condition* that each node belonging to $\mathcal{P}_{\mathrm{c}}$ may come into contact with the rigid obstacle as a result of the deformation of the elastic body. A node subjected to a contact boundary condition is called a *contact candidate node*. A contact candidate node cannot interpenetrate the rigid obstacle. When a contact candidate node is on the obstacle surface, it can be pushed from the obstacle. This force, which is thus related to the non-penetration condition, is called the *reaction*. The *contact problem* is the problem to find the equilibrium state of an elastic body which involves the contact boundary conditions. Thus, in a contact problem, both

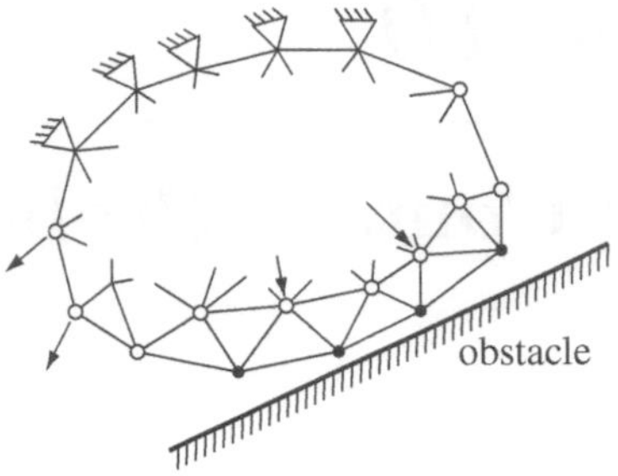

FIGURE 10.1: A discretized elastic body and a rigid obstacle. "•": the contact candidate node ($\in \mathcal{P}_{\rm c}$); "○": the node for which the displacement is prescribed ($\in \mathcal{P}_{\rm D}$); "△": the node for which the applied force is prescribed ($\in \mathcal{P}_{\rm N}$).

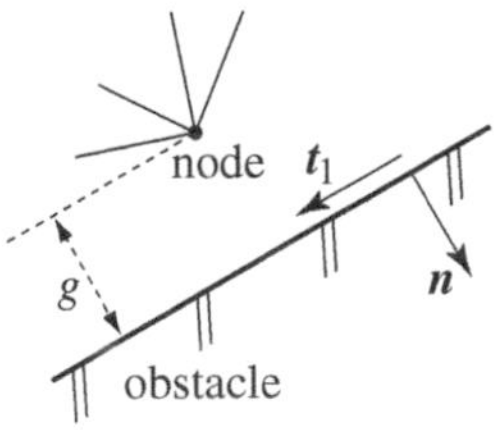

FIGURE 10.2: The local coordinate system for a contact candidate node.

the displacement and the reaction are considered to be unknown variables for each contact candidate node.

Throughout this chapter, we assume that a surface of the obstacle with which each contact candidate node can possibly contact is identified in the sense that the surface is locally regarded as a plane, in accordance with the assumptions of the small deformation and the flatness of the obstacle surface.

In the remainder of this section, we consider a *friction law*, which relates the reaction to the displacement of the contact candidate node when the node contacts with the obstacle. The friction law is also called the tangential contact law. The normal contact law, which complements the contact boundary condition, is dealt with in section 10.2.2; see also section 3.3.4 for the normal contact law.

10.1.1 Coulomb's law

Consider a contact candidate node and the corresponding obstacle surface, as illustrated in Figure 10.2. We introduce a orthonormal reference frame $(\boldsymbol{t}_1, \boldsymbol{t}_2, \boldsymbol{n})$ such that $\boldsymbol{n} \in \mathbb{R}^3$ is the normal vector inward to the obstacle, and $\boldsymbol{t}_1$, $\boldsymbol{t}_2 \in \mathbb{R}^3$ are in the tangential directions of the obstacle surface. The

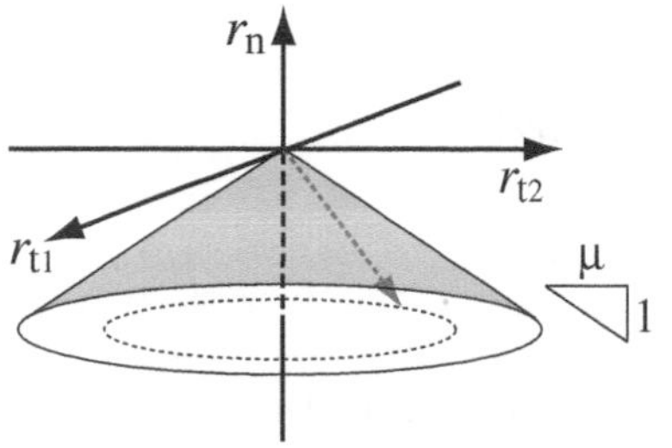

FIGURE 10.3: Friction cone $\{(r_n, \boldsymbol{r}_t) \in \mathbb{R} \times \mathbb{R}^2 \mid -\mu r_n \geq \|\boldsymbol{r}_t\|\}$.

reaction vector $\boldsymbol{r} \in \mathbb{R}^3$ at the contact candidate node can be decomposed with respect to the coordinate system $(\boldsymbol{t}_1, \boldsymbol{t}_2, \boldsymbol{n})$ as

$$\boldsymbol{r} = r_{t1}\boldsymbol{t}_1 + r_{t2}\boldsymbol{t}_2 + r_n\boldsymbol{n}, \tag{10.1}$$

and we then write $\boldsymbol{r}_t = \begin{bmatrix} r_{t1} \\ r_{t2} \end{bmatrix} \in \mathbb{R}^2$ and $\boldsymbol{r} = \begin{bmatrix} \boldsymbol{r}_t \\ r_n \end{bmatrix} \in \mathbb{R}^3$. Similarly, the displacement vector $\boldsymbol{u} \in \mathbb{R}^3$ of the node is decomposed into three components as

$$\boldsymbol{u} = u_{t1}\boldsymbol{t}_1 + u_{t2}\boldsymbol{t}_2 + u_n\boldsymbol{n}, \tag{10.2}$$

and we write $\boldsymbol{u}_t = \begin{bmatrix} u_{t1} \\ u_{t2} \end{bmatrix} \in \mathbb{R}^2$ and $\boldsymbol{u} = \begin{bmatrix} \boldsymbol{u}_t \\ u_n \end{bmatrix} \in \mathbb{R}^3$. Hereafter, the subscripts "t" and "n" denote the quantities with respect to the frame $(\boldsymbol{t}_1, \boldsymbol{t}_2)$ and the axis $\boldsymbol{n}$, respectively.

Let $\mu > 0$ be a scalar coefficient, called the *coefficient of friction*, at the point under consideration. The Coulomb friction law consists of two elements: the inclusion of the reaction and the disjunction of the stick and slip states, which are written as

$$-\mu r_n \geq \|\boldsymbol{r}_t\|, \tag{10.3a}$$

$$\begin{cases} -\mu r_n > \|\boldsymbol{r}_t\| & \Rightarrow \quad \dot{\boldsymbol{u}}_t = \boldsymbol{0}, \\ -\mu r_n = \|\boldsymbol{r}_t\| > 0 & \Rightarrow \quad \dot{\boldsymbol{u}}_t = -\hat{\gamma}\boldsymbol{r}_t \ (\hat{\gamma} \geq 0), \end{cases} \tag{10.3b}$$

where $\dot{\boldsymbol{u}}_t$ denotes the (right-hand) time derivative of $\boldsymbol{u}_t$. The constraint (10.3a) defines the admissible set of the reaction force $\boldsymbol{r}(r_n, \boldsymbol{r}_t)$, while (10.3b) describes the evolution law of the tangential velocity $\dot{\boldsymbol{u}}_t$.

Although the disjunctive condition in (10.3b) is adequate for understanding the mechanical meaning, the case of $r_n = \|\boldsymbol{r}_t\| = 0$ and $\dot{\boldsymbol{u}}_t \neq \boldsymbol{0}$ is not stated clearly. This technical difficulty is resolved by stating the Coulomb friction law as

$$-\mu r_{\rm n} \geq \|\boldsymbol{r}_{\rm t}\|, \tag{10.4a}$$

$$\begin{cases} -\mu r_{\rm n} > \|\boldsymbol{r}_{\rm t}\| & \Rightarrow \quad \dot{\boldsymbol{u}}_{\rm t} = \boldsymbol{0}, \\ \dot{\boldsymbol{u}}_{\rm t} \neq \boldsymbol{0} & \Rightarrow \quad -\mu r_{\rm n} = \|\boldsymbol{r}_{\rm t}\|, \\ & \qquad \exists \gamma \geq 0 : \ \boldsymbol{r}_{\rm t} = -\gamma \dot{\boldsymbol{u}}_{\rm t}. \end{cases} \tag{10.4b}$$

The set of $\boldsymbol{r}$ satisfying (10.4a) is called the *friction cone*, which is illustrated in Figure 10.3. Concerning the tangential velocity, the first case of (10.4b), say, $\dot{\boldsymbol{u}}_{\rm t} = \boldsymbol{0}$, corresponds to the *stick* state, while the second case, say, $\dot{\boldsymbol{u}}_{\rm t} \neq \boldsymbol{0}$, is the *slip* state. As an intermediate situation, the case of $-\mu r_{\rm n} = \|\boldsymbol{r}_{\rm t}\|$ and $\dot{\boldsymbol{u}}_{\rm t} = \boldsymbol{0}$ is sometimes called the *impending slip* state, although it is a stick state by the definition above. The disjunctive condition (10.4b) leads to the following two statements:

(i) A slip $\dot{\boldsymbol{u}}_{\rm t} \neq \boldsymbol{0}$ can take place only if the reaction is on the boundary of the friction cone.

(ii) In a slip state, the tangential reaction $\boldsymbol{r}_{\rm t}$ should be parallel with $\dot{\boldsymbol{u}}_{\rm t}$ in the opposite direction.

In the next section, the Coulomb friction law (10.4) is reformulated as a complementarity condition over the second-order cone.

10.1.2 Second-order cone complementarity formulation

To reformulate (10.4) by using the second-order cone constraints, we first recall that the complementarity condition, between $\boldsymbol{x}$ and $\boldsymbol{s}$, over the second-order cone is written as

$$x_0 \geq \|\boldsymbol{x}_1\|, \quad s_0 \geq \|\boldsymbol{s}_1\|, \quad \begin{bmatrix} x_0 \\ \boldsymbol{x}_1 \end{bmatrix} \cdot \begin{bmatrix} s_0 \\ \boldsymbol{s}_1 \end{bmatrix} = 0. \tag{10.5}$$

From the investigation in section 1.4.3 (for the concise consequence, see Table 1.4), the condition (10.5) is satisfied if and only if any one of the following six cases is true:

(i) $x_0 > \|\boldsymbol{x}_1\|$, $(s_0, \boldsymbol{s}_1) = \boldsymbol{0}$.

(ii) $(x_0, \boldsymbol{x}_1) = \boldsymbol{0}$, $s_0 > \|\boldsymbol{s}_1\|$.

(iii) $x_0 = \|\boldsymbol{x}_1\| \neq 0$, $s_0 = \|\boldsymbol{s}_1\| \neq 0$, and there exists $\gamma > 0$ such that $\boldsymbol{x}_1 = -\gamma \boldsymbol{s}_1$.

(iv) $x_0 = \|\boldsymbol{x}_1\| \neq 0$, $(s_0, \boldsymbol{s}_1) = \boldsymbol{0}$.

(v) $(x_0, \boldsymbol{x}_1) = \boldsymbol{0}$, $s_0 = \|\boldsymbol{s}_1\| \neq 0$.

(vi) $(x_0, \boldsymbol{x}_1) = \boldsymbol{0}$, $(s_0, \boldsymbol{s}_1) = \boldsymbol{0}$.

Roughly speaking, case (i) corresponds to the first case in (10.4b), which asserts that $-\mu r_{\rm n} > \|\boldsymbol{r}_{\rm t}\|$ means the stick state, say, $\dot{\boldsymbol{u}}_{\rm t} = \boldsymbol{0}$. On the other hand, case (iii) means that $\boldsymbol{x}_1$ is parallel with $-\boldsymbol{s}_1$. This condition corresponds to the second case in (10.4b), which requires that in the sliding state $\boldsymbol{r}_{\rm t}$ is parallel with $-\dot{\boldsymbol{u}}_{\rm t}$. Thus condition (10.4b), together with the inclusion (10.4a), can be embedded into the form of (10.5), as stated more precisely below.

Proposition 10.1.1. *$\dot{\boldsymbol{u}}_{\rm t} \in \mathbb{R}^2$ and $(\boldsymbol{r}_{\rm t}, r_{\rm n}) \in \mathbb{R}^2 \times \mathbb{R}$ satisfy the Coulomb friction law* (10.4) *if and only if there exists a $\lambda_{\rm n} \in \mathbb{R}$ satisfying*

$$\lambda_{\rm n} \geq \|\dot{\boldsymbol{u}}_{\rm t}\|, \quad -\mu r_{\rm n} \geq \|\boldsymbol{r}_{\rm t}\|, \quad \begin{bmatrix} \lambda_{\rm n} \\ \dot{\boldsymbol{u}}_{\rm t} \end{bmatrix} \cdot \begin{bmatrix} -\mu r_{\rm n} \\ \boldsymbol{r}_{\rm t} \end{bmatrix} = 0. \tag{10.6}$$

Proof. The friction cone constraint (10.4a) is included in condition (10.6). Let

$$x_0 = -\mu r_{\rm n}, \quad \boldsymbol{x}_1 = \boldsymbol{r}_{\rm t}, \tag{10.7a}$$
$$s_0 = \lambda_{\rm n}, \quad \boldsymbol{s}_1 = \dot{\boldsymbol{u}}_{\rm t}. \tag{10.7b}$$

Condition (10.6) is then reduced to (10.5).

It follows from Fact 1.4.7 that $\boldsymbol{x}$ and $\boldsymbol{s}$ satisfy (10.5) if and only if any one of cases (i)–(vi) above holds. From (i)–(vi), for $\boldsymbol{x}$ and $\boldsymbol{s}$ satisfying (10.5) we have the implication

$$x_0 > \|\boldsymbol{x}_1\| \quad \Rightarrow \quad (s_0, \boldsymbol{s}_1) = \boldsymbol{0}.$$

By substituting (10.7), this implication is reduced to the first case in (10.4b). Furthermore, from (ii), (iii), and (v) we see that (i)–(vi) also imply that

$$\|\boldsymbol{s}_1\| \neq 0 \quad \Rightarrow \quad x_0 = \|\boldsymbol{x}_1\|, \ \exists \gamma \geq 0 : \ \boldsymbol{x}_1 = -\gamma \boldsymbol{s}_1$$

holds. This implication is, in turn, reduced to the second case in (10.4b) by substituting (10.7).[1] An alternative proof of Proposition 10.1.1 is given in section 10.3.2.1 based on the SOCP formulation of the maximum dissipation principle for the friction law. □

In (10.6), the Coulomb friction law is expressed as a complementarity condition over the second-order cone, $\mathbb{L}^3$, in the three-dimensional space. As a distinguished feature of this formulation, it is noted that the constraints on the kinematic variables $(\lambda_{\rm n}, \dot{\boldsymbol{u}}_{\rm t})$ and the static variables $(-\mu r_{\rm n}, \boldsymbol{r}_{\rm t})$ are represented in a same form. Certainly, this feature stems from the self-duality of the second-order cone. Figure 10.4 illustrates the vectors $(\lambda_{\rm n}, \dot{\boldsymbol{u}}_{\rm t})$ and $(-\mu r_{\rm n}, \boldsymbol{r}_{\rm t})$ included in $\mathbb{L}^3$. If both of these two vectors are nonzero, then they are on the boundary of $\mathbb{L}^3$, and $\dot{\boldsymbol{u}}_{\rm t}$ is placed in parallel with $\boldsymbol{r}_{\rm t}$ in the opposite direction,

[1] If $r_{\rm n} = 0$, then (10.4a) implies $\boldsymbol{r}_{\rm t} = \boldsymbol{0}$. In this case condition (10.4b) is automatically satisfied, because $-\gamma \dot{\boldsymbol{u}}_{\rm t} = \boldsymbol{0}$ holds for any $\gamma \geq 0$ and $\dot{\boldsymbol{u}}_{\rm t}$. This situation corresponds to case (ii); i.e., $(-\mu r_{\rm n}, \boldsymbol{r}_{\rm t}) = \boldsymbol{0}$ and $\lambda_{\rm n} > \|\dot{\boldsymbol{u}}_{\rm t}\|$. From a mechanical point of view, the absence of the normal reaction means vanishing friction, and hence the tangential velocity $\dot{\boldsymbol{u}}_{\rm t}$ is not subjected to any constraint.

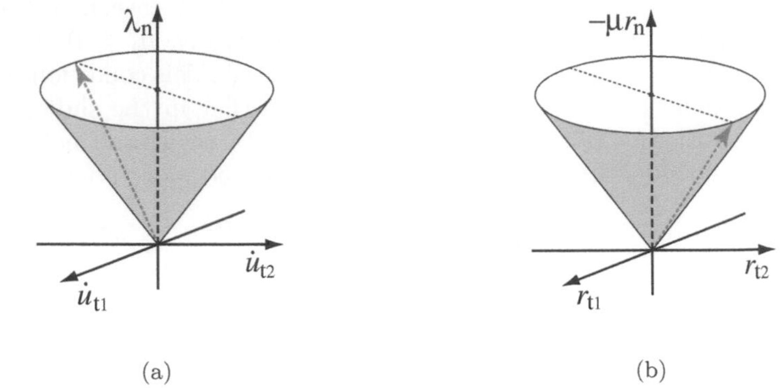

FIGURE 10.4: Second-order cone constraints for the Coulomb law in (10.6). (a) the constraint on the kinematic variables vector $(\lambda_n, \dot{\boldsymbol{u}}_t) \in \mathbb{R}^3$; (b) the constraint on the static variables vector $(-\mu r_n, \boldsymbol{r}_t) \in \mathbb{R}^3$.

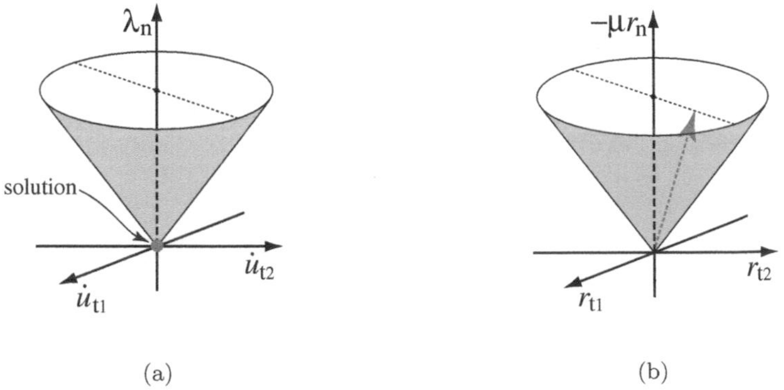

FIGURE 10.5: Second-order cone constraints (10.6) and the contact variables in a stick state.

as shown in Figure 10.4. This situation corresponds to the slip state. Alternatively, if $-\mu r_n > \|\boldsymbol{r}_t\|$, then $(-\mu r_n, \boldsymbol{r}_t)$ is in the interior of $\mathbb{L}^3$ as illustrated in Figure 10.5(b). Then (10.6) is satisfied if and only if $(\lambda_n, \dot{\boldsymbol{u}}_t) = (0, \boldsymbol{0})$. Thus the situation in Figure 10.5 corresponds to the stick state.

Remark 10.1.2. The formulation (10.6) is valid even for $\mu = 0$, although in section 10.1.1 we restricted ourselves to $\mu > 0$. In the case of $\mu = 0$, condition (10.4) leads to a wrong conclusion that only $\dot{\boldsymbol{u}}_t = \boldsymbol{0}$ is allowed. In contrast, in (10.6) we see that $\mu = 0$ implies $(-\mu r_n, \boldsymbol{r}_t) = \boldsymbol{0}$. Therefore, for any $\dot{\boldsymbol{u}}_t$, there always exists λ_n satisfying (10.6), which correctly reflects the physical phenomenon.

The quasistatic problem investigated in section 10.2, which is based on (10.6), is hence valid for any $\mu \geq 0$. ■

Remark 10.1.3. The formulation (10.6) newly obtained for the Coulomb law includes an extra variable $\lambda_{\rm n}$. The meaning of $\lambda_{\rm n}$, from the mechanical point of view, is obtained as follows.

In cases (i)–(vi) listed for the second-order cone complementarity, for $\boldsymbol{x}$ and $\boldsymbol{s}$ satisfying (10.5) we have that

$$x_0 > 0 \quad \Leftrightarrow \quad s_0 = \|\boldsymbol{s}_1\|. \tag{10.8}$$

By substituting (10.7) into this equivalence, $\lambda_{\rm n}$ is determined as

$$\begin{cases} \lambda_{\rm n} = \|\dot{\boldsymbol{u}}_{\rm t}\| & \text{if } r_{\rm n} > 0, \\ \lambda_{\rm n} \geq \|\dot{\boldsymbol{u}}_{\rm t}\| & \text{if } r_{\rm n} = 0. \end{cases} \tag{10.9}$$

Thus the auxiliary variable $\lambda_{\rm n}$ coincides with the tangential slip, so far as $r_{\rm n} > 0$. If $r_{\rm n} = 0$, which means $\|\boldsymbol{r}_{\rm t}\| = 0$, then $\lambda_{\rm n}$ is not determined uniquely; any $\lambda_{\rm n}$ satisfying $\lambda_{\rm n} \geq \|\dot{\boldsymbol{u}}_{\rm t}\|$ is feasible for (10.6). ■

Remark 10.1.4. For planar contact problems, the second-order cone inequalities involved in (10.6) are reduced to linear inequalities.

We introduced the three-dimensional coordinate system $(\boldsymbol{t}_1, \boldsymbol{t}_2, \boldsymbol{n})$ for the expression of $\boldsymbol{r}$ and $\boldsymbol{u}$ in (10.1) and (10.2). In the planar case, we consider the two-dimensional coordinate system $(\boldsymbol{t}, \boldsymbol{n})$ consisting of the normal and tangential direction of the obstacle surface. Then $\boldsymbol{r}$ and $\boldsymbol{u}$ are decomposed into two parts as

$$\boldsymbol{r} = r_{\rm t}\boldsymbol{t} + r_{\rm n}\boldsymbol{n}, \qquad \boldsymbol{u} = u_{\rm t}\boldsymbol{t} + u_{\rm n}\boldsymbol{n}.$$

Therefore, the variables $\boldsymbol{r}_{\rm t}$ and $\dot{\boldsymbol{u}}_{\rm t}$ in (10.6) are replaced with scalar variables $r_{\rm t}$ and $\dot{u}_{\rm t}$, respectively. Then the inequality $\lambda_{\rm n} \geq |\dot{u}_{\rm t}|$ involved in (10.6) is equivalently rewritten as $-\lambda_{\rm n} \leq \dot{u}_{\rm t} \leq \lambda_{\rm n}$. Similarly, $-\mu r_{\rm n} \geq |r_{\rm t}|$ is equivalent to $\mu r_{\rm n} \leq r_{\rm t} \leq -\mu r_{\rm n}$. Thus in the planar case the second-order cone constraints in (10.6) are reduced to the linear inequality constraints. ■

10.2 Incremental Problem

The objective of this section is the investigation of the *incremental problem* of frictional contacts. In the *quasistatic analysis*, the inertia forces are supposed to be negligible, and the evolution of the solution is regarded as a sequence of equilibrium states corresponding to the given loading steps. Such an equilibrium path is numerically traced by solving a sequence of incremental problems. The governing equations for the incremental problem are obtained by considering the equilibrium equations of the external, internal, and reaction forces and by replacing the time derivative in the friction law with a finite difference within a time interval. We present a formulation of the incremental problem, which is in the form of the complementarity problem over the second-order cone.[2]

[2]See section 1.2.2 for the definition of the complementarity problem over a convex cone.

10.2.1 Friction law in incremental problems

The Coulomb friction law, i.e., (10.6) of Proposition 10.1.1, primarily formulated with respect to the velocities is reduced to an incremental form.

Suppose that the time interval $[0, T]$ under consideration is subdivided into finitely many intervals. For a specific subinterval, denoted $[\tau_l, \tau_{l+1}]$, the backward Euler time discretization yields for $\dot{\boldsymbol{u}}_{\rm t}$ the approximation

$$\dot{\boldsymbol{u}}_{\rm t} \simeq \frac{\boldsymbol{u}_{\rm t}(\tau_{l+1}) - \boldsymbol{u}_{\rm t}(\tau_l)}{\tau_{l+1} - \tau_l}.$$

Then condition (10.6) is approximated for the reaction $\boldsymbol{r}$ at τ_{l+1} as

$$\lambda_{\rm n} \geq \left\| \frac{\Delta \boldsymbol{u}_{\rm t}}{\Delta \tau} \right\|, \quad -\mu r_{\rm n} \geq \|\boldsymbol{r}_{\rm t}\|, \quad \begin{bmatrix} \lambda_{\rm n} \\ \frac{\Delta \boldsymbol{u}_{\rm t}}{\Delta \tau} \end{bmatrix} \cdot \begin{bmatrix} -\mu r_{\rm n} \\ \boldsymbol{r}_{\rm t} \end{bmatrix} = 0, \tag{10.10}$$

where

$$\Delta \tau = \tau_{l+1} - \tau_l, \quad \Delta \boldsymbol{u}_{\rm t} = \boldsymbol{u}_{\rm t}(\tau_{l+1}) - \boldsymbol{u}_{\rm t}(\tau_l).$$

Thus the friction law does not dependent on the time increment $\tau_{l+1} - \tau_l$, which reflects the property referred to as the *rate-independence* of Coulomb's law. By introducing a variable $\Delta\lambda_{\rm n}$ by

$$\Delta\lambda_{\rm n} = \lambda_n \Delta\tau$$

we can rewrite (10.10) simply

$$\Delta\lambda_{\rm n} \geq \|\Delta \boldsymbol{u}_{\rm t}\|, \quad -\mu r_{\rm n} \geq \|\boldsymbol{r}_{\rm t}\|, \quad \begin{bmatrix} \Delta\lambda_{\rm n} \\ \Delta \boldsymbol{u}_{\rm t} \end{bmatrix} \cdot \begin{bmatrix} -\mu r_{\rm n} \\ \boldsymbol{r}_{\rm t} \end{bmatrix} = 0. \tag{10.11}$$

Note again that $\boldsymbol{r}_{\rm t}$ and $r_{\rm n}$ in (10.11) represent the reaction forces at τ_{l+1}, i.e., $(\boldsymbol{r}_{\rm t}, r_{\rm n}) = (\boldsymbol{r}_{\rm t}(\tau_{l+1}), r_{\rm n}(\tau_{l+1}))$.

Suppose that the equilibrium state at τ_l has been already found and that the one at τ_{l+1} is to be computed; i.e., $\Delta \boldsymbol{u}$, $\boldsymbol{r}$, and $\Delta\lambda_n$ are the unknown variables. The problem for finding $\Delta \boldsymbol{u}$, $\boldsymbol{r}$, and $\Delta\lambda_{\rm n}$ consists of (10.11) and the incremental equilibrium equations. Such a problem is called the *incremental problem* for frictional contacts, and a sequence of incremental problems is to be solved for finding the equilibrium path of the quasistatic problem. Hereafter, we simply write, e.g., $\boldsymbol{u}^l$ instead of $\boldsymbol{u}(\tau_l)$.

10.2.2 Contact kinematics

Consider an elastic body that is loaded and may possibly make contact with a fixed rigid obstacle, as illustrated in Figure 10.1. The body is discretized into finite elements so that its deformed state is represented by the displacement vector $\boldsymbol{u} \in \mathbb{R}^d$, where d is the number of degrees of freedom of displacements.

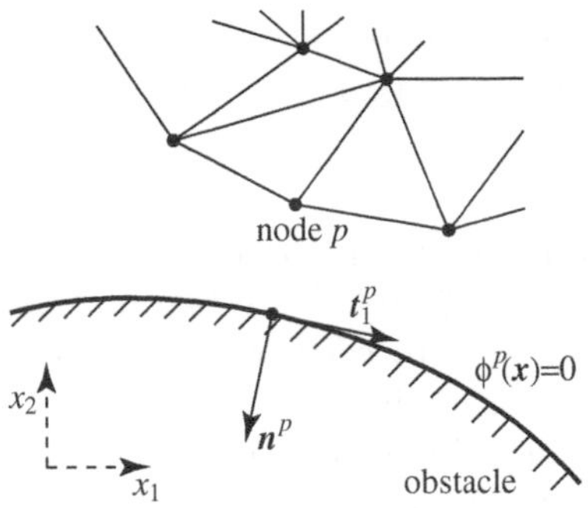

FIGURE 10.6: Contact candidate node p and the obstacle.

Note that the the prescribed displacements are not involved in $\boldsymbol{u}$. In other words, $\boldsymbol{u}$ consists of the nodal displacements of the nodes belonging to $\mathcal{P}_{\rm c}$ and $\mathcal{P}_{\rm N}$ in Figure 10.1.

Let $\mathcal{P}_{\rm c}$ denote the set of indices of contact candidate nodes, and let $n_{\rm c} = |\mathcal{P}_{\rm c}|$. We assume that the obstacle, or rigid surface, corresponding to each contact candidate node is known. Specifically, for each $p \in \mathcal{P}_{\rm c}$, the interior of the corresponding obstacle in the three-dimensional space is identified by $\{\boldsymbol{x} \in \mathbb{R}^3 \mid \phi^p(\boldsymbol{x}) > 0\}$, where $\boldsymbol{x}$ is the position vector with respect to a fixed orthonormal reference coordinate system, and $\phi^p : \mathbb{R}^3 \to \mathbb{R}$ is assumed to be sufficiently smooth (see Figure 10.6). We deal only with possible contact between a candidate node and the corresponding obstacle. In other words, the self-contact of the elastic body is not considered.

At the deformed state corresponding to $\boldsymbol{u}$, we denote by $\boldsymbol{x}^p(\boldsymbol{u}) \in \mathbb{R}^3$ the position vector of the pth node with respect to the reference frame. Then the admissible set of $\boldsymbol{u}$ is given by

$$\left\{\boldsymbol{u} \in \mathbb{R}^d \mid \phi^p(\boldsymbol{x}^p(\boldsymbol{u})) \le 0 \ (p \in \mathcal{P}_{\rm c})\right\}.$$

We may assume without loss of generality that ϕ^p satisfies $\|\nabla\phi^p(\boldsymbol{x})\| = 1$ on $\{\boldsymbol{x} \in \mathbb{R}^3 \mid \phi^p(\boldsymbol{x}) = 0\}$. Then, at a point $\boldsymbol{x}$ on the obstacle the vector

$$\boldsymbol{n}^p(\boldsymbol{x}) = \nabla\phi^p(\boldsymbol{x})$$

corresponds to the unit inner normal vector as illustrated in Figure 10.6. Define $\boldsymbol{\phi} : \mathbb{R}^d \to \mathbb{R}^{n_{\rm c}}$ by

$$\boldsymbol{\phi}(\boldsymbol{u}) = \left(\phi^1(\boldsymbol{x}^1(\boldsymbol{u})), \dots, \phi^{n_{\rm c}}(\boldsymbol{x}^{n_{\rm c}}(\boldsymbol{u}))\right)^{\rm T}.$$

Then the *gap* g_p between the pth node and the obstacle is given by

$$g_p = -\phi^p(\boldsymbol{u}),$$

which should be nonnegative so that the non-penetration condition is satisfied. Introduce a vector $\boldsymbol{g}$ by $\boldsymbol{g} = (g_p) \in \mathbb{R}^{n_c}$ to see

$$\boldsymbol{g} = -\boldsymbol{\phi}(\boldsymbol{u}). \tag{10.12}$$

Let $\boldsymbol{r}_p \in \mathbb{R}^3$ denote the reaction force, which acts from the obstacle to the pth node. We denote by $\boldsymbol{t}_1^p$ and $\boldsymbol{t}_2^p$ the two unit vectors normal to $\boldsymbol{n}^p$, where $\boldsymbol{t}_1^p$ and $\boldsymbol{t}_2^p$ are also normal to each other. In other words, $\boldsymbol{t}_1^p$ and $\boldsymbol{t}_2^p$ correspond to the two tangential directions of the obstacle. Thus the orthonormal basis $\{\boldsymbol{t}_1^p, \boldsymbol{t}_2^p, \boldsymbol{n}^p\}$ is introduced for each $p \in \mathcal{P}_{\mathrm{c}}$. Concerning this basis, the decomposition of $\boldsymbol{r}_p$ results in

$$\boldsymbol{r}_p = r_{\mathrm{t}1p}\boldsymbol{t}_1^p(\boldsymbol{x}^p(\boldsymbol{u})) + r_{\mathrm{t}2p}\boldsymbol{t}_2^p(\boldsymbol{x}^p(\boldsymbol{u})) + r_{\mathrm{n}p}\boldsymbol{n}^p(\boldsymbol{x}^p(\boldsymbol{u})). \tag{10.13}$$

Similarly, the nodal displacement vector $\boldsymbol{u}_p \in \mathbb{R}^3$ of the pth node is decomposed as

$$\boldsymbol{u}_p = u_{\mathrm{t}1p}\boldsymbol{t}_1^p(\boldsymbol{x}^p(\boldsymbol{u})) + u_{\mathrm{t}2p}\boldsymbol{t}_2^p(\boldsymbol{x}^p(\boldsymbol{u})) + u_{\mathrm{n}p}\boldsymbol{n}^p(\boldsymbol{x}^p(\boldsymbol{u})). \tag{10.14}$$

In the small deformation theory the function $\boldsymbol{\phi}$ can be regarded as being affine. Accordingly, $C_{\mathrm{n}} \in \mathbb{R}^{n_c \times d}$ defined by

$$C_{\mathrm{n}} = \left(\frac{\partial \phi_p}{\partial u_j}(\boldsymbol{u})\right) = \begin{bmatrix} \nabla\phi^1(\boldsymbol{x}^1(\boldsymbol{u}))^{\mathrm{T}} \\ \vdots \\ \nabla\phi^{n_c}(\boldsymbol{x}^{n_c}(\boldsymbol{u}))^{\mathrm{T}} \end{bmatrix}$$

is a constant matrix, and thence the kinematic relation in (10.12) is reduced to

$$\boldsymbol{g} = \boldsymbol{g}^0 - C_{\mathrm{n}}\boldsymbol{u}, \tag{10.15}$$

where $\boldsymbol{g} = (g_p^0) \in \mathbb{R}^{n_c}$ is the vector of the initial gaps. Similarly, the remaining basis vectors $\boldsymbol{t}_1^p$ and $\boldsymbol{t}_2^p$ are also regarded as to be constant. Denoting by $C_{\mathrm{t}} \in \mathbb{R}^{2n_c \times d}$ the transformation matrix from the generalized coordinate system to the frame $\{\boldsymbol{t}_1^1, \boldsymbol{t}_2^1, \dots, \boldsymbol{t}_1^{n_c}, \boldsymbol{t}_2^{n_c}\}$, the total reaction with respect to the generalized coordinate, denoted $\boldsymbol{r}^{\mathrm{global}} \in \mathbb{R}^d$, is obtained from (10.13) as

$$\boldsymbol{r}^{\mathrm{global}} = \begin{bmatrix} C_{\mathrm{t}}^{\mathrm{T}} \ C_{\mathrm{n}}^{\mathrm{T}} \end{bmatrix} \begin{bmatrix} \boldsymbol{r}_{\mathrm{t}} \\ \boldsymbol{r}_{\mathrm{n}} \end{bmatrix}, \tag{10.16}$$

where

$$\boldsymbol{r}_{\mathrm{t}p} = \begin{bmatrix} r_{\mathrm{t}1p} \\ r_{\mathrm{t}2p} \end{bmatrix}, \quad \boldsymbol{r}_{\mathrm{t}} = \begin{bmatrix} \boldsymbol{r}_{\mathrm{t}1} \\ \vdots \\ \boldsymbol{r}_{\mathrm{t}n_c} \end{bmatrix}, \quad \boldsymbol{r}_{\mathrm{n}} = \begin{bmatrix} r_{\mathrm{n}1} \\ \vdots \\ r_{\mathrm{n}n_c} \end{bmatrix}.$$

Thus the reaction $\boldsymbol{r}_p$ with respect to the reference coordinate system is related to the vector $\boldsymbol{r}^{\mathrm{global}}$ with respect to the global coordinate system.

The nodal displacement $\boldsymbol{u}_p \in \mathbb{R}^3$, defined in (10.14) with respect to the reference coordinate system, is related to the displacement vector $\boldsymbol{u} \in \mathbb{R}^d$ with respect to the global coordinate system as follows. Let $C_{\mathrm{n}p} \in \mathbb{R}^{1\times d}$ and $C_{\mathrm{t}p} \in \mathbb{R}^{2\times d}$ denote the rows C_{n} and C_{t}, respectively, as

$$C_{\mathrm{n}} = \begin{bmatrix} C_{\mathrm{n}1} \\ \vdots \\ C_{\mathrm{n}n_{\mathrm{c}}} \end{bmatrix}, \quad C_{\mathrm{t}} = \begin{bmatrix} C_{\mathrm{t}1} \\ \vdots \\ C_{\mathrm{t}n_{\mathrm{c}}} \end{bmatrix}.$$

Then the normal and tangential displacement of the pth node are written as

$$\begin{bmatrix} u_{\mathrm{t}1p} \\ u_{\mathrm{t}2p} \end{bmatrix} = C_{\mathrm{t}p}\boldsymbol{u}, \quad u_{\mathrm{n}p} = C_{\mathrm{n}p}\boldsymbol{u}.$$

10.2.3 Problem formulation

The incremental problem for frictional contacts is formulated as an SOCCP (*second-order cone complementarity problem*), which consists of the complementarity conditions, second-order cone constraints, and linear equality constraints.

10.2.3.1 Equilibrium equation

Recall that the deformed configuration of the finitely discretized elastic body is represented by the displacement vector $\boldsymbol{u} \in \mathbb{R}^d$, where d is the number of degrees of freedom of displacements. Notice again that $\boldsymbol{u}$ consists of the nodal displacements of the nodes that are contact candidates ($p \in \mathcal{P}_{\mathrm{c}}$) and the nodes for which the applied forces are prescribed ($p \in \mathcal{P}_{\mathrm{N}}$). Let $K \in \mathbb{S}^d$ denote the stiffness matrix. We denote by $\boldsymbol{f} \in \mathbb{R}^d$ and $\boldsymbol{r}^{\mathrm{global}} \in \mathbb{R}^d$ the external and reaction force vectors, respectively, with respect to the generalized coordinate system. Then the equilibrium equation is written as

$$K\boldsymbol{u} = \boldsymbol{f} + \boldsymbol{r}^{\mathrm{global}}. \tag{10.17}$$

In the quasistatic evolution, we focus on a subinterval $[\tau_l, \tau_{l+1}]$ of the time interval $[0, T]$ under consideration, as in section 10.2.1. Each time τ_l is called a loading step for convenience.

Suppose that the equilibrium state, say, the pair $\boldsymbol{u}_l$ and $\boldsymbol{r}_l$, at the loading step τ_l has been already found. We in turn aim to find the equilibrium state at the loading step τ_{l+1}. The current loading step τ_{l+1} is characterized by the specified external load vector $\boldsymbol{f}^{l+1} := \boldsymbol{f}(\tau_{l+1})$, which is written as[3]

$$\boldsymbol{f}^{l+1} = \boldsymbol{f}^l + \Delta\boldsymbol{f}.$$

[3]For simplicity, we suppose that only the prescribed external forces change, and the displacements prescribed for the constrained degrees of freedom are assumed to be fixed in the course of loading under consideration.

Here, $\boldsymbol{f}^l = \boldsymbol{f}(\tau_l)$ is the external load vector specified at the previous loading step τ_l, and $\Delta\boldsymbol{f}$ is the specified increment between τ_l and τ_{l+1}. Then the incremental problem is to find the displacement $\boldsymbol{u}^{l+1}$ and the reaction $\boldsymbol{r}^{l+1}$ at the equilibrium state corresponding to $\boldsymbol{f}^{l+1}$.

We write $\Delta\boldsymbol{u} = \boldsymbol{u}^{l+1} - \boldsymbol{u}^l$ in accordance with the notation in section 10.2.1. In what follows, the unknown $\boldsymbol{r}^{l+1}$ at τ_{l+1} is written as $\boldsymbol{r}$, and the pair of $\Delta\boldsymbol{u}$ and $\boldsymbol{r}$ are considered to be unknown variables of the incremental problem.[4]

By using (10.16) for the expression of $\boldsymbol{r}^{\mathrm{global}}$, the equilibrium equation (10.17) at the current loading step τ_{l+1} is written as

$$K(\boldsymbol{u}^l + \Delta\boldsymbol{u}) = C^{\mathrm{T}}\boldsymbol{r} + (\boldsymbol{f}^l + \Delta\boldsymbol{f}), \tag{10.18}$$

where

$$C = \begin{bmatrix} C_{\mathrm{t}} \\ C_{\mathrm{n}} \end{bmatrix}, \quad \boldsymbol{r} = \begin{bmatrix} \boldsymbol{r}_{\mathrm{t}} \\ \boldsymbol{r}_{\mathrm{n}} \end{bmatrix}.$$

On the other hand, at the previous loading step the equilibrium equation reads

$$K\boldsymbol{u}^l = C^{\mathrm{T}}\boldsymbol{r}^l + \boldsymbol{f}^l,$$

where $\boldsymbol{u}^l$ and $\boldsymbol{r}^l$ are considered to be known. Substitution of this relation into (10.18) results in

$$K\Delta\boldsymbol{u} = C^{\mathrm{T}}\boldsymbol{r} + (\Delta\boldsymbol{f} - C^{\mathrm{T}}\boldsymbol{r}^l), \tag{10.19}$$

which is the equilibrium equation for the unknown $\Delta\boldsymbol{u}$ and $\boldsymbol{r}$.

10.2.3.2 Unilateral contacts

In this section, the unilateral contact law is formulated in the incremental form.[5]

In section 10.2.2 we define the rigid obstacle by using the function ϕ^p, with which the gap g_p between the contact candidate node and the obstacle is given by (10.12). In the small deformation theory $\boldsymbol{g}$ can be written as (10.15). At the loading step τ_{l+1}, (10.15) is evaluated as $\boldsymbol{g} = \boldsymbol{g}^0 - C_{\mathrm{n}}(\boldsymbol{u}^l + \Delta\boldsymbol{u})$. Thus the vector of gaps is written in terms of the incremental displacements as

$$\boldsymbol{g} = \boldsymbol{g}^l - C_{\mathrm{n}}\Delta\boldsymbol{u}, \tag{10.20}$$

where $\boldsymbol{g}^l := \boldsymbol{g}^0 - C_{\mathrm{n}}\boldsymbol{u}^l$.

The *Signorini law* of unilateral normal contact consists of the following three conditions:

---

[4]It should be clear that the incremental problem is formulated for the *unknown incremental displacement* $\Delta\boldsymbol{u}$ and the *unknown total reaction* $\boldsymbol{r}$ (and not for the *unknown incremental reaction*). This is because the friction law (10.11) is written in terms of the incremental displacement and the total reaction.

[5]See also section 3.3 for more detailed exposition of the unilateral contact law.

(i) a kinematic constraint of the non-penetration on the contact gap g_p

(ii) a static constraint of non-adhesion (or no-tension) on the normal contact reaction $r_{\mathrm{n}p}$

(iii) a complementarity condition between g_p and $r_{\mathrm{n}p}$ asserting that the reaction cannot act on the node with nonzero gap

Thus, for each $p \in \mathcal{P}_{\mathrm{c}}$, the unilateral contact condition is written as

$$g_p \geq 0 \quad : \text{[non-penetration]}, \tag{10.21a}$$
$$r_{\mathrm{n}p} \leq 0 \quad : \text{[non-adhesion]}, \tag{10.21b}$$
$$g_p r_{\mathrm{n}p} = 0 \quad : \text{[complementarity]}. \tag{10.21c}$$

The complementarity condition (10.21) allows the two alternative situations: $g_p > 0$ implies $r_{\mathrm{n}p} = 0$, which is called the *free* state; $r_{\mathrm{n}p} > 0$ implies $g = 0$, which is said to be in *contact* with nonzero reaction. The intermediate situation, i.e., $g_p = r_{\mathrm{n}p} = 0$, is called the *grazing contact* state.

10.2.3.3 SOCCP formulation

We now state the incremental problem for the quasistatic analysis.

Recall that the Coulomb friction law is introduced in Proposition 10.1.1 and that its incremental form is written as (10.11). The unilateral contact condition in terms of the incremental displacements is given by (10.20) and (10.21). The equilibrium equation in the incremental form is presented in (10.19).

Consequently, the incremental problem is formulated as the following complementarity problem:

$$\left.\begin{array}{l}
\text{find } (\Delta \boldsymbol{u}, \boldsymbol{r}, \boldsymbol{\lambda}_{\mathrm{n}}) \in \mathbb{R}^d \times \mathbb{R}^{3n_{\mathrm{c}}} \times \mathbb{R}^{n_{\mathrm{c}}} \\
\text{s.t. } K\Delta \boldsymbol{u} = \begin{bmatrix} C_{\mathrm{t}}^{\mathrm{T}} \; C_{\mathrm{n}}^{\mathrm{T}} \end{bmatrix} \begin{bmatrix} \boldsymbol{r}_{\mathrm{t}} \\ \boldsymbol{r}_{\mathrm{n}} \end{bmatrix} + (\Delta \boldsymbol{f} - C^{\mathrm{T}} \boldsymbol{r}^l), \\
\quad \boldsymbol{g} = \boldsymbol{g}^l - C_{\mathrm{n}} \Delta \boldsymbol{u}, \quad \Delta \boldsymbol{u}_{\mathrm{t}} = C_{\mathrm{t}} \Delta \boldsymbol{u}, \\
\quad g_p \geq 0, \ -r_{\mathrm{n}p} \geq 0, \ g_i r_{\mathrm{n}p} = 0 \qquad\qquad (\forall p \in \mathcal{P}_{\mathrm{c}}), \\
\quad \lambda_{\mathrm{n}p} \geq \|\Delta \boldsymbol{u}_{\mathrm{t}p}\|, \ -\mu r_{\mathrm{n}p} \geq \|\boldsymbol{r}_{\mathrm{t}p}\|, \ \begin{bmatrix} \lambda_{\mathrm{n}p} \\ \Delta \boldsymbol{u}_{\mathrm{t}p} \end{bmatrix} \cdot \begin{bmatrix} -\mu r_{\mathrm{n}p} \\ \boldsymbol{r}_{\mathrm{t}p} \end{bmatrix} = 0 \quad (\forall p \in \mathcal{P}_{\mathrm{c}}).
\end{array}\right\} \tag{10.22}$$

Here, $g_p \in \mathbb{R}$ and $\Delta \boldsymbol{u}_{\mathrm{t}p} \in \mathbb{R}^2$ are also unknown variables, which correspond to the gap and the incremental tangential displacement, respectively. Problem (10.22) is the complementarity problem involving the second-order cone constraints as well as the linear inequality constraints.

Remark 10.2.1. In problem (10.22), provided $\mu > 0$, the constraint $-r_{\mathrm{n}p} \geq 0$ is redundant, because this condition is a consequence of $-\mu r_{\mathrm{n}p} \geq \|\boldsymbol{r}_{\mathrm{t}p}\|$. However, we leave it in formulation (10.22) (and also in the sequel) to emphasize the complementarity relation between $r_{\mathrm{n}p}$ and g_p. ∎

10.2.3.4 Formulation involving only contact variables

Problem (10.22) is reduced to the form of the complementarity problem involving only the variables associated with the contact candidate nodes.

Let $\boldsymbol{u}^{\mathrm{C}} = (\boldsymbol{u}_p \mid p \in \mathcal{P}_{\mathrm{c}})$ and $\boldsymbol{f}^{\mathrm{C}} = (\boldsymbol{f}_p \mid p \in \mathcal{P}_{\mathrm{c}})$, where $\boldsymbol{u}_p \in \mathbb{R}^3$ and $\boldsymbol{f}_p \in \mathbb{R}^3$ are the nodal displacement vector and the external force vector, respectively, of the pth node. Similarly, $\boldsymbol{u}^{\mathrm{N}}$ and $\boldsymbol{f}^{\mathrm{N}}$ are defined as the vectors associated with the nodes belonging to $\mathcal{P}_{\mathrm{N}}$ (i.e., the nodes for which the forces are prescribed). Without loss of generality we suppose that $\boldsymbol{u}$ and $\boldsymbol{f}$ are partitioned as

$$\boldsymbol{u} = \begin{bmatrix} \boldsymbol{u}^{\mathrm{N}} \\ \boldsymbol{u}^{\mathrm{C}} \end{bmatrix}, \quad \boldsymbol{f} = \begin{bmatrix} \boldsymbol{f}^{\mathrm{N}} \\ \boldsymbol{f}^{\mathrm{C}} \end{bmatrix}.$$

Corresponding to this block-partitioned expression, the matrices C_{t} and C_{n} introduced in section 10.2.3.3 have the forms of $C_{\mathrm{t}} = \begin{bmatrix} O \ \hat{C}_{\mathrm{t}} \end{bmatrix}$ and $C_{\mathrm{n}} = \begin{bmatrix} O \ \hat{C}_{\mathrm{n}} \end{bmatrix}$ so that

$$C_{\mathrm{n}} \Delta \boldsymbol{u} = \hat{C}_{\mathrm{n}} \Delta \boldsymbol{u}^{\mathrm{C}}, \quad C_{\mathrm{t}} \Delta \boldsymbol{u} = \hat{C}_{\mathrm{t}} \Delta \boldsymbol{u}^{\mathrm{C}}, \tag{10.23}$$

where $\hat{C}_{\mathrm{n}} \in \mathbb{R}^{n_{\mathrm{c}} \times 3n_{\mathrm{c}}}$ and $\hat{C}_{\mathrm{t}} \in \mathbb{R}^{2n_{\mathrm{c}} \times 3n_{\mathrm{c}}}$. As a consequence, the equilibrium equation (10.19) can be written in a block-matrix form as

$$\begin{bmatrix} K_{\mathrm{NN}} \ K_{\mathrm{NC}} \\ K_{\mathrm{CN}} \ K_{\mathrm{CC}} \end{bmatrix} \begin{bmatrix} \Delta \boldsymbol{u}^{\mathrm{N}} \\ \Delta \boldsymbol{u}^{\mathrm{C}} \end{bmatrix} = \begin{bmatrix} \boldsymbol{0} \\ \hat{C}^{\mathrm{T}} \boldsymbol{r} \end{bmatrix} + \begin{bmatrix} \Delta \boldsymbol{f}^{\mathrm{N}} \\ \Delta \boldsymbol{f}^{\mathrm{C}} - \hat{C}^{\mathrm{T}} \boldsymbol{r}^l \end{bmatrix}, \tag{10.24}$$

where $\hat{C} \in \mathbb{R}^{3n_{\mathrm{c}} \times 3n_{\mathrm{c}}}$ is defined by

$$\hat{C} := \begin{bmatrix} \hat{C}_{\mathrm{t}} \\ \hat{C}_{\mathrm{n}} \end{bmatrix}. \tag{10.25}$$

In what follows we eliminate the variables $\Delta \boldsymbol{u}^{\mathrm{N}}$ from the problem (10.22).

Since K is assumed to be positive definite, the first row-block of (10.24) can be solved as

$$\Delta \boldsymbol{u}^{\mathrm{N}} = K_{\mathrm{NN}}^{-1} (\Delta \boldsymbol{f}^{\mathrm{N}} - K_{\mathrm{NC}} \Delta \boldsymbol{u}^{\mathrm{C}}),$$

from which the second row-block of (10.24) is reduced to

$$K_{\mathrm{S}} \Delta \boldsymbol{u}^{\mathrm{C}} = \hat{C}^{\mathrm{T}} \boldsymbol{r} + (\Delta \boldsymbol{f}^{\mathrm{C}} - \hat{C}^{\mathrm{T}} \boldsymbol{r}^l - K_{\mathrm{CN}} K_{\mathrm{NN}}^{-1} \Delta \boldsymbol{f}^{\mathrm{N}}), \tag{10.26}$$

where $K_{\mathrm{S}} = K_{\mathrm{CC}} - K_{\mathrm{CN}} K_{\mathrm{NN}}^{-1} K_{\mathrm{NC}}$ is the Schur complement of K_{NN} in K.

The expressions in (10.23) motivate us to consider a basis transformation

$$\begin{bmatrix} \Delta \boldsymbol{u}_{\mathrm{t}} \\ \Delta \boldsymbol{u}_{\mathrm{n}} \end{bmatrix} = \begin{bmatrix} \hat{C}_{\mathrm{t}} \\ \hat{C}_{\mathrm{n}} \end{bmatrix} \Delta \boldsymbol{u}^{\mathrm{C}}.$$

Here, we easily see that $\hat{C}$ in (10.25) is nonsingular. In accordance with this transformation, define $\hat{K} \in \mathbb{S}^{3n_c}$ by

$$\hat{K} = \hat{C}^{-\mathrm{T}} K_{\mathrm{S}} \hat{C}^{-1}, \tag{10.27}$$

which is then in the block-partitioned form as

$$\hat{K} = \begin{bmatrix} \hat{K}_{\mathrm{tt}} & \hat{K}_{\mathrm{tn}} \\ \hat{K}_{\mathrm{nt}} & \hat{K}_{\mathrm{nn}} \end{bmatrix}.$$

As a consequence, (10.26) is rewritten as

$$\begin{bmatrix} \hat{K}_{\mathrm{tt}} & \hat{K}_{\mathrm{tn}} \\ \hat{K}_{\mathrm{nt}} & \hat{K}_{\mathrm{nn}} \end{bmatrix} \begin{bmatrix} \Delta \boldsymbol{u}_{\mathrm{t}} \\ \Delta \boldsymbol{u}_{\mathrm{n}} \end{bmatrix} = \boldsymbol{r} + \hat{\boldsymbol{f}}', \tag{10.28}$$

where $\hat{\boldsymbol{f}}' = \hat{C}^{-\mathrm{T}}(\Delta \boldsymbol{f}^{\mathrm{C}} - \hat{C}^{\mathrm{T}} \boldsymbol{r}^l - K_{\mathrm{CN}} K_{\mathrm{NN}}^{-1} \Delta \boldsymbol{f}^{\mathrm{N}})$. Substituting (10.20) into (10.28) yields

$$\begin{bmatrix} \hat{K}_{\mathrm{tt}} & \hat{K}_{\mathrm{tn}} \\ \hat{K}_{\mathrm{nt}} & \hat{K}_{\mathrm{nn}} \end{bmatrix} \begin{bmatrix} \Delta \boldsymbol{u}_{\mathrm{t}} \\ \boldsymbol{g}^l - \boldsymbol{g} \end{bmatrix} = \begin{bmatrix} \boldsymbol{r}_{\mathrm{t}} \\ \boldsymbol{r}_{\mathrm{n}} \end{bmatrix} + \hat{\boldsymbol{f}}'. \tag{10.29}$$

The constant vector $\boldsymbol{g}^l$ is eliminated from the left-hand side of (10.29) as

$$\begin{bmatrix} \hat{K}_{\mathrm{tt}} & \hat{K}_{\mathrm{tn}} \\ \hat{K}_{\mathrm{nt}} & \hat{K}_{\mathrm{nn}} \end{bmatrix} \begin{bmatrix} \Delta \boldsymbol{u}_{\mathrm{t}} \\ \boldsymbol{g} \end{bmatrix} = \begin{bmatrix} \boldsymbol{r}_{\mathrm{t}} \\ -\boldsymbol{r}_{\mathrm{n}} \end{bmatrix} + \begin{bmatrix} \hat{\boldsymbol{f}}_{\mathrm{t}} \\ -\hat{\boldsymbol{f}}_{\mathrm{n}} \end{bmatrix}, \tag{10.30}$$

where $\hat{\boldsymbol{f}}_{\mathrm{t}}$ and $\hat{\boldsymbol{f}}_{\mathrm{n}}$ are defined by

$$\begin{bmatrix} \hat{\boldsymbol{f}}_{\mathrm{t}} \\ \hat{\boldsymbol{f}}_{\mathrm{n}} \end{bmatrix} = \hat{\boldsymbol{f}}' - \begin{bmatrix} \hat{K}_{\mathrm{tn}} \boldsymbol{g}^l \\ \hat{K}_{\mathrm{nn}} \boldsymbol{g}^l \end{bmatrix} = \hat{C}^{-\mathrm{T}}(\Delta \boldsymbol{f}^{\mathrm{C}} - \hat{C}^{\mathrm{T}} \boldsymbol{r}^l - K_{\mathrm{CN}} K_{\mathrm{NN}}^{-1} \Delta \boldsymbol{f}^{\mathrm{N}}) - \begin{bmatrix} \hat{K}_{\mathrm{tn}} \boldsymbol{g}^l \\ \hat{K}_{\mathrm{nn}} \boldsymbol{g}^l \end{bmatrix}.$$

Consequently, from (10.23) and (10.30), the incremental problem in (10.22) is reformulated only in terms of the contact variables as

$$\left.\begin{array}{ll} \text{find } (\Delta \boldsymbol{u}_{\mathrm{t}}, \boldsymbol{g}, \boldsymbol{r}, \boldsymbol{\lambda}_{\mathrm{n}}) \in \mathbb{R}^{2n_c} \times \mathbb{R}^{n_c} \times \mathbb{R}^{3n_c} \times \mathbb{R}^{n_c} & \\ \text{s.t. } \begin{bmatrix} \hat{K}_{\mathrm{tt}} & \hat{K}_{\mathrm{tn}} \\ \hat{K}_{\mathrm{nt}} & \hat{K}_{\mathrm{nn}} \end{bmatrix} \begin{bmatrix} \Delta \boldsymbol{u}_{\mathrm{t}} \\ \boldsymbol{g} \end{bmatrix} = \begin{bmatrix} \boldsymbol{r}_{\mathrm{t}} \\ -\boldsymbol{r}_{\mathrm{n}} \end{bmatrix} + \begin{bmatrix} \hat{\boldsymbol{f}}_{\mathrm{t}} \\ -\hat{\boldsymbol{f}}_{\mathrm{n}} \end{bmatrix}, & \\ \quad g_p \geq 0, \ -r_{\mathrm{n}p} \geq 0, \ g_p r_{\mathrm{n}p} = 0 & (\forall p \in \mathcal{P}_{\mathrm{c}}), \\ \quad \lambda_{\mathrm{n}p} \geq \|\Delta \boldsymbol{u}_{\mathrm{t}p}\|, \ -\mu r_{\mathrm{n}p} \geq \|\boldsymbol{r}_{\mathrm{t}p}\|, \ \begin{bmatrix} \lambda_{\mathrm{n}p} \\ \Delta \boldsymbol{u}_{\mathrm{t}p} \end{bmatrix} \cdot \begin{bmatrix} -\mu r_{\mathrm{n}p} \\ \boldsymbol{r}_{\mathrm{t}p} \end{bmatrix} = 0 & (\forall p \in \mathcal{P}_{\mathrm{c}}). \end{array}\right\} \tag{10.31}$$

This is the formulation as the complementarity problem (CP) defined by Definition 1.2.1, where the canonical form of (CP) is characterized by $\mathcal{C}$ and G; Remark 10.2.2 confirms for problem (10.31) that $\mathcal{C} = \mathcal{C}^*$ corresponds to the direct product of the second-order cones and the nonnegative orthant and that G is affine. For this reason we call (10.31) the *second-order cone linear complementarity problem* (SOCLCP). It is noted that the SOCLCP may be solved numerically by using the algorithms introduced in Chapter 6.

Remark 10.2.2. It is shown here that problem (10.31) is encoded into the form of (CP) in Definition 1.2.1 with a convex cone $\mathcal{C} \subseteq \mathbb{R}^{4n_c}$ defined by

$$\mathcal{C} = \left\{ \boldsymbol{z} = \begin{bmatrix} \boldsymbol{z}^{(1)} \\ \boldsymbol{z}^{(2)} \\ \boldsymbol{z}^{(3)} \end{bmatrix} \in \mathbb{R}^{4n_c} \mid z_i^{(1)} \geq \|\boldsymbol{z}_i^{(2)}\|,\ z_i^{(3)} \geq 0\ (i = 1, \dots, n_c) \right\},$$

where $z_i^{(1)} \in \mathbb{R}$, $\boldsymbol{z}_i^{(2)} \in \mathbb{R}^2$, and $z_i^{(3)} \in \mathbb{R}$ $(i = 1, \dots, n_c)$. Note that $\mathcal{C}$ is a direct product of some second-order cones and a nonnegative orthant, and hence it is certainly a self-dual cone, i.e., $\mathcal{C}^* = \mathcal{C}$. We next define $\boldsymbol{x}$ by

$$\boldsymbol{x} = \begin{bmatrix} \boldsymbol{\lambda}_{\rm n} \\ \Delta \boldsymbol{u}_{\rm t} \\ \boldsymbol{g} \end{bmatrix},$$

which is a vector of kinematic variables. Then the kinematic constraints, $\lambda_{{\rm n}p} \geq \|\Delta \boldsymbol{u}_{{\rm t}p}\|$ and $g_p \geq 0$, in (10.31) are written as $\boldsymbol{x} \in \mathcal{C}$. Turning to static variables, it follows from (10.30) that the relation

$$\begin{bmatrix} -\mu \boldsymbol{r}_{\rm n} \\ \boldsymbol{r}_{\rm t} \\ -\boldsymbol{r}_{\rm n} \end{bmatrix} = \begin{bmatrix} O & \mu \hat{K}_{\rm nt} & \mu \hat{K}_{\rm nn} \\ O & \hat{K}_{\rm tt} & \hat{K}_{\rm tn} \\ O & \hat{K}_{\rm nt} & \hat{K}_{\rm nn} \end{bmatrix} \begin{bmatrix} \boldsymbol{\lambda}_{\rm n} \\ \Delta \boldsymbol{u}_{\rm t} \\ \boldsymbol{g} \end{bmatrix} - \begin{bmatrix} -\mu \hat{\boldsymbol{f}}_{\rm n} \\ \hat{\boldsymbol{f}}_{\rm t} \\ -\hat{\boldsymbol{f}}_{\rm n} \end{bmatrix}$$

holds. Therefore, by introducing $\boldsymbol{y} = \check{K}\boldsymbol{x} - \check{\boldsymbol{f}}$ with $\check{K}$ and $\check{\boldsymbol{f}}$ defined by

$$\check{K} = \begin{bmatrix} O & \mu \hat{K}_{\rm nt} & \mu \hat{K}_{\rm nn} \\ O & \hat{K}_{\rm tt} & \hat{K}_{\rm tn} \\ O & \hat{K}_{\rm nt} & \hat{K}_{\rm nn} \end{bmatrix}, \quad \check{\boldsymbol{f}} = \begin{bmatrix} -\mu \hat{\boldsymbol{f}}_{\rm n} \\ \hat{\boldsymbol{f}}_{\rm t} \\ -\hat{\boldsymbol{f}}_{\rm n} \end{bmatrix},$$

the static constraints, $-\mu r_{{\rm n}p} \geq \|\boldsymbol{r}_{{\rm t}p}\|$ and $-r_{{\rm n}p} \geq 0$, in (10.31) are written as $\boldsymbol{y} \in \mathcal{C}$. Consequently, problem (10.31) is rewritten simply as

$$\left.\begin{array}{l} \text{find } \boldsymbol{x}, \boldsymbol{y} \in \mathbb{R}^{4n_c} \\ \text{s.t. } \boldsymbol{y} = \check{K}\boldsymbol{x} - \check{\boldsymbol{f}}, \\ \qquad \mathcal{C} \ni \boldsymbol{x} \perp \boldsymbol{y} \in \mathcal{C}, \end{array}\right\} \tag{10.32}$$

which is in the form of (CP) in Definition 1.2.1. ■

Remark 10.2.3. In formulation (10.32) we can clearly see that all the nonlinear (and also nonsmooth) properties of the frictional contact are condensed to the last line of the constraint, while the equality constraints, $\boldsymbol{y} = \check{K}\boldsymbol{x} - \check{\boldsymbol{f}}$, are affine. This is a particular feature of the second-order cone formulation of the frictional contact. See also various other formulations presented in section 10.3.2–section 10.3.4. ■

Remark 10.2.4. In a planar case the incremental problem can be formulated as an LCP (linear complementarity problem), because the Coulomb friction law (10.4) is reduced to an LCP as follows.

As mentioned in Remark 10.1.4, for a planar problem the variables $r_{\rm t}$ and $u_{\rm t}$ are scalar variables. In the Coulomb law in (10.4) (in the incremental form), we first focus on the conditions

$$-\mu r_{\rm n} \geq |r_{\rm t}|, \quad \begin{cases} -\mu r_{\rm n} > |r_{\rm t}| \ \Rightarrow \ \Delta u_{\rm t} = 0, \\ \Delta u_{\rm t} \neq 0 \ \Rightarrow \ -\mu r_{\rm n} = |r_{\rm t}|. \end{cases} \tag{10.33}$$

Write $\lambda_{\rm n} \geq |\Delta u_{\rm t}|$ to see that $\lambda_{\rm n} = 0$ implies $\Delta u_{\rm t} = 0$. Thus (10.33) holds if and only if there exists a $\lambda_{\rm n}$ satisfying

$$\lambda_{\rm n} \geq |\Delta u_{\rm t}|, \quad -\mu r_{\rm n} \geq |r_{\rm t}|, \quad \lambda_{\rm n}(-\mu r_{\rm n} - |r_{\rm t}|) = 0. \tag{10.34}$$

The remaining condition to be considered in (10.4b) is

$$\Delta u_{\rm t} \neq 0 \quad \Rightarrow \quad \exists \gamma \geq 0 : \; r_{\rm t} = -\gamma \Delta u_{\rm t}.$$

Since $r_{\rm t}$ and $\Delta u_{\rm t}$ are scalar variables, this condition is equivalently rewritten as

$$\begin{cases} \Delta u_{\rm t} > 0 & \Rightarrow \quad r_{\rm t} \leq 0, \\ \Delta u_{\rm t} < 0 & \Rightarrow \quad r_{\rm n} \geq 0. \end{cases} \tag{10.35}$$

For dealing with (10.35), we introduce an additive decomposition of the tangential reaction $r_{\rm t}$ as

$$r_{\rm t} = r_{\rm t}^{+} - r_{\rm t}^{-}, \quad r_{\rm t}^{+} \geq 0, \quad r_{\rm t}^{-} \geq 0. \tag{10.36}$$

From the first inequality constraint in (10.34) we obtain "$\Delta u_{\rm t} > 0 \Rightarrow \lambda_n + \Delta u_{\rm t} > 0$." Therefore, the condition

$$\lambda_{\rm n} + \Delta u_{\rm t} \geq 0, \quad r_{\rm t}^{+} \geq 0, \quad r_{\rm t}^{+}(\lambda_{\rm n} + \Delta u_{\rm t}) = 0 \tag{10.37}$$

implies that "$\Delta u_{\rm t} > 0 \Rightarrow r_{\rm t}^{+} = 0$," and thence using (10.36) we conclude that $r_{\rm t} \leq 0$, as required in (10.35). Similarly, the latter case of (10.36) (i.e., the case of $\Delta u_{\rm t} < 0$) is derived by considering the condition

$$\lambda_{\rm n} - \Delta u_{\rm t} \geq 0, \quad r_{\rm t}^{-} \geq 0, \quad r_{\rm t}^{-}(\lambda_{\rm n} - \Delta u_{\rm t}) = 0. \tag{10.38}$$

Consequently, the Coulomb law in (10.33) and (10.35) is equivalently rewritten as (10.34), (10.36), (10.37), and (10.38), which are summed up as

$$r_{\rm t} = r_{\rm t}^{+} - r_{\rm t}^{-}, \tag{10.39a}$$

$$r_{\rm t}^{+} \geq 0, \qquad \lambda_{\rm n} + \Delta u_{\rm t} \geq 0, \qquad r_{\rm t}^{+}(\lambda_{\rm n} + \Delta u_{\rm t}) = 0, \tag{10.39b}$$

$$r_{\rm t}^{-} \geq 0, \qquad \lambda_{\rm n} - \Delta u_{\rm t} \geq 0, \qquad r_{\rm t}^{-}(\lambda_{\rm n} - \Delta u_{\rm t}) = 0, \tag{10.39c}$$

$$\lambda_{\rm n} \geq 0, \qquad -\mu r_{\rm n} \geq |r_{\rm t}^{+} - r_{\rm t}^{-}|, \quad \lambda_{\rm n}(-\mu r_{\rm n} - |r_{\rm t}^{+} - r_{\rm t}^{-}|) = 0. \tag{10.39d}$$

Notice here that $\lambda_{\rm n} \geq |\Delta u_{\rm t}|$ in (10.34) is satisfied if $\lambda_{\rm n} + \Delta u_{\rm t} \geq 0$ and $\lambda_{\rm n} - \Delta u_{\rm t} \geq 0$ hold; hence, $\lambda_{\rm n} \geq 0$ in (10.39d) is redundant. However, we leave it in order to explicitly show the complementarity relation between $\lambda_{\rm n}$ and $-\mu r_{\rm n} - |r_{\rm t}^{+} - r_{\rm t}^{-}|$ in (10.39d).

In (10.36), suppose $r_{\rm t}^{+} r_{\rm t}^{-} = 0$. Then we obtain

$$|r_{\rm t}^{+} - r_{\rm t}^{-}| = r_{\rm t}^{+} + r_{\rm t}^{-}.$$

Alternatively, suppose $r_{\rm t}^{+} > 0$ and $r_{\rm t}^{-} > 0$ in (10.36). Then (10.39b) and (10.39c) imply $\lambda_{\rm n} = \Delta u_{\rm t} = 0$. Since this is a stick state, only the constraint $-\mu r_{\rm n} \geq |r_{\rm t}|$ is required to be satisfied. For any fixed $r_{\rm n} \leq 0$, one can see that

$$\{r_{\rm t} \mid -\mu r_{\rm n} \geq |r_{\rm t}|\} = \{r_{\rm t}^{+} - r_{\rm t}^{-} \mid -\mu r_{\rm n} \geq r_{\rm t}^{+} + r_{\rm t}^{-}, \; r_{\rm t}^{+} \geq 0, \; r_{\rm t}^{-} \geq 0\}$$

holds, from which we see that the inequality $-\mu r_\mathrm{n} \geq |r_\mathrm{t}^+ - r_\mathrm{t}^-|$ is equivalent to $-\mu r_\mathrm{n} \geq r_\mathrm{t}^+ + r_\mathrm{t}^-$ and (10.36). Consequently, we can replace $|r_\mathrm{t}^+ - r_\mathrm{t}^-|$ in (10.39d) with $r_\mathrm{t}^+ + r_\mathrm{t}^-$ without changing the solution set of (10.39). We thus conclude that the Coulomb friction law for the planar case is equivalent to

$$
\begin{aligned}
&r_\mathrm{t} = r_\mathrm{t}^+ - r_\mathrm{t}^-, && && \quad (10.40\mathrm{a})\\
&r_\mathrm{t}^+ \geq 0, && \lambda_\mathrm{n} + \Delta u_\mathrm{t} \geq 0, && r_\mathrm{t}^+(\lambda_\mathrm{n} + \Delta u_\mathrm{t}) = 0, \quad (10.40\mathrm{b})\\
&r_\mathrm{t}^- \geq 0, && \lambda_\mathrm{n} - \Delta u_\mathrm{t} \geq 0, && r_\mathrm{t}^-(\lambda_\mathrm{n} - \Delta u_\mathrm{t}) = 0, \quad (10.40\mathrm{c})\\
&\lambda_\mathrm{n} \geq 0, && -\mu r_\mathrm{n} - r_\mathrm{t}^+ - r_\mathrm{t}^- \geq 0, && \lambda_\mathrm{n}(-\mu r_\mathrm{n} - r_\mathrm{t}^+ - r_\mathrm{t}^-) = 0, \quad (10.40\mathrm{d})
\end{aligned}
$$

which is an LCP formulation. ■

Remark 10.2.5. Based on the result in Remark 10.2.4, we here show that, in the planar case, the incremental problem (10.31) is reduced to an LCP in the standard form.

Let $\hat{A} := \hat{K}^{-1}$ denote the flexibility matrix, where $\hat{K}$ is the stiffness matrix defined by (10.27). By pre-multiplying $\hat{A}$ with (10.30), the equilibrium equation is rewritten as

$$
\begin{bmatrix} \Delta \boldsymbol{u}_\mathrm{t} \\ \boldsymbol{g} \end{bmatrix} = \begin{bmatrix} \hat{A}_\mathrm{tt} & \hat{A}_\mathrm{tn} \\ \hat{A}_\mathrm{nt} & \hat{A}_\mathrm{nn} \end{bmatrix} \begin{bmatrix} \boldsymbol{r}_\mathrm{t} \\ -\boldsymbol{r}_\mathrm{n} \end{bmatrix} + \begin{bmatrix} \hat{\boldsymbol{u}}_\mathrm{t} \\ -\hat{\boldsymbol{u}}_\mathrm{n} \end{bmatrix},
$$

where $\begin{bmatrix} \hat{\boldsymbol{u}}_\mathrm{t} \\ -\hat{\boldsymbol{u}}_\mathrm{n} \end{bmatrix} := \hat{A} \begin{bmatrix} \hat{\boldsymbol{f}}_\mathrm{t} \\ -\hat{\boldsymbol{f}}_\mathrm{n} \end{bmatrix}$. Applying the decomposition (10.40a) to each $r_{\mathrm{t}p}$ yields

$$
\begin{bmatrix} -\mu \boldsymbol{r}_\mathrm{n} - \boldsymbol{r}_\mathrm{t}^+ - \boldsymbol{r}_\mathrm{t}^- \\ \boldsymbol{\lambda}_\mathrm{n} - \Delta \boldsymbol{u}_\mathrm{t} \\ \boldsymbol{\lambda} + \Delta \boldsymbol{u}_\mathrm{t} \\ \boldsymbol{g} \end{bmatrix} = \begin{bmatrix} O & -I & -I & \mu I \\ I & \hat{A}_\mathrm{tt} & -\hat{A}_\mathrm{tt} & -\hat{A}_\mathrm{tn} \\ I & -\hat{A}_\mathrm{tt} & \hat{A}_\mathrm{tt} & \hat{A}_\mathrm{tn} \\ O & -\hat{A}_\mathrm{nt} & \hat{A}_\mathrm{nt} & \hat{A}_\mathrm{nn} \end{bmatrix} \begin{bmatrix} \boldsymbol{\lambda}_\mathrm{n} \\ \boldsymbol{r}_\mathrm{t}^- \\ \boldsymbol{r}_\mathrm{t}^+ \\ -\boldsymbol{r}_\mathrm{n} \end{bmatrix} + \begin{bmatrix} \boldsymbol{0} \\ -\hat{\boldsymbol{u}}_\mathrm{t} \\ \hat{\boldsymbol{u}}_\mathrm{t} \\ -\hat{\boldsymbol{u}}_\mathrm{n} \end{bmatrix}. \quad (10.41)
$$

Thus all the terms subjected to the complementarity relations in (10.40) appear in (10.41). Write

$$
M = \begin{bmatrix} O & -I & -I & \mu I \\ I & \hat{A}_\mathrm{tt} & -\hat{A}_\mathrm{tt} & -\hat{A}_\mathrm{tn} \\ I & -\hat{A}_\mathrm{tt} & \hat{A}_\mathrm{tt} & \hat{A}_\mathrm{tn} \\ O & -\hat{A}_\mathrm{nt} & \hat{A}_\mathrm{nt} & \hat{A}_\mathrm{nn} \end{bmatrix}, \quad \boldsymbol{q} = \begin{bmatrix} \boldsymbol{0} \\ -\hat{\boldsymbol{u}}_\mathrm{t} \\ \hat{\boldsymbol{u}}_\mathrm{t} \\ -\hat{\boldsymbol{u}}_\mathrm{n} \end{bmatrix}, \quad \tilde{\boldsymbol{x}} = \begin{bmatrix} \boldsymbol{\lambda}_\mathrm{n} \\ \boldsymbol{r}_\mathrm{t}^- \\ \boldsymbol{r}_\mathrm{t}^+ \\ -\boldsymbol{r}_\mathrm{n} \end{bmatrix}
$$

to see that the incremental problem, consisting of (10.40) and (10.41), is encoded to the following LCP:

$$
\left.
\begin{aligned}
&\text{find } \tilde{\boldsymbol{x}}, \tilde{\boldsymbol{y}} \in \mathbb{R}^{4n_\mathrm{c}} \\
&\text{s.t. } \tilde{\boldsymbol{y}} = M\tilde{\boldsymbol{x}} + \boldsymbol{q}, \\
&\qquad \tilde{\boldsymbol{x}} \geq \boldsymbol{0}, \quad \tilde{\boldsymbol{y}} \geq \boldsymbol{0}, \quad \tilde{\boldsymbol{x}}^\mathrm{T}\tilde{\boldsymbol{y}} = 0.
\end{aligned}
\right\} \quad (10.42)
$$

As mentioned by Klarbring [227], formulation (10.42) is crucial for solution analysis of the planar incremental problem of planar frictional contacts. One can show that the matrix M is copositive (see [227, Lemma 1]) and that the implication

$$
\tilde{\boldsymbol{x}} \geq \boldsymbol{0},\ M\tilde{\boldsymbol{x}} \geq \boldsymbol{0},\ \tilde{\boldsymbol{x}}^\mathrm{T} M \tilde{\boldsymbol{x}} = 0 \quad \Rightarrow \quad \boldsymbol{q}^\mathrm{T}\tilde{\boldsymbol{x}} \geq 0
$$

holds. From these properties of M and $\boldsymbol{q}$ it can be guaranteed that the incremental problem has a solution, because Theorem 3.8.6 in Cottle–Pang–Stone [96] can be applied to problem (10.42).[6] Moreover, it is also shown that the Lemke method can find such a solution under a standard non-degeneracy assumption (see Cottle–Pang–Stone [96, section 4.4]). ■

10.3 Discussions on Various Complementarity Forms

Because of its long history of research, formulations of the frictional contact problems have diverged. In this section we discuss some that are important from the viewpoints both of nonsmooth mechanics and modern optimization. We begin with a variant of the second-order cone formulation in section 10.3.1. Formulations derived as the optimality conditions of certain optimization problems are presented in section 10.3.2. Section 10.3.3 deals with the formulation using the projection operator. The relation between the normality rule in the classical plasticity theory and the frictional contact condition is investigated in section 10.3.4.[7]

10.3.1 On auxiliary variables

In section 10.1.2 we showed that the Coulomb friction law (10.4) is expressed in the form of second-order cone complementarity as follows (see Proposition 10.1.1).

$$\exists \lambda_{\mathrm{n}} :$$

$$\lambda_{\mathrm{n}} \geq \|\dot{\boldsymbol{u}}_{\mathrm{t}}\|, \quad -\mu r_{\mathrm{n}} \geq \|\boldsymbol{r}_{\mathrm{t}}\|, \tag{10.43a}$$

$$\begin{bmatrix} \lambda_{\mathrm{n}} \\ \dot{\boldsymbol{u}}_{\mathrm{t}} \end{bmatrix} \cdot \begin{bmatrix} -\mu r_{\mathrm{n}} \\ \boldsymbol{r}_{\mathrm{t}} \end{bmatrix} = 0. \tag{10.43b}$$

As shown in Remark 10.1.3, the auxiliary variable λ_{n} satisfies

$$\begin{cases} r_{\mathrm{n}} > 0 & \Rightarrow \quad \lambda_{\mathrm{n}} = \|\dot{\boldsymbol{u}}_{\mathrm{t}}\|, \\ r_{\mathrm{n}} = 0 & \Rightarrow \quad \lambda_{\mathrm{n}} \geq \|\dot{\boldsymbol{u}}_{\mathrm{t}}\|. \end{cases}$$

Hence, without loss of generality we can restrict $\lambda_{\mathrm{n}} = \|\dot{\boldsymbol{u}}_{\mathrm{t}}\|$ in (10.43), which yields the following alternative formulation of the Coulomb law.

[6] See also Theorem 5 in Klarbring [227].

[7] See section 11.1 for the normality rule in plasticity.

$$-\mu r_{\mathrm{n}} \geq \|\boldsymbol{r}_{\mathrm{t}}\|, \tag{10.44a}$$

$$\begin{bmatrix} \|\dot{\boldsymbol{u}}_{\mathrm{t}}\| \\ \dot{\boldsymbol{u}}_{\mathrm{t}} \end{bmatrix} \cdot \begin{bmatrix} -\mu r_{\mathrm{n}} \\ \boldsymbol{r}_{\mathrm{t}} \end{bmatrix} = 0. \tag{10.44b}$$

In this way we can eliminate the auxiliary variable λ_n. As a consequence, however, this formulation includes a nonsmooth equation (10.44b). In contrast, the complementarity condition (10.43b) involved in the SOCCP formulation is a smooth equation.

Remark 10.3.1. Here is an alternative proof of the equivalence of (10.44) and the Coulomb evolution law in (10.3), say,

$$-\mu r_{\mathrm{n}} > \|\boldsymbol{r}_{\mathrm{t}}\| \quad \Rightarrow \quad \dot{\boldsymbol{u}}_{\mathrm{t}} = \boldsymbol{0}, \tag{10.45a}$$

$$-\mu r_{\mathrm{n}} = \|\boldsymbol{r}_{\mathrm{t}}\| \quad \Rightarrow \quad \exists \hat{\gamma} \geq 0 : \ \dot{\boldsymbol{u}}_{\mathrm{t}} = -\hat{\gamma} \boldsymbol{r}_{\mathrm{t}}. \tag{10.45b}$$

Note that we assume $\boldsymbol{r}_{\mathrm{t}} \neq \boldsymbol{0}$ for simplicity.

We begin by considering the case of (10.45a), i.e., $-\mu r_{\mathrm{n}} > \|\boldsymbol{r}_{\mathrm{t}}\|$. Assume $\dot{\boldsymbol{u}}_{\mathrm{t}} \neq \boldsymbol{0}$ for contradiction. Then we obtain the inequality

$$(-\mu r_{\mathrm{n}})\|\dot{\boldsymbol{u}}_{\mathrm{t}}\| > \|\boldsymbol{r}_{\mathrm{t}}\|\|\dot{\boldsymbol{u}}_{\mathrm{t}}\|,$$

and hence if (10.44b) holds, then we have that

$$\begin{aligned} 0 &= (-\mu r_{\mathrm{n}})\|\dot{\boldsymbol{u}}_{\mathrm{t}}\| + \boldsymbol{r}_{\mathrm{t}} \cdot \dot{\boldsymbol{u}}_{\mathrm{t}} \\ &> \|\boldsymbol{r}_{\mathrm{t}}\|\|\dot{\boldsymbol{u}}_{\mathrm{t}}\| + \boldsymbol{r}_{\mathrm{t}} \cdot \dot{\boldsymbol{u}}_{\mathrm{t}} \\ &\geq 0, \end{aligned}$$

where the last inequality follows from the Cauchy–Schwarz inequality. Thus the contradiction is obtained, and hence $\dot{\boldsymbol{u}}_{\mathrm{t}} = \boldsymbol{0}$. Conversely, $\dot{\boldsymbol{u}}_{\mathrm{t}} = \boldsymbol{0}$ implies (10.44b).

Second, consider the case of (10.45b), i.e., $-\mu r_{\mathrm{n}} = \|\boldsymbol{r}_{\mathrm{t}}\|$. If (10.44b) is satisfied, then we obtain

$$\begin{aligned} 0 &= (-\mu r_{\mathrm{n}})\|\dot{\boldsymbol{u}}_{\mathrm{t}}\| + \boldsymbol{r}_{\mathrm{t}} \cdot \dot{\boldsymbol{u}}_{\mathrm{t}} \\ &= \|\boldsymbol{r}_{\mathrm{t}}\|\|\dot{\boldsymbol{u}}_{\mathrm{t}}\| + \boldsymbol{r}_{\mathrm{t}} \cdot \dot{\boldsymbol{u}}_{\mathrm{t}}. \end{aligned} \tag{10.46}$$

It follows from the Cauchy–Schwarz inequality that (10.46) holds if and only if there exists a $\hat{\gamma} \geq 0$ such that $\dot{\boldsymbol{u}}_{\mathrm{t}} = -\hat{\gamma} \boldsymbol{r}_{\mathrm{t}}$. Conversely, it is straightforward to see that (10.45b) implies (10.44b). ■

10.3.2 Maximum dissipation law and its optimality conditions

The *maximum dissipation law* is a fundamental postulate in the plasticity theory. This postulate relates the plastic strain rate and stress when the plastic deformation takes place as follows. For a given plastic strain rate $\dot{\boldsymbol{\varepsilon}}_{\mathrm{p}}$, the corresponding stress $\boldsymbol{\sigma}$ is the one that maximizes the (rate of) plastic work $\boldsymbol{\sigma} : \dot{\boldsymbol{\varepsilon}}_{\mathrm{p}}$ among all admissible stresses. Note that the plastic work corresponds to the dissipated energy during elastoplastic deformation, and hence, in other words, the actual stress $\boldsymbol{\sigma}$ maximizes the energy dissipation. The set of admissible stresses is defined by the yield condition; see section 11.1. As a consequence of the maximum dissipation law, the plastic strain rate $\dot{\boldsymbol{\varepsilon}}_{\mathrm{p}}$ is normal to the tangent hyperplane at the point $\boldsymbol{\sigma}$ to the boundary of the admissible set of stresses. This relation is called the *normality rule*. When the normality rule holds, the plastic strain rate $\dot{\boldsymbol{\varepsilon}}_{\mathrm{p}}$, which is also called the plastic flow, is said to satisfy the *associated flow rule*.

Frictional contact problem of an elastic body is also a dissipative system; the work done by the friction is dissipative. The Coulomb friction law (10.4) can also be expressed in the form of the maximum dissipation law, as far as only the evolution of the tangential displacement $\boldsymbol{u}_{\mathrm{t}}$ is considered.[8] In this section we first state the maximum dissipation law for the tangential contact law by analogical reasoning from the plasticity theory and then show that this law is equivalent to the evolution law of the tangential displacement $\boldsymbol{u}_{\mathrm{t}}$ in the Coulomb friction law. Throughout this section we assume $\mu > 0$ and $r_{\mathrm{n}} < 0$ for simplicity.

For a given $r_{\mathrm{n}} < 0$, the admissible set of the tangential reactions is given by (10.4a), i.e.,

$$\{\boldsymbol{r}_{\mathrm{t}} \mid -\mu r_{\mathrm{n}} \geq \|\boldsymbol{r}_{\mathrm{t}}\|\}. \tag{10.47}$$

The energy dissipation is due to the friction force $\boldsymbol{r}_{\mathrm{t}}$. Therefore, for the given $\dot{\boldsymbol{u}}_{\mathrm{t}}$, the (rate of) dissipated energy is written as $-\dot{\boldsymbol{u}}_{\mathrm{t}}^{\mathrm{T}}\boldsymbol{r}_{\mathrm{t}}$, where the negative sign is due to the fact that $\boldsymbol{r}_{\mathrm{t}}$ is in the opposite direction to $\dot{\boldsymbol{u}}_{\mathrm{t}}$. Then the maximum dissipation law is stated as the postulate that the actual tangential reaction $\boldsymbol{r}_{\mathrm{t}}$ is the one that maximizes $-\dot{\boldsymbol{u}}_{\mathrm{t}}^{\mathrm{T}}\boldsymbol{r}_{\mathrm{t}}$ among all tangential reactions belonging to the set (10.47). More precisely, for the given $r_{\mathrm{n}} < 0$ and $\dot{\boldsymbol{u}}_{\mathrm{t}}$, the actual $\boldsymbol{r}_{\mathrm{t}}$ is an optimal solution of the following optimization problem in the variables $\check{\boldsymbol{r}}_{\mathrm{t}} = (\check{r}_{\mathrm{t}1}, \check{r}_{\mathrm{t}2})^{\mathrm{T}}$:

$$\boldsymbol{r}_{\mathrm{t}} \in \arg \max_{\check{\boldsymbol{r}}_{\mathrm{t}} \in \mathbb{R}^2} \left\{ -\dot{\boldsymbol{u}}_{\mathrm{t}}^{\mathrm{T}} \check{\boldsymbol{r}}_{\mathrm{t}} \mid -\mu r_{\mathrm{n}} \geq \|\check{\boldsymbol{r}}_{\mathrm{t}}\| \right\}. \tag{10.48}$$

A proof of the equivalence of (10.48) and the Coulomb friction law is given in section 10.3.2.1 based on SOCP. Then some equivalent complementarity formulations are derived from (10.48) in section 10.3.2.2 and section 10.3.2.3.

10.3.2.1 An SOCP formulation

The optimization problem (10.48) can be perceived as an SOCP problem, because its constraint is regarded as a second-order cone in three-dimensional

[8] If we consider the normal contact law together with the tangential one, then we cannot state the Coulomb friction law in the form of the maximum dissipation law (and hence the normality rule). In section 10.3.4, we shall see that with an appropriate modification the Coulomb friction law can be expressed in a form similar to the normality rule.

space. From this perspective we here show that the optimality condition of (10.48) is equivalent to the second-order cone complementarity formulation (10.6) of the Coulomb friction law.

Proposition 10.3.2. *For given* $\dot{\boldsymbol{u}}_{\rm t}$ *and* $r_{\rm n} < 0$, $\boldsymbol{r}_{\rm t}$ *satisfies* (10.48) *if and only if there exists a* $\lambda_{\rm n}$ *satisfying* (10.6).

Proof. Since (10.48) is a convex optimization with an interior feasible solution,[9] its optimality condition can be derived within the framework of the Fenchel duality investigated in section 2.2.5; we particularly make use of the results established in Proposition 2.2.9. To this end, we begin by rewriting problem (10.48) as the minimization problem

$$\boldsymbol{r}_{\rm t} \in \arg \min_{\check{\boldsymbol{r}}_{\rm t} \in \mathbb{R}^2} \left\{ \dot{\boldsymbol{u}}_{\rm t}^{\rm T} \check{\boldsymbol{r}}_{\rm t} \mid -\mu r_{\rm n} \geq \|\check{\boldsymbol{r}}_{\rm t}\| \right\}, \tag{10.49}$$

which can be encoded to the form of $(\mathrm{P_F})$ in (2.28) as follows.

Let $\check{\boldsymbol{r}} = \begin{bmatrix} \check{r}_0 \\ \check{\boldsymbol{r}}_{\rm t} \end{bmatrix} \in \mathbb{R}^3$, and define $f,\, g : \mathbb{R}^3 \to \mathbb{R} \cup \{+\infty\}$ and $\Lambda : \mathbb{R}^3 \to \mathbb{R}^3$ by

$$f(\check{\boldsymbol{r}}) = \begin{cases} \dot{\boldsymbol{u}}_{\rm t}^{\rm T} \check{\boldsymbol{r}}_{\rm t} & \text{if } \check{r}_0 = -\mu r_{\rm n}, \\ +\infty & \text{otherwise}, \end{cases} \tag{10.50}$$

$$g(\check{\boldsymbol{r}}) = \begin{cases} 0 & \text{if } \check{r}_0 \geq \|\check{\boldsymbol{r}}_{\rm t}\|, \\ +\infty & \text{otherwise}, \end{cases} \tag{10.51}$$

$$\Lambda \check{\boldsymbol{r}} = \check{\boldsymbol{r}}. \tag{10.52}$$

Then problem (10.49) is embedded in the form of (2.28), i.e., the primal problem of the Fenchel duality framework. Hereafter we derive explicit expressions of the conditions in Proposition 2.2.9 (b) for this problem.

An explicit form of the conjugate function of f is obtained as

$$\begin{aligned} f^*(\check{\boldsymbol{r}}^*) &= \sup_{\check{\boldsymbol{r}}} \{ (\check{\boldsymbol{r}}^*)^{\rm T} \check{\boldsymbol{r}} - f(\check{\boldsymbol{r}}) \mid \check{\boldsymbol{r}} \in \mathbb{R}^3 \} \\ &= \sup_{\check{\boldsymbol{r}}} \{ (\check{\boldsymbol{r}}^*)^{\rm T} \check{\boldsymbol{r}} - \dot{\boldsymbol{u}}_{\rm t}^{\rm T} \check{\boldsymbol{r}}_{\rm t} \mid \check{r}_0 = -\mu r_{\rm n} \} \\ &= \begin{cases} -(\mu r_{\rm n}) \check{r}_0^* & \text{if } \check{\boldsymbol{r}}_{\rm t}^* = \dot{\boldsymbol{u}}_{\rm t}, \\ +\infty & \text{otherwise}, \end{cases} \end{aligned} \tag{10.53}$$

where $\check{\boldsymbol{r}}^* = \begin{bmatrix} \check{r}_0^* \\ \check{\boldsymbol{r}}_{\rm t}^* \end{bmatrix} \in \mathbb{R}^3$. From Fact 1.4.4 (ii), we see that the conjugate function of g is obtained as

$$\begin{aligned} g^*(\check{\boldsymbol{r}}^*) &= \sup_{\check{\boldsymbol{r}}} \{ (\check{\boldsymbol{r}}^*)^{\rm T} \check{\boldsymbol{r}} \mid \check{r}_0 \geq \|\check{\boldsymbol{r}}_{\rm t}\| \} \\ &= \begin{cases} 0 & \text{if } -\check{r}_0^* \geq \| -\check{\boldsymbol{r}}_{\rm t}^* \|, \\ +\infty & \text{otherwise}. \end{cases} \end{aligned} \tag{10.54}$$

[9] Recall that we assume $\mu > 0$ and $-r_{\rm n} < 0$ throughout section 10.3.2.

It follows from (10.50) and (10.53) that $f(\check{\boldsymbol{r}}) + f^*(\check{\boldsymbol{r}}^*)$ is finite if and only if $\check{r}_0 = -\mu r_{\rm n}$ and $\check{\boldsymbol{r}}^* = \dot{\boldsymbol{u}}_{\rm t}$. If this is the case the equality $f(\check{\boldsymbol{r}}) + f^*(\check{\boldsymbol{r}}^*) = \langle \check{\boldsymbol{r}}^*, \check{\boldsymbol{r}} \rangle$ is reduced to

$$-(\mu r_{\rm n})\check{r}_0^* + \dot{\boldsymbol{u}}_{\rm t}^{\rm T} \check{\boldsymbol{r}}_{\rm t} = 0.$$

From (10.51) and (10.54), we see that $g(\check{\boldsymbol{r}}) + g^*(-\check{\boldsymbol{r}}^*)$ is finite if and only if $\check{r}_0 \geq \|\check{\boldsymbol{r}}_{\rm t}\|$ and $\check{r}_0^* \geq \|\check{\boldsymbol{r}}_{\rm t}^*\|$. If this is the case the equality $g(\check{\boldsymbol{r}}) + g^*(-\check{\boldsymbol{r}}^*) = \langle -\check{\boldsymbol{r}}^*, \check{\boldsymbol{r}} \rangle$ always holds. Thus, by putting $\check{r}_0^* = \lambda_{\rm n}$, the conditions in Proposition 2.2.9 (b) are shown to be equivalent to (10.6). □

Although variables involved in (10.48) are $\check{\boldsymbol{r}}_{\rm t} \in \mathbb{R}^2$, we introduce an extra variable $\check{r}_0$ fixed as $-\mu r_{\rm n}$ in the proof of Proposition 10.3.2, to enjoy the self-duality property of the second-order cone in $\mathbb{R}^3$. In contrast, the investigation worked out in the next section is based on the admissible set of $\boldsymbol{r}_{\rm t} \in \mathbb{R}^2$, say, $\{\boldsymbol{r}_{\rm t} \in \mathbb{R}^2 \mid -\mu r_{\rm n} \geq \|\boldsymbol{r}_{\rm t}\|\}$.

10.3.2.2 Geometry of friction cone

Although we introduced an auxiliary variable $\check{r}_0$ in Proposition 10.3.2 to enjoy the self-dual property of the second-order cone, problem (10.48) essentially involves only two variables, say, $(\check{r}_{\rm t1}, \check{r}_{\rm t2})$. We here study the optimality condition of (10.48) without using additional variables, which yields for the Coulomb law an alternative formulation in the form of the inclusion.

For a given $r_{\rm n} < 0$, define $\tilde{\mathcal{K}}(r_{\rm n}) \subset \mathbb{R}^2$ by

$$\tilde{\mathcal{K}}(r_{\rm n}) = \{\boldsymbol{r}_{\rm t} \mid -\mu r_{\rm n} \geq \|\boldsymbol{r}_{\rm t}\|\}, \tag{10.55}$$

which is the admissible set of tangential reactions in the friction cone. The indicator function of $\tilde{\mathcal{K}}(r_{\rm n})$, denoted $\delta_{\tilde{\mathcal{K}}(r_{\rm n})} : \mathbb{R}^2 \to \mathbb{R} \cup \{+\infty\}$, is given by

$$\delta_{\tilde{\mathcal{K}}(r_{\rm n})}(\boldsymbol{r}_{\rm t}) = \begin{cases} 0 & \text{if } \boldsymbol{r}_{\rm t} \in \tilde{\mathcal{K}}(r_{\rm n}), \\ +\infty & \text{otherwise.} \end{cases} \tag{10.56}$$

Then problem (10.48) is regarded as the minimization of $\dot{\boldsymbol{u}}_{\rm t}^{\rm T} \boldsymbol{r}_{\rm t} + \delta_{\tilde{\mathcal{K}}(r_{\rm n})}(\boldsymbol{r}_{\rm t})$,[10] and thence its optimal solution satisfies $\boldsymbol{0} \in \partial(\dot{\boldsymbol{u}}_{\rm t}^{\rm T} \boldsymbol{r}_{\rm t} + \delta_{\tilde{\mathcal{K}}(r_{\rm n})}(\boldsymbol{r}_{\rm t}))$. This assertion is formally stated as follows.

Proposition 10.3.3. *For given* $\dot{\boldsymbol{u}}_{\rm t}$ *and* $r_{\rm n} < 0$, $\boldsymbol{r}_{\rm t}$ *satisfies* (10.48) *if and only if*

$$-\dot{\boldsymbol{u}}_{\rm t} \in \partial \delta_{\tilde{\mathcal{K}}(r_{\rm n})}(\boldsymbol{r}_{\rm t}), \tag{10.57}$$

[10] Because it is equivalent to the minimization problem in (10.49).

where

$$\partial \delta_{\tilde{\mathcal{K}}(r_{\rm n})}(\boldsymbol{r}_{\rm t}) = \begin{cases} \{\boldsymbol{0}\} & \text{if } \boldsymbol{r}_{\rm t} \in \operatorname{int} \tilde{\mathcal{K}}(r_{\rm n}), \\ \{\alpha \boldsymbol{r}_{\rm t} \mid \alpha \geq 0\} & \text{if } \boldsymbol{r}_{\rm t} \in \operatorname{bd} \tilde{\mathcal{K}}(r_{\rm n}), \\ \emptyset & \text{if } \boldsymbol{r}_{\rm t} \notin \tilde{\mathcal{K}}(r_{\rm n}). \end{cases} \tag{10.58}$$

Proof. To clarify the difference from the analysis in Proposition 10.3.2, we again employ the framework of the Fenchel duality. Specifically, we rewrite problem (10.48) as the minimization problem (10.49), and apply the results in Proposition 2.2.9 to derive the optimality condition.

Define f, $g : \mathbb{R}^2 \to \mathbb{R} \cup \{+\infty\}$ and $\Lambda : \mathbb{R}^2 \to \mathbb{R}^2$ by

$$f(\check{\boldsymbol{r}}_{\rm t}) = \dot{\boldsymbol{u}}_{\rm t}^{\rm T} \check{\boldsymbol{r}}_{\rm t},$$
$$g(\check{\boldsymbol{r}}_{\rm t}) = \delta_{\tilde{\mathcal{K}}(r_{\rm n})}(\check{\boldsymbol{r}}_{\rm t}),$$
$$\Lambda \check{\boldsymbol{r}}_{\rm t} = \check{\boldsymbol{r}}_{\rm t}.$$

Then problem (10.49) is embedded in the form of (2.28), i.e., the primal problem of the Fenchel duality framework. Since $\partial f(\check{\boldsymbol{r}}_{\rm t}) = \{\dot{\boldsymbol{u}}_{\rm t}\}$, the optimality condition established in Proposition 2.2.9 (c) is immediately reduced to (10.57).

To see (10.58), by definition of subgradient, we have that

$$\begin{aligned} &\partial \delta_{\tilde{\mathcal{K}}(r_{\rm n})}(\boldsymbol{r}_{\rm t}) \\ &\quad = \left\{ \dot{\boldsymbol{u}}_{\rm t} \in \mathbb{R}^2 \mid \delta_{\tilde{\mathcal{K}}(r_{\rm n})}(\boldsymbol{r}_{\rm t}') \geq \delta_{\tilde{\mathcal{K}}(r_{\rm n})}(\boldsymbol{r}_{\rm t}) + \langle \dot{\boldsymbol{u}}_{\rm t}, \boldsymbol{r}_{\rm t}' - \boldsymbol{r}_{\rm t} \rangle \ (\forall \boldsymbol{r}_{\rm t}' \in \mathbb{R}^2) \right\} \\ &\quad = \left\{ \dot{\boldsymbol{u}}_{\rm t} \in \mathbb{R}^2 \mid 0 \geq \delta_{\tilde{\mathcal{K}}(r_{\rm n})}(\boldsymbol{r}_{\rm t}) + \langle \dot{\boldsymbol{u}}_{\rm t}, \boldsymbol{r}_{\rm t}' - \boldsymbol{r}_{\rm t} \rangle \ (\forall \boldsymbol{r}_{\rm t}' \in \tilde{\mathcal{K}}(r_{\rm n})) \right\}. \end{aligned} \tag{10.59}$$

If $\boldsymbol{r}_{\rm t} \notin \tilde{\mathcal{K}}(r_{\rm n})$, then $\delta_{\tilde{\mathcal{K}}(r_{\rm n})}(\boldsymbol{r}_{\rm t}) = +\infty$ and hence $\partial \delta_{\tilde{\mathcal{K}}(r_{\rm n})}(\boldsymbol{r}_{\rm t}) = \emptyset$. If $\boldsymbol{r}_{\rm t} \in \tilde{\mathcal{K}}(r_{\rm n})$, then (10.59) is reduced to

$$\partial \delta_{\tilde{\mathcal{K}}(r_{\rm n})}(\boldsymbol{r}_{\rm t}) = \left\{ \dot{\boldsymbol{u}}_{\rm t} \in \mathbb{R}^2 \mid \langle \dot{\boldsymbol{u}}_{\rm t}, \boldsymbol{r}_{\rm t} \rangle \geq \langle \dot{\boldsymbol{u}}_{\rm t}, \boldsymbol{r}_{\rm t}' \rangle \ (\forall \boldsymbol{r}_{\rm t}' \in \tilde{\mathcal{K}}(r_{\rm n})) \right\}. \tag{10.60}$$

Concerning the inequality involved in (10.60), the Cauchy–Schwarz inequality yields

$$\begin{aligned} \max_{\boldsymbol{r}_{\rm t}'} \{ \langle \dot{\boldsymbol{u}}_{\rm t}, \boldsymbol{r}_{\rm t}' \rangle \mid \boldsymbol{r}_{\rm t}' \in \tilde{\mathcal{K}}(r_{\rm n}) \} &= \| \dot{\boldsymbol{u}}_{\rm t} \| \max_{\boldsymbol{r}_{\rm t}'} \{ \| \boldsymbol{r}_{\rm t}' \| \mid \boldsymbol{r}_{\rm t}' \in \tilde{\mathcal{K}}(r_{\rm n}) \} \\ &= \| \dot{\boldsymbol{u}}_{\rm t} \| (-\mu r_{\rm n}). \end{aligned} \tag{10.61}$$

If $\boldsymbol{r}_{\rm t} \in \operatorname{int} \tilde{\mathcal{K}}(r_{\rm n})$, then $\langle \dot{\boldsymbol{u}}_{\rm t}, \boldsymbol{r}_{\rm t} \rangle < \| \dot{\boldsymbol{u}}_{\rm t} \| (-\mu r_{\rm n})$ for $\dot{\boldsymbol{u}}_{\rm t} \neq \boldsymbol{0}$, from which and (10.61) we see that the inequality on the right-hand side of (10.60) is satisfied only by $\dot{\boldsymbol{u}}_{\rm t} = \boldsymbol{0}$, which means that $\partial \delta_{\tilde{\mathcal{K}}(r_{\rm n})}(\boldsymbol{r}_{\rm t}) = \{\boldsymbol{0}\}$. If $\boldsymbol{r}_{\rm t} \in \operatorname{bd} \tilde{\mathcal{K}}(r_{\rm n})$ (i.e., if $\| \boldsymbol{r}_{\rm t} \| = -\mu r_{\rm n}$), then $\langle \dot{\boldsymbol{u}}_{\rm t}, \boldsymbol{r}_{\rm t} \rangle = \| \dot{\boldsymbol{u}}_{\rm t} \| (-\mu r_{\rm n})$ is satisfied with $\dot{\boldsymbol{u}}_{\rm t} = \lambda \boldsymbol{r}_{\rm t}$ for any $\lambda \geq 0$, and from (10.61) such a $\dot{\boldsymbol{u}}_{\rm t}$ satisfies the inequality in (10.60), which means $\partial \delta_{\tilde{\mathcal{K}}(r_{\rm n})}(\boldsymbol{r}_{\rm t}) = \{\lambda \boldsymbol{r}_{\rm t} \mid \lambda \geq 0\}$. □

The assertion of Proposition 10.3.3 is illustrated in Figure 10.7. If the tangential reaction $\boldsymbol{r}_{\rm t}$ is on the boundary of the admissible set $\tilde{\mathcal{K}}(r_{\rm n})$, then

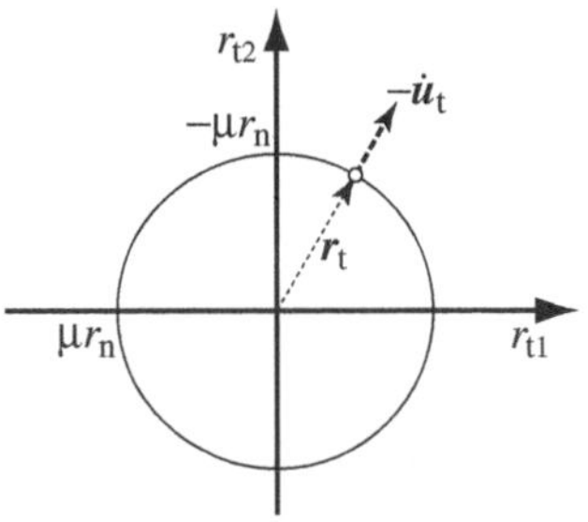

FIGURE 10.7: The normality rule (10.57) for the tangential velocity $\dot{\boldsymbol{u}}_t$ associated with the tangential reaction $\boldsymbol{r}_t$.

the tangential velocity $-\dot{\boldsymbol{u}}_t$ should be normal to the boundary of $\tilde{\mathcal{K}}(r_n)$ at the point $\boldsymbol{r}_t$. Thus, concerning the tangential contact variables, the Coulomb friction law can be stated in the form of the normality rule (10.57), which is a consequence of the maximum dissipation law (10.48). Note again that we here consider only the evolution of tangential displacement $\boldsymbol{u}_t$, and the evolution of the normal displacement g is not considered. If we incorporate the evolution law of g, which is the unilateral contact law, then the maximum dissipation law and the normality rule do not hold; see section 10.3.4.

We next derive the inverse law of (10.57). The conjugate function of $\delta_{\tilde{\mathcal{K}}(r_n)}$, denoted $\delta^*_{\tilde{\mathcal{K}}(r_n)}$, is obtained as

$$\begin{aligned}
\delta^*_{\tilde{\mathcal{K}}(r_n)}(\dot{\boldsymbol{u}}_t) &= \sup_{\boldsymbol{r}_t}\{\langle \dot{\boldsymbol{u}}_t, \boldsymbol{r}_t\rangle - \delta_{\tilde{\mathcal{K}}(r_n)}(\boldsymbol{r}_t)\} \\
&= \sup_{\boldsymbol{r}_t}\{\langle \dot{\boldsymbol{u}}_t, \boldsymbol{r}_t\rangle \mid \|\boldsymbol{r}_t\| \le -\mu r_n\} \\
&= \dot{\boldsymbol{u}}_t^{\mathrm{T}}\left[(-\mu r_n)\frac{\dot{\boldsymbol{u}}_t}{\|\dot{\boldsymbol{u}}_t\|}\right] \\
&= (-\mu r_n)\|\dot{\boldsymbol{u}}_t\|.
\end{aligned} \tag{10.62}$$

The subdifferential of $\delta^*_{\tilde{\mathcal{K}}(r_n)}$ is then given by[11]

$$\partial\delta^*_{\tilde{\mathcal{K}}(r_n)}(\dot{\boldsymbol{u}}_t) = \begin{cases} \left\{-\mu r_n \dfrac{\dot{\boldsymbol{u}}_t}{\|\dot{\boldsymbol{u}}_t\|}\right\} & \text{if } \dot{\boldsymbol{u}}_t \neq \boldsymbol{0}, \\ \{\boldsymbol{r}_t \mid \|\boldsymbol{r}_t\| \le -\mu r_n\} & \text{if } \dot{\boldsymbol{u}}_t = \boldsymbol{0}. \end{cases} \tag{10.63}$$

[11]Although we suppose throughout this section that $\mu > 0$ and $r_n < 0$, (10.62) and (10.63) hold even when $\mu r_n = 0$. If that is the case, $\delta_{\tilde{\mathcal{K}}(r_n)}(\boldsymbol{r}_t) = 0$ for $\boldsymbol{r}_t = \boldsymbol{0}$, while $\delta_{\tilde{\mathcal{K}}(r_n)}(\boldsymbol{r}_t) = +\infty$ for $\boldsymbol{r}_t \neq \boldsymbol{0}$. Simple calculations then result in $\delta^*_{\tilde{\mathcal{K}}(r_n)}(\dot{\boldsymbol{u}}_t) = 0$ $(\forall \dot{\boldsymbol{u}}_t)$ and $\partial\delta^*_{\tilde{\mathcal{K}}(r_n)}(\dot{\boldsymbol{u}}_t) = \{\boldsymbol{0}\}$.

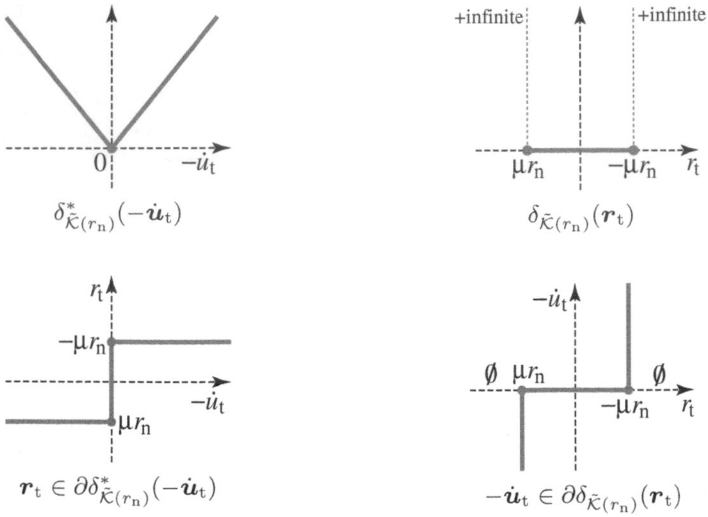

FIGURE 10.8: Schematic exposition of the inclusion forms of the Coulomb friction law. $\delta^*_{\tilde{\mathcal{K}}(r_n)}$ and $\delta_{\tilde{\mathcal{K}}(r_n)}$ defined by (10.62) and (10.56) serve as the potential energy and complementary energy functions, respectively; $\boldsymbol{r}_t \in \partial\delta^*_{\tilde{\mathcal{K}}(r_n)}$ and $-\dot{\boldsymbol{u}}_t \in \partial\delta_{\tilde{\mathcal{K}}(r_n)}$ with (10.63) and (10.58) serve as the constitutive relation and its inversion, respectively.

It follows from Proposition 2.1.12 that inclusion (10.57) is equivalent to

$$\delta_{\tilde{\mathcal{K}}(r_n)}(\boldsymbol{r}_t) + \delta^*_{\tilde{\mathcal{K}}(r_n)}(-\dot{\boldsymbol{u}}_t) = \langle -\dot{\boldsymbol{u}}_t, \boldsymbol{r}_t \rangle,$$

which yields the following alternative inclusion formulation:

$$\boldsymbol{r}_t \in \partial\delta^*_{\tilde{\mathcal{K}}(r_n)}(-\dot{\boldsymbol{u}}_t). \tag{10.64}$$

It is noted that (10.64) and (10.57) are regarded as the "constitutive law" and its inversion between $\boldsymbol{r}_t$ and $\dot{\boldsymbol{u}}_t$ for the Coulomb friction, as illustrated in Figure 10.8. Then $\delta^*_{\tilde{\mathcal{K}}(r_n)}$ and $\delta_{\tilde{\mathcal{K}}(r_n)}$ in (10.62) and (10.56) serve as the "strain energy" function and the corresponding "complementary strain energy" function, respectively.

10.3.2.3 A QCQP formulation

In section 10.3.2.1 the optimality condition of the maximum dissipation law, given in (10.48), has been studied from the perspective of SOCP. Since we assume $r_n < 0$ and $\mu > 0$, problem (10.48) can be rewritten equivalently as

$$\boldsymbol{r}_t \in \arg\max_{\check{\boldsymbol{r}}_t \in \mathbb{R}^2} \left\{ -\dot{\boldsymbol{u}}_t^{\mathrm{T}} \check{\boldsymbol{r}}_t \mid \mu^2 r_n^2 - \|\check{\boldsymbol{r}}_t\|^2 \geq 0 \right\}. \tag{10.65}$$

Problem (10.65) is then a QCQP problem.[12] Note that the constraint of (10.65) is entirely smooth, while that of (10.48) has a nonsmooth point. Thus the Lagrangian $L : \mathbb{R}^2 \times \mathbb{R} \to \mathbb{R} \cup \{+\infty\}$ of (10.65) is defined in a usual manner as[13]

$$L(\check{\boldsymbol{r}}_{\mathrm{t}}, \lambda_{\mathrm{Q}}) = \begin{cases} -\dot{\boldsymbol{u}}_{\mathrm{t}}^{\mathrm{T}} \check{\boldsymbol{r}}_{\mathrm{t}} + \lambda_{\mathrm{Q}}(\mu^2 r_{\mathrm{n}}^2 - \check{\boldsymbol{r}}_{\mathrm{t}}^{\mathrm{T}} \check{\boldsymbol{r}}_{\mathrm{t}}) & \text{if } \lambda_{\mathrm{Q}} \geq 0, \\ +\infty & \text{otherwise,} \end{cases} \tag{10.66}$$

where $\lambda_{\mathrm{Q}} \in \mathbb{R}$ is the Lagrange multiplier. Consequently, $\boldsymbol{r}_{\mathrm{t}}$ satisfies (10.65) if and only if there exists a $\bar{\lambda}_{\mathrm{Q}}$ satisfying the following KKT conditions:

$$\mu^2 r_{\mathrm{n}}^2 - \boldsymbol{r}_{\mathrm{t}}^{\mathrm{T}} \boldsymbol{r}_{\mathrm{t}} \geq 0, \tag{10.67a}$$

$$\bar{\lambda}_{\mathrm{Q}} \geq 0, \tag{10.67b}$$

$$\bar{\lambda}_{\mathrm{Q}}(\mu^2 r_{\mathrm{n}}^2 - \boldsymbol{r}_{\mathrm{t}}^{\mathrm{T}} \boldsymbol{r}_{\mathrm{t}}) = 0, \tag{10.67c}$$

$$\dot{\boldsymbol{u}}_{\mathrm{t}} + 2\bar{\lambda}_{\mathrm{Q}} \boldsymbol{r}_{\mathrm{t}} = \boldsymbol{0}, \tag{10.67d}$$

which are derived as the stationarity condition of L in (10.66). By putting $\lambda = 2\bar{\lambda}_{\mathrm{Q}}$ for simplicity, (10.67) is reduced to

$$\exists \lambda \geq 0 :$$

$$\dot{\boldsymbol{u}}_{\mathrm{t}} = -\lambda \boldsymbol{r}_{\mathrm{t}}, \tag{10.68a}$$

$$\mu^2 r_{\mathrm{n}}^2 \geq \boldsymbol{r}_{\mathrm{t}}^{\mathrm{T}} \boldsymbol{r}_{\mathrm{t}}, \tag{10.68b}$$

$$\lambda(\mu^2 r_{\mathrm{n}}^2 - \boldsymbol{r}_{\mathrm{t}}^{\mathrm{T}} \boldsymbol{r}_{\mathrm{t}}) = 0. \tag{10.68c}$$

Thus we obtain an alternative formulation of the Coulomb friction law. In (10.68) it is noted that the complementarity condition, say, (10.68c), is given in terms of λ and $\mu^2 r_{\mathrm{n}}^2 - \boldsymbol{r}_{\mathrm{t}}^{\mathrm{T}} \boldsymbol{r}_{\mathrm{t}}$, which means that $\boldsymbol{r}_{\mathrm{t}} \in \operatorname{int} \tilde{\mathcal{K}}(r_{\mathrm{n}})$ implies $\lambda = 0$, and that $\lambda > 0$ implies $\boldsymbol{r}_{\mathrm{t}} \in \operatorname{bd} \tilde{\mathcal{K}}(r_{\mathrm{n}})$. Then (10.68a) requires that $\dot{\boldsymbol{u}}_{\mathrm{t}}$ should be parallel with $\boldsymbol{r}_{\mathrm{t}}$ with the opposite sign. This is a physical interpretation given to (10.68). As another feature, all the functions involved in (10.68) are smooth, in contrast to, e.g., (10.43). It is noted, on the other hand, that the parallelism constraint (10.68a) is not required to be imposed explicitly in the case of the second-order cone complementarity formulation.[14]

Remark 10.3.4. Condition (10.68) is also derived from the inclusion form (10.57) of the Coulomb law as follows. In (10.58), the first two cases can be put together as

$$\partial \delta_{\tilde{\mathcal{K}}(r_{\mathrm{n}})}(\boldsymbol{r}_{\mathrm{t}}) = \begin{cases} \{\lambda \boldsymbol{r}_{\mathrm{t}} \mid \lambda \geq 0,\ \lambda(\mu r_{\mathrm{n}} + \|\boldsymbol{r}_{\mathrm{t}}\|) = 0\} & \text{if } \boldsymbol{r}_{\mathrm{t}} \in \tilde{\mathcal{K}}(r_{\mathrm{n}}), \\ \emptyset & \text{if } \boldsymbol{r}_{\mathrm{t}} \notin \tilde{\mathcal{K}}(r_{\mathrm{n}}). \end{cases}$$

[12]See section 1.6.2 for the definitions of QP (quadratic programming) and QCQP (quadratically constrained quadratic programming).

[13]Since (10.65) is a maximization problem, it corresponds to ($\mathrm{P}_{\mathrm{L}}^*$) in (2.35) with L in (10.66).

[14]See also Remark 10.2.3.

By using this expression, condition (10.57) is explicitly written as

$$\exists \lambda \geq 0 :$$

$$\dot{\boldsymbol{u}}_{\mathrm{t}} = -\lambda \boldsymbol{r}_{\mathrm{t}}, \tag{10.69a}$$

$$-\mu r_{\mathrm{n}} \geq \|\boldsymbol{r}_{\mathrm{t}}\|, \tag{10.69b}$$

$$\lambda(\mu r_{\mathrm{n}} + \|\boldsymbol{r}_{\mathrm{t}}\|) = 0. \tag{10.69c}$$

It is easy to see that (10.68c) is equivalent to (10.69c), and thence (10.69) can be reduced to (10.68). ■

Remark 10.3.5. For deriving (10.68) we used the stationarity condition of the Lagrangian (10.66) associated with the QCQP problem (10.65). Similarly, formulation (10.43) can also be obtained from the Lagrangian for the SOCP problem (10.48).

By using the self-duality of the second-order cone, the Lagrangian $L_{\mathrm{SOCP}} : \mathbb{R}^2 \times \mathbb{R}^3 \to \mathbb{R} \cup \{+\infty\}$ of (10.48) is written as

$$L_{\mathrm{SOCP}}(\check{\boldsymbol{r}}_{\mathrm{t}}, \check{\boldsymbol{\lambda}}) = \begin{cases} -\dot{\boldsymbol{u}}_{\mathrm{t}}^{\mathrm{T}} \check{\boldsymbol{r}}_{\mathrm{t}} + (\check{\lambda}_{\mathrm{n}}, \check{\boldsymbol{\lambda}}_{\mathrm{t}}) \cdot (\mu r_{\mathrm{n}}, \check{\boldsymbol{r}}_{\mathrm{t}}) & \text{if } \check{\lambda}_{\mathrm{n}} \geq \|\check{\boldsymbol{\lambda}}_{\mathrm{t}}\|, \\ +\infty & \text{otherwise,} \end{cases}$$

where $\check{\boldsymbol{\lambda}} = (\check{\lambda}_{\mathrm{n}}, \check{\boldsymbol{\lambda}}_{\mathrm{t}}) \in \mathbb{R} \times \mathbb{R}^2$ is the Lagrange multiplier vector. The SOCP formulation, (10.43), of the Coulomb friction law is then recovered from the stationarity condition of L_{SOCP}. Thus the difference between (10.68) and (10.43) is originated from the difference of the Lagrangian formulations. ■

By eliminating λ from (10.68), we can obtain an alternative expression below for Coulomb's law (see Remark 10.3.6 for a proof).

$$-\mu r_{\mathrm{n}} \geq \|\boldsymbol{r}_{\mathrm{t}}\|, \tag{10.70a}$$

$$(\mu r_{\mathrm{n}})\dot{\boldsymbol{u}}_{\mathrm{t}} = \|\dot{\boldsymbol{u}}_{\mathrm{t}}\|\boldsymbol{r}_{\mathrm{t}}. \tag{10.70b}$$

Remark 10.3.6. Condition (10.70) is derived from (10.68) as follows. By multiplying λ with (10.68c), we obtain

$$(\lambda \mu r_{\mathrm{n}})^2 = (\lambda \boldsymbol{r}_{\mathrm{t}})^{\mathrm{T}}(\lambda \boldsymbol{r}_{\mathrm{t}}). \tag{10.71}$$

Substituting (10.68a) into (10.71) results in $(\lambda \mu r_{\mathrm{n}})^2 = \dot{\boldsymbol{u}}_{\mathrm{t}}^{\mathrm{T}} \dot{\boldsymbol{u}}_{\mathrm{t}}$, which is reduced to

$$-\mu r_{\mathrm{n}} \lambda = \|\dot{\boldsymbol{u}}_{\mathrm{t}}\|, \tag{10.72}$$

because $\lambda \geq 0$ and $r_{\mathrm{n}} \leq 0$. On the other hand, by multiplying μr_{n} with (10.68a) we obtain

$$(\mu r_{\mathrm{n}})\dot{\boldsymbol{u}}_{\mathrm{t}} = -\mu r_{\mathrm{n}} \lambda \boldsymbol{r}_{\mathrm{t}}. \tag{10.73}$$

Substituting (10.72) into (10.73) then yields (10.70b). ■

10.3.3 A formulation using projection operator

For a closed convex set $\mathcal{S} \subseteq \mathbb{R}^n$, let $\mathbf{p}_{\mathcal{S}} : \mathbb{R}^n \to \mathbb{R}^n$ be the *projection* on $\mathcal{S}$ defined by[15]

$$\mathbf{p}_{\mathcal{S}}(\boldsymbol{x}) = \arg\min_{\boldsymbol{y}}\{\|\boldsymbol{y} - \boldsymbol{x}\| \mid \boldsymbol{y} \in \mathcal{S}\}. \tag{10.74}$$

In other words, $\mathbf{p}_{\mathcal{S}}(\boldsymbol{x})$ is the *nearest point* to $\boldsymbol{x}$ in $\mathcal{S}$. For $\tilde{\mathcal{K}}(r_{\mathrm{n}})$ defined by (10.55), in particular, the associated projection operator is explicitly written as

$$\mathbf{p}_{\tilde{\mathcal{K}}(r_{\mathrm{n}})}(\check{\boldsymbol{r}}_{\mathrm{t}}) = \min\left\{-\frac{\mu r_{\mathrm{n}}}{\|\check{\boldsymbol{r}}_{\mathrm{t}}\|}, 1\right\}\check{\boldsymbol{r}}_{\mathrm{t}}, \tag{10.75}$$

where we use the convention $0/0 = 1$. We show that the Coulomb friction law can be expressed by using $\mathbf{p}_{\tilde{\mathcal{K}}(r_{\mathrm{n}})}$.

To see this, recall the maximum dissipation law (10.48). As a variational characterization, $\check{\boldsymbol{r}}_{\mathrm{t}} \in \tilde{\mathcal{K}}(r_{\mathrm{n}})$ satisfies (10.48) if and only if

$$\langle -\dot{\boldsymbol{u}}_{\mathrm{t}}, \check{\boldsymbol{r}}_{\mathrm{t}} - \boldsymbol{r}_{\mathrm{t}}\rangle \leq 0, \quad \forall \check{\boldsymbol{r}}_{\mathrm{t}} \in \tilde{\mathcal{K}}(r_{\mathrm{n}})$$

holds. Add and subtract $\langle \boldsymbol{r}_{\mathrm{t}}, \check{\boldsymbol{r}}_{\mathrm{t}} - \boldsymbol{r}_{\mathrm{t}}\rangle$ to the left-hand side to obtain

$$\langle (\boldsymbol{r}_{\mathrm{t}} - \dot{\boldsymbol{u}}_{\mathrm{t}}) - \boldsymbol{r}_{\mathrm{t}}, \check{\boldsymbol{r}}_{\mathrm{t}} - \boldsymbol{r}_{\mathrm{t}}\rangle \leq 0, \quad \forall \check{\boldsymbol{r}}_{\mathrm{t}} \in \tilde{\mathcal{K}}(r_{\mathrm{n}}). \tag{10.76}$$

It can be shown (see Remark 10.3.8 below for a proof) that the variational inequality (10.76) is satisfied if and only if $\boldsymbol{r}_{\mathrm{t}}$ is the projection of $\boldsymbol{r}_{\mathrm{t}} - \dot{\boldsymbol{u}}_{\mathrm{t}}$ on $\tilde{\mathcal{K}}(r_{\mathrm{n}})$. Thus the maximum dissipation law (10.48) can be rewritten as follows:

$$\boldsymbol{r}_{\mathrm{t}} = \mathbf{p}_{\tilde{\mathcal{K}}(r_{\mathrm{n}})}(\boldsymbol{r}_{\mathrm{t}} - \dot{\boldsymbol{u}}_{\mathrm{t}}). \tag{10.77}$$

It is noted that condition (10.77) does not involve inequalities but is given in the form of nonsmooth equations.[16]

Remark 10.3.7. The unilateral contact law (10.21) can also be written by using a projection.

It follows from Proposition 2.1.12 that the unilateral contact law (10.21) is rewritten into the inclusion form as[17]

$$r_{\mathrm{n}} \in \partial\delta_{\mathbb{R}_+}(g). \tag{10.78}$$

By using the projection on $\mathbb{R}_+$, (10.78) is equivalently rewritten as

$$g = \mathbf{p}_{\mathbb{R}_+}(g + r_{\mathrm{n}}). \tag{10.79}$$

[15]See section 6.2.1 (specifically (6.39) and (6.52)) for more on the projection.

[16]The nonsmoothness property stems from the norm-operator and the min-operator in (10.75).

[17]See section 3.3 for more detail on the equivalence of (10.21) and (10.78).

Indeed, from Fact 2.1.7 we see that $r_{\rm n} \le 0$ satisfies (10.78) if and only if $\langle r_{\rm n}, \breve{g}-g\rangle \le 0$ $(\forall \breve{g} \in \mathbb{R}_+)$ holds, which is equivalent to

$$\langle (g + r_{\rm n}) - g, \breve{g} - g\rangle \le 0, \quad \forall \breve{g} \in \mathbb{R}_+.$$

This variational inequality is equivalent to (10.79), as shown Remark 10.3.8.

Alternatively, (10.79) is also written as

$$r_{\rm n} = \mathbf{p}_{\mathbb{R}_-}(r_{\rm n} + \rho g), \tag{10.80}$$

where $\mathbb{R}_- = \{x \in \mathbb{R} \mid x \le 0\}$, and $\rho > 0$ is a constant. In (10.80), we may recognize $r_{\rm n}+\rho g$ as the *augmented normal reaction* with a *penalty* factor ρ. Thus the projection law (10.80) serves as a fundamental tool for the augmented Lagrangian method for contact problems. An augmented Lagrangian method for frictional contacts, essentially based on (10.77) and (10.80), is in Alart–Curnier [6]. ■

Remark 10.3.8. We here show a general variational characteristic of the projection; i.e., for a given closed convex set $\mathcal{S}$, $\bar{\boldsymbol{y}} \in \mathcal{S}$ satisfies $\bar{\boldsymbol{y}} = \mathbf{p}_{\mathcal{S}}(\boldsymbol{x})$ if and only if

$$\langle \boldsymbol{x} - \bar{\boldsymbol{y}}, \boldsymbol{y} - \bar{\boldsymbol{y}}\rangle \le 0, \quad \forall \boldsymbol{y} \in \mathcal{S}. \tag{10.81}$$

Suppose that $\bar{\boldsymbol{y}}$ is the optimal solution of problem (10.74). For any $\boldsymbol{y} \in \mathcal{S}$ $(\boldsymbol{y} \neq \bar{\boldsymbol{y}})$, a point on the (open) line segment connecting $\boldsymbol{y}$ and $\bar{\boldsymbol{y}}$ is written as $\bar{\boldsymbol{y}} + \alpha(\boldsymbol{y} - \bar{\boldsymbol{y}})$ $(\alpha \in]0,1[)$. From the optimality of $\bar{\boldsymbol{y}}$ for (10.74), we obtain

$$\|\bar{\boldsymbol{y}} - \boldsymbol{x}\|^2 \le \|[\bar{\boldsymbol{y}} + \alpha(\boldsymbol{y} - \bar{\boldsymbol{y}})] - \boldsymbol{x}\|^2,$$

which is simplified as

$$0 \le \alpha\langle \boldsymbol{x} - \bar{\boldsymbol{y}}, \boldsymbol{y} - \bar{\boldsymbol{y}}\rangle + \alpha^2\|\boldsymbol{y} - \bar{\boldsymbol{y}}\|^2.$$

Then dividing by α and taking $\alpha \searrow 0$, we obtain (10.81).

Conversely, suppose that $\bar{\boldsymbol{y}} \in \mathcal{S}$ satisfies (10.81). Then, for any $\boldsymbol{y} \in \mathcal{S}$, we obtain

$$\begin{aligned} 0 &\ge \langle \boldsymbol{x} - \bar{\boldsymbol{y}}, \boldsymbol{y} - \bar{\boldsymbol{y}}\rangle \\ &= \langle \boldsymbol{x} - \bar{\boldsymbol{y}}, (\boldsymbol{x} - \bar{\boldsymbol{y}}) + (\boldsymbol{y} - \boldsymbol{x})\rangle \\ &= \|\boldsymbol{x} - \bar{\boldsymbol{y}}\|^2 + \langle \boldsymbol{x} - \bar{\boldsymbol{y}}, \boldsymbol{y} - \boldsymbol{x}\rangle \\ &\ge \|\boldsymbol{x} - \bar{\boldsymbol{y}}\|(\|\boldsymbol{x} - \bar{\boldsymbol{y}}\| - \|\boldsymbol{x} - \boldsymbol{y}\|), \end{aligned}$$

where the last inequality follows from the Cauchy–Schwarz inequality. If $\bar{\boldsymbol{y}} \neq \boldsymbol{x}$, divide by $\|\boldsymbol{x} - \bar{\boldsymbol{y}}\|$ to see $\bar{\boldsymbol{y}} = \mathbf{p}_{\mathcal{S}}(\boldsymbol{x})$ in (10.74). If $\bar{\boldsymbol{y}} = \boldsymbol{x}$, then $\bar{\boldsymbol{y}}$ is certainly the nearest point to $\boldsymbol{x}$. ■

10.3.4 Friction law and normality rule

We saw in section 10.3.2 that the evolution property of the tangential displacement $\boldsymbol{u}_{\rm t}$ in the Coulomb friction law can be stated in the form of the maximum dissipation law, which is a fundamental postulate of the classical plasticity theory. Hence, the normality rule is true with the evolution of $\boldsymbol{u}_{\rm t}$; i.e., $-\dot{\boldsymbol{u}}_{\rm t}$ should be normal to the boundary of $\widetilde{\mathcal{K}}(r_{\rm n})$ at the point $\boldsymbol{r}_{\rm t}$. Thus

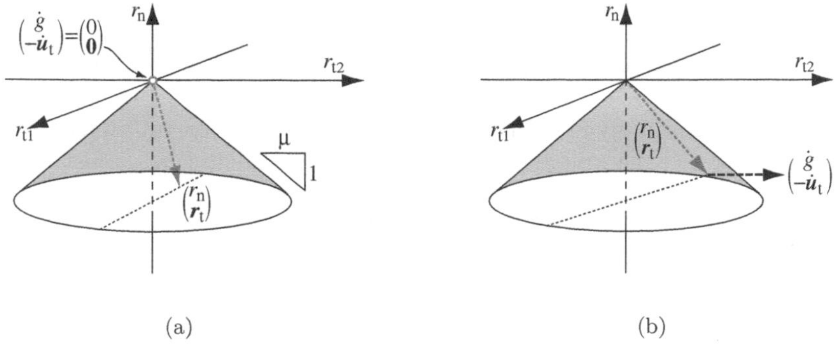

FIGURE 10.9: Non-association property of the friction law. (a) $-\mu r_n > \|\boldsymbol{r}_t\|$ implies $(g, -\dot{\boldsymbol{u}}_t) = \mathbf{0}$; (b) $\dot{\boldsymbol{u}}_t \neq \mathbf{0}$ implies $-\mu r_n = \|\boldsymbol{r}_t\|$, but the velocity vector $(\dot{g}, -\dot{\boldsymbol{u}}_t)$ is not normal to the surface of the friction cone.

the evolution of $\boldsymbol{u}_t$ is related to $\boldsymbol{r}_t$, by using the basic concept of the plasticity theory. On the other hand, the normal displacement g and the normal reaction r_n are also the state variables of the contact candidate node; it seems to be useful to describe the evolution of the normal displacement g and the tangential displacement $\boldsymbol{u}_t$ at the same time. We here show, unfortunately, that the normality rule (and hence the maximum dissipation law also) does not hold if we consider the evolution of g and $\boldsymbol{u}_t$ simultaneously.

Suppose that $r_n < 0$, which implies $g = 0$. If $-\mu r_n > \|\boldsymbol{r}_t\|$, no tangential slip can take place, i.e., $\dot{\boldsymbol{u}}_t = \mathbf{0}$. Moreover, in the presence of the reaction, the gap opening is not allowed, and hence $\dot{g} = 0$. Thus, as shown in Figure 10.9(a), if $\boldsymbol{r}$ is in the interior of the friction cone $\mathcal{K} = \{(r_n, \boldsymbol{r}_t) \mid -\mu r_n \geq \|\boldsymbol{r}_t\|\}$, then only $(\dot{g}, -\dot{\boldsymbol{u}}_t) = (0, \mathbf{0})$ is admissible. Alternatively, if $-\mu r_n = \|\boldsymbol{r}_t\| > 0$, then $-\dot{\boldsymbol{u}}_t$ may take place in parallel with $\boldsymbol{r}_t$. Because of the presence of the reaction, however, the gap opening is not allowed, and hence $\dot{g} = 0$. Thus, if $\boldsymbol{r}$ is on the boundary of $\mathcal{K}$, then $(\dot{g}, -\dot{\boldsymbol{u}}_t) = (0, \alpha \boldsymbol{r}_t)$ with any $\alpha \geq 0$ is admissible. As shown in Figure 10.9(b), such a $(\dot{g}, -\dot{\boldsymbol{u}}_t)$ is *not* normal to the boundary of $\mathcal{K}$. Thus, the evolution of $(g, -\boldsymbol{u}_t)$ in the Coulomb friction law cannot be described as the normality rule. For this reason, the Coulomb friction law is said to be a *non-associated flow rule*.

The lack of the normality between $(\dot{g}, -\dot{\boldsymbol{u}}_t)$ and the boundary surface of $\mathcal{K}$ at the point $(r_n, \boldsymbol{r}_t)$ means that there exists no potential corresponding to the Coulomb friction law; see [109] for a proof of the non-existence of a potential. In other words, the incremental problem of a contact problem with the Coulomb friction, e.g., (10.22), cannot be reformulated as an optimization problem. This characteristic of the Coulomb friction law makes the frictional contact problem somewhat complicated compared with the other problems considered in this book. This is one of the reasons the formulations of frictional contacts diverge, as already seen in the preceding sections; no almighty formulation is known up to now.

In the following, we explore some formulations that relate the contact displacement vector $(g, \boldsymbol{u}_{\mathrm{t}})$ to the reaction $(r_{\mathrm{n}}, \boldsymbol{r}_{\mathrm{t}})$ by somewhat extending the concepts commonly used in mechanics; the first two formulations, investigated in section 10.3.4.1 and section 10.3.4.2, are regarded as a sophisticated modification of the normality rule, while the one in section 10.3.4.3 is based on an extension of the potential energy.

10.3.4.1 Friction law coupled with contact law

In all the formulations presented above, the Coulomb friction law and the unilateral contact law are treated separatedly. Therefore, each contact candidate node is subjected to a pair of complementarity conditions, as seen in, e.g., (10.22). In contrast, the following formulation expresses both of the Coulomb friction law and unilateral contact law simultaneously as a single complementarity condition.[18]

$$g \geq 0, \tag{10.82a}$$

$$-\mu r_{\mathrm{n}} \geq \|\boldsymbol{r}_{\mathrm{t}}\|, \tag{10.82b}$$

$$\begin{bmatrix} r_{\mathrm{n}} \\ \boldsymbol{r}_{\mathrm{t}} \end{bmatrix} \cdot \begin{bmatrix} -g - \mu\|\dot{\boldsymbol{u}}_{\mathrm{t}}\| \\ \dot{\boldsymbol{u}}_{\mathrm{t}} \end{bmatrix} = 0. \tag{10.82c}$$

Proposition 10.3.9. *$(g, \dot{\boldsymbol{u}}_{\mathrm{t}})$ and $(r_{\mathrm{n}}, \boldsymbol{r}_{\mathrm{t}})$ satisfy* (10.82) *if and only if they satisfy the Coulomb friction law* (10.4) *and the unilateral contact law* (10.21).

Proof. We prove the assertion by showing that (10.82) is equivalent to (10.21) and (10.44).

We firstly observe that (10.82b) implies

$$r_{\mathrm{n}} \leq 0. \tag{10.83}$$

The condition (10.82c) is reduced to

$$\begin{aligned}
0 &= \begin{bmatrix} r_{\mathrm{n}} \\ \boldsymbol{r}_{\mathrm{t}} \end{bmatrix} \cdot \begin{bmatrix} -g - \mu\|\dot{\boldsymbol{u}}_{\mathrm{t}}\| \\ \dot{\boldsymbol{u}}_{\mathrm{t}} \end{bmatrix} \\
&= -r_{\mathrm{n}} g - \mu r_{\mathrm{n}} \|\dot{\boldsymbol{u}}_{\mathrm{t}}\| + \boldsymbol{r}_{\mathrm{n}}^{\mathrm{T}} \dot{\boldsymbol{u}}_{\mathrm{t}} \\
&\geq -r_{\mathrm{n}} g + \|\boldsymbol{r}_{\mathrm{n}}\| \|\dot{\boldsymbol{u}}_{\mathrm{t}}\| + \boldsymbol{r}_{\mathrm{n}}^{\mathrm{T}} \dot{\boldsymbol{u}}_{\mathrm{t}} && \text{[from (10.82b)]} \\
&\geq -r_{\mathrm{n}} g && \text{[from Schwarz's inequality]} \\
&\geq 0. && \text{[from (10.82a) and (10.83)]}
\end{aligned} \tag{10.84}$$

[18] In (10.82), the tangential contact velocity $\dot{\boldsymbol{u}}_{\mathrm{t}}$ and (not $\dot{g}$ but) the normal contact displacement g are related to the reaction $(r_{\mathrm{n}}, \boldsymbol{r}_{\mathrm{t}})$, because the unilateral contact law (10.21) gives the relation between g and r_{n}.

Therefore, all the inequalities in (10.84) are satisfied with the equalities, and hence we obtain

$$r_{\rm n} g = 0. \tag{10.85}$$

Suppose that $g > 0$. Then (10.82b) and (10.85) imply $(r_{\rm n}, \boldsymbol{r}_{\rm t}) = (0, \boldsymbol{0})$, and hence (10.82c) holds for any $\dot{\boldsymbol{u}}_{\rm t}$ as expected. Alternatively, suppose that $r_{\rm n} > 0$. Then (10.85) implies $g = 0$, and (10.82c) is then reduced to

$$\begin{bmatrix} -\mu r_{\rm n} \\ \boldsymbol{r}_{\rm t} \end{bmatrix} \cdot \begin{bmatrix} \|\dot{\boldsymbol{u}}_{\rm t}\| \\ \dot{\boldsymbol{u}}_{\rm t} \end{bmatrix} = 0,$$

which coincides with the complementarity condition (10.44b) involved in the formulation (10.44) of the Coulomb friction law. □

At first glance (10.82) is not a complementarity condition, but, in fact, we can restate (10.82) in the form of the complementarity condition as follows. For a given $\mu > 0$, let $\mathcal{K} \subset \mathbb{R}^3$ denote the friction cone, i.e.,

$$\mathcal{K} = \{(r_{\rm n}, \boldsymbol{r}_{\rm t}) \mid -\mu r_{\rm n} \geq \|\boldsymbol{r}_{\rm t}\|\}. \tag{10.86}$$

The dual cone of $\mathcal{K}$, denoted by $\mathcal{K}^* \subset \mathbb{R}^3$, is explicitly obtained as

$$\mathcal{K}^* = \{(v_{\rm n}, \boldsymbol{v}_{\rm t}) \mid -v_{\rm n} \geq \mu\|\boldsymbol{v}_{\rm t}\|\}, \tag{10.87}$$

because the second-order cone $\mathbb{L}^3$ is self-dual, i.e., $(\mathbb{L}^3)^* = \mathbb{L}^3$ (see Fact 1.3.17). Then we have that

$$\begin{bmatrix} -g - \mu\|\dot{\boldsymbol{u}}_{\rm t}\| \\ \dot{\boldsymbol{u}}_{\rm t} \end{bmatrix} \in \mathcal{K}^* \quad \Leftrightarrow \quad g \geq 0,$$

and hence condition (10.82) is equivalently rewritten as

$$\mathcal{K} \ni \begin{bmatrix} r_{\rm n} \\ \boldsymbol{r}_{\rm t} \end{bmatrix} \perp \begin{bmatrix} -g - \mu\|\dot{\boldsymbol{u}}_{\rm t}\| \\ \dot{\boldsymbol{u}}_{\rm t} \end{bmatrix} \in \mathcal{K}^*. \tag{10.88}$$

This is clearly in a form of the complementarity condition.

We next show that (10.88) is rewritten as a form *similar* to the normality rule. More specifically, we consider the relation between the reaction vector $(r_{\rm n}, \boldsymbol{r}_{\rm t})$ and the *modified* kinematic variable vector $(-g - \mu\|\dot{\boldsymbol{u}}_{\rm t}\|, \dot{\boldsymbol{u}}_{\rm t})$. To this end, recall the definitions of dual cone, indicator function, and conjugate function (see (1.1), (2.1), and (2.13)) to see that for a cone $C \subseteq \mathbb{R}^n$ we have that

$$\delta_C^*(\boldsymbol{s}) = \delta_{C^*}(-\boldsymbol{s}). \tag{10.89}$$

One can easily see that the complementarity condition (10.88) is equivalent to the equation

$$\delta_{\mathcal{K}}(r_{\rm n}, \boldsymbol{r}_{\rm t}) + \delta_{\mathcal{K}^*}(-g - \mu\|\dot{\boldsymbol{u}}_{\rm t}\|, \dot{\boldsymbol{u}}_{\rm t}) = -\begin{bmatrix} r_{\rm n} \\ \boldsymbol{r}_{\rm t} \end{bmatrix} \cdot \begin{bmatrix} -g - \mu\|\dot{\boldsymbol{u}}_{\rm t}\| \\ \dot{\boldsymbol{u}}_{\rm t} \end{bmatrix},$$

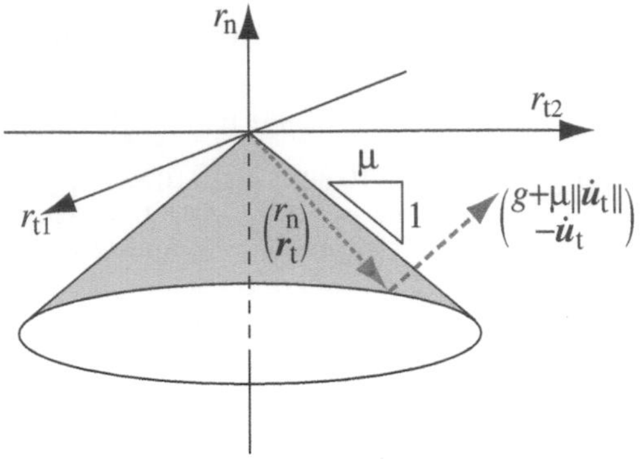

FIGURE 10.10: Schematic representation of the "modified" normality rule in (10.90) for the Coulomb friction law together with the unilateral contact law.

which can further be reduced to

$$\delta_{\mathcal{K}}(r_n, r_t) + \delta^*_{\mathcal{K}}(g + \mu\|\dot{\boldsymbol{u}}_t\|, -\dot{\boldsymbol{u}}_t) = \begin{bmatrix} r_n \\ \boldsymbol{r}_t \end{bmatrix} \cdot \begin{bmatrix} g + \mu\|\dot{\boldsymbol{u}}_t\| \\ -\dot{\boldsymbol{u}}_t \end{bmatrix}.$$

From Proposition 2.1.12, this equation is equivalent to the inclusion

$$\begin{bmatrix} g + \mu\|\dot{\boldsymbol{u}}_t\| \\ -\dot{\boldsymbol{u}}_t \end{bmatrix} \in \partial\delta_{\mathcal{K}}(r_n, \boldsymbol{r}_t). \tag{10.90}$$

Moreover, it follows from Fact 2.1.7 that (10.90) is equivalent to

$$\begin{bmatrix} r_n \\ \boldsymbol{r}_t \end{bmatrix} \in \arg\max_{r_n, \boldsymbol{r}_t} \left\{ \begin{bmatrix} g + \mu\|\dot{\boldsymbol{u}}_t\| \\ -\dot{\boldsymbol{u}}_t \end{bmatrix} \cdot \begin{bmatrix} r_n \\ \boldsymbol{r}_t \end{bmatrix} \mid -\mu r_n \geq \|\boldsymbol{r}_t\| \right\}. \tag{10.91}$$

The inclusion (10.90) can be regarded as being similar to the normality rule. Indeed, as illustrated in Figure 10.10, (10.90) means that, if the reaction $(r_n, \boldsymbol{r}_t)$ is on the boundary of the friction cone $\mathcal{K}$, the "modified" contact kinematic variable vector $(g + \mu\|\dot{\boldsymbol{u}}_t\|, -\dot{\boldsymbol{u}}_t)$ should be normal to the boundary of $\mathcal{K}$ at $(r_n, \boldsymbol{r}_t)$. Thus, in (10.90), *all* the kinematic contact variables are related to the reaction. Similarly, (10.91) can be regarded as a modification of the maximum dissipation law, which describes the normal and tangential contact laws simultaneously.

10.3.4.2 Modified normality rule for Coulomb friction law

We again consider only the Coulomb friction law. This means that we suppose $g = 0$ and that the unilateral contact condition should be treated separately. In this situation, we here aim to relate the evolution of $\boldsymbol{u}_t$ to the reaction $(r_n, \boldsymbol{r}_t)$ in a manner similar to the normality rule. Since $\dot{\boldsymbol{u}}_t \in \mathbb{R}^2$ and

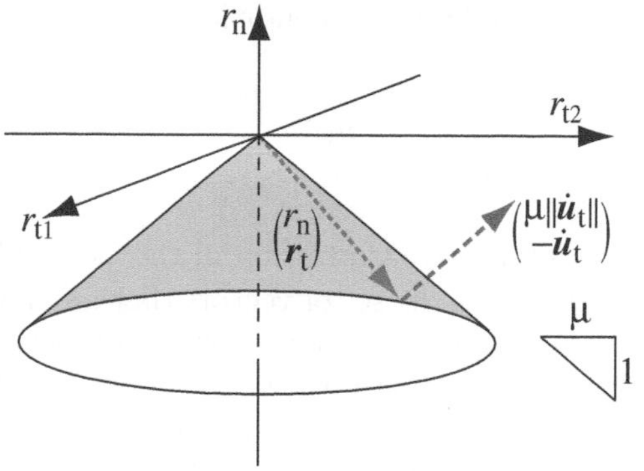

FIGURE 10.11: Schematic representation of the "modified" normality rule in (10.93) for the Coulomb friction law.

$(r_n, \boldsymbol{r}_t) \in \mathbb{R}^3$, we introduce a three-dimensional "modified" contact velocity vector below.

Retain $\mathcal{K}$ and $\mathcal{K}^*$ defined by (10.86) and (10.87), respectively, and recall from (10.89) that $\delta_{\mathcal{K}}$ and $\delta_{\mathcal{K}}^*$ are given by

$$\delta_{\mathcal{K}}(\boldsymbol{r}) = \begin{cases} 0 & \text{if } -\mu r_n \geq \|\boldsymbol{r}_t\|, \\ +\infty & \text{otherwise}, \end{cases}$$

$$\delta_{\mathcal{K}}^*(\boldsymbol{v}) = \begin{cases} 0 & \text{if } v_n \geq \mu\|-\boldsymbol{v}_t\|, \\ +\infty & \text{otherwise}. \end{cases}$$

Some direct calculations yield[19]

$$\partial\delta_{\mathcal{K}}(\boldsymbol{r}) = \begin{cases} \{\boldsymbol{0}\} & \text{if } -\mu r_n > \|\boldsymbol{r}_t\|, \\ \left\{\alpha \begin{bmatrix} \mu\|\boldsymbol{r}_t\| \\ \boldsymbol{r}_t \end{bmatrix} \mid \alpha \geq 0\right\} & \text{if } -\mu r_n = \|\boldsymbol{r}_t\| \neq 0, \\ \left\{\begin{bmatrix} v_n \\ \boldsymbol{v}_t \end{bmatrix} \mid v_n \geq \mu\|\boldsymbol{v}_t\|\right\} & \text{if } -\mu r_n = \|\boldsymbol{r}_t\| = 0, \\ \emptyset & \text{otherwise}. \end{cases} \tag{10.92}$$

[19]It follows from Proposition 2.1.12 that $\boldsymbol{v} \in \partial\delta_{\mathcal{K}}(\boldsymbol{r})$ if and only if $\delta_{\mathcal{K}}(\boldsymbol{r}) + \delta_{\mathcal{K}}^*(\boldsymbol{v}) = \boldsymbol{r}^{\mathrm{T}}\boldsymbol{v}$, i.e.,

$$-\mu r_n \geq \|\boldsymbol{r}_t\|, \quad v_n \geq \mu\|-\boldsymbol{v}_t\|, \quad \begin{bmatrix} -\mu r_n \\ \boldsymbol{r}_t \end{bmatrix} \cdot \begin{bmatrix} v_n \\ -\mu\boldsymbol{v}_t \end{bmatrix} = 0.$$

Thus, $(-\mu r_n, \boldsymbol{r}_t)$ and $(v_n, -\mu\boldsymbol{v}_t)$ are subjected to the complementarity condition over the second-order cone. In particular, if $-\mu r_n = \|\boldsymbol{r}_t\| \neq 0$, then $v_n = \mu\|-\boldsymbol{v}_t\|$ should hold, and $-\boldsymbol{v}_t$ should be parallel with $\boldsymbol{r}_t$ in the opposite direction. Thus we obtain (10.92).

By comparing (10.58) and (10.92), condition (10.57) in Proposition 10.3.3 can be rewritten as

$$\begin{bmatrix} \mu\|\dot{\boldsymbol{u}}_{\mathrm{t}}\| \\ -\dot{\boldsymbol{u}}_{\mathrm{t}} \end{bmatrix} \in \partial\delta_{\mathcal{K}}(\boldsymbol{r}). \tag{10.93}$$

Thus (10.93) is an alternative expression of the Coulomb friction law.

Geometrically, $\partial\delta_{\mathcal{K}}(\boldsymbol{r})$ is the set of vectors that are normal to the boundary surface of the friction cone $\mathcal{K}$. Therefore, as illustrated in Figure 10.11, condition (10.93) asserts that the vector $(\mu\|\dot{\boldsymbol{u}}_{\mathrm{t}}\|, -\dot{\boldsymbol{u}}_{\mathrm{t}}) \in \mathbb{R}^3$, which may be regarded as a *modified* contact velocity vector, should be normal to the boundary surface of the friction cone. Thus in (10.93) the evolution of $\boldsymbol{u}_{\mathrm{t}}$ is related to the reaction $(r_{\mathrm{n}}, \boldsymbol{r}_{\mathrm{t}})$ in a manner similar to the normality rule. Note again that, unlike the result in section 10.3.4.1, the unilateral contact law is not dealt with in (10.93).

10.3.4.3 Bi-potential for Coulomb friction law

Based on the result (10.93) in section 10.3.4.2, we derive a generalization of potential energy, called the bi-potential, for representing the contact law with the Coulomb friction in a variational form.

Recall that the unilateral contact law is originally given by (10.21), which is stated for the normal contact *displacement* g. In contrast, the unilateral contact law in the *velocity* level is written as

$$g = 0 \quad \Rightarrow \quad \dot{g} \geq 0,\ r_{\mathrm{n}} \leq 0,\ \dot{g} r_{\mathrm{n}} = 0. \tag{10.94}$$

In what follows, we suppose that $g = 0$. We aim to derive a "modified" potential energy for representing the Coulomb friction law together with the unilateral contact law (10.94) in the velocity level.[20]

We begin by rewriting (10.93) as

$$\begin{bmatrix} \dot{g} + \mu\|\dot{\boldsymbol{u}}_{\mathrm{t}}\| \\ -\dot{\boldsymbol{u}}_{\mathrm{t}} \end{bmatrix} \in \partial\delta_{\mathcal{K}}(\boldsymbol{r}). \tag{10.95}$$

One can easily verify from Figure 10.11 and (10.92) that (10.95) implies

$$\begin{aligned} \dot{g} > 0 \quad &\Rightarrow \quad \boldsymbol{r} = \boldsymbol{0}, \\ \boldsymbol{r} \neq \boldsymbol{0} \quad &\Rightarrow \quad \dot{g} = 0. \end{aligned}$$

Therefore, (10.95) expresses the Coulomb friction law and the unilateral contact law in the velocity level simultaneously.

[20] In contrast, in section 10.3.4.1 we considered the unilateral contact law in the *displacement* level, i.e., (10.21). Therefore, the result (10.95) is not restricted to the case of $g = 0$.

We next rewrite (10.95) as variational forms. It follows from the definition of the subdifferential that (10.95) is equivalent to

$$\delta_{\mathcal{K}}(\boldsymbol{r}') - \delta_{\mathcal{K}}(\boldsymbol{r}) \geq \begin{bmatrix} \dot{g} + \mu\|\dot{\boldsymbol{u}}_{\rm t}\| \\ -\dot{\boldsymbol{u}}_{\rm t} \end{bmatrix} \cdot \begin{bmatrix} r'_{\rm n} - r_{\rm n} \\ \boldsymbol{r}'_{\rm t} - \boldsymbol{r}_{\rm t} \end{bmatrix} \quad (\forall \boldsymbol{r}' \in \mathbb{R}^3),$$

i.e., for any $\boldsymbol{r}' \in \mathbb{R}^3$ the inequality

$$(\delta_{\mathcal{K}}(\boldsymbol{r}') - \mu r'_{\rm n}\|\dot{\boldsymbol{u}}_{\rm t}\|) - (\delta_{\mathcal{K}}(\boldsymbol{r}) - \mu r_{\rm n}\|\dot{\boldsymbol{u}}_{\rm t}\|) \geq \begin{bmatrix} \dot{g} \\ -\dot{\boldsymbol{u}}_{\rm t} \end{bmatrix} \cdot \begin{bmatrix} r'_{\rm n} - r_{\rm n} \\ \boldsymbol{r}'_{\rm t} - \boldsymbol{r}_{\rm t} \end{bmatrix} \tag{10.96}$$

holds. On the other hand, (10.95) is alternatively written as

$$\begin{bmatrix} r_{\rm n} \\ \boldsymbol{r}_{\rm t} \end{bmatrix} \in \partial\delta^*_{\mathcal{K}}(\dot{g} + \mu\|\dot{\boldsymbol{u}}_{\rm t}\|, -\dot{\boldsymbol{u}}_{\rm t}). \tag{10.97}$$

From the definition of the subdifferential and simple calculations, we see that (10.97) is satisfied if and only if, for any $(g', \dot{\boldsymbol{u}}'_{\rm t}) \in \mathbb{R}^3$, the inequality

$$(\delta_{\mathbb{R}_+}(\dot{g}') - \mu r_{\rm n}\|\dot{\boldsymbol{u}}'_{\rm t}\|) - (\delta_{\mathbb{R}_+}(\dot{g}) - \mu r_{\rm n}\|\dot{\boldsymbol{u}}_{\rm t}\|) \geq \begin{bmatrix} \dot{g}' - \dot{g} \\ -\dot{\boldsymbol{u}}'_{\rm t} + \dot{\boldsymbol{u}}_{\rm t} \end{bmatrix} \cdot \begin{bmatrix} r_{\rm n} \\ \boldsymbol{r}_{\rm t} \end{bmatrix} \tag{10.98}$$

holds. Define $b : \mathbb{R}^3 \times \mathbb{R}^3 \to \mathbb{R} \cup \{+\infty\}$ by

$$\begin{aligned} b(\dot{g}, -\dot{\boldsymbol{u}}_{\rm t}; \boldsymbol{r}) &= \delta^*_{\mathcal{K}}(\dot{g} + \mu\|\dot{\boldsymbol{u}}_{\rm t}\|, -\dot{\boldsymbol{u}}_{\rm t}) + \delta_{\mathcal{K}}(\boldsymbol{r}) - \mu r_{\rm n}\|\dot{\boldsymbol{u}}_{\rm t}\| \\ &= \delta_{\mathbb{R}_+}(\dot{g}) + \delta_{\mathcal{K}}(\boldsymbol{r}) - \mu r_{\rm n}\|\dot{\boldsymbol{u}}_{\rm t}\|. \end{aligned} \tag{10.99}$$

Then the left-hand side of (10.96) can be written as $b(\dot{g}, -\dot{\boldsymbol{u}}_{\rm t}; \boldsymbol{r}') - b(\dot{g}, -\dot{\boldsymbol{u}}_{\rm t}; \boldsymbol{r})$, while the left-hand side of (10.98) can be written as $b(\dot{g}', -\dot{\boldsymbol{u}}'_{\rm t}; \boldsymbol{r}) - b(\dot{g}, -\dot{\boldsymbol{u}}_{\rm t}; \boldsymbol{r})$. Thus we have that

(a): (10.95) $\Leftrightarrow$

$$b(\dot{g}, -\dot{\boldsymbol{u}}_{\rm t}; \boldsymbol{r}') - b(\dot{g}, -\dot{\boldsymbol{u}}_{\rm t}; \boldsymbol{r}) \geq \begin{bmatrix} \dot{g} \\ -\dot{\boldsymbol{u}}_{\rm t} \end{bmatrix} \cdot \begin{bmatrix} r'_{\rm n} - r_{\rm n} \\ \boldsymbol{r}'_{\rm t} - \boldsymbol{r}_{\rm t} \end{bmatrix} \quad (\forall \boldsymbol{r}'), \tag{10.100}$$

(b): (10.97) $\Leftrightarrow$

$$b(\dot{g}', -\dot{\boldsymbol{u}}'_{\rm t}; \boldsymbol{r}) - b(\dot{g}, -\dot{\boldsymbol{u}}_{\rm t}; \boldsymbol{r}) \geq \begin{bmatrix} \dot{g}' - \dot{g} \\ -\dot{\boldsymbol{u}}'_{\rm t} + \dot{\boldsymbol{u}}_{\rm t} \end{bmatrix} \cdot \begin{bmatrix} r_{\rm n} \\ \boldsymbol{r}_{\rm t} \end{bmatrix} \quad (\forall \dot{g}',\ \dot{\boldsymbol{u}}'_{\rm t}). \tag{10.101}$$

Since (10.100) and (10.101) are considered as a generalization of the extremal relation expressing the constitutive law (see Remark 10.3.10 for detail), b defined by (10.99) is called *the bi-potential* for the contact condition with the Coulomb friction [109].

Remark 10.3.10. The relation between the (conventional) potentials and the bi-potential b is understood as follows. In general, the inequalities in (10.100) and (10.101) for $b : \mathbb{R}^n \times \mathbb{R}^n \to \mathbb{R} \cup \{+\infty\}$ are written as

$$b(\boldsymbol{x}'; \boldsymbol{s}) - b(\boldsymbol{x}; \boldsymbol{s}) \geq \langle \boldsymbol{s}, \boldsymbol{x}' - \boldsymbol{x} \rangle, \quad \forall \boldsymbol{x}' \in \mathbb{R}^n, \tag{10.102}$$

$$b(\boldsymbol{x}; \boldsymbol{s}') - b(\boldsymbol{x}; \boldsymbol{s}) \geq \langle \boldsymbol{s}' - \boldsymbol{s}, \boldsymbol{x} \rangle, \quad \forall \boldsymbol{s}' \in \mathbb{R}^n. \tag{10.103}$$

As a particular case, suppose that b can be additively decomposed as

$$b(\boldsymbol{x}, \boldsymbol{s}) = w(\boldsymbol{x}) + w^*(\boldsymbol{s}), \tag{10.104}$$

which is *not* the case with b in (10.99). If b can be written as (10.104), then (10.102) is reduced to $w(\boldsymbol{x}') - w(\boldsymbol{x}) \geq \langle \boldsymbol{s}, \boldsymbol{x}' - \boldsymbol{x} \rangle$ $(\forall \boldsymbol{x}' \in \mathbb{R}^n)$, i.e.,

$$\boldsymbol{s} \in \partial w(\boldsymbol{x}).$$

Similarly, (10.102) is reduced to

$$\boldsymbol{x} \in \partial w^*(\boldsymbol{s}).$$

Thus, if b is in the form of (10.104), then the extremal conditions (10.102) and (10.103) are reduced to the constitutive law, and thence b corresponds to the sum of strain energy and complementary strain energy. Moreover, the pair w and w^* satisfies the Fenchel–Young inequality (Fact 2.1.9), which is written by using (10.104) as

$$b(\boldsymbol{x}, \boldsymbol{s}) \geq \langle \boldsymbol{s}, \boldsymbol{x} \rangle, \quad \forall (\boldsymbol{x}, \boldsymbol{s}) \in \mathbb{R}^n \times \mathbb{R}^n. \tag{10.105}$$

The function b defined by (10.99) satisfies (10.105), although it cannot be represented in the form of (10.104). ∎

10.4 Notes

1. There have been many books concerning contact mechanics. Among them, one may mention Frémond [136], Laursen [253], and Wriggers [456] as contemporary ones, from both theoretical and numerical points of view. Some recent review articles are Barber–Ciavarella [29], Mijar–Arora [309], Nguyen [336], Pfeiffer–Glocker [364], and Popp [374]. A concise survey in Klarbring [227] is also excellent.

In this book, we do not deal at all with the variational inequality treatment of frictional contact problems (particularly in the continuum theory), although such an analysis is another major stream of research regarding contact problems; see Han–Sofonea [160], Haslinger [165], Hlaváček–Haslinger–Nečas–Lovíšek [183], Kikuchi–Oden [221], Shillor–Sofonea–Telega [410], and Sofonea–Matei [412].

2. Concerning the unilateral contact condition (Signorini condition), historical remarks are provided by Curnier [101], Gladwell [148], Shillor–Sofonea–Telega [410, Chap. 1], and Wriggers [456, Chap. 1]. The first authors to write about variational inequality formulations for contact problems are Duvaut–Lions [120]. See section 3.4 for more references on the unilateral contact condition.

The existence and the regularity of a solution for frictional contact problems is a very important topic in nonsmooth mechanics, although it is not discussed in this book at all; see Andersson [12, 13], Cocu–Pratt–Raous [91], Cocu–Rocca [92], Klarbring–Mikelić–Shillor [228, 229], Martins–Oden [300], Oden–Martins [342], Rocca [387], and Rochdi–Shillor–Sofonea [388] for contributions on the existence results, Andersson–Klarbring [14] for a survey article, and Shillor–Sofonea–Telega [410] and Haslinger [165] for more information.

3. The formulation of Coulomb's law based on the the complementarity condition over the second-order cones, i.e., the result of Proposition 10.1.1, is in Kanno–Martins–Pinto da Costa [202]. The proof has been adapted to the context of the present work. Problem (10.31) was solved in [202] by using the smoothing method for the second-order cone complementarity problems from Hayashi–Yamashita–Fukushima [170]; see section 6.2.3 for fundamentals of the algorithm.

Formulation (10.44) in section 10.3.1 is from Duvaut–Lions [120, section III 5.2]. The maximum dissipation law (10.65) in section 10.3.2, together with its optimality condition (10.65), is dealt with in Christensen–Pang [88] and Klarbring–Pang [230].

The projection operator, (10.77) in section 10.3.3, was shown to be B-differentiable (B for Bouligand) by Strömberg [422], in which the fretting problem was solved numerically by using the B-differentiable Newton method from Pang [357]. Later, this formulation was applied to the quasistatic frictional contact problems by Christensen–Klarbring–Pang–Strömberg [87]; see also Strömberg [423].

Formulation (10.82) in section 10.3.4.1 can be found in Martins–Pinto da Costa–Simões [301]. Use of this formulation for studying the stability of deformable bodies subjected to the frictional contact conditions can be found in Martins–Pinto da Costa [299] and Pinto da Costa–Martins–Figueiredo–Júdice [368].

The bi-potential formulation, say, (10.99), (10.100), and (10.101) in section 10.3.4.2, is in de Saxcé–Feng [109]. More numerical aspects based on the bi-potential formulation can be found in Fortin–Millet–de Saxće [133] and Hjiaj–Feng–de Saxcé–Mróz [181].

4. In section 10.2 we have treated only the incremental problem, which is to be solved to analyze quasistatic problems. For the planar problem, the LCP formulation (10.42) presented in Remark 10.2.4 is important for the solution analysis of the incremental problem of frictional contacts. This formulation is from Klarbring [227], which was inspired by the formulation in Stewart–Trinkle [421] for the dynamic contact problems of rigid bodies.

Another problem of importance is the *rate problem*, with which we can investigate the bifurcation and instability along the equilibrium path; see Chateau–Nguyen [73], Klarbring [225, 226], Klarbring–Mikelić–Shillor [228], Martins–Barbarin–Raous–Pinto da Costa [298], and Pinto da Costa–Martins [366, 367]. The rate problem does not necessarily have a solution, while the incremental problem always has (but possibly multiple) solutions; see Ballard [28], Cocu–Pratt–Raous [91], Hassani–Hild–Ionescu [166], and Klarbring [224, 225].

As an interesting consequence of non-uniqueness of the equilibrium path under the friction, even for the small deformation, a linear elastic structure in contact with a fixed rigid obstacle may possibly have nontrivial (i.e., with the nonzero internal forces) equilibrium states; such an equilibrium configuration is called the *wedged configuration*, while a problem for finding wedged configurations is called the *wedging problem*. The wedging problem has received recent attention, initiated by Barber–Hild [30, 31]. A genetic algorithm for finding the critical value of the friction coefficient, for which wedged configurations can exist, was proposed by Hassani–Ionescu–Oudet [167], while an enumeration algorithm of all the wedged configurations is in Fujita–Kanno [139].

Chapter 11

Plasticity

Orientation As the final topic in nonsmooth mechanics, this chapter takes up the elastoplastic problem, which was briefly mentioned in the Preface of this book. Specifically, we shall see that the von Mises yield criterion can be represented by using a second-order cone constraint and that the associated flow rule can be formulated as a complementarity condition over the second-order cone. As a consequence, the incremental problem for quasistatic analysis can be formulated as a primal-dual pair of the SOCP (second-order cone programming) problems.

All the technical results in this chapter can be shown using the Fenchel duality theory, like in Chapter 8. In this chapter, we adopt, however, the Lagrangian duality theory introduced in section 2.2.6 to show its usage.

11.1 Fundamentals of Plasticity

Elastoplastic response of solids consists of elastic and plastic deformations, which are completely distinguishable from each other: An elastic deformation is reversible, while a plastic deformation is irreversible. Transition from a reversible deformation to a deformation involving an irreversible process, and vice versa, is thus the resource of nonsmoothness property in an elastoplastic problem, as seen below.

The elastoplastic property is governed by the yield condition and the flow rule. To describe the yield condition, we begin by introducing the principal invariants of a 3 × 3 symmetric matrix. Note that a second-order symmetric tensor in the three-dimensional space is identified with a 3 × 3 symmetric matrix, as is done in Chapter 8.

The *characteristic polynomial* of $\boldsymbol{\tau} \in \mathbb{S}^3$ is written as

$$\det(\boldsymbol{\tau} - \lambda \boldsymbol{I}) = -\lambda^3 + \imath_1 \lambda^2 + \imath_2 \lambda + \imath_3,$$

where $\boldsymbol{I} \in \mathbb{S}^3$ is the unit tensor. The *principal invariants* of $\boldsymbol{\tau}$ are the three coefficients of this polynomial, i.e., $\imath_1(\boldsymbol{\tau})$, $\imath_2(\boldsymbol{\tau})$, and $\imath_3(\boldsymbol{\tau})$. It follows immediately from the definition that the principal invariants are explicitly written as

$$\imath_1(\boldsymbol{\tau}) = \mathrm{tr}(\boldsymbol{\tau}), \tag{11.1a}$$

$$\imath_2(\boldsymbol{\tau}) = \frac{1}{2}\left[\mathrm{tr}(\boldsymbol{\tau}^2) - (\mathrm{tr}(\boldsymbol{\tau}))^2\right], \tag{11.1b}$$

$$\imath_3(\boldsymbol{\tau}) = \det \boldsymbol{\tau}. \tag{11.1c}$$

Let λ_1, λ_2, λ_3 denote the eigenvalues of $\boldsymbol{\tau}$, i.e., the solutions of the characteristic equation $\det(\boldsymbol{\tau} - \lambda \boldsymbol{I}) = 0$. By using the relation between roots and coefficients of a polynomial, (11.1) is alternatively written in terms of the eigenvalues as

$$\imath_1(\boldsymbol{\tau}) = \lambda_1 + \lambda_2 + \lambda_3, \tag{11.2a}$$

$$\imath_2(\boldsymbol{\tau}) = -\lambda_1\lambda_2 - \lambda_2\lambda_3 - \lambda_3\lambda_1, \tag{11.2b}$$

$$\imath_3(\boldsymbol{\tau}) = \lambda_1\lambda_2\lambda_3. \tag{11.2c}$$

It is clearly seen in (11.2) that the principal invariants are preserved through the orthogonal transformation of $\boldsymbol{\tau}$, because the eigenvalues do not change through the orthogonal transformation. Note that (11.1b) is also rewritten in terms of the components of $\boldsymbol{\tau}$ as

$$\imath_2(\boldsymbol{\tau}) = \tau_{12}^2 + \tau_{23}^2 + \tau_{31}^2 - \tau_{11}\tau_{22} - \tau_{22}\tau_{33} - \tau_{33}\tau_{11}, \tag{11.3}$$

which is an expression often found in the plasticity theory literature.

The *deviator* of $\boldsymbol{\tau} \in \mathbb{S}^3$, denoted $\mathrm{dev}(\boldsymbol{\tau})$, is defined by

$$\mathrm{dev}(\boldsymbol{\tau}) = \boldsymbol{\tau} - \frac{1}{3}\mathrm{tr}(\boldsymbol{\tau})\boldsymbol{I}. \tag{11.4}$$

From this definition, we immediately see that $\mathrm{dev}(\mathrm{dev}(\boldsymbol{\tau})) = \mathrm{dev}(\boldsymbol{\tau})$ for any $\boldsymbol{\tau} \in \mathbb{S}^3$. Particularly, for the stress tensor $\boldsymbol{\sigma}$, $\mathrm{dev}(\boldsymbol{\sigma}) \in \mathbb{S}^3$ is called the *deviatoric stress tensor*, while $\frac{1}{3}\mathrm{tr}(\boldsymbol{\sigma})\boldsymbol{I}$ is called the *volumetric stress tensor* (or *mean normal stress tensor*). Moreover, $\frac{1}{3}\mathrm{tr}(\boldsymbol{\sigma}) = \frac{1}{3}\imath_1(\boldsymbol{\sigma})$ is sometimes called the *mean normal stress*. Accordingly, a stress tensor $\boldsymbol{\sigma}$ is additively decomposed to the deviatoric and volumetric stress tensors as

$$\boldsymbol{\sigma} = \mathrm{dev}(\boldsymbol{\sigma}) + \frac{1}{3}\mathrm{tr}(\boldsymbol{\sigma})\boldsymbol{I}.$$

Let $\jmath_r(\boldsymbol{\tau})$ $(r = 1, 2, 3)$ denote the principal invariants of the deviator of $\boldsymbol{\tau} \in \mathbb{S}^3$, i.e.,

$$\jmath_r(\boldsymbol{\tau}) = \imath_r(\mathrm{dev}(\boldsymbol{\tau})), \quad r = 1, 2, 3. \tag{11.5}$$

By substituting (11.4) into (11.5) and by using (11.1), we obtain

$$\jmath_1(\boldsymbol{\tau}) = 0, \tag{11.6a}$$

$$\jmath_2(\boldsymbol{\tau}) = \frac{1}{2}\mathrm{dev}(\boldsymbol{\tau}) : \mathrm{dev}(\boldsymbol{\tau}), \tag{11.6b}$$

$$\jmath_3(\boldsymbol{\tau}) = \det \mathrm{dev}(\boldsymbol{\tau}). \tag{11.6c}$$

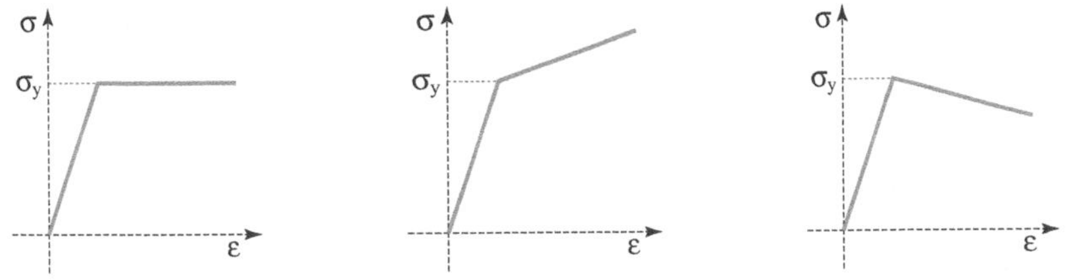

(a) Perfect plasticity. (b) Strain hardening. (c) Strain softening.

FIGURE 11.1: The stress–strain curve of a one-dimensional model. (a) a perfect plasticity model; (b) a strain-hardening model; (c) a strain-softening model.

It is common in plasticity theory to postulate that the stress tensor $\boldsymbol{\sigma}$ is always included in a closed convex set, which is written as

$$\Sigma_{\mathrm{ad}} = \{\boldsymbol{\sigma} \in \mathbb{S}^3 \mid f(\boldsymbol{\sigma}) \leq 0\}. \tag{11.7}$$

The function $f : \mathbb{S}^3 \to \mathbb{R}$ defining Σ_{ad} is called the *yield function*. The stress tensor $\boldsymbol{\sigma}$ is said to be an *admissible stress* if $\boldsymbol{\sigma} \in \Sigma_{\mathrm{ad}}$. The inequality $f(\boldsymbol{\sigma}) \leq 0$ involved in (11.7) is sometimes called the *yield condition*, although it rather defines the admissible set of stress tensors.

The boundary of Σ_{ad}, i.e.,

$$\operatorname{bd} \Sigma_{\mathrm{ad}} = \{\boldsymbol{\sigma} \in \mathbb{S}^3 \mid f(\boldsymbol{\sigma}) = 0\}, \tag{11.8}$$

is called the *yield surface*. The *purely elastic behavior* is characterized by $\boldsymbol{\sigma} \in \operatorname{int} \Sigma_{\mathrm{ad}}$, while the *plastic loading* can take place only if $\boldsymbol{\sigma} \in \operatorname{bd} \Sigma_{\mathrm{ad}}$.[1] More precisely, the material behavior is said to be plastic loading if $\boldsymbol{\sigma}$ is on $\operatorname{bd} \Sigma_{\mathrm{ad}}$ and continues to stay on $\operatorname{bd} \Sigma_{\mathrm{ad}}$ during the loading step under consideration. In contrast, if $\boldsymbol{\sigma} \in \operatorname{bd} \Sigma_{\mathrm{ad}}$ moves toward $\operatorname{int} \Sigma_{\mathrm{ad}}$, then the corresponding behavior is called *elastic unloading*.

Figure 11.1 depicts typical stress–strain curves of a one-dimensional elasto-plastic model. For a while, σ and ε are considered to be one-dimensional variables. In the case of Figure 11.1(a), the yield function is defined by

$$f(\sigma) = |\sigma| - \sigma_{\mathrm{y}} \tag{11.9}$$

with a constant $\sigma^{\mathrm{y}} > 0$ called the *flow stress*. It is assumed in Figure 11.1(a) that the yield function is not affected by plastic deformation at all and is always given by (11.9). This is the case of *perfect plasticity*. In various metals it is experimentally observed that the stress increases in the course of plastic

[1] It should be clear that $\boldsymbol{\sigma} \in \Sigma_{\mathrm{ad}}$ must be satisfied even after the plastic loading occurs.

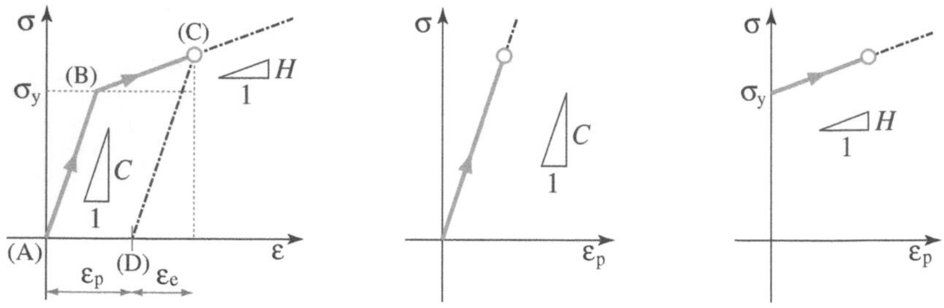

(a) Stress–strain curve. (b) Elastic strain. (c) Plastic strain.

FIGURE 11.2: Decomposition of the strain $\boldsymbol{\varepsilon}$ into the elastic part $\boldsymbol{\varepsilon}_{\rm e}$ and the plastic part $\boldsymbol{\varepsilon}_{\rm p}$. "o": the current point.

loading, as illustrated in Figure 11.1(b). This phenomenon is called the *strain hardening* behavior. It should be clear that some parameters involved in the yield function f vary due to the strain hardening, but the stress is still required to satisfy $f(\sigma) \le 0$. In materials such as soil and concrete, it is often that the stress decreases with the progress of plastic deformation, as shown in Figure 11.1(c). This behavior is known as *strain softening*.

As a simple model of strain hardening, consider the yield function given by

$$f(\sigma) = |\sigma - \beta| - R, \tag{11.10}$$

instead of (11.9). Here, $R > 0$ represents the radius of the yield surface, while the center of the yield surface is shifted by β, which is called the *back stress*. We can describe the response in Figure 11.1(b) by considering R and β increase in the course of plastic loading, where we set $R = \sigma_{\rm y}$ and $\beta = 0$ at the initial state. Suppose that the evolution of β can be ignored, and only the increase of R is considered. This strain-hardening rule is known as the *isotropic hardening*, in which the center of the yield surface remains at the origin. In contrast, when R is assumed to be constant and the evolution of β is considered, the material is said to undergo the *kinematic hardening*.

Consider a three-dimensional model again. The strain tensor $\boldsymbol{\varepsilon} \in \mathbb{S}^3$ is additively decomposed to the elastic part $\boldsymbol{\varepsilon}_{\rm e} \in \mathbb{S}^3$ and the plastic part $\boldsymbol{\varepsilon}_{\rm p} \in \mathbb{S}^3$, corresponding to the reversible and irreversible parts, respectively. In addition, we postulate that the stress depends linearly on the elastic strain as

$$\boldsymbol{\sigma} = \boldsymbol{C} : \boldsymbol{\varepsilon}_{\rm e}, \tag{11.11}$$

where $\boldsymbol{C}$ is the elasticity tensor. To fix the meaning of this decomposition, suppose that the stress–strain curve is given as shown in Figure 11.2(a), and consider the decomposition of the strain at the state (C). The material undergoes the purely elastic behavior ((A)→(B)) and the plastic loading ((B)→(C)).

If the stress decreases from the point (C), then the elastic unloading occurs, indicated as (C)→(D). The residual strain at the state (D), at which the stress becomes equal to zero, coincides with the plastic strain $\boldsymbol{\varepsilon}_{\mathrm{p}}$ at the state (C). The elastic strain $\boldsymbol{\varepsilon}_{\mathrm{e}}$ at the state (C) is then determined by $\boldsymbol{\varepsilon}_{\mathrm{e}} = \boldsymbol{\varepsilon} - \boldsymbol{\varepsilon}_{\mathrm{p}}$, as illustrated in Figure 11.2.

We next consider a postulate on the evolution rule of $\boldsymbol{\varepsilon}_{\mathrm{p}}$. Let $\dot{\boldsymbol{\varepsilon}}_{\mathrm{p}} \in \mathbb{S}^3$ denote the rate of plastic strain tensor (or the plastic flow), where the standard dot notation for the time derivative is used. As mentioned before, if $\boldsymbol{\sigma} \in \operatorname{int} \Sigma_{\mathrm{ad}}$, then $\dot{\boldsymbol{\varepsilon}}_{\mathrm{p}} = \boldsymbol{o}$. For a given $\boldsymbol{\sigma} \in \operatorname{bd} \Sigma_{\mathrm{ad}}$, a postulate that determines the direction of the plastic strain rate is known as a *flow rule*. For most metals, the so-called *associated flow rule* can be accepted, which is stated as

$$\exists \alpha \geq 0: \quad \dot{\boldsymbol{\varepsilon}}_{\mathrm{p}} = \alpha \frac{\partial f(\boldsymbol{\sigma})}{\partial \boldsymbol{\sigma}}; \tag{11.12}$$

i.e., the direction of the flow $\boldsymbol{\varepsilon}_{\mathrm{p}}$ is determined associated with the yield function f. Condition (11.12) is also known as the *normality rule*, because it requires that $\dot{\boldsymbol{\varepsilon}}_{\mathrm{p}}$ is in the outer normal direction of the yield surface at $\boldsymbol{\sigma}$. Furthermore, we can also view (11.12) as a consequence of an alternative postulate, known as the *maximum dissipation law*. This is a postulate that, for a given $\dot{\boldsymbol{\varepsilon}}_{\mathrm{p}} \in \mathbb{S}^3$, the corresponding stress $\boldsymbol{\sigma}$ is the one that maximizes the rate of plastic work $\boldsymbol{\sigma} : \dot{\boldsymbol{\varepsilon}}_{\mathrm{p}}$ among admissible stresses. Formally, the maximum dissipation law is written as

$$\boldsymbol{\sigma} \in \arg\max_{\check{\boldsymbol{\sigma}} \in \mathbb{S}^3} \{ \check{\boldsymbol{\sigma}} : \dot{\boldsymbol{\varepsilon}}_{\mathrm{p}} \mid f(\check{\boldsymbol{\sigma}}) \leq 0 \}. \tag{11.13}$$

Provided that f is a differentiable convex function, one can show for $\dot{\boldsymbol{\varepsilon}}_{\mathrm{p}} \neq \boldsymbol{o}$ that (11.13) is equivalent to (11.12).

It is experimentally observed in most metals that the yield function $f(\boldsymbol{\sigma})$ does not depend on the mean normal stress. Moreover, for an isotropic material, $f(\boldsymbol{\sigma})$ should be invariant under the orthogonal transformation of $\boldsymbol{\sigma}$. Therefore, without loss of generality we can consider that $f(\boldsymbol{\sigma})$ depends solely on $\jmath_2(\boldsymbol{\sigma})$ and $\jmath_3(\boldsymbol{\sigma})$ defined by (11.6).

As a simple model of the yield function f, the *von Mises yield criterion* assumes that f is written as[2]

$$\begin{aligned} f(\boldsymbol{\sigma}) &= \sqrt{2\jmath_2(\boldsymbol{\sigma})} - \sqrt{\tfrac{2}{3}}\sigma_{\mathrm{y}} \\ &= \| \operatorname{dev}(\boldsymbol{\sigma}) \| - \sqrt{\tfrac{2}{3}}\sigma_{\mathrm{y}}, \end{aligned} \tag{11.14}$$

[2]In expression (11.14), σ_{y} corresponds to the yield stress subjected to the uniaxial load, because, for the stress tensor

$$\boldsymbol{\sigma}_{\mathrm{uni}} = \begin{bmatrix} \sigma_{\mathrm{y}} & 0 & 0 \\ 0 & 0 & 0 \\ 0 & 0 & 0 \end{bmatrix}$$

corresponding to the uniaxial loading, we have that $\| \operatorname{dev}(\boldsymbol{\sigma}_{\mathrm{uni}}) \| = \sqrt{2/3}\sigma_{\mathrm{y}}$.

where $\|\mathrm{dev}(\boldsymbol{\sigma})\|$ is the Frobenius norm of $\mathrm{dev}(\boldsymbol{\sigma}) \in \mathbb{S}^3$, i.e., $\|\mathrm{dev}(\boldsymbol{\sigma})\| = \sqrt{\mathrm{dev}(\boldsymbol{\sigma}) : \mathrm{dev}(\boldsymbol{\sigma})}$. Since $f(\boldsymbol{\sigma})$ in (11.14) depend solely on $\jmath_2(\boldsymbol{\sigma})$, the elastoplastic model based on the associated flow rule (11.12) and the von Mises yield criterion (11.14) is sometimes called the J_2-flow theory.

The effect of strain hardening is incorporated in the von Mises yield criterion by modifying (11.14) slightly as

$$f(\boldsymbol{\sigma}) = \|\boldsymbol{\eta}\| - \sqrt{\tfrac{2}{3}} R, \tag{11.15a}$$

$$\boldsymbol{\eta} = \mathrm{dev}(\boldsymbol{\sigma}) - \boldsymbol{\beta}, \tag{11.15b}$$

where $R > 0$ represents the radius of the von Mises yield surface, and the *back stress* tensor $\boldsymbol{\beta} \in \mathbb{S}^3$ defines the center of the von Mises yield surface. From a physical point of view, the back stress tensor $\boldsymbol{\beta} \in \mathbb{S}^3$ is regarded as an internal static variable representing the effect of the internal restructuring that takes place due to plastic deformation.

In what follows we investigate second-order cone complementarity formulations of the plasticity laws with the von Mises yield criterion; in section 11.2 we deal with the perfect plasticity theory, while the isotropic and kinematic strain-hardening effects are the subjects of section 11.3 and section 11.4, respectively.

11.2 Perfect Plasticity

Perfect plasticity is an idealization of the elastoplastic behavior in which we assume that the yield function does not change even after the material undergoes plastic deformation. In this section we show that the associated flow rule for the perfect plasticity with the von Mises criterion can be formulated as a complementarity condition over second-order cone.

11.2.1 Classical formulation of flow rule in perfect plasticity

We here present a conventional formulation of the perfect plasticity assuming the von Mises yield criterion and the associated flow rule.

Recall that the von Mises yield criterion is defined by (11.15), which involves the parameters R and $\boldsymbol{\beta}$ related to the strain-hardening effect. The perfect plasticity theory ignores the evolution of R and $\boldsymbol{\beta}$, i.e.,

$$\boldsymbol{\beta} = \boldsymbol{o}, \quad R = \sigma_{\mathrm{y}}. \tag{11.16}$$

Therefore, from (11.15b) we obtain

$$\boldsymbol{\eta} = \mathrm{dev}(\boldsymbol{\sigma}). \tag{11.17}$$

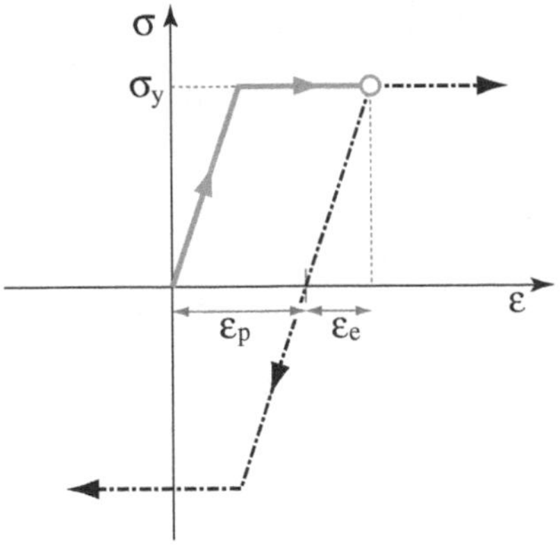

FIGURE 11.3: Constitutive relation of a one-dimensional model of perfect plasticity. "o": the current point.

Although we suppose (11.16) throughout section 11.2, we use expression (11.15a) of f with R and $\boldsymbol{\eta}$ to provide general formulations that work in the subsequent sections. Note that the evolution of R is dealt with in section 11.3, while the evolution of $\boldsymbol{\beta}$ is considered in section 11.4. Figure 11.3 shows a schematic representation of the response of a one-dimensional model of the perfect plasticity.

In accordance with definition (11.7) of the admissible set of stress tensors, the inequality

$$f(\boldsymbol{\sigma}) \leq 0 \tag{11.18}$$

should be satisfied. The distinctive condition between the purely elastic behavior and the plastic loading is then given by

$$\begin{cases} f(\boldsymbol{\sigma}) < 0 & \Rightarrow \quad \dot{\boldsymbol{\varepsilon}}_{\mathrm{p}} = \boldsymbol{o}, \\ \dot{\boldsymbol{\varepsilon}}_{\mathrm{p}} \neq \boldsymbol{o} & \Rightarrow \quad f(\boldsymbol{\sigma}) = 0. \end{cases} \tag{11.19}$$

To obtain the (classical) complementarity condition concerning the yield law, define $\dot{\gamma}$ by

$$\gamma = \|\boldsymbol{\varepsilon}_{\mathrm{p}}\|;$$

i.e., γ is the modulus of the plastic strain. The time derivative of γ is then written as[3]

$$\dot{\gamma} = \|\dot{\boldsymbol{\varepsilon}}_{\mathrm{p}}\|, \tag{11.20}$$

[3] Accumulation of $\sqrt{\frac{2}{3}}\dot{\gamma}$, i.e., $\displaystyle\int_0^{\tau} \sqrt{\frac{2}{3}} \|\dot{\boldsymbol{\varepsilon}}_{\mathrm{p}}\| \mathrm{d}\tau$, is called the *equivalent plastic strain* or the *effective plastic strain*.

from which we obtain the inclusion

$$\dot{\gamma} \geq 0. \tag{11.21}$$

By using (11.20) and (11.21), condition (11.19) can be reduced to

$$\begin{cases} f(\boldsymbol{\sigma}) < 0 & \Rightarrow \quad \dot{\gamma} = 0, \\ \dot{\gamma} > 0 & \Rightarrow \quad f(\boldsymbol{\sigma}) = 0. \end{cases} \tag{11.22}$$

Consequently, the yield condition given in (11.18), (11.21), and (11.22) is equivalently rewritten as the complementarity condition between $f(\boldsymbol{\sigma})$ and $\dot{\gamma}$, i.e.,[4]

$$\|\boldsymbol{\eta}\| - \sqrt{\tfrac{2}{3}}R \leq 0, \tag{11.23a}$$

$$\dot{\gamma} \geq 0, \tag{11.23b}$$

$$\dot{\gamma}\left(\|\boldsymbol{\eta}\| - \sqrt{\tfrac{2}{3}}R\right) = 0. \tag{11.23c}$$

The associated flow rule, given by (11.12), is explicitly written for the von Mises yield criterion by using (11.15a) and (11.20) as[5]

$$\dot{\boldsymbol{\varepsilon}}_{\mathrm{p}} = \dot{\gamma}\frac{\boldsymbol{\eta}}{\|\boldsymbol{\eta}\|}, \tag{11.24}$$

where $\dot{\gamma}$ should satisfy (11.23).[6] Note that we use the convention $\boldsymbol{o}/\boldsymbol{o} = \boldsymbol{o}$ in (11.24), because from $R = \sigma_{\mathrm{y}} > 0$ we see that $\boldsymbol{\sigma} \in \operatorname{bd} \Sigma_{\mathrm{ad}}$ implies $\boldsymbol{\eta} \neq \boldsymbol{o}$. Thus, as well known in the plasticity theory literature (e.g., [413]), the perfect plasticity model is formulated as (11.23) and (11.24). In expression (11.24), the variable $\dot{\gamma}$ is often called the *consistency parameter*.

11.2.2 Second-order cone complementarity formulation

In this section, we show that the conventional formulation of the perfect plasticity in (11.23) and (11.24) can be restated as a complementarity condition over second-order cone. A key idea of this reduction is to see in (11.24)

[4]Note again that we suppose (11.16) and (11.17) in section 11.2 but use R and $\boldsymbol{\eta}$ for consistency with the subsequent sections.

[5]To see this, use (11.15) and (11.16) to rewrite f as

$$f(\boldsymbol{\sigma}) = \|\operatorname{dev}(\boldsymbol{\sigma})\| = \left[(\boldsymbol{\sigma} - \tfrac{1}{3}\operatorname{tr}(\boldsymbol{\sigma})) : (\boldsymbol{\sigma} - \tfrac{1}{3}\operatorname{tr}(\boldsymbol{\sigma}))\right]^{1/2} = \left[\boldsymbol{\sigma} : \boldsymbol{\sigma} - \tfrac{1}{3}(\boldsymbol{I} : \boldsymbol{\sigma})^2\right]^{1/2},$$

from which we obtain

$$\frac{\partial}{\partial \boldsymbol{\sigma}} f(\boldsymbol{\sigma}) = \frac{\boldsymbol{\sigma} - \frac{1}{3}(\boldsymbol{I} : \boldsymbol{\sigma})\boldsymbol{I}}{\left[\boldsymbol{\sigma} : \boldsymbol{\sigma} - \frac{1}{3}(\boldsymbol{I} : \boldsymbol{\sigma})\right]^{1/2}} = \frac{\operatorname{dev}(\boldsymbol{\sigma})}{\|\operatorname{dev}(\boldsymbol{\sigma})\|}.$$

[6]Condition (11.12), together with (11.19), is alternatively written as $\dot{\boldsymbol{\varepsilon}}_{\mathrm{p}} \in \partial\delta_{\Sigma_{\mathrm{ad}}}(\boldsymbol{\sigma})$, where $\delta_{\Sigma_{\mathrm{ad}}} : \mathbb{S}^3 \rightarrow \mathbb{R} \cup \{+\infty\}$ is the indicator function of Σ_{ad}.

that $\dot{\varepsilon}_{\rm p}$ is required to be parallel with $\boldsymbol{\eta}$, while a similar parallelism condition is implied for two vectors subjected to a complementarity condition over second-order cone as investigated in section 1.4.3. Recall that $(x_0, \boldsymbol{x}_1) \in \mathbb{R} \times \mathbb{R}^{n-1}$ and $(s_0, \boldsymbol{s}_1) \in \mathbb{R} \times \mathbb{R}^{n-1}$ satisfy

$$x_0 \geq \|\boldsymbol{x}_1\|, \quad s_0 \geq \|\boldsymbol{s}_1\|, \quad x_0 s_0 + \boldsymbol{x}_1^{\rm T} \boldsymbol{s}_1 = 0 \qquad \text{(see (1.23))}$$

if and only if they satisfy (1.24) and (1.25) and that condition (1.25c) corresponds to the parallelism condition of $\boldsymbol{x}_1$ and $\boldsymbol{s}_1$.

Conventionally the second-order cone constraint is defined as an inequality between a scalar x_0 and the Euclidean norm $\|\boldsymbol{x}_1\|$ of a vector $\boldsymbol{x}_1$. To treat the plasticity, we slightly generalize the notion of second-order cone constraint for a tensor $\boldsymbol{\xi}_1$ instead of a vector $\boldsymbol{x}_1$. Specifically, for $(x_0, \boldsymbol{\xi}_1) \in \mathbb{R} \times \mathbb{S}^3$ and $(z_0, \boldsymbol{\zeta}_1) \in \mathbb{R} \times \mathbb{S}^3$, we consider[7]

$$x_0 \geq \|\boldsymbol{\xi}_1\|, \quad z_0 \geq \|\boldsymbol{\zeta}_1\|, \quad x_0 z_0 + \boldsymbol{\xi}_1 : \boldsymbol{\zeta}_1 = 0, \tag{11.25}$$

where $\|\boldsymbol{\xi}_1\|$ is the Frobenius norm of the tensor (or the 3×3 matrix) $\boldsymbol{\xi}_1$, i.e., $\|\boldsymbol{\xi}_1\| = (\boldsymbol{\xi}_1 : \boldsymbol{\xi}_1)^{1/2} = \sqrt{\mathrm{tr}(\boldsymbol{\xi}_1^{\rm T} \boldsymbol{\xi}_1)}$ (see section 1.3.2). Throughout this chapter, we mean (11.25) by the complementarity condition over second-order cone, unless otherwise stated. In a manner similar to the conventional second-order cone inequality (see Fact 1.4.7 in section 1.4.3), the solution set of (11.25) is explicitly described as follows.[8]

Fact 11.2.1. $(x_0, \boldsymbol{\xi}_1) \in \mathbb{R} \times \mathbb{S}^3$ and $(z_0, \boldsymbol{\zeta}_1) \in \mathbb{R} \times \mathbb{S}^3$ satisfy (11.25) if and only if any one of the following six cases is true:

(i) $x_0 > \|\boldsymbol{\xi}_1\|$, $(z_0, \boldsymbol{\zeta}_1) = (0, \boldsymbol{o})$.

(ii) $(x_0, \boldsymbol{\xi}_1) = (0, \boldsymbol{o})$, $z_0 > \|\boldsymbol{\zeta}_1\|$.

(iii) $x_0 = \|\boldsymbol{\xi}_1\| \neq 0$, $z_0 = \|\boldsymbol{\zeta}_1\| \neq 0$, and there exists $\alpha > 0$ such that $\boldsymbol{\xi}_1 = -\alpha \boldsymbol{\zeta}_1$.

(iv) $x_0 = \|\boldsymbol{x}_1\| \neq 0$, $(z_0, \boldsymbol{\zeta}_1) = (0, \boldsymbol{o})$.

(v) $(x_0, \boldsymbol{\xi}_1) = (0, \boldsymbol{o})$, $z_0 = \|\boldsymbol{\zeta}_1\| \neq 0$.

(vi) $(x_0, \boldsymbol{\xi}_1) = (0, \boldsymbol{o})$, $(z_0, \boldsymbol{\zeta}_1) = (0, \boldsymbol{o})$. ■

---

[7]In fact, we need not restrict ourselves to symmetric matrices; the result of Fact 11.2.1 is true for $\boldsymbol{\xi}_1 \in \mathbb{R}^{3\times 3}$ and $\boldsymbol{\zeta}_1 \in \mathbb{R}^{3\times 3}$.

[8]To see the relation between Fact 1.4.7 and Fact 11.2.1, observe that the inner product of $\boldsymbol{\xi}_1 = (\xi_{ij}) \in \mathbb{R}^{3\times 3}$ and $\boldsymbol{\zeta}_1 = (\zeta_{ij}) \in \mathbb{R}^{3\times 3}$ is identified with the inner product of nine-dimensional vectors $\boldsymbol{x}_1 = (\xi_{11}, \xi_{12}, \dots, \xi_{33})^{\rm T}$ and $\boldsymbol{z}_1 = (\zeta_{11}, \zeta_{12}, \dots, \zeta_{33})^{\rm T}$, and thence the Frobenius norm $\|\boldsymbol{\xi}_1\|$ of the matrix $\boldsymbol{\xi}_1$ is identified with the Euclidean norm $\|\boldsymbol{x}_1\|$ of the vector $\boldsymbol{x}_1$.

Using Fact 11.2.1, the associated perfect plasticity law is reformulated as a complementarity condition over second-order cone as follows.

Proposition 11.2.2. *Let $R > 0$ be given. Then $(\dot{\gamma}, \dot{\boldsymbol{\varepsilon}}_{\rm p}) \in \mathbb{R} \times \mathbb{S}^3$ and $\boldsymbol{\eta} \in \mathbb{S}^3$ satisfy* (11.23) *and* (11.24) *if and only if they satisfy*

$$\dot{\gamma} \geq \|-\dot{\boldsymbol{\varepsilon}}_{\rm p}\|, \tag{11.26a}$$

$$\sqrt{\tfrac{2}{3}}R \geq \|\boldsymbol{\eta}\|, \tag{11.26b}$$

$$\dot{\gamma}\left(\sqrt{\tfrac{2}{3}}R\right) + (-\dot{\boldsymbol{\varepsilon}}_{\rm p}) : \boldsymbol{\eta} = 0. \tag{11.26c}$$

Proof. Among (11.23) and (11.24), observe that the admissibility condition of the stress, say, (11.23a), is explicitly included in (11.26) as (11.26b). Moreover, let

$$x_0 = \dot{\gamma}, \qquad \boldsymbol{\xi}_1 = -\dot{\boldsymbol{\varepsilon}}_{\rm p},$$
$$z_0 = \sqrt{\tfrac{2}{3}}R, \quad \boldsymbol{\zeta}_1 = \boldsymbol{\eta}$$

to see that condition (11.26) is encoded in the form of (11.25).

Note that $(x_0, \boldsymbol{\xi}_1)$ and $(z_0, \boldsymbol{\zeta}_1)$ satisfy (11.25) if and only if any one of the cases (i)–(vi) of Fact 11.2.1 holds. Since $z_0 = \sqrt{2/3}R > 0$ is constant, cases (i), (iv), and (vi) can never occur. Therefore, it suffices to show that (11.23) and (11.24) are satisfied if and only if any one of the following holds:

(ii) $\dot{\gamma} = \|-\dot{\boldsymbol{\varepsilon}}_{\rm p}\| = 0$, $\sqrt{\tfrac{2}{3}}R > \|\boldsymbol{\eta}\|$.

(iii) $\dot{\gamma} = \|-\dot{\boldsymbol{\varepsilon}}_{\rm p}\| \neq 0$, $\sqrt{\tfrac{2}{3}}R = \|\boldsymbol{\eta}\| \neq 0$, and there exists $\alpha > 0$ such that $-\dot{\boldsymbol{\varepsilon}}_{\rm p} = -\alpha\boldsymbol{\eta}$.

(v) $\dot{\gamma} = \|-\dot{\boldsymbol{\varepsilon}}_{\rm p}\| = 0$, $\sqrt{\tfrac{2}{3}}R = \|\boldsymbol{\eta}\| \neq 0$.

Here, situation (iii) can be rewritten equivalently as

$$\|\boldsymbol{\eta}\| - \sqrt{\tfrac{2}{3}}R = 0, \quad \dot{\gamma} \neq 0, \quad \dot{\boldsymbol{\varepsilon}}_{\rm p} = \dot{\gamma}\frac{\boldsymbol{\eta}}{\|\boldsymbol{\eta}\|}. \tag{11.27}$$

Then we can easily see that (11.23) and (11.24) are satisfied if and only if any one of (ii), (v), and (11.27) holds. □

Figure 11.4 gives a schematic interpretation of the assertion of Proposition 11.2.2, where $(\dot{\gamma}, -\dot{\boldsymbol{\varepsilon}}_{\rm p})$ and $(\sqrt{2/3}R, \boldsymbol{\eta})$ are included in the second-order cone. Suppose $\dot{\gamma} \neq 0$. Then, the "inner product" $\dot{\gamma}(\sqrt{2/3}R) + (-\dot{\boldsymbol{\varepsilon}}_{\rm p}) : \boldsymbol{\eta}$ vanishes if and only if (i) $(\dot{\gamma}, -\dot{\boldsymbol{\varepsilon}}_{\rm p})$ and $(\sqrt{2/3}R, \boldsymbol{\eta})$ are on the boundary of the second-order cone, and (ii) $-\dot{\boldsymbol{\varepsilon}}_{\rm p}$ is parallel with $\boldsymbol{\eta}$ and has the opposite direction. Condition (ii) means the parallelism condition (11.24) of the associated flow rule, which is shown in Figure 11.5. Thus the associated flow rule is satisfied at the solution of (11.26).

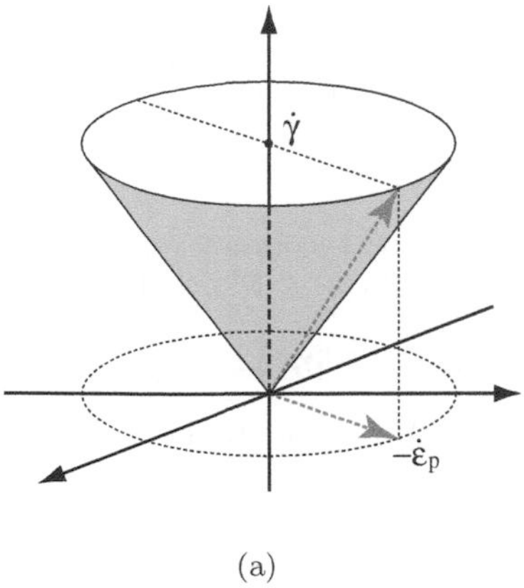

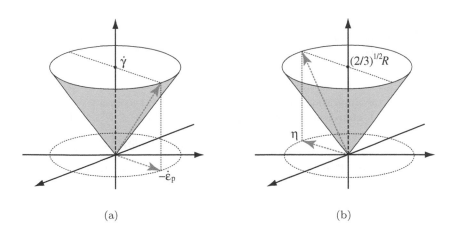

FIGURE 11.4: Second-order cone constraints for the flow rule in (11.26). (a) the constraint on the kinematic variables $\dot{\gamma}$ and $\dot{\boldsymbol{\varepsilon}}_\mathrm{p}$; (b) the constraint on the static variables $\boldsymbol{\eta}$, where $R = \sigma_\mathrm{y}$ and $\boldsymbol{\eta} = \mathrm{dev}(\boldsymbol{\sigma})$.

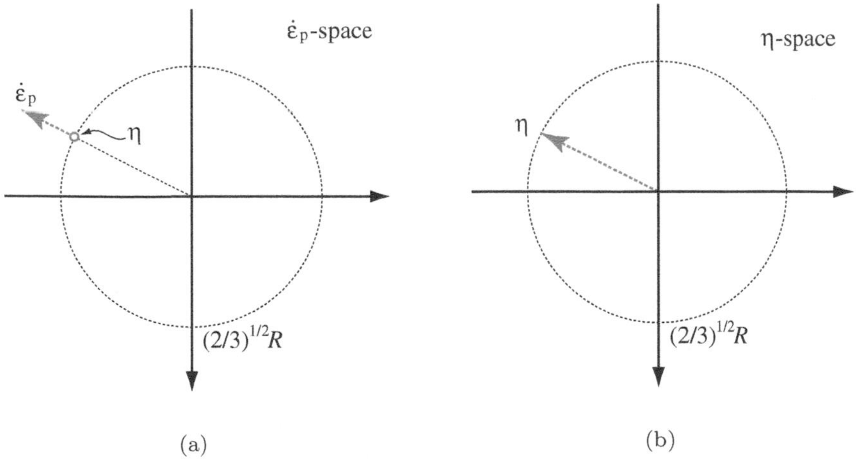

FIGURE 11.5: The associated flow rule in (11.24). (a) $\dot{\boldsymbol{\varepsilon}}_\mathrm{p}$ is in the normal outer direction of the yield surface at $\boldsymbol{\eta}$; (b) $\boldsymbol{\eta}$ is on the yield surface if $\dot{\boldsymbol{\varepsilon}}_\mathrm{p} \neq \boldsymbol{o}$.

Remark 11.2.3. The difference of formulation (11.26) given in Proposition 11.2.2 from the classical formulation in the plasticity theory is recognized further by comparing their inclusion forms as follows. Let $R = \sigma_{\rm y}$, and define $\Sigma_{\boldsymbol{\eta}} \subset \mathbb{S}^3$ by

$$\Sigma_{\boldsymbol{\eta}} = \left\{ \boldsymbol{\eta} \in \mathbb{S}^3 \mid \sqrt{\tfrac{2}{3}} R \geq \|\boldsymbol{\eta}\| \right\},$$

which is the admissible set of $\boldsymbol{\eta}$. We denote by $\delta_{\Sigma_{\boldsymbol{\eta}}}$ the indicator function of $\Sigma_{\boldsymbol{\eta}}$. Then the normality rule (11.12) can be rewritten as the inclusion

$$\dot{\boldsymbol{\varepsilon}}_{\rm p} \in \partial \delta_{\Sigma_{\boldsymbol{\eta}}}(\boldsymbol{\eta}), \tag{11.28}$$

which is the conventional formulation in the plasticity theory; see, e.g., [159, Chap. 4]. In contrast, concerning expression (11.26), define $\Sigma_{\rm SOC} \subset \mathbb{R} \times \mathbb{S}^3$ by

$$\Sigma_{\rm SOC} = \{(\rho, \boldsymbol{\eta}) \in \mathbb{R} \times \mathbb{S}^3 \mid \rho \geq \|\boldsymbol{\eta}\|\}.$$

By using the self-duality property, i.e., $\Sigma_{\rm SOC}^* = \Sigma_{\rm SOC}$, we obtain[9]

$$\delta_{\Sigma_{\rm SOC}}^*(-\dot{\gamma}, \dot{\boldsymbol{\varepsilon}}_{\rm p}) = \begin{cases} 0 & \text{if } \dot{\gamma} \geq \|-\dot{\boldsymbol{\varepsilon}}_{\rm p}\|, \\ +\infty & \text{otherwise.} \end{cases}$$

Hence, (11.26) can be rewritten as

$$\partial \delta_{\Sigma_{\rm SOC}}(\sqrt{\tfrac{2}{3}} R, \boldsymbol{\eta}) + \partial \delta_{\Sigma_{\rm SOC}}^*(-\dot{\gamma}, \dot{\boldsymbol{\varepsilon}}_{\rm p}) = \begin{bmatrix} \sqrt{\frac{2}{3}} R \\ \boldsymbol{\eta} \end{bmatrix} \cdot \begin{bmatrix} -\dot{\gamma} \\ \dot{\boldsymbol{\varepsilon}}_{\rm p} \end{bmatrix}. \tag{11.29}$$

Applying Proposition 2.1.12 to (11.29) yields for (11.26) the following alternative expression:

$$\begin{bmatrix} -\dot{\gamma} \\ \dot{\boldsymbol{\varepsilon}}_{\rm p} \end{bmatrix} \in \partial \delta_{\Sigma_{\rm SOC}}(\sqrt{\tfrac{2}{3}} R, \boldsymbol{\eta}). \tag{11.30}$$

Formulation (11.30) is to be compared with the classical one (11.28). ■

11.3 Plasticity with Isotropic Hardening

We next introduce the *strain-hardening* effect, which is the increase of the modulus of the stress in the course of plastic loading. In continuation with the preceding section, the von Mises yield criterion (11.23) and the associated flow rule (11.24) are assumed.

[9]To see the detail of this assertion, recall the definitions of the dual cone (in (1.1)) and the conjugate function (in (2.13)) to see for a cone C that

$$\delta_C^*(z) = \sup_x\{\langle z, x\rangle - \delta_C(x)\} = \sup_x\{\langle z, x\rangle \mid x \in C\} = -\inf_x\{\langle -z, x\rangle \mid x \in C\} = \delta_{C^*}(-z).$$

Therefore, for a self-dual cone satisfying $C^* = C$, we obtain $\delta_C^*(z) = \delta_C(-z)$.

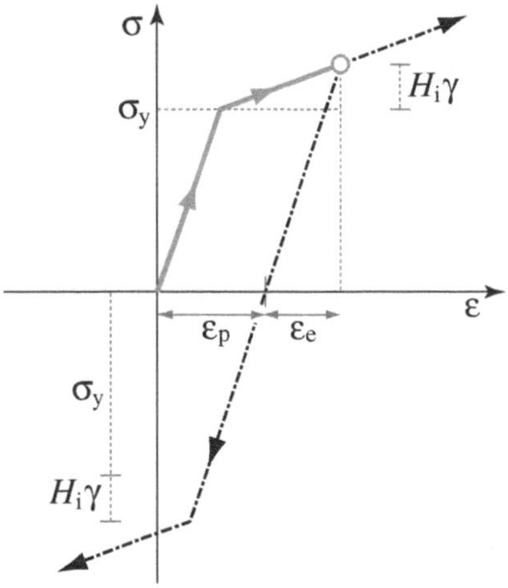

FIGURE 11.6: Constitutive relation of a one-dimensional elastoplastic model with linear isotropic hardening. "o": the current point.

This section treats isotropic hardening, which is an expansion of the set of admissible stresses (11.23a) in terms of the increase of its radius R. In contrast, the center of the yield surface is assumed to remain at the origin; i.e., we put

$$\boldsymbol{\beta} = \boldsymbol{o}$$

in (11.15b). The evolution of $\boldsymbol{\beta}$, which is known as kinematic hardening, is the subject of section 11.4. For a one-dimensional model, Figure 11.6 depicts the stress–strain response including the linear isotropic hardening effect.

11.3.1 Linear isotropic hardening law

Suppose that we examine response of the elastoplastic body within the time interval $[0, T]$, which is to be subdivided into finitely many intervals. For a specific subinterval, denoted $[\tau_l, \tau_{l+1}]$, the response at time τ_{l+1} is found by applying the standard backward Euler time discretization. With the subscript $(\ )_l$ we represent the values of various variables at time τ_l (e.g., $\boldsymbol{u}_l$ for the displacement), and with $\dot{(\ )}$ we represent the increments between time τ_l and τ_{l+1} (e.g., $\dot{\boldsymbol{u}}$ for the incremental displacement).

The isotropic hardening means the increase of the radius R in (11.23a) of the yield surface. We denote by $\dot{s}_{\rm i} \geq 0$ the increment of the radius of the yield surface in the time interval $[\tau_l, \tau_{l+1}]$. In the linear isotropic hardening model, we postulate that the hardening is linear in $\dot{\gamma}$, i.e.,

$$\dot{s}_{\rm i} = \sqrt{\tfrac{2}{3}} H_{\rm i} \dot{\gamma}, \tag{11.31}$$

where the constant $H_{\mathrm{i}} > 0$ is called the *isotropic hardening modulus*.

At time τ_{l+1}, the von Mises yield criterion (11.23) and the associated flow rule (11.24) are written as follows. Define R_l by

$$R_l = \sigma_{\mathrm{y}} + s_{\mathrm{i}l} = \sigma_{\mathrm{y}} + \int_0^{\tau_l} \dot{s}_{\mathrm{i}} \mathrm{d}\tau \tag{11.32}$$

for simplicity. Then the yield condition (11.23a) at time τ_{l+1} is given by $\|\boldsymbol{\eta}\| - \sqrt{\frac{2}{3}}R \leq 0$, where

$$R = R_l + \dot{s}_{\mathrm{i}}. \tag{11.33}$$

Accordingly, concerning the incremental problem, the flow rule in (11.26) is to be updated at time τ_{l+1} as

$$\|\boldsymbol{\eta}\| - \sqrt{\tfrac{2}{3}}R \leq 0, \quad \dot{\gamma} \geq 0, \quad \dot{\gamma}\Big(\|\boldsymbol{\eta}\| - \sqrt{\tfrac{2}{3}}R\Big) = 0. \tag{11.34}$$

Note that $\dot{\gamma}$ corresponds to the amount of the plastic flow (see (11.24)):

$$\dot{\boldsymbol{\varepsilon}}_{\mathrm{p}} = \dot{\gamma}\frac{\boldsymbol{\eta}}{\|\boldsymbol{\eta}\|}. \tag{11.35}$$

It is clear from expression (11.33) that $s_{\mathrm{i}l}$ is a static variable, while $\dot{\gamma}$ in (11.35) is a kinematic variable. These variables are related by (11.31).

11.3.2 Second-order cone complementarity formulation

The classical formulation of the isotropic hardening plasticity was introduced in section 11.3.1, where the flow rule in conjunction with the linear isotropic hardening is given by (11.31)–(11.35). We here present an alternative formulation of this rule in terms of the complementarity condition subjected to self-dual cone constraints. We first state in Proposition 11.3.1 that the associated flow rule, say, (11.34) and (11.35), can be reformulated as the complementarity condition over second-order cone, an intuitive interpretation of which is given by Figure 11.4.

Proposition 11.3.1. *Suppose* $R > 0$. *Then* $(\dot{\gamma}, \dot{\boldsymbol{\varepsilon}}_{\mathrm{p}}) \in \mathbb{R} \times \mathbb{S}^3$ *and* $\boldsymbol{\eta} \in \mathbb{S}^3$ *satisfy* (11.34) *and* (11.35) *if and only if they satisfy*

$$\dot{\gamma} \geq \|-\dot{\boldsymbol{\varepsilon}}_{\mathrm{p}}\|, \tag{11.36a}$$

$$\sqrt{\tfrac{2}{3}}R \geq \|\boldsymbol{\eta}\|, \tag{11.36b}$$

$$\dot{\gamma}\Big(\sqrt{\tfrac{2}{3}}R\Big) + (-\dot{\boldsymbol{\varepsilon}}_{\mathrm{p}}) : \boldsymbol{\eta} = 0. \tag{11.36c}$$

Proof. Analogous to the proof of Proposition 11.2.2. □

Recall that the evolution of R is defined by (11.32) and (11.33), where the initial flow stress $\sigma_{\mathrm{y}} > 0$ is a constant. It follows from (11.31) and (11.34) that $\dot{s}_{\mathrm{i}} \geq 0$. Therefore, from (11.33) we see that the hypothesis of Proposition 11.3.1, say, $R > 0$, is always satisfied.

Thus we saw in Proposition 11.3.1 that the associated flow rule, say, (11.35) with (11.34), is reformulated as a complementarity condition over the second-order cone in terms of the kinematic variables $(\dot{\gamma}, \dot{\boldsymbol{\varepsilon}}_{\mathrm{p}})$ and the static variables $(R, \boldsymbol{\eta})$.

We next consider the evolution law (11.31) of the static internal state variable $\dot{s}_{\mathrm{i}}$ due to the isotropic hardening effect. In Proposition 11.3.2, this evolution rule, together with the flow rule, is reformulated as follows[10]:

$$(\dot{s}_{\mathrm{i}}, \boldsymbol{\eta}) \in \arg\max_{\check{s}_{\mathrm{i}}, \check{\boldsymbol{\eta}}} \left\{ \dot{\boldsymbol{\varepsilon}}_{\mathrm{p}} : \check{\boldsymbol{\eta}} - \frac{1}{2H_{\mathrm{i}}} \check{s}_{\mathrm{i}}^2 \;\middle|\; \sqrt{\tfrac{2}{3}}(R_l + \check{s}_{\mathrm{i}}) \geq \|\check{\boldsymbol{\eta}}\| \right\}. \tag{11.37}$$

Note that the SOCP problem on the right-hand side of (11.37) can be regarded as the maximum dissipation law for the linear isotropic hardening.

Proposition 11.3.2. *Suppose that* $\dot{\boldsymbol{\varepsilon}}_{\mathrm{p}} \in \mathbb{S}^3$ *and* $R_l > 0$ *are given. Then* $(\dot{s}_{\mathrm{i}}, \boldsymbol{\eta}) \in \mathbb{R} \times \mathbb{S}^3$ *satisfies* (11.37) *if and only if there exists* $\dot{\gamma} \in \mathbb{R}$ *satisfying* (11.31), (11.33), (11.34), *and* (11.35).

Proof. We prove the assertion by deriving the optimality condition of the problem on the right-hand side of (11.37) based on the Lagrangian duality introduced in section 2.2.6. The Lagrangian of this maximization problem is defined by[11]

$$L(x_0, \boldsymbol{\xi}_1; \check{s}_{\mathrm{i}}, \check{\boldsymbol{\eta}}) = \begin{cases} \dot{\boldsymbol{\varepsilon}}_{\mathrm{p}} : \check{\boldsymbol{\eta}} - \dfrac{1}{2H_{\mathrm{i}}} \check{s}_{\mathrm{i}}^2 \\ \qquad + \left[x_0 \sqrt{\tfrac{2}{3}}(R_l + \check{s}_{\mathrm{i}}) + \boldsymbol{\xi}_1 : \check{\boldsymbol{\eta}} \right] & \text{if } x_0 \geq \|\boldsymbol{\xi}_1\|, \\ +\infty & \text{otherwise}, \end{cases} \tag{11.38}$$

where $x_0 \in \mathbb{R}$ and $\boldsymbol{\xi}_1 \in \mathbb{S}^3$ are the Lagrangian multipliers. For reverting to the situation of the Lagrangian duality studied in section 2.2.6, we set

$$x = (x_0, \boldsymbol{\xi}_1), \quad z^* = (\check{s}_{\mathrm{i}}, \check{\boldsymbol{\eta}}),$$
$$\mathbb{V} = \mathbb{Y} = \mathbb{R} \times \mathbb{S}^3.$$

Note that the variables $\check{s}_{\mathrm{i}}$ and $\check{\boldsymbol{\eta}}$ are treated as dual variables in (11.38), because the optimization problem in (11.37) is a maximization problem, while in section 2.2.6 the dual problem (2.35) in the Lagrangian duality theory has been defined as a maximization problem. Therefore, we here

[10] We write $\check{s}_{\mathrm{i}}$ instead of $\check{\dot{s}}_{\mathrm{i}}$ to avoid a too complicated notation.

[11] See Definition 2.2.13 for the definition of the Lagrangian, but note that here we consider a maximization problem.

treat the problem in (11.37) as a dual problem for consistency with section 2.2.6. It then follows from Proposition 2.2.17 that the optimal solution of the problem in (11.37) is characterized as the saddle point of L, i.e.,

$$L(\bar{x}; z^*) \leq L(\bar{x}; \bar{z}^*) \leq L(x; \bar{z}^*), \quad \forall x, \ \forall z^*. \tag{11.39}$$

In what follows we shall show that (11.39) coincides with (11.31)–(11.35).

For (11.39) to hold, the leftmost term should satisfy $L(\bar{x}; z^*) < +\infty$, and hence from (11.38) $\bar{x} = (\bar{x}_0, \bar{\boldsymbol{\xi}}_1)$ should satisfy

$$\bar{x}_0 \geq \|\bar{\boldsymbol{\xi}}_1\|. \tag{11.40}$$

Moreover, the rightmost term in (11.39) should satisfy $L(x; \bar{z}^*) > -\infty$. Regarding the terms inside the square brackets of (11.39), Fact 1.4.4 (ii) yields

$$\begin{aligned} &\inf_{x_0, \boldsymbol{\xi}_1} \left\{ x_0 \sqrt{\tfrac{2}{3}}(R_l + \check{s}_{\mathrm{i}}) + \boldsymbol{\xi}_1 : \check{\boldsymbol{\eta}} \mid x_0 \geq \|\boldsymbol{\xi}_1\| \right\} \\ &\quad = \begin{cases} 0 & \text{if } \sqrt{\tfrac{2}{3}}(R_l + \check{s}_{\mathrm{i}}) \geq \|\check{\boldsymbol{\eta}}\|, \\ -\infty & \text{otherwise.} \end{cases} \end{aligned} \tag{11.41}$$

If $\bar{z}^* = (\bar{s}_{\mathrm{i}}, \bar{\boldsymbol{\eta}})$ does not satisfy

$$\sqrt{\tfrac{2}{3}}(R_l + \bar{s}_{\mathrm{i}}) \geq \|\bar{\boldsymbol{\eta}}\|, \tag{11.42}$$

then (11.41) means that we can choose $\{x_k\}$ such that $L(x_k; \bar{z}^*) \to -\infty$. Hence, $\bar{z}^*$ is required to satisfy (11.42). For $\bar{x}$ and $\bar{z}^*$ satisfying (11.40) and (11.42), it follows from (11.41) that the sum of the terms inside the square brackets of (11.38) takes the minimum value of 0. Therefore, the second inequality of (11.39) requires

$$\bar{x}_0 \sqrt{\tfrac{2}{3}}(R_l + \bar{s}_{\mathrm{i}}) + \bar{\boldsymbol{\xi}}_1 : \bar{\boldsymbol{\eta}} = 0. \tag{11.43}$$

On the other hand, the first inequality of (11.39) means that $\bar{z}^*$ corresponds to the maximum point of $L(\bar{x}; z^*)$, which is characterized by the stationarity conditions of $L(\bar{x}; z^*)$ with respect to $\check{s}_{\mathrm{i}}$ and $\check{\boldsymbol{\eta}}$, i.e.,

$$\bar{\dot{\boldsymbol{\varepsilon}}}_{\mathrm{p}} + \bar{\boldsymbol{\xi}}_1 = \boldsymbol{o}, \tag{11.44}$$

$$-\frac{1}{H_{\mathrm{i}}}\bar{s}_{\mathrm{i}} + \sqrt{\tfrac{2}{3}}\bar{x}_0 = 0. \tag{11.45}$$

Let

$$\dot{\gamma} = \bar{x}_0 \tag{11.46}$$

to see that (11.45) is reduced to (11.31), while (11.40) and (11.42)–(11.45) are reduced to (11.33) and (11.36). Moreover, Proposition 11.3.1 asserts that (11.36) is equivalent to (11.34) and (11.35), which concludes the proof. □

The relation (11.37) means that, for the given plastic strain rate $\dot{\boldsymbol{\varepsilon}}_{\mathrm{p}}$, the actual static variable maximizes the plastic dissipation. This alternative postulate to the associative flow rule is called the *maximum plastic-dissipation law*; see Simo–Hughes [413, section 2.6], and also Hill [178] and Lubliner [275] as classic work.

Recall that L defined by (11.38) in the proof of Proposition 11.3.2 serves as the Lagrangian of the optimization problem in (11.37). Since L is a concave function (i.e., $-L$ is a convex function), substituting (11.44) and (11.45) into (11.38) yields

$$\sup_{\check{s}_{\mathrm{i}},\check{\boldsymbol{\eta}}} L(x_0, \boldsymbol{\xi}_1; \check{s}_{\mathrm{i}}, \check{\boldsymbol{\eta}}) = \begin{cases} \sqrt{\frac{2}{3}} R_l x_0 + \dfrac{1}{3} H_{\mathrm{i}} x_0^2 & \text{if } x_0 \geq \|\boldsymbol{\xi}_1\|,\ \boldsymbol{\xi}_1 = -\dot{\boldsymbol{\varepsilon}}_{\mathrm{p}}, \\ +\infty & \text{otherwise.} \end{cases} \tag{11.47}$$

By rewriting the variable x_0 according to (11.46), the problem dual to (11.37) is obtained as[12]

$$\dot{\gamma} \in \arg\min_{\check{\gamma}} \left\{ \sqrt{\tfrac{2}{3}} R_l \check{\gamma} + \frac{1}{3} H_{\mathrm{i}} \check{\gamma}^2 \mid \check{\gamma} \geq \|-\dot{\boldsymbol{\varepsilon}}_{\mathrm{p}}\| \right\}. \tag{11.48}$$

It is easy to see that the right-hand side of (11.48) is equal to the singleton $\{\|-\dot{\boldsymbol{\varepsilon}}_{\mathrm{p}}\|\}$, which is compatible to definition (11.20) of $\dot{\gamma}$. Note that the dual principle (11.48) means no more than (11.20), because we consider that $\dot{\boldsymbol{\varepsilon}}_{\mathrm{p}}$ is given in the framework of the maximum dissipation law. In section 11.3.4 we shall use (11.48) and (11.37) to formulate the minimization problems of the potential energy and the complementary energy, respectively; see (PE) and (CE) in (11.59) and (11.61).

11.3.3 Incremental problem

The incremental problem for the associative plasticity with isotropic hardening is formulated within the geometrically linear theory.

The notation concerning the continuum mechanics is the same as that used in Chapters 8 and 9; see section 8.2.1 for a detailed introduction of notation. Consider the elastoplastic body occupying $\overline{\Omega} := \operatorname{cl} \Omega$, where $\Omega \subset \mathbb{R}^3$ is a bounded, open, and connected set with a sufficiently smooth boundary $\Gamma := \operatorname{bd} \Omega$. For a partition $\Gamma_{\mathrm{N}} \cup \Gamma_{\mathrm{D}}$ of Γ, we denote by $\underline{\boldsymbol{t}} : \Gamma_{\mathrm{N}} \to \mathbb{R}^3$ the specified traction per unit area in the reference configuration, while $\underline{\boldsymbol{u}} : \Gamma_{\mathrm{D}} \to \mathbb{R}^3$ denotes prescribed displacement. For Ω, we denote by $\underline{\boldsymbol{p}} : \Omega \to \mathbb{R}^3$ the body force per unit volume in the reference configuration.

[12]Since problem in (11.37) is a maximization problem, its Lagrangian dual is defined as the minimization problem given in (2.37).

As introduced in section 11.3.1, we consider the evolution of the solution among the time subinterval $[\tau_l, \tau_{l+1}]$. With the superscript $(\ )_l$ we represent the value of the corresponding variable at time τ_l, while with $(\dot{\ })$ we represent its increment between time τ_l and τ_{l+1}. It should be clear that $\underline{\boldsymbol{u}}$, $\underline{\boldsymbol{t}}$, and $\underline{\boldsymbol{p}}$ are the values at time τ_{l+1}.

The strain tensor is additively decomposed to the elastic and plastic parts as

$$\boldsymbol{\varepsilon} = \boldsymbol{\varepsilon}_{\mathrm{e}} + \boldsymbol{\varepsilon}_{\mathrm{p}}.$$

Then the compatibility condition within the infinitesimal deformation theory is written in terms of the incremental values as

$$\dot{\boldsymbol{\varepsilon}}_{\mathrm{e}} + \dot{\boldsymbol{\varepsilon}}_{\mathrm{p}} = \frac{1}{2}(\nabla \dot{\boldsymbol{u}}^{\mathrm{T}} + \nabla \dot{\boldsymbol{u}}). \tag{11.49}$$

According to (11.11), the elastic constitutive relation is given by

$$\boldsymbol{\sigma} = \boldsymbol{\sigma}_l + \boldsymbol{C} : \dot{\boldsymbol{\varepsilon}}_{\mathrm{e}}, \tag{11.50}$$

where $\boldsymbol{\sigma}_l = \boldsymbol{C} : \boldsymbol{\varepsilon}_{\mathrm{e}l}$ is the stress tensor at time τ_l. The force-balance equation is given by

$$-\operatorname{div} \boldsymbol{\sigma} = \underline{\boldsymbol{p}} \tag{11.51}$$

in Ω. The definition of $\boldsymbol{\eta}$ is given by (11.15b) with $\boldsymbol{\beta} = \boldsymbol{o}$ for the isotropic hardening. The evolution of R is defined by (11.31) and (11.33). As shown in Proposition 11.3.1, the associated flow rule is formulated as (11.36). Recall that (11.36) implies (11.24). From (11.24) we obtain $\dot{\boldsymbol{\varepsilon}}_{\mathrm{p}} = \operatorname{dev}(\dot{\boldsymbol{\varepsilon}}_{\mathrm{p}})$, because from definition (11.15b) we can easily see that $\boldsymbol{\eta}$ satisfies $\boldsymbol{\eta} = \operatorname{dev}(\boldsymbol{\eta})$. Therefore, without loss of generality we can replace (11.49) with

$$\dot{\boldsymbol{\varepsilon}}_{\mathrm{e}} + \operatorname{dev}(\dot{\boldsymbol{\varepsilon}}_{\mathrm{p}}) = \frac{1}{2}(\nabla \dot{\boldsymbol{u}}^{\mathrm{T}} + \nabla \dot{\boldsymbol{u}}).$$

By summing up these equations, the incremental problem is formulated as

$$\boldsymbol{\sigma} = \boldsymbol{\sigma}_l + \boldsymbol{C} : \dot{\boldsymbol{\varepsilon}}_{\mathrm{e}} \quad \text{in } \Omega, \tag{11.52a}$$

$$\dot{\boldsymbol{\varepsilon}}_{\mathrm{e}} + \operatorname{dev}(\dot{\boldsymbol{\varepsilon}}_{\mathrm{p}}) = \frac{1}{2}(\nabla \dot{\boldsymbol{u}}^{\mathrm{T}} + \nabla \dot{\boldsymbol{u}}) \quad \text{in } \Omega, \tag{11.52b}$$

$$-\operatorname{div} \boldsymbol{\sigma} = \underline{\boldsymbol{p}} \quad \text{in } \Omega, \tag{11.52c}$$

$$\dot{\boldsymbol{u}} = \dot{\underline{\boldsymbol{u}}} \quad \text{on } \Gamma_{\mathrm{D}}; \qquad \boldsymbol{t} = \underline{\boldsymbol{t}} \quad \text{on } \Gamma_{\mathrm{N}}, \tag{11.52d}$$

$$\boldsymbol{\eta} = \operatorname{dev}(\boldsymbol{\sigma}) \quad \text{in } \Omega, \tag{11.52e}$$

$$R = R_l + \sqrt{\tfrac{2}{3}} H_{\mathrm{i}} \dot{\gamma} \quad \text{in } \Omega, \tag{11.52f}$$

$$\dot{\gamma} \geq \|-\dot{\boldsymbol{\varepsilon}}_{\mathrm{p}}\|, \ \sqrt{\tfrac{2}{3}} R \geq \|\boldsymbol{\eta}\|, \ \dot{\gamma}\Big(\sqrt{\tfrac{2}{3}} R\Big) + (-\dot{\boldsymbol{\varepsilon}}_{\mathrm{p}}) : \boldsymbol{\eta} = 0 \quad \text{in } \Omega. \tag{11.52g}$$

In what follows, we suppose that the elastoplastic body is discretized into finite elements to avoid any technical difficulty arising from the treatment of infinite-dimensional optimization problems. As done in section 8.4.1, the domain Ω is subdivided into n^{E} finite elements according to the standard procedure. We employ the Gauss quadrature as usual by considering n^{G} evaluation point for each element. Accordingly, the total number of the Gauss points is $n^{\mathrm{P}} = n^{\mathrm{E}}n^{\mathrm{G}}$. The constraints, such as the yield condition and the compatibility relation, are imposed at each Gauss point. To simplify the notation, with the superscript $(\ )^r$ we represent the value of a state variable at the rth Gauss point. For example, we denote by $\boldsymbol{\varepsilon}_{\mathrm{p}}^r$ the plastic strain at the rth Gauss point. Without $(\ )^r$ we mean the assemblage of the values at all Gauss points; i.e., $\boldsymbol{\varepsilon}_{\mathrm{p}}$ denotes the assemblage of $\boldsymbol{\varepsilon}_{\mathrm{p}}^1, \dots, \boldsymbol{\varepsilon}_{\mathrm{p}}^{n^{\mathrm{P}}}$.

The field of the displacement vector $\boldsymbol{u}$ is discretized into the vector of the (generalized) nodal displacements, also denoted $\boldsymbol{u} \in \mathbb{R}^d$, where d is the number of degrees of freedom. Similarly, we denote by $\boldsymbol{t} \in \mathbb{R}^d$ the vector of the (generalized) nodal stresses. The set of indices of the degrees of freedom $\{1, \dots, d\}$ is partitioned as $\tilde{\Gamma}_{\mathrm{D}} \cup \tilde{\Gamma}_{\mathrm{N}}$, where the displacement u_j is prescribed for $j \in \tilde{\Gamma}_{\mathrm{D}}$, and the generalized stress t_j is prescribed for $j \in \tilde{\Gamma}_{\mathrm{N}}$.

We denote by B^* the discrete equilibrium operator, so that the force-balance equation (11.51) is discretized as

$$B^* \cdot \boldsymbol{\sigma} = \boldsymbol{t}, \tag{11.53}$$

where $\boldsymbol{\sigma}$ is the assemblage of $\boldsymbol{\sigma}^1, \dots, \boldsymbol{\sigma}^{n^{\mathrm{P}}}$. The discrete version of the compatibility relation (11.49) is then written as

$$\dot{\boldsymbol{\varepsilon}}_{\mathrm{e}} + \dot{\boldsymbol{\varepsilon}}_{\mathrm{p}} = B\dot{\boldsymbol{u}}, \tag{11.54}$$

where B is the conjugate operator of B^*, which satisfies $\langle B^* \cdot \boldsymbol{\sigma}, \boldsymbol{u}\rangle = \langle \boldsymbol{\sigma}, B\boldsymbol{u}\rangle$ for any $\boldsymbol{u}$ and $\boldsymbol{\sigma}$. The elastic constitutive relation (11.50) is written at each Gauss point as

$$\boldsymbol{\sigma}^r = \boldsymbol{\sigma}_l^r + \boldsymbol{C} : \dot{\boldsymbol{\varepsilon}}_{\mathrm{e}}^r, \quad r = 1, \dots, n^{\mathrm{P}},$$

which is simply written in the assembled form as

$$\boldsymbol{\sigma} = \boldsymbol{\sigma}_l + C : \dot{\boldsymbol{\varepsilon}}_{\mathrm{e}}. \tag{11.55}$$

By using the discretized forms in (11.53), (11.54), and (11.55), the incremental problem (11.52) can be discretized as

$$\boldsymbol{\sigma} = \boldsymbol{\sigma}_l + \boldsymbol{C} : \dot{\boldsymbol{\varepsilon}}_{\rm e}, \tag{11.56a}$$

$$\dot{\boldsymbol{\varepsilon}}_{\rm e} + \mathrm{dev}(\dot{\boldsymbol{\varepsilon}}_{\rm p}) = B\dot{\boldsymbol{u}}, \tag{11.56b}$$

$$B^* \cdot \boldsymbol{\sigma} = \underline{t}, \tag{11.56c}$$

$$\dot{u}_j = \underline{\dot{u}}_j \quad (j \in \tilde{\Gamma}_{\rm D}); \qquad t_j = \underline{t}_j \quad (j \in \tilde{\Gamma}_{\rm N}), \tag{11.56d}$$

$$\boldsymbol{\eta}^r = \mathrm{dev}(\boldsymbol{\sigma}^r) \quad (\forall r), \tag{11.56e}$$

$$R^r = R^r_l + \sqrt{\tfrac{2}{3}} H_{\rm i} \dot{\gamma}^r \quad (\forall r), \tag{11.56f}$$

$$\dot{\gamma}^r \geq \|-\dot{\boldsymbol{\varepsilon}}^r_{\rm p}\|, \ \sqrt{\tfrac{2}{3}} R^r \geq \|\boldsymbol{\eta}^r\|, \ \dot{\gamma}^r\left(\sqrt{\tfrac{2}{3}} R^r\right) + (-\dot{\boldsymbol{\varepsilon}}^r_{\rm p}) : \boldsymbol{\eta}^r = 0 \quad (\forall r). \tag{11.56g}$$

Note that, by using (11.31), (11.56f) can be alternatively written only in terms of static variables as

$$R^r = R^r_l + \dot{s}^r_{\rm i} \quad (\forall r). \tag{11.57}$$

In the subsequent section we show that (11.56) can be reduced to an SOCP problem.

11.3.4 SOCP formulation of incremental problem

For the incremental problem, the minimization problem of the potential energy is formulated as an optimization problem, which involves only kinematic variables, and the optimality condition of which coincides with (11.56). In contrast, the minimization problem of the complementary energy involves only static variables.

For simplicity, define $\hat{w}_{\rm e} : \mathbb{S}^3 \to \mathbb{R}$ by

$$\hat{w}_{\rm e}(\boldsymbol{\varepsilon}_{\rm e}) = \frac{1}{2} \boldsymbol{\varepsilon}_{\rm e} : \boldsymbol{C} : \boldsymbol{\varepsilon}_{\rm e}, \tag{11.58}$$

which is the elastic strain energy function. Then the minimization problem of the potential energy, corresponding to the incremental problem (11.56), is formulated as

$$\left.\begin{aligned} \text{(PE)} : \min_{\dot{\boldsymbol{u}}, \dot{\boldsymbol{\varepsilon}}_{\rm e}, \dot{\boldsymbol{\varepsilon}}_{\rm p}, \dot{\gamma}} \quad & \sum_{r=1}^{n^{\rm P}} \left(\hat{w}_{\rm e}(\dot{\boldsymbol{\varepsilon}}^r_{\rm e}) + \boldsymbol{\sigma}^r_l : \dot{\boldsymbol{\varepsilon}}^r_{\rm e} + \sqrt{\tfrac{2}{3}} R^r_l \dot{\gamma}^r + \frac{1}{3} H_{\rm i} (\dot{\gamma}^r)^2 \right) \\ & \quad - \sum_{j \in \tilde{\Gamma}_{\rm N}} \underline{t}_j \dot{u}_j \\ \text{s.t.} \quad & \dot{\boldsymbol{\varepsilon}}_{\rm e} + \mathrm{dev}(\dot{\boldsymbol{\varepsilon}}_{\rm p}) = B\dot{\boldsymbol{u}}, \\ & \dot{\gamma}^r \geq \|-\dot{\boldsymbol{\varepsilon}}^r_{\rm p}\|, \quad \forall r, \\ & \dot{u}_j = \underline{\dot{u}}_j, \quad \forall j \in \tilde{\Gamma}_{\rm D}. \end{aligned}\right\} \tag{11.59}$$

The variables to be optimized in (11.59) are $\dot{\boldsymbol{u}}$, $\dot{\boldsymbol{\varepsilon}}_{\rm e}$, $\dot{\boldsymbol{\varepsilon}}_{\rm p}$, and $\dot{\gamma}$, which are the increments of kinematic variables. Note that $\dot{\gamma}$ corresponds to the norm of the plastic strain increment $\dot{\varepsilon}_{\mathrm{p}}$; see (11.20). Problem (11.59) essentially consists of the following:

- the elastic part of the strain energy $\hat{w}_{\mathrm{e}}(\dot{\varepsilon}_{\mathrm{e}}^{r}) + \boldsymbol{\sigma}_{l}^{r} : \dot{\varepsilon}_{\mathrm{e}}^{r}$
- the plastic part of the strain energy $\sqrt{\frac{2}{3}} R_{l}^{r} \dot{\gamma}^{r} + \frac{1}{3} H_{\mathrm{i}} (\dot{\gamma}^{r})^{2}$ (see (11.48))
- the compatibility condition (see (11.56b))
- the second-order cone constraint on the kinematic variables in (11.56g)

For $\hat{w}_{\mathrm{e}}$ defined by (11.58), its conjugate function can be written as

$$\hat{w}_{\mathrm{e}}^{*}(\boldsymbol{\sigma}) = \frac{1}{2} \boldsymbol{\sigma} : \boldsymbol{C}^{-1} : \boldsymbol{\sigma}, \tag{11.60}$$

which is the complementary strain energy for the linear elasticity. The minimization problem of the complementary energy is formulated as an optimization problem in terms of the static variables involved in (11.56) as

$$\left.\begin{aligned} \text{(CE)} : \min_{\boldsymbol{\sigma}, \boldsymbol{t}, \dot{s}_{\mathrm{i}}} \quad & \sum_{r=1}^{n^{\mathrm{P}}} \left(\hat{w}_{\mathrm{e}}^{*}(\boldsymbol{\sigma}^{r} - \boldsymbol{\sigma}_{l}^{r}) + \frac{1}{2H_{\mathrm{i}}} (\dot{s}_{\mathrm{i}}^{r})^{2} \right) - \sum_{j \in \tilde{\Gamma}_{\mathrm{D}}} t_{j} \underline{\dot{u}}_{j} \\ \text{s.t.} \quad & B^{*} \cdot \boldsymbol{\sigma} = \boldsymbol{t}, \\ & \sqrt{\tfrac{2}{3}} (R_{l}^{r} + \dot{s}_{\mathrm{i}}^{r}) \geq \| \operatorname{dev}(\boldsymbol{\sigma}^{r}) \|, \quad \forall r, \\ & t_{j} = \underline{t}_{j}, \quad \forall j \in \tilde{\Gamma}_{\mathrm{N}}. \end{aligned}\right\} \tag{11.61}$$

The total stress tensor $\boldsymbol{\sigma}$ and the incremental variable $\dot{s}_{\mathrm{i}}$, which represents the extension of the yield surface, are the variables to be optimized in (11.61). Note that problem (11.61) consists of the following:

- the elastic part of the complementary strain energy $\hat{w}_{\mathrm{e}}^{*}(\boldsymbol{\sigma}^{r} - \boldsymbol{\sigma}_{l}^{r})$
- the plastic part of the complementary strain energy $\frac{1}{2H_{\mathrm{i}}} (\dot{s}_{\mathrm{i}}^{r})^{2}$ (see (11.37))
- the force-balance equation (see (11.56c))
- the second-order cone constraint on the static variables in (11.56g), with (11.56e) and (11.57)

It should be emphasized that problem (11.61) is dual to problem (11.59). Moreover, both problem (11.59) and problem (11.61) consist of (i) a convex quadratic objective function, (ii) linear equality constraints, and (iii) second-order cone constraints. Therefore, both of these problems are convex optimization problems and can be reduced to SOCP problems in a manner similar to Remark 8.4.2.[13]

Remark 11.3.3. It can be confirmed that (CE) in (11.61) is obtained as the Lagrangian dual problem of (PE) in (11.59) as follows.[14]

To simplify presentation, let $\Phi(\dot{\boldsymbol{u}}, \dot{\boldsymbol{\varepsilon}}_{\rm e}, \dot{\boldsymbol{\varepsilon}}_{\rm p}, \dot{\gamma})$ denote the objective function of (PE). Then the Lagrangian for (PE) is defined by

$$L(\dot{\boldsymbol{u}}, \dot{\boldsymbol{\varepsilon}}_{\rm e}, \dot{\boldsymbol{\varepsilon}}_{\rm p}, \dot{\gamma}; z_0^*, \boldsymbol{\zeta}_1^*, \boldsymbol{\zeta}_2^*) = \begin{cases} \Phi(\dot{\boldsymbol{u}}, \dot{\boldsymbol{\varepsilon}}_{\rm e}, \dot{\boldsymbol{\varepsilon}}_{\rm p}, \dot{\gamma}) \\ \quad - \sum_{r=1}^{n^{\rm P}} \boldsymbol{\zeta}_2^{r*} : (\dot{\boldsymbol{\varepsilon}}_{\rm e}^r + \mathrm{dev}(\dot{\boldsymbol{\varepsilon}}_{\rm p}^r) - B\dot{\boldsymbol{u}}) \\ \quad - \sum_{r=1}^{n^{\rm P}} (z_0^{r*}\dot{\gamma}^r - \boldsymbol{\zeta}_1^{r*} : \dot{\boldsymbol{\varepsilon}}_{\rm p}^r) - \sum_{j \in \tilde{\Gamma}_{\rm D}} t_j(\dot{u}_j - \underline{\dot{u}}_j) & \text{if } z_0^{r*} \geq \|\boldsymbol{\zeta}_1^{r*}\| \ (\forall r), \\ -\infty & \text{otherwise,} \end{cases} \tag{11.62}$$

where $z_0^{r*} \in \mathbb{R}$, $\boldsymbol{\zeta}_1^{r*} \in \mathbb{S}^3$, and $\boldsymbol{\zeta}_2^{r*} \in \mathbb{S}^3$ are the Lagrangian multipliers. For reverting to the situation of the general results of the Lagrangian duality in section 2.2.6, the primal and dual variables are set as

$$x = (\dot{\boldsymbol{u}}, \dot{\boldsymbol{\varepsilon}}_{\rm e}, \dot{\boldsymbol{\varepsilon}}_{\rm p}, \dot{\gamma}), \quad z^* = (z_0^*, \boldsymbol{\zeta}_1^*, \boldsymbol{\zeta}_2^*).$$

In (11.62), the term $\boldsymbol{\zeta}_2^{r*} : \mathrm{dev}(\dot{\boldsymbol{\varepsilon}}_{\rm p}^r)$ can be reduced by using (11.4) as

$$\begin{aligned} \boldsymbol{\zeta}_2^{r*} : \mathrm{dev}(\dot{\boldsymbol{\varepsilon}}_{\rm p}^r) &= \boldsymbol{\zeta}_2^{r*} : (\dot{\boldsymbol{\varepsilon}}_{\rm p}^r - \mathrm{tr}(\dot{\boldsymbol{\varepsilon}}_{\rm p}^r)\boldsymbol{I}) \\ &= \boldsymbol{\zeta}_2^{r*} : \dot{\boldsymbol{\varepsilon}}_{\rm p}^r - \mathrm{tr}(\boldsymbol{\zeta}_2^{r*})\,\mathrm{tr}(\dot{\boldsymbol{\varepsilon}}_{\rm p}^r) \\ &= \boldsymbol{\zeta}_2^{r*} : \dot{\boldsymbol{\varepsilon}}_{\rm p}^r - (\mathrm{tr}(\boldsymbol{\zeta}_2^{r*})\boldsymbol{I}) : \dot{\boldsymbol{\varepsilon}}_{\rm p}^r \\ &= \mathrm{dev}(\boldsymbol{\zeta}_2^{r*}) : \dot{\boldsymbol{\varepsilon}}_{\rm p}^r. \end{aligned} \tag{11.63}$$

By substituting (11.63) into (11.62), we see that when $z_0^{r*} \geq \|\boldsymbol{\zeta}_1^{r*}\|$ $(\forall r)$ is satisfied, L takes the value

$$\begin{aligned} L = &\sum_{r=1}^{n^{\rm P}} [\hat{w}_{\rm e}(\dot{\boldsymbol{\varepsilon}}_{\rm e}^r) + (\boldsymbol{\sigma}_l^r - \boldsymbol{\zeta}_2^{r*}) : \dot{\boldsymbol{\varepsilon}}_{\rm e}^r] + \sum_{r=1}^{n^{\rm P}} (\sqrt{\tfrac{2}{3}} R_l^r \dot{\gamma}^r + \frac{1}{3} H_{\rm i}(\dot{\gamma}^r)^2 - z_0^{r*}\dot{\gamma}^r) \\ &+ \sum_{r=1}^{n^{\rm P}} (\boldsymbol{\zeta}_1^{r*} - \mathrm{dev}(\boldsymbol{\zeta}_2^{r*})) : \dot{\boldsymbol{\varepsilon}}_{\rm p}^r + (B^* \cdot \boldsymbol{\sigma} - \boldsymbol{t}) \cdot \dot{\boldsymbol{u}} + \sum_{j \in \tilde{\Gamma}_{\rm D}} t_j \underline{\dot{u}}_j \end{aligned}$$

[13]Note that in section 8.4.1 we have treated an optimization problem with (i) a convex quadratic objective function, (ii) linear equality constraints, and (iii) positive-semidefinite constraints; see problem (8.76), which has been reduced to the SDP problem in (8.79). The essential difference of problem (11.59) considered in this section from problem (8.76) for masonry structures is that problem (11.59) involves the second-order cone constraints instead of the positive-semidefinite constraints. From this observation we notice that problem (11.59) can be reduced to an SOCP problem.

[14]Recall that for problem (2.37) its Lagrangian dual is defined by (2.35).

with

$$t_j = \underline{t}_j, \quad j \in \tilde{\Gamma}_{\rm N}, \tag{11.64}$$

and otherwise $L = -\infty$. Consequently, by using the self-duality of the second-order cone in a manner similar to the proof of Proposition 11.3.2, we obtain

$$\inf_{\dot{\boldsymbol{u}}, \dot{\boldsymbol{\varepsilon}}_{\rm e}, \dot{\boldsymbol{\varepsilon}}_{\rm p}, \dot{\gamma}} L(\dot{\boldsymbol{u}}, \dot{\boldsymbol{\varepsilon}}_{\rm e}, \dot{\boldsymbol{\varepsilon}}_{\rm p}, \dot{\gamma}; z_0^*, \boldsymbol{\zeta}_1^*, \boldsymbol{\zeta}_2^*)$$

$$= \begin{cases} -\displaystyle\sum_{r=1}^{n^{\rm P}} \left[\hat{w}_{\rm e}^*(\boldsymbol{\zeta}_2^{r*} - \boldsymbol{\sigma}_l^r) + \frac{1}{2H_{\rm i}} \left(\sqrt{\tfrac{3}{2}} z_0^{r*} - R_l^r \right)^2 \right] & \\ \quad + \displaystyle\sum_{j \in \tilde{\Gamma}_{\rm D}} t_j \underline{\dot{u}}_j & \text{if } z_0^{r*} = \sqrt{\tfrac{2}{3}} R_l^r + \tfrac{2}{3} H_{\rm i} \dot{\gamma}^r, \\ & \quad \boldsymbol{\zeta}_1^{r*} = \mathrm{dev}(\boldsymbol{\zeta}_2^{r*}), \\ & \quad B^* \cdot \boldsymbol{\sigma} = \boldsymbol{t}, \\ & \quad z_0^{r*} \geq \|\boldsymbol{\zeta}_1^{r*}\|, \\ & \quad t_j = \underline{t}_j \ (j \in \tilde{\Gamma}_{\rm N}), \\ -\infty & \text{otherwise.} \end{cases}$$

By rewriting the variables as

$$\boldsymbol{\sigma}^r = \boldsymbol{\zeta}_2^{r*}, \quad \dot{s}_{\rm i}^r = \sqrt{\tfrac{3}{2}} z_0^{r*} - R_l^r,$$

and by eliminating $\boldsymbol{\zeta}_1^{r*}$, we can easily see that the Lagrangian dual problem

$$\max_{z_0^*, \boldsymbol{\zeta}_1^*, \boldsymbol{\zeta}_2^*} \ \inf_{\dot{\boldsymbol{u}}, \dot{\boldsymbol{\varepsilon}}_{\rm e}, \dot{\boldsymbol{\varepsilon}}_{\rm p}, \dot{\gamma}} L(\dot{\boldsymbol{u}}, \dot{\boldsymbol{\varepsilon}}_{\rm e}, \dot{\boldsymbol{\varepsilon}}_{\rm p}, \dot{\gamma}; z_0^*, \boldsymbol{\zeta}_1^*, \boldsymbol{\zeta}_2^*)$$

coincides with problem (11.61) as expected. ■

11.4 Plasticity with Kinematic Hardening

It is observed in many metals subjected to cyclic loading that the center of the yield surface shifts in the direction of the plastic flow. This phenomenon is called the *kinematic hardening* effect.

Recall that the von Mises yield criterion is given by (11.23) with (11.15b). The center of the von Mises yield surface is represented by an internal variable $\boldsymbol{\beta} \in \mathbb{S}^3$ called the back stress tensor. Figure 11.7 illustrates a stress–strain response of a one-dimensional elastoplastic model in the presence of the kinematic hardening effect.

In this section we consider the linear kinematic hardening law, while for simplicity the isotropic hardening is neglected; i.e., the radius of the yield surface is assumed to be constant as

$$R = \sigma_{\rm y}$$

in (11.23).

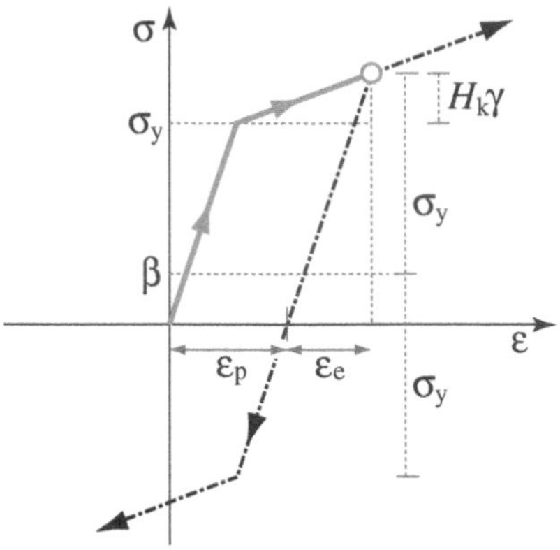

FIGURE 11.7: Constitutive relation of a one-dimensional elastoplastic model with linear kinematic hardening. "∘": the current point.

11.4.1 Linear kinematic hardening

Recall that the von Mises criterion is defined as the upper bound constraint of $\|\boldsymbol{\eta}\|$ (see (11.23a)), where $\boldsymbol{\eta}$ is defined by (see (11.15b))

$$\boldsymbol{\eta} = \mathrm{dev}(\boldsymbol{\sigma}) - \boldsymbol{\beta}. \tag{11.65}$$

The kinematic hardening is characterized by the evolution of the back stress tensor $\boldsymbol{\beta}$.

Let $s_{\mathrm{k}} = \|\boldsymbol{\beta}\|$, and we denote by $\dot{s}_{\mathrm{k}} \geq 0$ the increment of the modulus of the back stress among the time interval $[\tau_l, \tau_{l+1}]$. Within the framework of the associated flow rule, the evolution of the back stress tensor is assumed to be normal to the yield surface. Hence, we obtain

$$\dot{\boldsymbol{\beta}} = \sqrt{\tfrac{2}{3}}\dot{s}_{\mathrm{k}}\frac{\boldsymbol{\eta}}{\|\boldsymbol{\eta}\|}. \tag{11.66}$$

Making use of (11.24), we can rewrite (11.66) alternatively as

$$\dot{\boldsymbol{\beta}} = \sqrt{\tfrac{2}{3}}\dot{s}_{\mathrm{k}}\frac{\dot{\boldsymbol{\varepsilon}}_{\mathrm{p}}}{\|\dot{\boldsymbol{\varepsilon}}_{\mathrm{p}}\|}. \tag{11.67}$$

It is clearly seen in (11.67) that translation of the yield surface due to kinematic hardening is in the direction of the plastic flow. The amount of translation is then given by an evolution law of s_{k}. We here assume that $\dot{s}_{\mathrm{k}}$ depends linearly on $\dot{\gamma}$ as

$$\dot{s}_{\mathrm{k}} = \sqrt{\tfrac{2}{3}}H_{\mathrm{k}}\dot{\gamma}, \tag{11.68}$$

where $\dot{\gamma}$ represents the norm of the plastic flow as seen in (11.24), i.e.,

$$\dot{\gamma} = \|\dot{\boldsymbol{\varepsilon}}_{\mathrm{p}}\|. \tag{11.69}$$

The constant $H_{\mathrm{k}} > 0$ in (11.68) is called the *kinematic hardening modulus*.

11.4.2 Second-order cone complementarity formulation

In the kinematic hardening law, we take notice that the parallelism condition (11.66) between $\dot{\varepsilon}_{\rm p}$ and $\dot{\beta}$ is in the same form of (11.24) relating η and $\dot{\varepsilon}$, which has been reduced to a second-order cone complementarity condition in Proposition 11.2.2. This observation suggests that the linear kinematic hardening law represented by (11.67), (11.68), and (11.69) can also be rewritten as a complementarity condition over second-order cone, which is actually performed below.

Proposition 11.4.1. *Suppose that $\dot{\gamma}$ and $\dot{s}_{\rm k}$ satisfy* (11.68)*. Then $(\dot{\gamma}, \dot{\boldsymbol{\varepsilon}}_{\rm p}) \in \mathbb{R} \times \mathbb{S}^3$ and $(\dot{s}_{\rm k}, \dot{\boldsymbol{\beta}}) \in \mathbb{R} \times \mathbb{S}^3$ satisfy* (11.67) *and* (11.69) *if and only if they satisfy*

$$\dot{\gamma} \geq \|-\dot{\boldsymbol{\varepsilon}}_{\rm p}\|, \tag{11.70a}$$

$$\sqrt{\tfrac{2}{3}}\dot{s}_{\rm k} \geq \|\dot{\boldsymbol{\beta}}\|, \tag{11.70b}$$

$$\dot{\gamma}\Big(\sqrt{\tfrac{2}{3}}\dot{s}_{\rm k}\Big) + (-\dot{\boldsymbol{\varepsilon}}_{\rm p}) : \dot{\boldsymbol{\beta}} = 0. \tag{11.70c}$$

Proof. It is straightforward to see that (11.67) and (11.69) imply (11.70).

Conversely, suppose that (11.68) and (11.70) are satisfied. From (11.70a) and (11.70b) we obtain $\dot{\gamma} \geq 0$ and $\dot{s}_{\rm k} \geq 0$. Since $H_{\rm k} > 0$, (11.68) implies that $\dot{\gamma}$ and $\dot{s}_{\rm k}$ satisfy either (a) $\dot{\gamma} > 0$ and $\dot{s}_{\rm k} > 0$, or (b) $\dot{\gamma} = 0$ and $\dot{s}_{\rm k} = 0$. From this and Fact 11.2.1 that (11.70) holds if and only if any one of the following two cases is true:

(iii) $\dot{\gamma} = \|-\dot{\boldsymbol{\varepsilon}}_{\rm p}\| \neq 0$, $\sqrt{\frac{2}{3}}\dot{s}_{\rm k} = \|\dot{\boldsymbol{\beta}}\| \neq 0$, and there exists $\alpha > 0$ such that $-\dot{\boldsymbol{\varepsilon}}_{\rm p} = -\alpha\dot{\boldsymbol{\beta}}$.

(vi) $\dot{\gamma} = 0$, $-\dot{\boldsymbol{\varepsilon}}_{\rm p} = \boldsymbol{o}$, $\sqrt{\frac{2}{3}}\dot{s}_{\rm k} = 0$, $\dot{\boldsymbol{\beta}} = \boldsymbol{o}$.

Situation (iii) means that $\dot{\boldsymbol{\varepsilon}}_{\rm p}$ and $\dot{\boldsymbol{\beta}}$ are in the same direction. Therefore, this situation is equivalently rewritten as

$$\dot{\gamma} = \|-\dot{\boldsymbol{\varepsilon}}_{\rm p}\| \neq 0, \quad \sqrt{\tfrac{2}{3}}\dot{s}_{\rm k} = \|\dot{\boldsymbol{\beta}}\| \neq 0, \quad \dot{\boldsymbol{\beta}} = \|\dot{\boldsymbol{\beta}}\|\frac{\boldsymbol{\eta}}{\|\boldsymbol{\eta}\|},$$

and hence (11.67) and (11.69) are satisfied. In situation (vi), it is immediate to see that (11.67) and (11.69) are satisfied, which concludes the proof. □

Thus the direction of $\dot{\beta}$ is determined by (11.70). Meanwhile, the modulus of $\dot{\beta}$ is defined by (11.68). The following proposition, in a manner similar to Proposition 11.3.2, provides us with an SOCP problem that determines the direction and modulus of $\dot{\beta}$ simultaneously.

Proposition 11.4.2. *Suppose that* $\dot{\boldsymbol{\varepsilon}}_{\rm p} \in \mathbb{S}^3$ *is given. Then there exists* $\dot{\gamma} \in \mathbb{R}$ *satisfying* (11.67), (11.68), *and* (11.69) *if and only if* $(\dot{s}_{\rm k}, \dot{\boldsymbol{\beta}}) \in \mathbb{R}\times\mathbb{S}^3$ *satisfies*[15]

$$(\dot{s}_{\rm k}, \dot{\boldsymbol{\beta}}) \in \arg\max_{\check{s}_{\rm k}, \check{\boldsymbol{\beta}}} \left\{ \dot{\boldsymbol{\varepsilon}}_{\rm p} : \check{\boldsymbol{\beta}} - \frac{1}{2H_{\rm k}} \check{s}_{\rm k}^2 \mid \sqrt{\tfrac{2}{3}} \check{s}_{\rm k} \geq \|\check{\boldsymbol{\beta}}\| \right\}. \tag{11.71}$$

Proof. The assertion can be shown in a manner similar to Proposition 11.3.2. The Lagrangian for the maximization problem in (11.71) can be defined by

$$L(x_0, \boldsymbol{\xi}_1; \check{s}_{\rm k}, \check{\boldsymbol{\beta}}) = \begin{cases} \dot{\boldsymbol{\varepsilon}}_{\rm p} : \check{\boldsymbol{\beta}} - \dfrac{1}{2H_{\rm k}} \check{s}_{\rm k}^2 + \left(x_0 \sqrt{\tfrac{2}{3}} \check{s}_{\rm k} + \boldsymbol{\xi}_1 : \check{\boldsymbol{\beta}} \right) & \text{if } x_0 \geq \|\boldsymbol{\xi}_1\|, \\ +\infty & \text{otherwise,} \end{cases} \tag{11.72}$$

where $x_0 \in \mathbb{R}$ and $\boldsymbol{\xi}_1 \in \mathbb{S}^3$ are the Lagrangian multipliers. Note that $x = (x_0, \boldsymbol{\xi}_1)$ and $z^* = (\check{s}_{\rm k}, \check{\boldsymbol{\beta}})$ correspond to the primal and dual variables, respectively, of the notation of the Lagrangian duality theory in section 2.2.6, because we have treated the maximization problem (2.35) as a dual problem in the duality theory.

It follows from Proposition 2.2.17 that the optimal solution of the problem in (11.71) corresponds to a saddle point of L. In the same manner as the proof of Proposition 11.3.2, the saddle-point condition for L is obtained explicitly as

$$\dot{\boldsymbol{\varepsilon}}_{\rm p} + \bar{\boldsymbol{\xi}}_1 = \boldsymbol{o}, \tag{11.73a}$$

$$-\frac{1}{H_{\rm k}} \bar{\check{s}}_{\rm k} + \sqrt{\tfrac{2}{3}} \bar{x}_0 = 0, \tag{11.73b}$$

$$\sqrt{\tfrac{2}{3}} \bar{\check{s}}_{\rm k} \geq \|\bar{\check{\boldsymbol{\beta}}}\|, \quad \bar{x}_0 \geq \|\bar{\boldsymbol{\xi}}_1\|, \quad \bar{x}_0 \sqrt{\tfrac{2}{3}} \bar{\check{s}}_{\rm k} + \bar{\boldsymbol{\xi}}_1 : \bar{\check{\boldsymbol{\beta}}} = 0. \tag{11.73c}$$

By putting

$$x_0 = \dot{\gamma}, \tag{11.74}$$

(11.73b) is reduced to (11.68). Moreover, (11.73a) and (11.73c) are reduced to (11.70), which is further equivalent to (11.67) and (11.69) as shown in Proposition 11.4.1. □

The condition (11.71) is regarded as the maximum dissipation law for linear kinematic hardening. We shall make use of (11.71) to formulate the minimization problem of the complementary energy in section 11.4.3 (see (CE) in (11.77)).

Substitute (11.73a) and (11.73b) to see that

$$\sup_{\check{s}_{\rm k}, \check{\boldsymbol{\beta}}} L(x_0, \boldsymbol{\xi}_1; \check{s}_{\rm k}, \check{\boldsymbol{\beta}}) = \begin{cases} \dfrac{1}{3} H_{\rm k} x_0^2 & \text{if } x_0 \geq \|\boldsymbol{\xi}_1\|, \ \boldsymbol{\xi}_1 = -\dot{\boldsymbol{\varepsilon}}_{\rm p}, \\ +\infty & \text{otherwise.} \end{cases}$$

[15] We write $\check{s}_{\rm i}$ and $\check{\boldsymbol{\beta}}$ instead of $\check{s}_{\rm i}$ and $\check{\boldsymbol{\beta}}$, respectively, to avoid too complicated notation.

By rewriting the variable x_0 according to (11.74), we obtain the problem dual to (11.71) as

$$\dot{\gamma} \in \arg\min_{\check{\gamma}} \left\{ \frac{1}{3} H_{\mathrm{k}} \check{\gamma}^2 \mid \check{\gamma} \geq \|-\dot{\boldsymbol{\varepsilon}}_{\mathrm{p}}\| \right\}. \tag{11.75}$$

The right-hand side of (11.75) is equal to the singleton $\{\|-\dot{\boldsymbol{\varepsilon}}_{\mathrm{p}}\|\}$, which agrees with (11.69). In section 11.4.3 we shall use the objective function of (11.75) to formulate the minimization problem of the potential energy (see (PE) in (11.76)).

11.4.3 SOCP formulation of incremental problem

The incremental problem for linear kinematic hardening is formulated as a primal-dual pair of conic optimization problems. See section 11.3.3 for the notation required to state the incremental problem.

In section 11.4.2 we showed that the linear kinematic hardening law can be reformulated as a second-order cone complementarity condition. However, unlike the case of isotropic hardening dealt with in section 11.3, the obtained condition (11.70) does not involve the associated flow rule; in (11.70), only the evolution rule of $\dot{\boldsymbol{\beta}}$ is treated, and the evolution rule of $\dot{\boldsymbol{\varepsilon}}_{\mathrm{p}}$ is not considered. Therefore, besides (11.70), we also have to consider (11.26) in Proposition 11.2.2, which is a second-order cone complementarity formulation of the associated flow rule. Thus the associative plasticity with linear kinematic hardening is described by a pair of second-order cone complementarity conditions, while in contrast the one with linear isotropic hardening is represented as a single second-order cone complementarity condition as shown in Proposition 11.3.1.

Recall that we denote by $\hat{w}_{\mathrm{e}}$ the strain energy function for the elastic part, as defined in (11.58). By using (11.26) and (11.75), the minimization problem of the potential energy for the incremental problem is formulated as

$$\left.\begin{aligned}
\text{(PE)}: \quad & \min_{\dot{\boldsymbol{u}}, \dot{\boldsymbol{\varepsilon}}_{\mathrm{e}}, \dot{\boldsymbol{\varepsilon}}_{\mathrm{p}}, \dot{\gamma}, \dot{\gamma}_{\mathrm{k}}} \quad \sum_{r=1}^{n^{\mathrm{P}}} \Big(\hat{w}_{\mathrm{e}}(\dot{\boldsymbol{\varepsilon}}_{\mathrm{e}}^r) + \boldsymbol{\sigma}_l^r : \dot{\boldsymbol{\varepsilon}}_{\mathrm{e}}^r + \boldsymbol{\beta}_l^r : \dot{\boldsymbol{\varepsilon}}_{\mathrm{p}}^r \\
& \qquad\qquad + \sqrt{\tfrac{2}{3}} R \dot{\gamma}^r + \frac{1}{3} H_{\mathrm{k}} (\dot{\gamma}_{\mathrm{k}}^r)^2 \Big) - \sum_{j \in \tilde{\varGamma}_{\mathrm{N}}} \underline{t}_j \dot{u}_j \\
\text{s.t.} \quad & \dot{\boldsymbol{\varepsilon}}_{\mathrm{e}} + \operatorname{dev}(\dot{\boldsymbol{\varepsilon}}_{\mathrm{p}}) = B \dot{\boldsymbol{u}}, \\
& \dot{\gamma}^r \geq \|-\dot{\boldsymbol{\varepsilon}}_{\mathrm{p}}^r\|, \quad \forall r, \\
& \dot{\gamma}_{\mathrm{k}}^r \geq \|-\dot{\boldsymbol{\varepsilon}}_{\mathrm{p}}^r\|, \quad \forall r, \\
& \dot{u}_j = \underline{\dot{u}}_j, \quad \forall j \in \tilde{\varGamma}_{\mathrm{D}}.
\end{aligned}\right\} \tag{11.76}$$

We can see that problem (11.76) consists of the following:

- the elastic part of the strain energy $\hat{w}_{\mathrm{e}}(\dot{\boldsymbol{\varepsilon}}_{\mathrm{e}}^r) + \boldsymbol{\sigma}_l^r : \dot{\boldsymbol{\varepsilon}}_{\mathrm{e}}^r$
- the plastic part of the strain energy $\boldsymbol{\beta}_l^r : \dot{\boldsymbol{\varepsilon}}_{\mathrm{p}}^r + \sqrt{\frac{2}{3}} R \dot{\gamma}^r + \frac{1}{3} H_{\mathrm{k}} (\dot{\gamma}_{\mathrm{k}}^r)^2$

- the compatibility condition (see (11.56b))
- the second-order cone constraints on the kinematic variables (see (11.26a) and (11.75))

Note that in (11.76) we use $\dot{\gamma}_{\rm k}^{r}$ for the conditions arising from (11.75) to distinguish it from the variable $\dot{\gamma}^{r}$ involved in the conditions arising from (11.26a).

Similarly, the minimization problem of the complementary energy is formulated by using (11.26) and (11.71) as

$$\left.\begin{array}{lll} (\mathrm{CE}): & \displaystyle\min_{\boldsymbol{\sigma},\boldsymbol{t},\dot{s}_{\rm k},\dot{\boldsymbol{\beta}}} & \displaystyle\sum_{r=1}^{n^{\rm P}}\left(\hat{w}_{\rm e}^{*}(\boldsymbol{\sigma}^{r}-\boldsymbol{\sigma}_{l}^{r})+\frac{1}{2}\frac{(\dot{s}_{\rm k}^{r})^{2}}{H_{\rm k}}\right)-\sum_{j\in\tilde{\Gamma}_{\rm D}}t_{j}\underline{\dot{u}}_{j} \\ & \mathrm{s.\,t.} & B^{*}\cdot\boldsymbol{\sigma}=\boldsymbol{t}, \\ & & \sqrt{\frac{2}{3}}R\geq\|\,\mathrm{dev}(\boldsymbol{\sigma}^{r})-(\boldsymbol{\beta}_{l}^{r}+\dot{\boldsymbol{\beta}}^{r})\|, \quad \forall r, \\ & & \sqrt{\frac{2}{3}}\dot{s}_{\rm k}^{r}\geq\|\dot{\boldsymbol{\beta}}^{r}\|, \quad \forall r, \\ & & t_{j}=\underline{t}_{j}, \quad \forall j\in\tilde{\Gamma}_{\rm N}, \end{array}\right\} \qquad (11.77)$$

where $\hat{w}_{\rm e}^{*}$ is the complementary strain energy of the elastic strain, as seen in (11.60). Note that problem (11.77) consists of the following:

- the elastic part of the complementary strain energy $\hat{w}_{\rm e}^{*}(\boldsymbol{\sigma}^{r}-\boldsymbol{\sigma}_{l}^{r})$
- the plastic part of the complementary strain energy $\frac{1}{2H_{\rm k}}(\dot{s}_{\rm k}^{r})^{2}$ (see (11.71))
- the force-balance equation (see (11.56c))
- the second-order cone constraints on the static variables (see (11.26b) and (11.71), together with (11.65))

Thus the incremental problem with the linear kinematic hardening rule is formulated as a prima-dual pair of conic optimization problems.[16] Like the isotropic hardening case discussed in section 11.3.4, (PE) in (11.76) has (i) a convex quadratic objective function, (ii) linear equality constraints, and (iii) second-order cone constraints. Therefore, (PE) can be reduced to a standard form of SOCP. See Yonekura–Kanno [462] for numerical experiments, in which (PE) is solved by using the prima-dual interior-point method for SOCP.

[16] Indeed, in a manner similar to Remark 11.3.3, the reader can confirm that (CE) in (11.77) can be derived as the Lagrangian dual of (PE) in (11.76).

11.5 Notes

1. The study of plasticity dates back to the late 19th century (Tresca and St. Venant) and the 1910s (von Mises). For more about the classic work on plasticity, see, e.g., Koiter [235] and Prager–Hodge [376].

2. The first comprehensive study of the boundary value problems in plasticity is in Duvaut–Lions [120, Chap. 5], in which the problem of the perfect plasticity is formulated as a variational inequality. An investigation of differential inclusion is in to Moreau [321].

Since the literature on plasticity has diverged, we do not aim to give a complete literature survey; see Han–Reddy [159] and Kojić–Bathe [237] for more information. We also refer to Simo–Hughes [413] for the radial return method (or the return-mapping algorithm), which is used very widely nowadays for elastoplastic analysis.

3. An introduction of optimization theory and algorithms to plasticity is in Maier [285, 286], in which the yield surface is approximated piecewise linearly. Accordingly, the admissible set of stress tensors is represented by finitely many linear inequalities, and hence the minimization problem of the complementary energy results in a quadratic programming (QP) problem.[17] To see the reduction to QP more clearly, consider (CE) in (11.61), and suppose that the second-order cone constraint $\sqrt{\frac{2}{3}}(R_l + \dot{s}_{\mathrm{i}}) \geq \|\operatorname{dev}(\boldsymbol{\sigma})\|$ is approximated by a polyhedral cone. Then the obtained problem involves only linear constraints, while the objective function is a convex quadratic function. Thus the approximated problem is a QP problem.

This QP-based approach, using a piecewise-linear approximation of the yield surface, has been extensively studied by Capurso–Maier [71] and Maier [285, 286, 287, 288, 289].

Subsequently, the optimization-based methods have been developed for more or less fine plasticity models. They include the augmented Lagrangian methods by Cuomo–Contrafatto [99] and Contrafatto–Ventura [93] and the parametric quadratic programming approaches by Liew–Wu–Zou–Ng [268], Zhang–He–Li–Wriggers [465], Zhang–Zhong–Wu–Liao [466], and Zhong–Sun [470]. The interior-point method for solving general nonlinear programming problems was applied by Krabbenhøft–Lyamin–Sloan–Wriggers [246] to the elastoplastic problems. For numerical solutions based on the complementarity problem, we may refer to Zhu [471], which is based on the linear complementarity problem, and Hu–Soh–Chen–Li–Lin [189], Tin-Loi–Pang [430], and

[17] By QP we mean that a minimization problem of a convex quadratic function under some linear inequality constraints; see (1.31) for the definition.

Tin-Loi–Xia [431, 432], which are based on the nonlinear complementarity problem. An approach based on the bi-potential, which was originally considered for frictional contact problems, was extended to non-associated plasticity by Hjiaj–Fortin–de Saxcé [180].

4. Recently, the Mohr–Coulomb yielding criterion, which is often used in the limit analysis of geo-materials, has been shown to be represented as a linear matrix inequality; see Bisboss–Pardalos [53] and Krabbenhøft–Lyamin–Sloan [244]. As a consequence, the limit analysis problem can be reduced to a semidefinite programming (SDP) problem (Krabbenhøft–Lyamin–Sloan [245]).

The von Mises yielding criterion is widely used for elastoplastic analysis of metals. The main results in this chapter are taken from Yonekura–Kanno [462], in which an SOCP formulation for a combined law of linear isotropic and kinematic hardening was also presented.

References

[1] Acary, V., Brogliato, B.: *Numerical Methods for Nonsmooth Dynamical Systems.* Springer-Verlag, Berlin (2008).

[2] Achtziger, W., Bendsøe, M.P., Ben-Tal, A., Zowe, J.: Equivalent displacement based formulations for maximum strength truss topology design. *Impact of Computing in Science and Engineering,* **4**, 315–345 (1992).

[3] Achtziger, W., Kočvara, M.: On the maximization of the fundamental eigenvalue in topology optimization. *Structural and Multidisciplinary Optimization,* **34**, 181–195 (2007).

[4] Achtziger, W., Kočvara, M.: Structural topology optimization with eigenvalues. *SIAM Journal on Optimization,* **18**, 1129–1164 (2007).

[5] Akita, T., Nakashino, K., Natori, M.C., Park, K.C.: A simple computer implementation of membrane wrinkle behaviour via a projection technique. *International Journal for Numerical Methods in Engineering,* **71**, 1231–1259 (2007).

[6] Alart, P., Curnier, A.: A mixed formulation for frictional contact problems prone to Newton like solution methods. *Computer Methods in Applied Mechanics and Engineering,* **92**, 353–375 (1991).

[7] Alfano, G., Rosati, L., Valoroso, N.: A numerical strategy for finite element analysis of no-tension materials. *International Journal for Numerical Methods in Engineering,* **48**, 317–350 (2000).

[8] Alizadeh, F., Goldfarb, D.: Second-order cone programming. *Mathematical Programming,* **95**, 3–51 (2003).

[9] Alizadeh, F., Haeberly, J.-P.A., Overton, M.L.: Complementarity and nondegeneracy in semidefinite programming. *Mathematical Programming,* **77**, 111–128 (1997).

[10] Allaire, G.: *Numerical Analysis and Optimization.* Oxford University Press, New York (2007).

[11] Anderson, E.J., Nash, P.: *Linear Programming in Infinite-Dimensional Spaces.* Wiley, Chichester (1987).

[12] Andersson, L.-E.: A quasistatic frictional problem with normal compliance. *Nonlinear Analysis,* **16**, 347–369 (1991).

[13] Andersson, L.-E.: Existence results for quasistatic contact problems with Coulomb friction. *Applied Mathematics and Optimization*, **42**, 169–202 (2000).

[14] Andersson, L.-E., Klarbring, A.: A review of the theory of static and quasi-static frictional contact problems in elasticity. *Philosophical Transactions of the Royal Society of London. Series A, Mathematical and Physical Sciences*, **359**, 2519–2539 (2001).

[15] Ando, K., Mitsugi, J., Senbokuya, Y.: Analyses of cable–membrane structure combined with deployable truss. *Computers & Structures*, **74**, 21–39 (2000).

[16] Angelillo, M.: Constitutive relations for no-tension materials. *Meccanica*, **28**, 195–202 (1993).

[17] Anjos, M.F.: Semidefinite optimization approaches for satisfiability and maximum-satisfiability problems. *Journal on Satisfiability, Boolean Modeling and Computation*, **1**, 1–47 (2005).

[18] Antman, S.S.: *Nonlinear Problems of Elasticity (2nd ed.)*. Springer, New York (2005).

[19] Anzellotti, G.: A class of convex non-coercive functionals and masonry-like materials. *Annales de l'Institut Henri Poincare*, **2**, 261–307 (1985).

[20] Argyris, J.H., Angelopoulos, T., Bichat, B.: A general method for the shape finding of lightweight tension structures. *Computer Methods in Applied Mechanics and Engineering*, **3**, 135–149 (1974).

[21] Argyris, J.H., Scharpf, D.W.: Large deflection analysis of prestressed networks. *Journal of Engineering Mechanics (ASCE)*, **98**, 633–654. (1972)

[22] Atai, A.A., Mioduchowski, A.: Equilibrium analysis of elasto-plastic cable nets. *Computers & Structures*, **66**, 163–171 (1998).

[23] Atai, A.A., Steigmann, D.J.: On the nonlinear mechanics of discrete networks. *Archive of Applied Mechanics*, **67**, 303–319 (1997).

[24] Aubin, J.-P., Ekeland, I.: *Applied Nonlinear Analysis.* John Wiley & Sons, New York (1984). Also: Dover Publications, Mineola (2006).

[25] Avriel, M.: *Nonlinear Programming.* Printice-Hall, Englewood Cliffs (1976). Also: Dover Publications, Mineola (2003).

[26] Awrejcewicz, J., Lamarque, C. H.: *Bifurcation and Chaos in Nonsmooth Mechanical Systems.* World Scientific Publishing, Singapore (2003).

[27] Baginski, F., Collier, W.: A mathematical model for the strained shape of a large scientific balloon at float altitude. *Journal of Applied Mechan-*

ics (ASME), **67**, 6–16 (2000).

[28] Ballard, P.: A counter-example to uniqueness in quasi-static elastic contact problems with small friction. *International Journal of Engineering Science*, **37**, 163–178 (1999).

[29] Barber, J.R., Ciavarella, M.: Contact mechanics. *International Journal of Solids and Structures*, **37**, 29–43 (2000).

[30] Barber, J.R., Hild, P.: Non-uniqueness, eigenvalue solutions and wedged configurations involving Coulomb friction. *Proceedings of ASME/STLE International Joint Tribology Conference*, TRIB2004-64368, Long Beach California (2004).

[31] Barber, J.R., Hild, P.: On wedged configurations with Coulomb friction. In: Wriggers, P., Nackenhorst, U. (eds.) *Analysis and Simulation of Contact Problems*, pp. 205–213, Springer-Verlag, Berlin (2006).

[32] Barnes, M.R.: Form-finding and analysis of prestressed nets and membranes. *Computers & Structures*, **30**, 685–695 (1988).

[33] Barnes, M.R.: Form finding and analysis of tension structures by dynamic relaxation. *International Journal of Space Structures*, **14**, 89–104 (1999).

[34] Barsotti, R., Ligarò, S.S., Royer-Carfagni, G.F.: The web bridge. *International Journal of Solids and Structures*, **38**, 8831–8850 (2001).

[35] Barvinok, A.: *A Course in Convexity.* Americal Mathematical Society, Providence (2002).

[36] Beckers, M., Fleury, C.: A primal-dual approach in truss topology optimization. *Computers & Structures*, **64**, 77-88 (1997).

[37] Bendsøe, M.P., Sigmund, O.: *Topology Optimization.* Springer-Verlag, Berlin (2003).

[38] Ben-Haim, Y.: *Information-gap Decision Theory: Decisions under Severe Uncertainty (2nd ed.).* Academic Press, London (2006).

[39] Ben-Haim, Y., Elishakoff, I.: *Convex Models of Uncertainty in Applied Mechanics.* Elsevier, New York (1990).

[40] Ben-Tal, A., Bendsøe, M.P.: A new method for optimal truss topology design. *SIAM Journal on Optimization*, **3**, 322–358 (1993).

[41] Ben-Tal, A., Jarre, F., Kočvara, M., Nemirovski, A., Zowe, J.: Optimal design of trusses under a nonconvex global buckling constraint. *Optimization and Engineering*, **1**, 189–213 (2000).

[42] Ben-Tal, A., Boyd, S., Nemirovski, A.: Extending scope of robust optimization: Comprehensive robust counterparts of uncertain problems.

Mathematical Programming, **B107**, 63–89 (2006).

[43] Ben-Tal, A., El Ghaoui, L., Nemirovski, A.: *Robust Optimization*. Princeton University Press, Princeton (2009).

[44] Ben-Tal, A., Nemirovski, A.: Robust truss topology design via semidefinite programming. *SIAM Journal on Optimization*, **7**, 991–1016 (1997).

[45] Ben-Tal, A., Nemirovski, A.: Robust convex optimization. *Mathematics of Operations Research*, **23**, 769–805 (1998).

[46] Ben-Tal, A., Nemirovski, A.: *Lectures on Modern Convex Optimization: Analysis, Algorithms, and Engineering Applications*. SIAM, Philadelphia (2001).

[47] Ben-Tal, A., Nemirovski, A.: Robust optimization—methodology and applications. *Mathematical Programming*, **B92**, 453–480 (2002).

[48] Ben-Tal, A., Nemirovski, A.: Selected topics in robust convex optimization. *Mathematical Programming*, **B112**, 125–158 (2008).

[49] Berto, L., Saetta, A., Scotta, R., Vitaliani, R.: An orthotropic damage model for masonry structures. *International Journal for Numerical Methods in Engineering*, **55**, 127–157 (2002).

[50] Bertóti, E.: On the principle of complementary virtual work in the non-linear theory of elasticity. *Journal of Computational and Applied Mechanics*, **1**, 109–122 (2000).

[51] Bertsimas, D., Popescu, I.: Optimal inequalities in probability theory: A convex optimization Approach. *SIAM Journal on Optimization*, **15**, 780–804 (2005).

[52] Birnstiel, C.: Analysis and design of cable structures. *Computers & Structures*, **2**, 817–831 (1972).

[53] Bisbos, C,D., Pardalos, P.M.: Second-order cone and semidefinite representations of material failure criteria. *Journal of Optimization Theory and Applications*, **134**, 275–301 (2007).

[54] Björkman, G.: Path following and critical points for contact problems. *Computational Mechanics*, **10**, 231–246 (1992).

[55] Bletzinger, K.-U., Ramm, E.: A general finite element approach to the form finding of tensile structures by the updated reference strategy. *International Journal of Space Structures*, **14**, 131–145 (1999).

[56] Bletzinger, K.-U., Wüchner, R., Daoud, F., Camprubí, N.: Computational methods for form finding and optimization of shells and membranes. *Computer Methods in Applied Mechanics and Engineering*, **194**, 3438–3452 (2005).

[57] Borchers, B.: CSDP 2.3 user's guide, a C library for semidefinite programming. *Optimization Methods and Software*, **11/12**, 597–611 (1999).

[58] Borchers, B., Young, J.G.: Implementation of a primal-dual method for SDP on a shared memory parallel architecture. *Computational Optimization and Applications*, **37**, 355–369 (2007).

[59] Borwein, J.M., Lewis, A.S.: *Convex Analysis and Nonlinear Optimization (2nd ed.)*. Springer, New York (2006).

[60] Boyd, S., El Ghaoui, L., Feron, E., Balakrishnan, V.: *Linear Matrix Inequalities in System and Control Theory.* SIAM, Philadelphia (1994).

[61] Boyd, S., Vandenberghe, L.: *Convex Optimization.* Cambridge University Press, Cambridge (2004).

[62] Bradshaw, R., Campbell, D., Gargari, M., Mirmiran, A., Tripeny, P.: Special structures: Past, Present, and Future. *Journal of Structural Engineering (ASCE)*, **128**, 691–709 (2002).

[63] Braides, A., Chiadò Piat, V.: Another brick in the wall. In: dal Maso, G., DeSimone, A., Tomarelli, F. (eds.) *Variational Problems in Material Science*, pp. 13–24, Birkhäuser, Basel (2006).

[64] Bratus, A.S., Seyranian, A.P.: Bimodal solutions in eigenvalue optimization problems. *Journal of Applied Mathematics and Mechanics*, **47**, 451–457 (1983).

[65] Bratus, A.S., Seyranian, A.P.: Sufficient conditions for an extremum in eigenvalue optimization problems. *Journal of Applied Mathematics and Mechanics*, **48**, 466–474 (1984).

[66] Brogliato, B.: *Nonsmooth Mechanics: Models, Dynamics and Control (2nd ed.)*. Springer-Verlag, London (1999).

[67] Calafiore, G., El Ghaoui, L.: Ellipsoidal bounds for uncertain linear equations and dynamical systems. *Automatica*, **40**, 773–787 (2004).

[68] Campos, L.T., Oden, J.T.: On the principle of stationary complementary energy in finite elastostatics. *International Journal of Engineering Science*, **23**, 57–63 (1985).

[69] Cannarozzi, M.: Stationary and extremum variational formulations for the elastostatics of cable networks. *Meccanica*, **20**, 136–143 (1985).

[70] Cannarozzi, M.: A minimum principle for tractions in the elastostatics of cable networks. *International Journal of Solids and Structures*, **23**, 551–568 (1987).

[71] Capurso, M., Maier, G.: Incremental elastoplastic analysis and quadratic optimization. *Meccanica*, **5**, 107–116 (1970).

[72] Cecchi, A., Di Marco, R.: Homogenization of masonry walls with a computational oriented procedure. Rigid or elastic block? *European Journal of Mechanics, A/Solids*, **19**, 535–546 (2000).

[73] Chateau, X., Nguyen, Q.S.: Buckling of elastic structures in unilateral contact with or without friction. *European Journal of Mechanics, A/Solids*, **10**, 71–89 (1991).

[74] Chen, B., Chen, X.: A global and local superlinear continuation-smoothing method for $P_0 + R_0$ and monotone NCP. *SIAM Journal Optimization*, **7**, 403–420 (1997).

[75] Chen, B., Chen, X., Kanzow, C.: A penalized Fishcer–Burmeister NCP functions. *Mathematical Programming*, **88**, 211–216 (2000).

[76] Chen, J.-S., Chen, X., Tseng, P.: Analysis of nonsmooth vector-valued functions associated with second-order cones. *Mathematical Programming*, **B101**, 95–117 (2004).

[77] Chen, X.D., Sun, D., Sun, J.: Complementarity functions and numerical experiments on some smoothing newton methods for second-order-cone complementarity problems. *Computational Optimization and Applications*, **25**, 39–56 (2003).

[78] Chen, X., Tseng, P.: Non-interior continuation methods for solving semidefinite complementarity problems. *Mathematical Programming*, **A95**, 431–474 (2003).

[79] Chen, J.-S., Tseng, P.: An unconstrained smooth minimization reformulation of the second-order cone complementarity problem. *Mathematical Programming*, **B104**, 293–327 (2005).

[80] Choi, K.K., Kim, N.H.: *Structural Sensitivity Analysis and Optimization. Volume I: Linear Systems; Volume II: Nonlinear Systems and Applications.* Springer, New York (2005).

[81] Chand, R., Haug, E.J., Rim, K.: Analysis of unbounded contact problems by means of quadratic programming. *Journal of Optimization Theory and Applications*, **20**, 171–189 (1976).

[82] Chen, M.X., Sun, Q.P., Wu, Z., Yuen, M.M.F.: A wrinkled membrane model for cloth draping with multigrid acceleration. *Journal of Manufacturing Science and Engineering (ASME)*, **121**, 695–700 (1999).

[83] Chisaliţa, A.: Finite deformation analysis of cable networks. *Journal of Engineering Mechanics (ASCE)*, **110**, 207–223 (1984).

[84] Christensen, P.W.: A nonsmooth Newton method for elastoplastic problems. *Computer Methods in Applied Mechanics and Engineering*, **191**, 1189–1219 (2002).

[85] Christensen, P.W.: A semi-smooth Newton method for elasto-plastic contact problems *International Journal of Solids and Structures*, **39**, 2323–2341 (2002).

[86] Christensen, P.W., Klarbring, A.: *An Introduction to Structural Optimization.* Springer (2009).

[87] Christensen, P.W., Klarbring, A., Pang, J.-S., Strömberg, N.: Formulation and comparison of algorithms for frictional contact problems. *International Journal for Numerical Methods in Engineering*, **42**, 145–173 (1998).

[88] Christensen, P.W., Pang, J.-S.: Frictional contact algorithms based on semismooth Newton methods. In: Fukushima, M., Qi, L. (eds.) *Reformulation—Nonsmooth, Piecewise Smooth, Semismooth and Smoothing Methods*, pp. 81–116, Kluwer Academic Publishers, Boston (1998).

[89] Chvátal, V.: *Linear Programming.* W.H. Freeman and Company, New York (1983).

[90] Ciarlet, P.G.: *Mathematical Elasticity. Volume I: Three-Dimensional Elasticity.* Elsevier, Amsterdam (1988).

[91] Cocu, M., Pratt, E., Raous, M.: Formulation and approximation of quasistatic frictional contact. *International Journal of Engineering Science*, **34**, 783–798 (1996).

[92] Cocu, M., Rocca, R.: Existence results for unilateral quasistatic contact problems with friction and adhesion. *Mathematical Modelling and Numerical Analysis*, **34**, 981–1001 (2000).

[93] Contrafatto, L., Ventura, G.: Numerical analysis of augmented Lagrangian algorithm in complementary elastoplasticity. *International Journal for Numerical Methods in Engineering*, **60**, 2263–2287 (2004).

[94] Contri, P., Schrefler, B.A.: A geometrically nonlinear finite element analysis of wrinkled membrane surfaces by a no-compression material model. *Communications in Applied Numerical Methods*, **4**, 5–15 (1988).

[95] Contro, R., Maier, G., Zavelani, A.: Inelastic analysis of suspension structures by nonlinear programming. *Computer Methods in Applied Mechanics and Engineering*, **5**, 127–143 (1975).

[96] Cottle, R.W., Pang, J.-S., Stone, R.E.: *The Linear Complementarity Problem.* Academic Press, San Diego (1992).

[97] Cox, S.J., Overton M.L.: On the optimal design of columns against buckling. *SIAM Journal on Mathematical Analysis*, **23**, 287–325 (1992).

[98] Coyette, J.P., Guisset, P.: Cable network analysis by a nonlinear pro-

gramming technique. *Engineering Structures*, **10**, 41–46 (1988).

[99] Cuomo, M., Contrafatto, L.: Stress rate formulation for elastoplastic models with internal variables based on augmented Lagrangian regularization. *International Journal of Solids and Structures*, **37**, 3935–3964 (2000).

[100] Cuomo, M., Ventura, G.: A complementary energy formulation of no tension masonry-like solids. *Computer Methods in Applied Mechanics and Engineering*, **189**, 313–339 (2000).

[101] Curnier, A.: Unilateral contact: mechanical modelling. In: Wriggers, P, Panagiotopoulos, P. (eds.) *New Developments in Contact Problems*, pp. 1–54, Springer-Verlag, Wien (1999).

[102] Danielson, D.A., Natarajan, S.: Tension field theory and the stress in stretched skin. *Journal of Biomechanics*, **8**, 135–142 (1975).

[103] Dantzig, G.B., Thapa, M.N.: *Linear Programming. Volume I: Introduction; Volume II: Theory and Extensions.* Springer, New York (1997, 2003).

[104] Day, A.S., Bunce, J.: Analysis of cable networks by dynamic relaxation. *Civil Engineering and Public Works Review*, **4**, 383–386 (1970).

[105] Deimling, K.: *Multivalued Differential Equations.* Walter de Gruyter, Berlin (1992).

[106] Del Piero, G.: Constitutive equation and compatibility of the external loads for linear elastic masonry-like materials. *Meccanica*, **24**, 150–162 (1989).

[107] Del Piero, G.: Limit analysis and no-tension materials. *International Journal of Plasticity*, **14**, 259–271 (1998).

[108] Den, H., Jiang, Q.F., Kwan, A.S.K.: Shape finding of incomplete cable-strut assemblies containing slack and prestressed elements. *Computers & Structures*, **83**, 1767–1779 (2005).

[109] de Saxcé, G., Feng, Z.-Q.: The bipotential method: a constructive approach to design the complete contact law with friction and improved numerical algorithms. *Mathematical and Computer Modelling*, **28**, 225–245 (1998).

[110] de Saxcé, G., Nguyen-Dang, H.: Dual analysis of frictionless problems by displacement and equilibrium finite elements. *Engineering Structures*, **6**, 26–32 (1984).

[111] Díaz, A.R., Kikuchi, N.: Solutions to shape and topology eigenvalue optimization problems using a homogenization method. *International Journal for Numerical Methods in Engineering*, **35**, 1487–1502 (1992).

[112] Diestel, R.: *Graph Theory (3rd ed.)*. Springer-Verlag, Heidelberg (2005).

[113] De Luca, T., Facchinei, F., Kanzow C.: A semismooth equation approach to the solution of nonlinear complementarity problems. *Mathematical Programming*, **75**, 407–439 (1996).

[114] De Luca, T., Facchinei, F., Kanzow C.: A theoretical and numerical comparison of some semismooth algorithms for complementarity problems. *Computational Optimization and Applications*, **16**, 173–205 (2000).

[115] Demkowicz, L.: On some results concerning the reciprocal formulation for the Signorini's problem. *Computers & Structures*, **8**, 57–74 (1982).

[116] di Bernardo, M., Budd, C.J., Champneys, A.R., Kowalczyk, P.: *Piecewise-Smooth Dynamical Systems*. Springer, London (2008).

[117] Ding, H., Yang, B.: The modeling and numerical analysis of wrinkled membranes. *International Journal for Numerical Methods in Engineering*, **58**, 1785–1801 (2003).

[118] Di Pasquale, S.: New trends in the analysis of masonry structures. *Meccanica*, **27**, 173–184 (1992).

[119] Dirkse, S.P., Ferris, M.C.: The PATH solver: A non-monotone stabilization scheme for mixed complementarity problems. *Optimization Methods and Software*, **5**, 123–156 (1995).

[120] Duvaut, G., Lions, J.L.: *Inequalities in Mechanics and Physics*. Springer-Verlag, Berlin (1976).

[121] Ekeland, I., Temam, R: *Convex Analysis and Variational Problems*. North-Holland, Amsterdam (1976); SIAM, Philadelphia (1999).

[122] Ekeland, I., Turnbull, T.: *Infinite-Dimensional Optimization and Convexity*. The University of Chicago Press, London (1983).

[123] Epstein, M.: On the wrinkling of anisotropic elastic membranes. *Journal of Elasticity*, **55**, 99–109 (1999).

[124] Epstein, M., Forcinito, M.A.: Anisotropic membrane wrinkling: theory and analysis. *International Journal of Solids and Structures*, **38**, 5253–5272 (2001).

[125] Facchinei, F., Pang, J.-S.: *Finite-Dimensional Variational Inequalities and Complementarity Problems. Volumes I & II*. Springer-Verlag, New York (2003).

[126] Faraut, U., Korányi, A.: *Analysis on Symmetric Cones*. Oxford University Press, New York (1994).

[127] Faybusovich, L., Moore, J.B.: Infinite-dimensional quadratic optimiza-

tion: Interior-point methods and control applications. *Applied Mathematics and Optimization*, **36**, 43–66 (1997).

[128] Faybusovich, L., Tsuchiya, T.: Primal-dual algorithms and infinite-dimensional Jordan algebras of finite rank. *Mathematical Programming*, **97**, 471–493 (2003).

[129] Feng, Q., Tu, J.: Modeling and algorithm on a class of mechanical systems with unilateral constraints. *Archive of Applied Mechanics*, **76**, 103–116 (2006).

[130] Ferris, M.C., Pang, J.-S.: Engineering and economic applications of complementarity problems. *SIAM Review*, **39**, 669–713 (1997).

[131] Fiacco, A.V., McCormick, G.P.: *Nonlinear Programming: Sequential Unconstrained Minimization Techniques*. John Wiley & Sons, New York (1968). Also: SIAM, Philadelphia (1990).

[132] Forsgren, A., Gill, P.E., Wright, M.H.: Interior methods for nonlinear optimization. *SIAM Review*, **44**, 525–597 (2002).

[133] Fortin, J., Millet, O., de Saxće, G.: Numerical simulation of granular materials by an improved discrete element method. *International Journal for Numerical Methods in Engineering*, **62**, 639–663 (2004).

[134] Fraeijs de Veubeke, B.: A new variational principle for finite elastic displacements. *International Journal of Engineering Science*, **10**, 745–763 (1972).

[135] Freire, A.M.S., Negrão, J.H.O., Lopes, A.V.: Geometrical nonlinearities on the static analysis of highly flexible steel cable-stayed bridges. *Computers & Structures*, **84**, 2128–2140 (2006).

[136] Frémond, M.: *Non-Smooth Thermomechanics*. Springer-Verlag, Berlin (2002).

[137] Fried, I.: Large deformation static and dynamic finite element analysis of extensible cables. *Computers & Structures*, **15**, 315–319 (1982).

[138] Fujikake, M., Kojima, O., Fukushima, S.: Analysis of fabric tension structures. *Computers & Structures*, **32**, 537–547 (1989).

[139] Fujita, R., Kanno, Y.: Enumeration of all wedged equilibrium configurations in contact problem with Coulomb friction. *Computer Methods in Applied Mechanics and Engineering*, **199**, 1202–1215 (2010).

[140] Fukushima, M., Luo, Z.-Q., Tseng, P.: Smoothing functions for second-order cone complementarity problems. *SIAM Journal on Optimization*, **12**, 436–460 (2002).

[141] Fuschi, P., Giambanco, G., Rizzo, S.: Nonlinear finite element analysis

of no-tension masonry structures. *Meccanica*, **30**, 233–249 (1995).

[142] Gao, D.Y.: Bi-complementarity and duality: A framework in nonlinear equilibria with applications to the contact problem of elastoplastic beam. *Journal of Mathematical Analysis and Applications*, **221**, 672–697 (1998).

[143] Gao, D.Y.: Pure complementary energy principle and triality theory in finite elasticity. *Mechanics Research Communications*, **26**, 31–37 (1999).

[144] Gao, D.Y.: General analytic solutions and complementary variational principles for large deformation nonsmooth mechanics. *Meccanica*, **34**, 167–196 (1999).

[145] Giaquinta, M. Giusti, E: Researches on the equilibrium of masonry structures. *Archive for Rational Mechanics and Analysis*, **88**, 359–392 (1985).

[146] Gil, A.J., Bonet, J.: Finite element analysis of partly wrinkled reinforced prestressed membranes. *Computational Mechanics*, **40**, 595–615 (2007).

[147] Giorgi, G., Guerraggio, A., Thierfelder, J.: *Mathematics of Optimization: Smooth and Nonsmooth Case*. Elsevier, Amsterdam (2004).

[148] Gladwell, G.M.L.: *Contact Problems in the Classical Theory of Elasticity*. Sijthoff & noordhoff, Alphen aan den Rijn (1980).

[149] Glocker, C.: *Set-Valued Force Laws*. Springer-Verlag, Berlin (2001).

[150] Goeleven, D., Motreanu, D., Dumont, Y., Rochdi, M.: *Variational and Hemivariational Inequalities: Theory, Methods and Applications. Volume I: Unilateral Analysis and Unilateral Mechanics; Volume II: Unilateral Problems*. Kluwer Academic Publishers, Dordrecht (2003).

[151] Goemans, M.X.: Semidefinite programming in combinatorial optimization. *Mathematical Programming*, **79**, 143–161 (1997).

[152] Goemans, M.X., Williamson, D.P.: Improved approximation algorithms for maximum cut and satisfiability problems using semidefinite programming. *Journal of the Association for Computing Machinery*, **42**, 1115–1145 (1995).

[153] Gould, N.I.M., Orban, D., Toint, P.L.: Numerical methods for large-scale nonlinear optimization. *Acta Numerica*, **14**, 229–361 (2005).

[154] Greenberg, D.P.: Inelastic analysis of suspension roof-structures. *Journal of the Structural Division (ASCE)*, **96**, 905–930 (1970).

[155] Grinold, R.C.: Symmetric duality for continuous linear programs. *SIAM Journal on Applied Mathematics*, **18**, 84–97. (1970).

[156] Guo, Z.-H.: Unified theory of variation principles in non-linear theory

of elasticity. *Applied Mathematics and Mechanics*, **1**, 1–22 (1980).

[157] Guo, X., Bai, W., Zhang, W.: Confidence extremal structural response analysis of truss structures under static load uncertainty via SDP relaxation. *Computers & Structures*, **87**, 246–253 (2009).

[158] Guo, X., Bai, W., Zhang, W., Gao, X.: Confidence structural robust design and optimization under stiffness and load uncertainties. *Computer Methods in Applied Mechanics and Engineering*, **98**, 3378–3399 (2009).

[159] Han, W., Reddy, B.D.: *Plasticity.* Springer-Verlag, New York (1999).

[160] Han, W., Sofonea, M.: *Quasistatic Contact Problems in Viscoelasticity and Viscoplasticity.* Americal Mathematical Society, Providence (2002).

[161] Harker, P.T., Pang, J.-S.: Finite-dimensional variational inequality and nonlinear complementarity problems: A survey of theory, algorithms and applications. *Mathematical Programming*, **48**, 161–220 (1990).

[162] Harris, J.B., Li, K.P.-K.: *Masted Structures in Architecture.* Architecture Press, Oxford (1996).

[163] Haseganu, E.M., Steigmann, D.J.: Analysis of partly wrinkled membranes by the method of dynamic relaxation. *Computational Mechanics*, **14**, 596–614 (1994).

[164] Haseganu, E.M., Steigmann, D.J.: Equilibrium analysis of finitely deformed elastic networks. *Computational Mechanics*, **17**, 359–373 (1996).

[165] Haslinger, J.: Approximation of variational inequalities of elliptic type: Applications to contact problems with friction. In: Haslinger, J., Stavroulakis, G.E. (eds.) *Nonsmooth Mechanics of Solids*, pp. 103–166, Springer-Verlag, Wien (2006).

[166] Hassani, R., Hild, P., Ionescu, I.R.: Sufficient conditions of nonuniqueness for the Coulomb friction problem. *Mathematical Methods in the Applied Sciences*, **27**, 47–67 (2004).

[167] Hassani, R., Ionescu, I.R., Oudet, E.: Critical friction for wedged configurations. *International Journal of Solids and Structures*, **44**, 6187–6200 (2007).

[168] Haug, E., Chand, R., Pan, K.: Multibody elastic contact analysis by quadratic programming. *Journal of Optimization Theory and Applications*, **21**, 189–198 (1977).

[169] Haug, E.J., Choi, K.K.: Systematic occurrence of repeated eigenvalues in structural optimization. *Journal of Optimization Theory and Applications*, **38**, 251–274 (1982).

[170] Hayashi, S., Yamashita, N., Fukushima, M.: A combined smoothing and

regularization method for monotone second-order cone complementarity problems. *SIAM Journal on Optimization*, **15**, 593-615 (2005).

[171] Hellinger, E.: Die Allgemeine Ansätze der Mechanik der Kontinua. *Wissenshafter IV*, **4**, 602–694 (1914).

[172] Helmberg, C.: Semidefinite programming. *European Journal of Operational Research*, **137**, 461–482 (2002).

[173] Helmberg, C., Rendl, F., Vanderbei, R.J., Wolkowicz, H.: An interior-point method for semidefinite programming. *SIAM Journal on Optimization*, **6**, 342–361 (1996)

[174] Henrion, D., Lasserre, J.B.: Convergent relaxations of polynomial matrix inequalities and static output feedback. *IEEE Transactions on Automatic Control*, **51**, 192–202 (2006).

[175] Hernández-Montes, E., Jurado-Piña, R., Bayo, E.: Topological mapping for tension structures. *Journal of Structural Engineering (ASCE)*, **132**, 970–977 (2006).

[176] Heyman, J.: The stone skeleton. *International Journal of Solids and Structures*, **2**, 249–256 (1966).

[177] Heyman, J.: The safety of masonry arches. *International Journal of Mechanical Sciences*, **11**, 363–385 (1969).

[178] Hill, R.: A variational principle of maximum plastic work in classical plasticity. *The Quarterly Journal of Mechanics and Applied Mathematics*, **1**, 18–28 (1948).

[179] Hiriart-Urruty, J.B., Lemaréchal, C.: *Convex Analysis and Minimization Algorithms. Volumes I & II.* Springer-Verlag, Berlin (1993).

[180] Hjiaj, M., Fortin, J., de Saxcé, G.: A complete stress update algorithm for the non-associated Drucker–Prager model including treatment of the apex. *International Journal of Engineering Science*, **41**, 1109–1143 (2003).

[181] Hjiaj, M., Feng, Z.-Q., de Saxcé, G., Mróz, Z.: Three-dimensional finite element computations for frictional contact problems with non-associated sliding rule. *International Journal for Numerical Methods in Engineering*, **60**, 2045–2076 (2004).

[182] Hlaváček, I., Chleboun, J., Babuška, I.: *Uncertain Input Data Problems and the Worst Scenario Method.* Elsevier, Amsterdam (2004).

[183] Hlaváček, I., Haslinger, J., Nečas, J., Lovíšek, J.: *Solution of Variational Inequalities in Mechanics.* Springer-Verlag, New York (1988).

[184] Holzapfel, G.A.: *Nonlinear Solid Mechanics.* John Wiley & Sons, Chich-

ester (2000).

[185] Holzapfel, G.A., Eberlein, R., Wriggers, P., Weizsäcker, H.W.: Large strain analysis of soft biological membranes: Formulation and finite element analysis. *Computer Methods in Applied Mechanics and Engineering*, **132**, 45–61 (1996).

[186] Horn, R.A., Johnson, C.R.: *Matrix Analysis.* Cambridge University Press, Cambridge (1985).

[187] Hornig, J., Schoop, H.: Closed form analysis of wrinkled membranes with linear stress-strain relation. *Computational Mechanics*, **30**, 259–264 (2003).

[188] Hornig, J., Schoop, H.: Wrinkling analysis of membranes with elastic-plastic material behavior. *Computational Mechanics*, **35**, 153–160 (2005).

[189] Hu, Z.Q., Soh, A.-K., Chen, W.J., Li, X.W., Lin, G.: Non-smooth nonlinear equations methods for solving 3D elastoplastic frictional contact problems. *Computational Mechanics*, **39**, 849–858 (2007).

[190] Jahn, J.: *Introduction to the Theory of Nonlinear Optimization (3rd ed.).* Springer-Verlag, Berlin (2007).

[191] Jarasjarunngkiat, A., Wüchner, R., Bletzinger, K.-U.: A wrinkling model based on material modification for isotropic and orthotropic membranes. *Computer Methods in Applied Mechanics and Engineering*, **197**, 773–788 (2008).

[192] Jarasjarunngkiat, A., Wüchner, R., Bletzinger, K.-U.: Efficient subgrid scale modeling of membrane wrinkling by a projection method. *Computer Methods in Applied Mechanics and Engineering*, **198**, 1097–1116 (2009).

[193] Jarre, F., Kočvara, M., Zowe, J.: Optimal truss design by interior-point methods. *SIAM Journal on Optimization*, **8**, 1084–1107 (1998).

[194] Jenkins, C.H.: Nonlinear dynamic response of membranes: State of the art—Update. *Applied Mechanics Review (ASME)*, **49**, S41–S48 (1996).

[195] Jenkins, C.H., Leonard, J.W.: Dynamic wrinkling of viscoelastic membranes. *Journal of Applied Mechanics (ASME)*, **60**, 575–582 (1993).

[196] Jennings, A.: Energy theorems in structural mechanics. *Journal of Engineering Mathematics*, **1**, 307–326 (1967).

[197] Jeong, D.G., Kwak, B.M.: Complementarity problem formulation for the wrinkled membrane and numerical implementation. *Finite Elements in Analysis and Design*, **12**, 91–104 (1992).

[198] Jonatowski, J.J., Birnstiel, C.: Inelastic stiffened suspension space structures. *Journal of the Structural Division*, **96**, 1143–1166 (1970).

[199] Jungnickel, D.: *Graphs, Networks and Algorithms (3rd ed.)*. Springer-Verlag, Berlin (2008).

[200] Kang, S., Im, S.: Finite element analysis of wrinkling membranes. *Journal of Applied Mechanics (ASME)*, **64**, 263–269 (1997).

[201] Kang, S., Im, S.: Finite element analysis of dynamic response of wrinkling membranes. *Computer Methods in Applied Mechanics and Engineering*, **173**, 227–240 (1999).

[202] Kanno, Y., Martins, J.A.C., Pinto da Costa, A.: Three-dimensional quasi-static frictional contact by using second-order cone linear complementarity problem. *International Journal for Numerical Methods in Engineering*, **65**, 62–83 (2006).

[203] Kanno, Y., Ohsaki, M.: Necessary and sufficient conditions for global optimality of eigenvalue optimization problems. *Structural and Multidisciplinary Optimization*, **22**, 248–252 (2001).

[204] Kanno, Y., Ohsaki, M.: Minimum principle of complementary energy of cable networks by using second-order cone programming. *International Journal of Solids and Structures*, **40**, 4437–4460 (2003).

[205] Kanno, Y., Ohsaki, M.: Semidefinite programming for finite element analysis of isotropic materials with unilateral constitutive laws. *AIS Research Report 04-02*, Department of Architecture and Architectural Systems, Kyoto University (October 2004).

[206] Kanno, Y., Ohsaki, M.: Minimum principle of complementary energy for nonlinear elastic cable networks with geometrical nonlinearities. *Journal of Optimization Theory and Applications*, **126**, 617–641 (2005).

[207] Kanno, Y., Ohsaki, M.: Contact analysis of cable networks by using second-order cone programming. *SIAM Journal on Scientific Computing*, **27**, 2032–2052 (2006).

[208] Kanno, Y., Ohsaki, M., Ito, J.: Large-deformation and friction analysis of nonlinear elastic cable networks by second-order cone programming. *International Journal for Numerical Methods in Engineering*, **55**, 1079–1114 (2002).

[209] Kanno, Y., Ohsaki, M., Katoh, N.: Sequential semidefinite programming for optimization of framed structures under multimodal buckling constraints. *International Journal of Structural Stability and Dynamics*, **1**, 585–602 (2001).

[210] Kanno, Y., Takewaki, I.: Confidence ellipsoids for static response of

trusses with load and structural uncertainties. *Computer Methods in Applied Mechanics and Engineering*, **196**, 393–403 (2006).

[211] Kanno, Y., Takewaki, I.: Sequential semidefinite program for maximum robustness design of structures under load uncertainties. *Journal of Optimization Theory and Applications*, **130**, 265–287 (2006).

[212] Kanno, Y., Takewaki, I.: Semidefinite programming for uncertain linear equations in static analysis of structures. *Computer Methods in Applied Mechanics and Engineering*, **198**, 102–115 (2008).

[213] Kanno, Y., Takewaki, I.: Semidefinite programming for dynamic steady-state analysis of structures under uncertain harmonic loads. *Computer Methods in Applied Mechanics and Engineering*, **198**, 3239–3261 (2009).

[214] Kanzow, C.: Some noninterior continuation methods for linear complementarity problems. *SIAM Journal on Matrix Analysis and Applications*, **17**, 851–868 (1996).

[215] Kanzow, K., Ferenczi, I., Fukushima, M.: On the local convergence of semismooth Newton methods for linear and nonlinear second-order cone programs without strict complementarity. *SIAM Journal on Optimization*, **20**, 297–320 (2009).

[216] Kanzow, C., Nagel, C.: Semidefinite programs: New search directions, smoothing-type methods, and numerical results. *SIAM Journal on Optimization*, **13**, 1–23 (2002); Corrigendum. *SIAM Journal on Optimization*, **14**, 936–937 (2003).

[217] Kanzow, C., Nagel, C.: Some structural properties of a Newton-type method for semidefinite programs. *Journal of Optimization Theory and Applications*, **122**, 219–226 (2004).

[218] Kanzow, K., Yamashita, N., Fukushima, M.: New NCP-functions and their properties. *Journal of Optimization Theory and Applications*, **94**, 115–135 (1997).

[219] Kawa, M., Pietruszczak, S., Shieh-Beygi, B.: Limit states for brick masonry based on homogenization approach. *International Journal of Solids and Structures*, **45**, 998–1016 (2008).

[220] Khot, N.S.: Optimization of structures with multiple frequency constraints. *Computers & Structures*, **20**, 869–876 (1985).

[221] Kikuchi, N., Oden, J.T.: *Contact Problems in Elasticity*. SIAM, Philadelphia (1988).

[222] Kinderlehrer, D., Stampacchia, G.: *An Introduction to Variational Inequalities and Their Applications*. Academic Press, New York (1980). Also: SIAM, Philadelphia (2000).

[223] Klarbring, A.: Quadratic programs in frictionless contact problems. *International Journal of Solids and Structures*, **24**, 1207–1217 (1986).

[224] Klarbring, A.: On discrete and discretized non-linear elastic structures in unilateral contact (stability, uniqueness and variational principles). *International Journal of Solids and Structures*, **24**, 459–479 (1988).

[225] Klarbring, A.: Examples of non-uniqueness and non-existence of solutions to quasistatic contact problems with friction. *Ingenieur-Archiv*, **60**, 529–541 (1990).

[226] Klarbring, A.: Derivation and analysis of rate boundary-value problems of frictional contact. *European Journal of Mechanics, A/Solids*, **9**, 53–85 (1990).

[227] Klarbring, A.: Contact, friction, discrete mechanical structures and mathematical programming. In: Wriggers, P, Panagiotopoulos, P. (eds.) *New Developments in Contact Problems*, pp. 55–100, Springer-Verlag, Wien (1999).

[228] Klarbring, A., Mikelić, A., Shillor, M.: Frictional contact problems with normal compliance. *International Journal of Engineering Science*, **26**, 811–832 (1988).

[229] Klarbring, A., Mikelić, A., Shillor, M.: On friction problems with normal compliance. *Nonlinear Analysis*, **13**, 935–955 (1989).

[230] Klarbring, A., Pang, J.-S.: Existence of solutions to discrete semicoercive frictional contact problems. *SIAM Journal on Optimization*, **8**, 414–442 (1998).

[231] Knudson, W.C.: Recent advances in the field of long span tension structures. *Engineering Structures*, **13**, 164–177 (1991).

[232] Koch, K.-M. (ed.): *Membrane Structures*. Prestel Verlag, Munich (2004).

[233] Kočvara, M.: On the modeling and solving of the truss design problem with global stability constraints. *Structural and Multidisciplinary Optimization*, **23**, 189–203 (2002). Also: Levy, R., Su, H.-H., Kočvara, M.: Discussion and author's reply. *Structural and Multidisciplinary Optimization*, **26**, 367–368 (2004).

[234] Kočvara, M., Stingl, M.: Solving nonconvex SDP problems of structural optimization with stability control. *Optimization Methods and Software*, **19**, 595–609 (2004).

[235] Koiter, W.T.: General theorems for elastic-plastic solids. In: Sneddon, I.N., Hill, R. (eds.) *Progress in Solid Mechanics*, pp. 167–221, North-Holland, Amsterdam (1960).

[236] Koiter, W.T.: On the complementary energy theorem in non-linear elasticity theory. In: Fichera, G. (ed.) *Trends in Applications of Pure Mathematics of Mechanics*, pp. 207–232, Pitman Publishing, London (1976).

[237] Kojić, M., Bathe, K.J.: *Inelastic Analysis of Solids and Structures*. Springer-Verlag, Berlin (2005).

[238] Kojima, M., Mizuno, S., Yoshise, A.: A primal-dual interior point algorithm for linear programming. In: Megiddo, N. (ed.) *Progress in Mathematical Programming: Interior-Point Algorithms and Related Methods*, pp. 29–47, Springer-Verlag, New York (1989).

[239] Kojima, M., Mizuno, S., Yoshise, A.: A polynomial-time algorithm for a class of linear complementarity problems. *Mathematical Programming*, **44**, 1–28 (1989).

[240] Kojima, M., Muramatsu, M.: An extension of sums of squares relaxations to polynomial optimization problems over symmetric cones. *Mathematical Programming*, **110**, 315–336 (2007).

[241] Kojima, M., Noma, T., Yoshise, A.: Global convergence in infeasible-interior-point algorithms. *Mathematical Programming*, **65**, 43–72 (1994).

[242] Kojima, M., Shida, M., Shindoh, S.: A note on the Nesterov–Todd and Kojima–Shindoh–Hara search directions in semidefinite programming. *Optimization Methods and Software*, **11/12**, 47–52 (1999).

[243] Kojima, M., Shindoh, S., Hara, S.: Interior-point methods for the monotone semidefinite linear complementarity problem in symmetric matrices. *SIAM Journal on Optimization*, **7**, 86–125 (1997).

[244] Krabbenhøft, K., Lyamin, A.V., Sloan, S.W.: Formulation and solution of some plasticity problems as conic programs. *International Journal of Solids and Structures*, **44**, 1533–1549 (2007).

[245] Krabbenhøft, K., Lyamin, A.V., Sloan, S.W.: Three-dimensional Mohr–Coulomb limit analysis using semidefinite programming. *Communications in Numerical Methods in Engineering*, **24**, 1107–1119 (2008).

[246] Krabbenhøft, K., Lyamin, A.V., Sloan, S.W., Wriggers, P.: An interior-point algorithm for elastoplasticity. *International Journal for Numerical Methods in Engineering*, **69**, 592–626 (2007).

[247] Kravchuk, A.S., Neittaanmäki, P.J.: *Variational and Quasi-Variational Inequalities in Mechanics*. Springer, Dordrecht (2007).

[248] Kwan, A.S.K.: A new approach to geometric nonlinearity of cable structures. *Computers & Structures*, **67**, 243–252 (1998).

[249] Labisch, F.K.: On the dual formulation of boundary value problems in

non-linear elastostatics. *International Journal of Engineering Science*, **20**, 413–431 (1982).

[250] Lasserre, J.B.: Global optimization with polynomials and the problems of moments. *SIAM Journal on Optimization*, **11**, 796–817 (2001).

[251] Lasserre, J.B.: Bounds on measures satisfying moment conditions. *The Annals of Applied Probability*, **12**, 1114–1137 (2002).

[252] Lasserre, J.B.: A semidefinite programming approach to the generalized problem of moments. *Mathematical Programming*, **112**, 65–92 (2008).

[253] Laursen, T.A.: *Computational Contact and Impact Mechanics*. Springer-Verlag, Berlin (2002).

[254] Lee, E.-S., Youn, S.-K.: Finite element analysis of wrinkling membrane structures with large deformations. *Finite Elements in Analysis and Design*, **42**, 780–791 (2006).

[255] Leine, R.I., Nijmeijer, H.: *Dynamics and Bifurcations of Non-Smooth Mechanical Systems*. Springer-Verlag, Berlin (2004).

[256] Leine, R.I., van de Wouw, N.: *Stability and Convergence of Mechanical Systems with Unilateral Constraints*. Springer-Verlag, Berlin (2008).

[257] Leung, A.Y.T., Chen, G., Chen, W.: Smoothing Newton method for solving two- and three-dimensional frictional contact problems. *International Journal for Numerical Methods in Engineering*, **41**, 1001–1027 (1998).

[258] Levinson, M.: The complementary energy theorem in finite elasticity. *Journal of Applied Mechanics (ASME)*, **32**, 826–828 (1965).

[259] Levinson, N.: A class of continuous linear programming problems. *Journal of Mathematical Analysis and Applications*, **16**, 73–83 (1966).

[260] Levy, R., Spillers, W.R.: Practical methods of shape-finding for membranes and cable nets. *Journal of Engineering Mechanics (ASCE)*, **124**, 466–468 (1998).

[261] Lewis, A.S., Overton, M.L.: Eigenvalue optimization. *Acta Numerica*, **5**, 149–190 (1996).

[262] Lewis, W.J.: The efficiency of numerical methods for the analysis of prestressed nets and pin-jointed frame structures. *Computers & Structures*, **33**, 791–800 (1989).

[263] Lewis, W.J.: *Tension Structures: Form and Behavior*. Thomas Telford Publishing, London (2003).

[264] Lewis, W.J., Jones, M.S., Rushton, K.R.: Dynamic relaxation analysis of the non-linear static response of pretensioned cable roofs. *Computers*

& *Structures*, **18**, 989–997 (1984).

[265] Lewis, W.J., Gosling, P.D.: Stable minimal surfaces in form-finding of lightweight tension structures. *International Journal of Space Structures*, **8**, 149–166 (1993).

[266] Li, D., Fukushima, M.: Smoothing Newton and quasi-Newton methods for mixed complementarity problems. *Computational Optimization and Applications*, **17**, 203–230 (2000).

[267] Libove, C.: Complementary energy method for finite deformations. *Journal of Engineering Mechanics (ASCE)*, **90(EM6)**, 49–71 (1964); **91(EM4)**, 203–209 (1965); **92(EM2)**, 279–290 (1966).

[268] Liew, K.M., Wu, Y.C., Zou, G.P., Ng, T.Y.: Elasto-plasticity revisited: numerical analysis via reproducing kernel particle method and parametric quadratic programming. *International Journal for Numerical Methods in Engineering*, **55**, 669–683 (2002).

[269] Linkwitz, K., Schek, H.-J.: Einige Bemerkungen zur Berechnung von vorgespannten Seilnetzkonstruktionen. *Ingenieur-Archiv*, **40**, 145–158 (1971).

[270] Lions, J.L., Magenes, E.: *Non-Homogeneous Boundary Value Problems and Applications. Volume I.* Springer-Verlag, Berlin (1972).

[271] Liu, X., Jenkins, C.H., Schur, W.W.: Large deflection analysis of pneumatic envelopes using a penalty parameter modified material model. *Finite Elements in Analysis and Design*, **37**, 233–251 (2001).

[272] Lobo, M.S., Vandenberghe, L., Boyd, S., Lebret, H.: Applications of second-order cone programming. *Linear Algebra and its Applications*, **284**, 193–238 (1998).

[273] Lourenço, P.B., Rots, J.G., Blaauwendraad, J.: Continuum model for masonry: parameter estimation and validation. *Journal of Structural Engineering (ASCE)*, **124**, 643–652 (1998).

[274] Lu, K., Accorsi, M., Leonard, J.: Finite element analysis of membrane wrinkling. *International Journal for Numerical Methods in Engineering*, **50**, 1017–1038 (2001).

[275] Lubliner, J.: A maximum-dissipation principle in generalized plasticity. *Acta Mechanica*, **52**, 225–237 (1984).

[276] Lucchesi, M., Padovani, C., Pagni, A.: A numerical method for solving equilibrium problems of masonry-like solids. *Meccanica*, **29**, 175–193 (1994).

[277] Lucchesi, M., Padovani, C., Pasquinelli, G.: On the numerical solution of equilibrium problems for elastic solids with bounded tensile strength.

Computer Methods in Applied Mechanics and Engineering, **127**, 37–56 (1995).

[278] Lucchesi, M., Padovani, C., Pasquinelli, G.: Thermodynamics of no-tension materials. *International Journal of Solids and Structures*, **37**, 6581–6604 (2000).

[279] Lucchesi, M., Padovani, C., Zani, N.: Masonry-like solids with bounded compressive strength. *International Journal of Solids and Structures*, **33**, 1961–1994 (1996).

[280] Lucchesi, M., Šilhavý, M., Zani, N.: Integration of measures and admissible stress fields for masonry bodies. *Journal of Mechanics of Materials and Structures*, **3**, 675–696 (2008).

[281] Luciano, R., Sacco, E.: Homogenization technique and damage model for old masonry material. *International Journal of Solids and Structures*, **34**, 3191–3208 (1997).

[282] Luenberger, D.G.: *Optimization by Vector Space Methods.* Wiley, New York (1969).

[283] Luo, Z.-Q., Sturm, J.F., Zhang, S.: Duality and self-duality for conic convex programming. Report 9620/A, Econometric Institute, Erasmus University Rotterdam (1996).

[284] Luo, Z.-Q., Sturm, J.F., Zhang, S.: Conic convex programming and self-dual embedding. *Optimization Methods and Software*, **14**, 169–218 (2000).

[285] Maier, G.: A quadratic programming approach for certain classes of non-linear structural problems. *Meccanica*, **3**, 121–130 (1968).

[286] Maier, G.: Quadratic programming and theory of elastic-perfectly plastic structures. *Meccanica*, **3**, 265–273 (1968).

[287] Maier, G.: Complementary plastic work theorems in piecewise-linear elastoplasticity. *International Journal of Solids and Structures*, **5**, 261–270 (1969).

[288] Maier, G.: Incremental plastic analysis in the presence of large displacements and physical instabilizing effects. *International Journal of Solids and Structures*, **7**, 345–372 (1971).

[289] Maier, G.: Mathematical programming applications to structural mechanics: some introductory thoughts. *Engineering Structures*, **6**, 2–6 (1984).

[290] Maier, G., Contro, R.: Energy approach to inelastic cable-structure analysis. *Journal of Engineering Mechanics (ASCE)*, **101**, 531–548 (1975).

[291] Maier, G., Nappi, A.: A theory of no-tension discretized structural systems. *Engineering Structures*, **12**, 227–234 (1990).

[292] Mangasarian, O.L.: Equivalence of the complementarity problem to a system of nonlinear equations. *SIAM Journal on Applied Mathematics*, **31**, 89–92 (1976).

[293] Mansfield, E.H.: Load transfer via a wrinkled membrane. *Proceedings of the Royal Society of London. Series A, Mathematical and Physical Sciences*, **316**, 269–289 (1970).

[294] Mansfield, E.H.: Analysis of wrinkled membranes with anisotropic and nonlinear elastic properties. *Proceedings of the Royal Society of London. Series A, Mathematical and Physical Sciences*, **353**, 475–498 (1977).

[295] Marsden, J.E., Hughes, T.J.R.: *Mathematical Foundations of Elasticity*. Printice-Hall, Englewood Cliffs (1983). Also: Dover Publications, Mineola (1994).

[296] Marfia, S., Sacco, E.: Modeling of reinforced masonry elements. *International Journal of Solids and Structures*, **38**, 4177–4198 (2001).

[297] Marfia, S., Sacco, E.: Numerical procedure for elasto-plastic no-tension model. *International Journal for Computational Methods in Engineering Science and Mechanics*, **6**, 187–199 (2005).

[298] Martins, J.M.C., Barbarin, S., Raous, M., Pinto da Costa, A.: Dynamic stability of finite dimensional linearly elastic systems with unilateral contact and Coulomb friction. *Computer Methods in Applied Mechanics and Engineering*, **177**, 289–328 (1999).

[299] Martins, J.M.C., Pinto da Costa, A.: Stability of finite-dimensional nonlinear elastic systems with unilateral contact and friction. *International Journal of Solids and Structures*, **37**, 2519–2564 (2000).

[300] Martins, J.A.C., Oden, J.T.: Existence and uniqueness results for dynamic contact problems with nonlinear normal and friction interface laws. *Nonlinear Analysis*, **11**, 407–428 (1987).

[301] Martins, J.A.C., Pinto da Costa, A., Simões, F.M.F.: Some notes on friction and instabilities. In: Martins, J.A.C., Raous, M. (eds.) *Friction and Instabilities*, pp. 65–136, Springer-Verlag, Wien (2002).

[302] Massabò, R., Gambarotta, L.: Wrinkling of plane isotropic biological membranes. *Journal of Applied Mechanics (ASME)*, **74**, 550–559 (2007).

[303] Masur, E.F.: Optimal structural design under multiple eigenvalue constraints. *International Journal of Solids and Structures*. **20**, 211–231 (1984).

[304] Mateus, H.C., Rodrigues, H.C., Mota Soares, C.M., Mota Soares, C.A.:

Sensitivity analysis and optimization of thin laminated structures with a nonsmooth eigenvalue based criterion. *Structural Optimization*, **14**, 219–224 (1997).

[305] MathWorks: *Using MATLAB*. The MathWorks, Inc., Natick (2009).

[306] Maurin, B., Motro, R.: The surface stress density method as a form-finding tool for tensile membranes. *Engineering Structures*, **20**, 712–719 (1998).

[307] Megiddo, N.: Pathways to the optimal set in linear programming. In: Megiddo, N. (ed.) *Progress in Mathematical Programming: Interior-Point Algorithms and Related Methods*, pp. 131–158, Springer-Verlag, New York (1989).

[308] Mehrotra, S.: On the implementation of a primal-dual interior point method. *SIAM Journal on Optimization*, **2**, 575–601 (1992).

[309] Mijar, A.R., Arora, J.S.: Review of formulations for elastostatic frictional contact problems. *Structural and Multidisciplinary Optimization*, **20**, 167–189 (2000).

[310] Mikkola, M.J.: Complementary energy theorem in geometrically non-linear structural problems. *International Journal of Non-Linear Mechanics*, **24**, 499–508 (1989).

[311] Miller, R.K., Hedgepeth, J.M., Weingarten, V.I., Das, P., Kahyai, S.: Finite element analysis of partly wrinkled membranes. *Computers & Structures*, **20**, 631–639 (1985).

[312] Miyamura, T.: Wrinkling on stretched circular membrane under in-plane torsion: Bifurcation analyses and experiments. *Engineering Structures*, **23**, 1407–1425 (2000).

[313] Miyazaki, Y.: Wrinkle/slack model and finite element dynamics of membrane. *International Journal for Numerical Methods in Engineering*, **66**, 1179–1209 (2006).

[314] Monteiro, R.D.C.: Primal-dual path-following algorithms for semidefinite programming. *SIAM Journal on Optimization*, **7**, 663–678 (1997).

[315] Monteiro, R.D.C., Adler, I.: Interior path following primal-dual algorithms. Part II: Convex quadratic programming. *Mathematical Programming*, **44**, 43–66 (1989).

[316] Monteiro, R.D.C., Adler, I., Resende, M.G.C.: A polynomial-time primal-dual affine scaling algorithm for linear and convex quadratic programming and its power series extension. *Mathematics of Operations Research*, **15**, 191–214 (1990).

[317] Monteiro, R.D.C., Tsuchiya, T.: Polynomial convergence of primal-dual

algorithms for the second-order cone program based on the MZ-family of directions. *Mathematical Programming*, **88**, 61–83 (2000).

[318] Monteiro Marques, M.D.P.: *Differential Inclusions in Nonsmooth Mechanical Problems*. Birkhäuser, Basel (1993).

[319] Moreau, J.J.: Quadratic programming in mechanics: dynamics of one-sided constraints. *SIAM Journal on Control*, **4**, 153–158 (1966).

[320] Moreau, J.J.: On unilateral constraints, friction and plasticity. In: Capriz, G., Stampacchia, G. (eds.) *New Variational Techniques in Mathematical Physics*, pp. 173-322, Springer-Veralg, Berlin (1973).

[321] Moreau, J.J.: Evolution problem associated with a moving convex set in a Hilbert space. *Journal of Differential Equations*, **26**, 347–374 (1977).

[322] Moreau, J.J.: Unilateral contact and dry friction in finite freedom dynamics. In: Moreau, J.J., Panagiotopoulos, P.D. (eds.) *Nonsmooth Mechanics and Applications*, pp. 1–82, Springer-Verlag, Wien (1988).

[323] Moreau, J.J.: Numerical aspects of the sweeping process. *Computer Methods in Applied Mechanics and Engineering*, **177**, 329–349 (1999).

[324] Moreau, J.J.: An introduction to unilateral dynamics. In: Frémond, M., Maceri, F. (eds.) *Novel Approaches in Civil Engineering*, pp. 1–46, Springer-Verlag, Berlin (2004).

[325] Mosler, J.: A novel variational algorithmic formulation for wrinkling at finite strains based on energy minimization: Application to mesh adaption. *Computer Methods in Applied Mechanics and Engineering*, **197**, 1131–1146 (2007).

[326] Mosler, J., Cirak, F.: A variational formulation for finite deformation wrinkling analysis of inelastic membranes. *Computer Methods in Applied Mechanics and Engineering*, **198**, 2087–2098 (2009).

[327] Motro, R.: *Tensegrity*. Kogan Page Science, London (2003).

[328] Murray, T.A., Willems, N.: Analysis of inelastic suspension structures. *Journal of the Structural Division (ASCE)*, **97**, 2791–2806 (1971).

[329] Muttin, F.: A finite element for wrinkled curved elastic membranes, and its application to sails. *Communications in Numerical Methods in Engineering*, **12**, 775-785 (1996).

[330] Nakamura, T., Ohsaki, M.: Sequential optimal truss generator for frequency ranges. *Computer Methods in Applied Mechanics and Engineering*, **67**, 189–209 (1988).

[331] Naniewicz, Z., Panagiotopoulos, P.D.: *Mathematical Theory of Hemivariational Inequalities and Applications*. Marcel Dekker, New York

(1995).

[332] Narayanan, S.: *Space Structures: Principles and Practice. Volumes I & II.* Multi-Science Publishing, Brentwood (2006).

[333] Nesterov, Y., Nemirovski, A.: *Interior-Point Polynomial Methods in Convex Programming.* SIAM, Philadelphia (1994).

[334] Nesterov, Y., Todd, M.J.: Self-scaled barriers and interior-point methods in convex programming. *Mathematics of Operations Research*, **22**, 1–42 (1997).

[335] Nesterov, Y., Todd, M.J.: Primal-dual interior-point methods for self-scaled cones. *SIAM Journal on Optimization*, **8**, 324–364 (1998).

[336] Nguyen, Q.S.: Instability and friction. *Comptes Rendus Mecanique*, **331**, 99–112 (2003).

[337] Nguyen, V.-H., Duhamel, D., Nedjar, B.: A continuum model for granular materials taking into account the no-tension effect. *Mechanics of Materials*, **35**, 955-967 (2003).

[338] Nguyen-Dang, H., de Saxcé, G.: Frictionless contact of elastic bodies by finite element method and mathematical programming technique. *Computers & Structures*, **11**, 55–67 (1980).

[339] Nie, J., Demmel, J.W.: Minimum ellipsoid bounds for solutions of polynomial systems via sum of squares. *Journal of Global Optimization*, **33**, 511–525 (2005).

[340] Nocedal, J., Wright, S.J.: *Numerical Optimization (2nd ed.).* Springer, New York (2006).

[341] Oden, J.T., Demkowicz, L.F.: *Applied Functional Analysis.* CRC Press, Boca Raton (1996).

[342] Oden, J.T., Martins, J.A.C.: Models and computational methods for dynamic friction phenomena. *Computer Methods in Applied Mechanics and Engineering*, **52**, 527–634 (1985).

[343] Ohsaki, M., Fujisawa, K., Katoh, N., Kanno, Y.: Semi-definite programming for topology optimization of trusses under multiple eigenvalue constraints. *Computer Methods in Applied Mechanics and Engineering*, **180**, 203–217 (1999).

[344] Ohsaki, M., Kanno, Y.: Form-finding of cable domes with specified stresses by using nonlinear programming. *Proceedings of IASS-APCS 2003, International Association for Shell and Spatial Structures*, Taipei, Taiwan (2003).

[345] Olhoff, N., Rasmussen, S.H.: On single and bimodal optimum buckling

loads of clamped columns. *International Journal of Solids and Structures*, **13**, 605–614 (1977).

[346] Oñate, E.: *Structural Analysis with the Finite Element Method. Linear Statics. Volume 1: Basis and Solids.* Springer-Verlag (2009).

[347] Otter, J.R.H.: Computations for prestressed concrete reactor pressure vessels using dynamic relaxation. *Nuclear Structural Engineering*, **1**, 61–75 (1965).

[348] Otter, J.R.H.: Dynamic relaxation compared with other iterative finite difference methods. *Nuclear Engineering and Design*, **3**, 183–185 (1966).

[349] Padovani, C.: No-tension solids in the presence of thermal expansion: An explicit solution. *Meccanica*, **31**, 687–703 (1996).

[350] Padovani, C.: On a class of non-linear elastic materials. *International Journal of Solids and Structures*, **37**, 7787–7807 (2000).

[351] Padovani, C., Pasquinelli, G., Zani, N.: A numerical method for solving equilibrium problems of no-tension solids subjected to thermal loads. *Computer Methods in Applied Mechanics and Engineering*, **190**, 55–73 (2000).

[352] Pan S., Chen, J.-S.: A semismooth Newton method for SOCCPs based on a one-parametric class of SOC complementarity functions. *Computational Optimization and Applications*, **45**, 59–88 (2010).

[353] Panagiotopoulos, P.D.: Stress-unilateral analysis of discretized cable and membrane structure in the presence of large displacements. *Ingenieur-Archiv*, **44**, 291–300 (1975).

[354] Panagiotopoulos, P.D.: A nonlinear programming approach to the unilateral contact-, and friction-boundary value problem in the theory of elasticity. *Ingenieur-Archiv*, **44**, 421–432 (1975).

[355] Panagiotopoulos, P.D.: Convex analysis and unilateral static problems. *Ingenieur-Archiv*, **45**, 55–68 (1975).

[356] Panagiotopoulos, P.D.: A variational inequality approach to the inelastic stress-unilateral analysis of cable-structures. *Computers & Structures*, **6**, 133–139 (1976).

[357] Pang, J.-S.: Newtons's method for B-differentiable equations. *Mathematics of Operations Research*, **15**, 311–341 (1990).

[358] Panzeca, T., Polizzotto, C.: Constitutive equations for no-tension materials. *Meccanica*, **23**, 88–93 (1988).

[359] Pataki, G.: On the rank of extreme matrices in semidefinite programs and the multiplicity of optimal eigenvalues. *Mathematics of Operations*

Research, **23**, 339–358 (1998).

[360] Pauletti, R.M.O., Pimenta, P.M.: The natural force density method for the shape finding of taut structures. *Computer Methods in Applied Mechanics and Engineering*, **197**, 4419–4428 (2008).

[361] Pedersen, N.L.: Maximization of eigenvalues using topology optimization. *Structural and Multidisciplinary Optimization*, **20**, 2–11 (2000).

[362] Pellegrino, S.: A class of tensegrity domes. *International Journal of Space Structures*, **7**, 127–142 (1992).

[363] Pevrot, A.H., Goulois, A.M.: Analysis of cable structures. *Computers & Structures*, **10**, 805–813 (1979).

[364] Pfeiffer, F., Glocker, C.: Contacts in multibody systems. *Journal of Applied Mathematics and Mechanics*, **64**, 773–782 (2000).

[365] Pieraccini, S., Gasparo, M.G., Pasquali, A.: Global Newton-type methods and semismooth reformulations for NCP. *Applied Numerical Mathematics*, **44**, 367–384 (2003).

[366] Pinto da Costa, A., Martins, J.A.C.: The evolution and rate problems and the computation of all possible evolutions in quasi-static frictional contact. *Computer Methods in Applied Mechanics and Engineering*, **192**, 2791–2821 (2003).

[367] Pinto da Costa, A., Martins, J.A.C.: A numerical study on multiple rate solutions and onset of directional instability in quasi-static frictional contact problems. *Computers & Structures*, **82**, 1485–1494 (2004).

[368] Pinto da Costa, A., Martins, J.A.C., Figueiredo, I.N., Júdice, J.J.: The directional instability problem in systems with frictional contacts. *Computer Methods in Applied Mechanics and Engineering*, **193**, 357–384 (2004).

[369] Pipkin, A.C.: The relaxed energy density for isotropic elastic membranes. *IMA Journal of Applied Mathematics*, **36**, 85–99 (1986).

[370] Pipkin, A.C.: Convexity conditions for strain-dependent energy functions for membranes. *Archive for Rational Mechanics and Analysis*, **121**, 361–376 (1993).

[371] Pipkin, A.C.: Relaxed energy densities for small deformations of membranes. *IMA Journal of Applied Mathematics*, **50**, 225–237 (1993).

[372] Pipkin, A.C.: Relaxed energy densities for large deformations of membranes. *IMA Journal of Applied Mathematics*, **52**, 297–308 (1994).

[373] Pólik, I., Terlaky, T.: A survey of $\mathcal{S}$-lemma. *SIAM Review*, **49**, 371–418 (2007).

[374] Popp, K.: Non-smooth mechanical systems. *Journal of Applied Mathematics and Mechanics*, **64**, 765–772 (2000).

[375] Potra, F.A., Ye, Y.: Interior-point methods for nonlinear complementarity problems. *Journal of Optimization Theory and Applications*, **88**, 617–642 (1996).

[376] Prager, W., Hodge, P.G.: *Theory of Perfectly Plastic Solids.* Wiley, New York (1951).

[377] Qi, H.-H., Liao, L.-Z.: A smoothing Newton method for general nonlinear complementarity problems. *Computational Optimization and Applications*, **17**, 231–253 (2000).

[378] Qi, L., Sun, J.: A nonsmooth version of Newton's method. *Mathematical Programming*, **58**, 353–367 (1993).

[379] Qi, L., Sun, D.: Smoothing functions and smoothing Newton method for complementarity and variational inequality problems. *Journal of Optimization Theory and Applications*, **113**, 121–147 (2002).

[380] Raible, T., Tegeler, K., Löhnert, S., Wriggers, P.: Development of a wrinkling algorithm for orthotropic membrane materials. *Computer Methods in Applied Mechanics and Engineering*, **194**, 2550–2568 (2005).

[381] Ramana, M.V.: An exact duality theory for semidefinite programming and its complexity implications. *Mathematical Programming*, **77**, 129–162 (1997).

[382] Ramana, M.V., Tunçel, L., Wolkowicz, H.: Strong duality for semidefinite programming. *SIAM Journal on Optimization*, **7**, 641–662 (1997).

[383] Reddy, B.D.: *Introductory Functional Analysis.* Springer-Verlag, New York (1998).

[384] Reddy, J.N.: *Energy Principles and Variational Methods in Applied Mechanics (2nd ed.).* John Wiley & Sons, Hoboken (2002).

[385] Reissner, E.: Some considerations on the problem of torsion and flexure of prismatical beams. *International Journal of Solids and Structures*, **15**, 41–53 (1979).

[386] Reissner, E.: Some aspects of the variational principles problem in elasticity. *Computational Mechanics*, **1**, 3–9 (1986).

[387] Rocca, R.: Existence of a solution for a quasistatic problem of unilateral contact with local friction. *Comptes Rendus de l'Academie des Sciences. Series I: Mathematics*, **328**, 1253–1258 (1999).

[388] Rochdi, M., Shillor, M., Sofonea, M.: Quasistatic viscoelastic contact with normal compliance and friction. *Journal of Elasticity*, **51**,105–126

(1998).

[389] Rockafellar, R.T.: *Convex Analysis.* Princeton University Press, Princeton (1970).

[390] Rockafellar, R.T.: *Conjugate Duality and Optimization.* SIAM, Philadelphia (1974).

[391] Roddeman, D.G.: Finite-element analysis of wrinkling membranes. *Communications in Applied Numerical Methods*, **7**, 299–307 (1991).

[392] Roddeman, D.G., Drukker, J., Oomens, C.W.J., Janssen, J.D.: The wrinkling of thin membranes: Part I: Theory. *Journal of Applied Mechanics (ASME)*, **54**, 884–887 (1987).

[393] Roddeman, D.G., Drukker, J., Oomens, C.W.J., Janssen, J.D.: The wrinkling of thin membranes: Part II: Numerical analysis. *Journal of Applied Mechanics (ASME)*, **54**, 888–892 (1987).

[394] Rodorigues, H.C., Guedes, J.M., Bendsøe, M.P.: Necessary conditions for optimal design of structures with a nonsmooth eigenvalue based criterion. *Structural Optimization*, **9**, 52–56 (1995).

[395] Rossi, R., Lazzari, M., Vitaliani, R., Oñate, E.: Simulation of light-weight membrane structures by wrinkling model. *International Journal for Numerical Methods in Engineering*, **62**, 2127–2153 (2005).

[396] Rozvany, G.I.N., Bendsøe, M.P., Kirsch, U.: Layout optimization of structures. *Applied Mechanics Reviews*, **48**, 41–119 (1995).

[397] Sacco, E.: A nonlinear homogenization procedure for periodic masonry. *European Journal of Mechanics, A/Solids*, **28**, 209–222 (2009).

[398] Saitoh, M., Okada, A.: The role of string in hybrid string structure. *Engineering Structures*, **21**, 756–769 (1999).

[399] Sasakawa, T., Tsuchiya, T.: Optimal magnetic shield design with second-order cone programming. *SIAM Journal on Scientific Computing*, **24**, 1930–1950 (2003).

[400] Schek, H.-J.: The force density method for form finding and computation of general networks. *Computer Methods in Applied Mechanics and Engineering*, **3**, 115–134 (1974).

[401] Scherer, C., Gahinet, P., Chilali, M.: Multiobjective output-feedback control via LMI optimization. *IEEE Transactions on Automatic Control*, **42**, 896–906 (1997).

[402] Schlaich Bergermann unt Partner, `http://www.sbp.de/`.

[403] Schoop, H., Taenzer, L., Hornig, J.: Wrinkling of nonlinear membranes. *Computational Mechanics*, **29**, 68–74 (2002).

[404] Schrijver, A.: *Theory of Linear and Integer Programming.* John Wiley & Sons, Chichester (1986).

[405] Sextro, W.: *Dynamical Contact Problems with Friction.* Springer-Verlag, Berlin (2002).

[406] Seyranian, A.P.: Sensitivity analysis of multiple eigenvalues. *Mechanics Based Design of Structures and Machines*, **21**, 261–284 (1993).

[407] Seyranian, A.P., Lund, E., Olhoff, N.: Multiple eigenvalues in structural optimization problem. *Structural Optimization*, **8**, 207–227 (1994).

[408] Shapiro, A.: On duality theory of convex semi-infinite programming. *Optimization*, **54**, 535–543 (2005).

[409] Shapiro, A.: Semi-infinite programming, duality, discretization and optimality conditions. *Optimization*, **58**, 131–161 (2009).

[410] Shillor, M., Sofonea, M., Telega, J.J.: *Models and Analysis of Quasistatic Contact.* Springer-Verlag, Berlin (2004).

[411] Skelton, R.E., de Olivira, M.C.: *Tensegrity Systems.* Springer, New York (2009).

[412] Sofonea, M., Matei, A.: *Variational Inequalities with Applications.* Springer, New York (2009).

[413] Simo, J.C., Hughes, T.J.R.: *Computational Inelasticity.* Springer-Verlag, New York (1998).

[414] Stavroulakis, G.E.: Applied nonsmooth mechanics of deformable bodies. In: Haslinger, J., Stavroulakis, G.E. (eds.) *Nonsmooth Mechanics of Solids*, pp. 275–314, Springer-Verlag, Wien (2006).

[415] Stefanou, G.D. Moossavi, E., Bishop, S., Koliopoulos, P.: Conjugate gradients method for calculating the response of large cable nets to static loads. *Computers & Structures*, **49**, 843–848 (1993).

[416] Steigmann, D.J.: Tension-field theory. *Proceedings of the Royal Society of London. Series A, Mathematical and Physical Sciences*, **429**, 141–173 (1990).

[417] Steigmann, D.J.: Puncturing a thin elastic sheet. *International Journal of Non-Linear Mechanics*, **40**, 255–270 (2005).

[418] Steigmann, D.J., Pipkin, A.C.: Wrinkling of pressurized membranes. *Journal of Applied Mechanics (ASME)*, **56**, 624–628 (1989).

[419] Steigmann, D.J., Pipkin, A.C.: Axisymmetric tension fields. *Journal of Applied Mathematics and Physics (ZAMP)*, **40**, 526–542 (1989).

[420] Stewart, D.E.: Rigid-body dynamics with friction and impact. *SIAM*

Review, **42**, 3–39 (2000).

[421] Stewart, D.E., Trinkle, J.C.: An implicit time-stepping scheme for rigid body dynamics with inelastic collisions and Coulomb friction. *International Journal for Numerical Methods in Engineering*, **39**, 2673–2691 (1996).

[422] Strömberg, N.: An augmented Lagrangian method for fretting problems. *European Journal of Mechanics, A/Solids*, **16**, 573–593 (1997).

[423] Strömberg, N.: A Newton method for three-dimensional fretting problems. *International Journal of Solids and Structures*, **36**, 2075–2090 (1999).

[424] Sturm, J.F.: Using SeDuMi 1.02, a MATLAB toolbox for optimization over symmetric cones. *Optimization Methods and Software*, **11/12**, 625–653 (1999).

[425] Sturm, J.F.: Theory and algorithms for semidefinite programming. In: Frenk, H., Roos, K., Terlaky, T., Zhang, S. (eds.) *High Performance Optimization*, pp. 3–196. Kluwer Academic Publishers, Dordrecht (2000).

[426] Sturm, J.F.: Implementation of interior point methods for mixed semidefinite and second order cone optimization problems. *Optimization Methods and Software*, **17**, 1105–1154 (2002).

[427] Sun, D., Sun, J.: Strong semismoothness of the Fischer-Burmeister SDC and SOC complementarity functions. *Mathematical Programming*, **A103**, 575-581 (2005).

[428] Sun, D., Sun, J.: Löewner's operator and spectral functions in Euclidean Jordan algebras. *Mathematics of Operations Research*, **33**, 421–445 (2008).

[429] Tin-Loi, F., Misa, J.S.: Large displacement elastoplastic analysis of semirigid steel frames. *International Journal for Numerical Methods in Engineering*, **39**, 741–762 (1996).

[430] Tin-Loi, F., Pang, J.-S.: Elastoplastic analysis of structures with nonlinear hardening: A nonlinear complementarity approach. *Computer Methods in Applied Mechanics and Engineering*, **107**, 299–312 (1993).

[431] Tin-Loi, F., Xia, S.H.: Nonholonomic elastoplastic analysis involving unilateral frictionless contact as a mixed complementarity problem. *Computer Methods in Applied Mechanics and Engineering*, **190**, 4551–4568 (2001).

[432] Tin-Loi, F., Xia, S.H.: An iterative complementarity approach for elastoplastic analysis involving frictional contact. *International Journal of Mechanical Sciences*, **45**, 197–216 (2003).

[433] Todd, M.J.: A study of search directions in primal-dual interior-point methods for semidefinite programming. *Optimization Methods and Software*, **11**, 1–46 (1999).

[434] Todd, M.J.: Semidefinite optimization. *Acta Numerica*, **10**, 515–560 (2001).

[435] Toh, K.C., Todd, M.J., Tütüncü, R.H.: SDPT3—a MATLAB software package for semidefinite programming, version 1.3. *Optimization Methods and Software*, **11/12**, 545–581 (1999).

[436] Tseng, P.: An infeasible path-following method for monotone complementarity problems. *SIAM Journal on Optimization*, **7**, 386–402 (1997).

[437] Tseng, P.: Merit functions for semi-definite complementarity problems. *Mathematical Programming*, **83**, 159–185 (1998).

[438] Tsuchiya, T.: A convergence analysis of the scaling-invariant primal-dual path-following algorithms for second-order cone programming. *Optimization Methods and Software*, **11/12**, 141–182 (1999)

[439] Tütüncü, R.H., Toh, K.C., Todd, M.J.: Solving semidefinite-quadratic-linear programs using SDPT3. *Mathematical Programming*, **B95**, 189–217 (2003).

[440] Tuy, H.: *Convex Analysis and Global Optimization*. Kluwer Academic Publishers, Dordrecht (1998).

[441] Tyndall, W.F.: A duality theorem for a class of continuous linear programming problems. *Journal of the Society for Industrial and Applied Mathematics*, **13**, 644–666 (1965).

[442] Valdés, J.G., Miquel, J., Oñate, E.: Nonlinear finite element analysis of orthotropic and prestressed membrane structures. *Finite Elements in Analysis and Design*, **45**, 395–405 (2009).

[443] Valid, R.: Duality in nonlinear theory of shells. *International Journal of Engineering Science*, **37**, 1521–1547 (1999).

[444] Vandenberghe, L., Boyd, S.: Semidefinite programming. *SIAM Review*, **38**, 49–95 (1996).

[445] Vanderbei, R.J.: *Linear Programming: Foundations and Extensions (2nd ed.)*. Springer, New York (2001).

[446] Vanderbei, R.J., Shanno, D.F.: An interior-point algorithm for nonconvex nonlinear programming. *Computational Optimization and Applications*, **13**, 231–252 (1999).

[447] Vassart, N., Motro, R.: Multiparametered formfinding method: Application to tensegrity systems. *International Journal of Space Structures*,

14, 147–154 (1999).

[448] Volokh, K.Yu., Vilnay, O.: Why pre-tensioning stiffens cable systems. *International Journal of Solids and Structures*, **37**, 1809–1816 (2000).

[449] Wagner, H.: Flat sheet metal girder with very thin metal web. *Zeitschrift fur Flugtechnik Motorlurftschiffahrt*, **20**, 200–207, 227–231, 281–284, 306-314 (1929). English translation: Technical Memorandum, NACA-TM-604, 605, 606, National Advisory Committee for Aeronautics, Washington, D.C. (1931).

[450] Wakefield, D.S.: Engineering analysis of tension structures: Theory and practice. *Engineering Structures*, **21**, 680–690 (1999).

[451] Wang, T., Monteiro, R.D.C., Pang, J.-S.: An interior point potential reduction method for constrained equations. *Mathematical Programming*, **74**, 159–195 (1996).

[452] Washizu, K.: Note on the principle of stationary complementary energy applied to free vibration of an elastic body. *International Journal of Solids and Structures*, **2**, 27–35 (1966).

[453] Wolkowicz, H., Anjos, M.F.: Semidefinite programming for discrete optimization and matrix completion problems. *Discrete Applied Mathematics*, **123**, 513–577 (2002).

[454] Wolkowicz, H., Saigal, R., Vandenberghe, L. (eds.): *Handbook on Semidefinite Programming: Theory, Algorithms and Applications*. Kluwer Academic Publishers, Boston (2000).

[455] Wong, Y.W., Pellegrino, S.: Wrinkled membranes, Part III: Numerical simulations. *Journal of Mechanics of Materials and Structures*, **1**, 63–95 (2006).

[456] Wriggers, P.: *Computational Contact Mechanics (2nd ed.)*. Springer-Verlag, Berlin (2006).

[457] Wriggers, P.: *Nonlinear Finite Element Methods*. Springer-Verlag, Berlin (2008).

[458] Wu, C.H.: Nonlinear wrinkling of nonlinear membranes of revolution. *Journal of Applied Mechanics (ASME)*, **45**, 533–538 (1978).

[459] Yamashita, M., Fujisawa, K., Kojima, M.: Implementation and evaluation of SDPA6.0 (SemiDefinite Programming Algorithm 6.0). *Optimization Methods and Software*, **18**, 491–505 (2003).

[460] Ye, Y.: *Interior Point Algorithms: Theory and Analysis*. John Wiley & Sons, New York (1997).

[461] Yonekura, K., Kanno, Y.: Global optimization of robust truss topology

via mixed integer semidefinite programming. *Optimization and Engineering*, **11**, 355–379 (2010).

[462] Yonekura, K., Kanno, Y.: Second-order cone programming with warm start for elastoplastic analysis with von Mises yield criterion. METR 2010-12, Department of Mathematical Informatics, University of Tokyo, May 2010.

[463] Zak, M.: Statics of wrinkling films. *Journal of Elasticity*, **12**, 51–63 (1982).

[464] Zhang, H.W., He, S.Y., Li, X.S: Non-interior smoothing algorithm for frictional contact problems. *Applied Mathematics and Mechanics*, **25**, 47–58 (2004).

[465] Zhang, H.W., He, S.Y., Li, X.S., Wriggers, P.: A new algorithm for numerical solution of 3D elastoplastic contact problems with orthotropic friction law. *Computational Mechanics*, **34**, 1–14 (2004).

[466] Zhang, H.W., Zhong, W.X., Wu, C.H., Liao, A.H.: Some advances and applications in quadratic programming method for numerical modeling of elastoplastic contact problems. *International Journal of Mechanical Science*, **48**, 176–189 (2006).

[467] Zhang, J.Y., Ohsaki, M.: Adaptive force density method for form-finding problem of tensegrity structures. *International Journal of Solids and Structures*, **43**, 5658–5673 (2006).

[468] Zhang, X., Jiang H., Wang, Y.: A smoothing Newton-type method for generalized nonlinear complementarity problem. *Journal of Computational and Applied Mathematics*, **212**, 75–85 (2008).

[469] Zhao, Y.B., Li, D.: A new path-following algorithm for nonlinear P_* complementarity problems. *Computational Optimization and Applications*, **34**, 183–214 (2005).

[470] Zhong, W., Sun, S.: A finite element method for elasto-plastic structures and contact problems by parametric quadratic programming. *International Journal for Numerical Methods in Engineering*, **26**, 2723–2738 (1988).

[471] Zhu, C.: A finite element-mathematical programming method for elasto-plastic contact problems with friction. *Finite Elements in Analysis and Design*, **20**, 273–282 (1995).

[472] Zienkiewicz, O.C., Taylor, R.L.: *The finite Element Method. Volume 1 (5th ed.)*. Butterworth-Heinemann, Oxford (2000).

[473] Zubov, L.M.: The stationary principle of complementary work in nonlinear theory of elasticity. *Prikladnaya Matematika i Mekhanika*, **34**,

241–245 (1970). English translation: *Journal of Applied Mathematics and Mechanics*, **34**, 228–232 (1970).

Index

$]a,b[$, 40
$[a,b]$, 40
$\succ$, 13
$\succeq$, 12, 13
aff, 40
bd, 4
C^k, 42, 43
$\mathrm{C}^k(\overline{\Omega})$, 227
$\mathrm{cl}\,\mathcal{S}$, 4
$\mathrm{cl}\,\mathrm{co}\,f$, 48
$\mathrm{cl}\,f$, 48
co, 40
$\delta_{\mathcal{S}}$, 41
diag$(\cdot)$, 27
dom, 39
epi, 41
$H^{-1/2}(\Gamma)$, 226
$H^{1/2}(\Gamma)$, 226, 227
$H^1(\Omega)^3$, 226
$H^m(\Omega)$, 226
$H_0^1(\Omega)$, 227
int, 4
$L^2(\Omega)$, 225
$L^2(\Omega)^3$, 225
$L^2(\Omega)_{\mathbb{S}}^{3\times 3}$, 226
$\mathbb{L}^n$, 20
∇, 9, 41, 213
∇^2, 41
$(\mathrm{P_F})$, 55
$(\mathrm{P_F^*})$, 56
$\mathbb{R}_+^n$, 5
ri, 40
$\mathbb{S}^n$, 12
$\mathbb{S}_{++}^n$, 13
$\mathbb{S}_+^n$, 12
$\mathbb{Y}$, 27

adjoint operator, 55
affine hull, 40
affine scaling direction, *see* search direction
affine space, 39
Almansi strain, *see* strain
arc
  = edge, 114
arrow-shaped matrix, 28
associated flow rule, *see* flow rule
augmented Lagrangian method, 340
axial force, 78, 86

B-differentiable, 179
B-differentiable Newton method, 184
back stress, 373
barrier method, 167
barrier problem, 172
bi-potential, 347
biconjugate function, 48
Biot strain, *see* strain
Biot stress, *see* stress
block matrix, 324
block-diagonal matrix, 77
boundary, 4
boundary condition
  contact —, 311
  Dirichlet —, 109, 311
  Neumann —, 109, 311

cable, 100, 108
  slack, 100
  taut, 100
centering direction, 172
central path, 168, 171
  for LP, 171
  for SDP, 175
characteristic polynomial, 351
Cholesky factorization, 15

closed convex function, 44, 45
closed convex hull
 of function, 48
closed proper convex function, 44
closure
 of function, 48
 of set, 4
compatibility relation, 108, 113, 215
complementarity condition, 10, 106
complementarity problem
 = CP, 7
complementary energy, 88
 truly complementary form, 143
complementary stored energy
 — of cable, 102
complementary strain energy function, 231
compliance, 74
 — optimization, 76
concave
 concave function, 40
cone, 5
 convex cone, 5
 dual cone, 6
 pointed cone, 5
 positive-semidefinite cone, 13
 proper cone, 5
 second-order cone, 20
 self-dual cone, 6
 solid cone, 5
congruence transformation, 15
conic inequality, 5, 26
conic linear program, 29
conic optimization, 29
conjugate function, 47
conservative force, 108
consistency parameter, 358
constitutive law, 87, 215
 bi-linear —, *see* piecewise-linear law
 piecewise-linear law, 120
constraint
 convex quadratic —, 28
 hyperbolic —, 29
 kinematic —, 109
 linear inequality —, 27
 positive-semidefinite —, 27
 second-order cone —, 27, 116
 static —, 109
constraint qualification, 63
 Slater's —, 63
contact
 contact state, 323
 free state, 323
 grazing contact state, 323
 normal law, 312
 tangential law, 312
contact candidate node, 91
contact force, 91
contact problem, 311
continuous, 45
continuously differentiable, 42, 43
convex
 convex function, 40, 87
 convex hull, 40
 convex set, 4
 strictly convex function, 40
convex function, 45
convex quadratic constraint, 28
convex quadratic inequality
 = convex quadratic constraint, 242
convex quadratic programming
 = QP, 251
convex set, 45
convexity
 of strain energy, 44
Courant–Fischer theorem, 14
CP, 7
 LCP, 9
 NCP, 9
crack strain, 217
critical load, 79

deformation gradient, 264
deviator, 352
deviatoric stress, 352
diagonal matrix, 27
diagonalization
 of symmetric matrix, 13

digraph, 114
directed graph, *see* digraph
direction cosine, 114
directional derivative, 47
dissipation
maximum —, 330, 355, 367
divergence operator, 214
dual cone, *see* cone
dual problem, 51
in Fenchel duality, 56
of LP, 30
of SDP, 32
of SOCP, 34
duality
duality gap, 53
duality measure, 171, 175
strong duality, 53, 66
weak duality, 51, 53
duality pairing, *see* pairing
dynamic relaxation method, 128, 206

edge, 114
head of —, 114
tail of —, 114
effective domain, 39
eigenvalue, 13
fundamental —, 82
multiple —, 82
eigenvalue decomposition, 13
eigenvalue optimization, 82
eigenvector, 13
elastic, 87, 264
— deformation, 351
purely —, 353
elastic modulus
= Young's modulus, 76
elastic unloading, 121, 353
elasticity tensor, 240, 270
elongation, 86
elongation stiffness, *see* stiffness
energy
complementary, 88
stored —, 87
strain —, 87, 216
total potential energy, 216

epigraph, 41
Euclidean space, 3
extended valued function, *see* function

feasible set, 169
strictly —, 169
feasible solution, 169
strictly —, 66, 169
Fenchel transformation, 47, 89
Fenchel–Legendre transformation, *see* Fenchel transformation
fictitious body
= reference elastic body, 107
first Piola–Kirchhoff stress, *see* stress
flexibility matrix, 328
flow rule, 331, 355
associated —, 355
non-associated, 341
flow stress, 353
force density, 206
— method, 206
force-balance equation, 78
form-finding problem
of cable network, 200
free state, *see* contact
friction cone, 314
Frobenius norm, *see* norm
function, 39
extended valued —, 39
multifunction, 47
set-valued —, 47
fundamental eigenvalue, 82

gap, 91, 319
Gauss quadrature, 240
generalized inverse, 16
global coordinate system, 238
gradient, 9, 41, 114, 213
granular material, 250
graph
directed graph, 114
of function, 40
Green's formula, 229
Green–Lagrange strain, 117, *see* strain

half-space
closed half-space, 4
open half-space, 4
hardening
isotropic —, 354, 363
kinematic —, 354, 373
strain —, 354, 362
head, 114
hemi-variational inequality, 8
Hessian, 41
Hilbert space, 225
homogenization method, 250
Hooke's law, 271
hyperbolic constraint, 29

incidence matrix, 114
truncated—, 146
inclusion, 3, 93, 102
incremental problem, 317, 367
indicator function, 41
inequality
conic —, 5
convex quadratic —, 28
linear matrix —, 84
second-order cone —, 116
inner product of matrices, 16
interior, 4
interior-point method
path-following —, 172
interval, 40
irreversible, 351
isoparametric element, 238
isotropic hardening, 363
— modulus, 364

J_2-flow theory, 356
Jordan frame, 24
Jordan product, 22

Karush–Kuhn–Tucker conditions
= KKT conditions, 63
kinematic boundary condition, 311
kinematic constraint, 109
kinematic hardening, 373
kinematic variable, 108
kinematically admissible, 109
kinematically indeterminate, 76
Kirchhoff stress, 268
KKT condition, 63
for LP, 169
modified —, 171
Kronecker product, 114

Lagrange multiplier, 59
Lagrangian, 59, 69
Lamé modulus, 218, 271
LCP, 9
Legendre transformation, 49
linear complementarity problem
= LCP, 9
linear elastic material, 87
linear elasticity, 87
linear inequality
constraint, 27
linear matrix inequality, 27, 84
linear program
= LP, 30
linear SDP, 32
linear strain, 116
Lipschitzian domain, 214, 229
LMI
= linear matrix inequality, 27
loading
elastic unloading, 353
plastic —, 353
Lorentz cone
= second-order cone, 20
lower semicontinuous, 44, 45
LP, 30

map
= mapping, 39
mapping
= function, 39
point-to-set mapping, 47
set-valued mapping, 47
masonry-like material, *see* no-tension material
matrix
block-diagonal —, 77

maximum dissipation law, 221, 330, 355, 367
mean normal stress, 352
member elongation, 78
membrane
  slack, 256
  taut, 255
  wrinkled, 256
merit function, 179
MiCP, 7
minimal-surface problem, 203
minimization of complementary energy, 109
minimization of potential energy, 109
minimum principle of complementary energy, 129
mixed complementarity problem
  = MiCP, 7
modified KKT condition, *see* KKT condition
monotone, 41, 87
  strictly —, 41, 88
multifunction = set-valued mapping, *see* mapping

natural coordinate, 238, 292
NCP, 9
nearest point, *see* projection
Newton–Raphson method, 128, 250
NLP, 61
no-compression model, 100
no-tension material, 212, 215
non-associated flow rule, *see* flow rule
non-dissipative process, 105
nonlinear complementarity problem
  = NCP, 9
nonlinear program
  = NLP, 61
nonlinear SDP, *see* SDP
nonnegative orthant, 5
  constraint, 27
norm, 4
  Frobenius norm, 16
normality rule, 221, 331, 355

octant, 5
optimal value function, 51

pairing, 45, 50
partial derivative, 226
partial order, 5
path independent, 108
perturbed problem, 51
piecewise-linear
  constitutive law, 120
plastic
  — deformation, 351
  — loading, 353
plastic flow
  = plastic strain rate, 355
plasticity
  perfect —, 353
point-to-set mapping
  = set-valued mapping, *see* mapping
Poisson's ratio, 219
positive definite, 12, 14
positive semidefinite, 12
positive-semidefinite
  — constraint, 27
positive-semidefinite cone, 13
power set, 47
predictor-corrector method, 172
primal problem, 50
  in Fenchel duality, 55
  of LP, 30
  of SDP, 32
  of SOCP, 34
principal invariant, 351
principal stress, 217
product
  Jordan, 22
  Kronecker —, 114
projection, 178, 339
proper convex function, 41, 45
pseudo-inverse, 16
purely elastic behavior, 353

QCQP, 33, 118
QP, 32, 117, 251

quadrant, 5
quadratic cone
  = second-order cone, 20
quadratic form, 12
quadratic program
  = QP, 32
quadratically constrained quadratic program
  = QCQP, 33
quasistatic analysis, 317

rate problem, 349
Rayleigh quotient, 14
reaction, 91, 108, 311
real-valued function, *see* function
reference elastic body, 107, 218, 258
reference element, 238, 292
relative interior, 40
relative interior point, 40
response function, 87
return-mapping method, 237, 251
reversible, 105, 215, 256, 351
robust optimization, 79

S-lemma, 36
saddle point, 60
scalar product
  — of tensors, 213
  — of vectors, 213
Schur complement, 15, 76
SDP, 32, 119
  nonlinear SDP, 96
search direction, 170, 179, 181
  affine scaling direction, 170
second Piola–Kirchhoff stress, *see* stress
second-order cone, 20
  — complementarity function, 182
  — constraint, 27, 116
  — inequality, 116
second-order cone program
  = SOCP, 34
self-dual, 17
self-dual cone, *see* cone
self-duality, – of second-order cone132
self-equilibrium configuration, 202
semidefinite program
  = SDP, 32
sequential unconstrained minimization technique
  = SUMT, 167
set
  closed set, 4
  convex set, 4
  open set, 4
  power set, 47
  uncertainty set, 79
set-valued function, 47, 102
set-valued mapping, *see* mapping
shape functions, 239, 292
Signorini contact condition
  = unilateral contact, 92
Signorini law, 322
simultaneously diagonalizable, 18
singleton, 40
slack state, 100, 256
Slater's constraint qualification, *see* constraint qualification
sliding joint, 195
slip
  impending slip state, 314
  slip state, 314
  stick state, 314
smeared crack, 217
smeared wrinkle, 256
smoothing function, 180
Sobolev space, 226
SOCP, 34, 116
square-root of matrix, 14
St. Venant material, 271
St. Venant–Kirchhoff material
  = St. Venant material, 271
static boundary condition, 311
static constraint, 109
static variable, 109
statically admissible, 109
step size, 170
stick, *see* slip
stiffness
  elongation —, 87
  tangent —, 87

stiffness matrix, 74
stored energy, 87, 215
  — of cable, 100
strain
  Almansi —, 265
  Biot —, 115
  elastic —, 217, 256
  Green–Lagrange —, 117, 268
  inelastic, 217
  inelastic —, 256
  linear —, 116, 215
  strain energy, 216
  wrinkling —, 256, 266
strain energy, 87, 215
strain hardening, 362
strain softening, 354
stress
  back —, 354, 356, 373
  Biot —, 115
  Cauchy —, 264
  deviatoric —, 352
  first Piola–Kirchhoff —, 161, 268
  flow —, 353
  mean normal —, 352
  second Piola–Kirchhoff —, 161, 267
  volumetric —, 352
stress–strain relation, *see* constitutive law
strict complementarity
  over $\mathbb{L}^n$, 25
  over $\mathbb{R}^n_+$, 10
  over $\mathbb{S}^n_+$, 19
strictly convex function, *see* convex
strictly feasible set, *see* feasible set
strictly feasible solution, *see* feasible solution, *see* feasible solution
strong duality, *see* duality
structural optimization
  compliance optimization, 76
  eigenvalue optimization, 82
  robust optimization, 79
subdifferential, 46
subgradient, 46
SUMT, 167

tail, 114
tangent modulus, 121
tangent stiffness, 87, 102, 128
taut state, 100, 255
topology
  of truss, 76
  of cable network, 200
total potential energy, 216
trace operator, 227
trace theorem, 227, 229
truly complementary form, 143
truncated incidence matrix, 146

uncertainty set, 79
unilateral contact, 92
unilateral contact condition, 322
unit tensor, 218
unloading, 121
upper semicontinuous, 44

variational inequality
  = VI, 8
variational principle
  truly complementary form, 143
vector space, 3
vertex, 114
VI, 8
Voigt notation, 240
volumetric stress, 352
von Mises yield condition, 355

weak duality, *see* duality
wedged configuration, 350
wedging problem, 350
worst case, 79
  — detection, 79
worst scenario
  = worst case, 79
wrinkled state, 256

yield condition, 353
yield function, 353
yield stress, 121
yield surface, 353

Young's modulus, 76, 100, 219

About the Author

Yoshihiro Kanno is an associate professor in the Department of Mathematical Informatics at the University of Tokyo, Japan. Dr. Kanno received his PhD in structural engineering from Kyoto University, Japan, in 2002. He received the Maeda Prize in Engineering in 2005 for his dissertation "Energy Principles of Structures Based on Mathematical Programming over Symmetric Cones."

The author and coauthor of numerous professional articles on applied mechanics and optimization, Dr. Kanno's research interest is in the interface between mechanics and mathematics. He is a member of the International Society for Structural and Multidisciplinary Optimization, the Japan Society of Mechanical Engineers, the Architectural Institute of Japan, and the Operations Research Society of Japan.